Artificial Intelligence of Things for Smarter Eco-Cities

This book takes readers on a captivating journey into the transformative role of Artificial Intelligence (AI) and AI of Things (AIoT) technologies in reshaping sustainable urban development. By combining comprehensive theoretical analyses, synthesized empirical evidence, and practical case studies, it offers pioneering interdisciplinary insights and unifying frameworks. The book highlights the synergistic integration of Urban Brain (UB), Urban Digital Twin (UDT), Smart Urban Metabolism (SUM), and platform urbanism, underscored by their collaborative potential to revolutionize the environmental management, planning, and governance of smarter eco-cities and sustainable smart cities. It leverages cutting-edge technologies and data-driven approaches to optimize urban systems, resource efficiency, and resilience. This approach provides a holistic understanding of the rapidly evolving landscape of AI- and AIoT-driven sustainable urban development.

Targeting a broad and diverse audience across multiple disciplines and fields, the book aims to share state-of-the-art research, present innovative solutions, and forecast future trends in urban sustainability. As both a seminal reference and a valuable resource for researchers, practitioners, industry experts, and policymakers, it provides essential guidance for those engaged in driving technological innovation, steering urban transformation, promoting environmental sustainability, or working at the crossroads of these critical areas.

Dr. Simon Elias Bibri's book is a groundbreaking exploration of the transformative role of Artificial Intelligence and Artificial Intelligence of Things in sustainable urban development. It introduces pioneering frameworks for data-driven environmental management, planning, and governance, offering profound insights for researchers, practitioners, and policymakers. It not only bridges the gap between theory and practice but also sets a new standard for interdisciplinary research in the field. A must-read for those committed to creating and advancing smarter, more sustainable cities.
Tan Yigitcanlar, *Professor, Queensland University of Technology, Australia*

Dr. Simon Elias Bibri's latest book provides an important contribution to the rapidly growing field of AI-driven urbanism. He explains how cutting-edge urban artificial intelligences, such as city brains and AI-mediated urban platforms, can become a medium to achieve sustainable urban development goals. His insights are very timely and much needed to expand the current debate about the urban sustainability of AI.
Federico Cugurullo, *Professor, Trinity College Dublin, Ireland*

Artificial Intelligence of Things for Smarter Eco-Cities
Pioneering the Environmental Synergies of Urban Brain, Digital Twin, Metabolic Circularity, and Platform

Simon Elias Bibri

CRC Press
Taylor & Francis Group
Boca Raton London New York

CRC Press is an imprint of the
Taylor & Francis Group, an **informa** business

Designed cover image: Smart City und Internet der Dinge Verschiedene Kommunikationsgeräte Stockfoto und mehr Bilder von Internet der Dinge—iStock (istockphoto.com)

First edition published 2025
by CRC Press
2385 NW Executive Center Drive, Suite 320, Boca Raton FL 33431

and by CRC Press
4 Park Square, Milton Park, Abingdon, Oxon, OX14 4RN

CRC Press is an imprint of Taylor & Francis Group, LLC

© 2025 Simon Elias Bibri

Reasonable efforts have been made to publish reliable data and information, but the author and publisher cannot assume responsibility for the validity of all materials or the consequences of their use. The authors and publishers have attempted to trace the copyright holders of all material reproduced in this publication and apologize to copyright holders if permission to publish in this form has not been obtained. If any copyright material has not been acknowledged, please write and let us know so we may rectify in any future reprint.

Except as permitted under U.S. Copyright Law, no part of this book may be reprinted, reproduced, transmitted, or utilized in any form by any electronic, mechanical, or other means, now known or hereafter invented, including photocopying, microfilming, and recording, or in any information storage or retrieval system, without written permission from the publishers.

For permission to photocopy or use material electronically from this work, access www.copyright. com or contact the Copyright Clearance Center, Inc. (CCC), 222 Rosewood Drive, Danvers, MA 01923, 978-750-8400. For works that are not available on CCC, please contact mpkbookspermissions @tandf.co.uk

Trademark notice: Product or corporate names may be trademarks or registered trademarks and are used only for identification and explanation without intent to infringe.

Library of Congress Cataloging-in-Publication Data
Names: Bibri, Simon Elias, author.
Title: Artificial intelligence of things for smarter eco-cities :
pioneering the environmental synergies of urban brain, digital twin,
metabolic circularity, and platform / Simon Elias Bibri.
Description: First edition. | Boca Raton, FL : CRC Press, [2025] |
Includes bibliographical references and index.
Identifiers: LCCN 2024037134 (print) | LCCN 2024037135 (ebook) |
ISBN 9781032881560 (hbk) | ISBN 9781032881553 (pbk) | ISBN 9781003536420 (ebk)
Subjects: LCSH: Smart cities. | Urban ecology (Sociology)—Technological
innovations. | Artificial intelligence—Industrial applications. |
Internet of things—Industrial applications.
Classification: LCC TD159.4 .B53 2025 (print) | LCC TD159.4 (ebook) |
DDC 307.1/416—dc23/eng/20241021
LC record available at https://lccn.loc.gov/2024037134
LC ebook record available at https://lccn.loc.gov/2024037135

ISBN: 978-1-032-88156-0 (hbk)
ISBN: 978-1-032-88155-3 (pbk)
ISBN: 978-1-003-53642-0 (ebk)

DOI: 10.1201/9781003536420

Typeset in Minion
by codeMantra

Contents

Preface, xiv

Acknowledgments, xviii

Author, xx

Chapter 1 ▪ Transformative Integration of Artificial Intelligence of Things and Cyber-Physical Systems of Systems: Advancing Smarter Eco-Cities and Sustainable Smart Cities	1
1.1 INTRODUCTION AND BACKGROUND	2
1.2 RESEARCH PROBLEM	15
1.3 RESEARCH GAP	16
1.4 RESEARCH AIM AND OBJECTIVES	16
1.5 RESEARCH MOTIVATION	17
1.6 RESEARCH SIGNIFICANCE	17
1.7 THE STRUCTURE AND CONTENT OF THE BOOK	18
1.8 THE NEED FOR THE BOOK	23
1.9 TARGET AUDIENCE AND READERSHIP	25
1.10 THE FIVE KEY FEATURES OF THE BOOK	27
1.11 AUTHOR'S CONTRIBUTION TO THE FIELD	28
REFERENCES	29

CHAPTER 2 ■ Smarter Eco-City Management, Planning, and Governance in the Era of Artificial Intelligence of Things: Theoretical and Conceptual Frameworks — 36

- 2.1 INTRODUCTION — 37
- 2.2 THEORETICAL AND CONCEPTUAL FOUNDATIONS AND FRAMEWORKS — 42
  - 2.2.1 Smarter Eco-Cities — 42
  - 2.2.2 Urban Governance, Planning, and Management: An Environmental Perspective — 46
  - 2.2.3 Urban Governance, Planning, and Management: Linkages, Dynamics, and Nuances — 53
  - 2.2.4 Environmental Sustainability and Climate Change — 55
  - 2.2.5 Artificial Intelligence — 58
  - 2.2.6 The Internet of Things — 62
  - 2.2.7 Artificial Intelligence of Things — 63
  - 2.2.8 Artificial Intelligence and Artificial Intelligence of Things: Commonalities and Differences — 66
  - 2.2.9 Machine Learning — 68
  - 2.2.10 Deep Learning, Machine Learning, and Generative AI as Subfields of Artificial Intelligence — 73
  - 2.2.11 City Brain — 78
  - 2.2.12 Smart Urban Metabolism — 80
  - 2.2.13 Urban Digital Twin — 81
  - 2.2.14 Platform Urbanism — 84
- 2.3 THE INTERCONNECTED CONCEPTS SHAPING THE DEVELOPMENT OF SMARTER ECO-CITIES — 85
- 2.4 A NOVEL CONCEPTUAL FRAMEWORK AND ITS ESSENTIAL COMPONENTS — 88
- 2.5 DISCUSSION — 90
- 2.6 CONCLUSION — 93
- REFERENCES — 94

CHAPTER 3 ▪ Artificial Intelligence and Artificial Intelligence of Things Solutions for Smarter Eco-Cities: Advancing Environmental Sustainability and Climate Change Strategies   111

  3.1   INTRODUCTION   113
  3.2   CONCEPTUAL DEFINITIONS AND DISCUSSIONS   117
      3.2.1   Artificial Intelligence of Things and Its Pillars   118
      3.2.2   Environmental Sustainability and Climate Change   120
  3.3   A SURVEY OF RELATED WORK   122
      3.3.1   Artificial Intelligence for Environmental Sustainability and Climate Change   123
      3.3.2   Artificial Intelligence and the Internet of Things for Smart Cities   124
  3.4   MATERIALS AND METHODS   127
  3.5   RESULTS: AI AND AIoT APPLICATIONS FOR ENVIRONMENTAL SUSTAINABILITY, CLIMATE CHANGE, AND SMART CITIES   129
      3.5.1   The First Period: 2017–2019   130
      3.5.2   The Second Period: 2020–2024   137
  3.6   A NOVEL CONCEPTUAL FRAMEWORK FOR SMARTER ECO-CITY DEVELOPMENT   161
  3.7   DISCUSSION   163
      3.7.1   Summary and Interpretation of Results/Findings   163
      3.7.2   Transformative Influence: Data-Driven Technologies of Smart Cities Shaping the Smart Eco-City Landscape and Dynamics   164
      3.7.3   Environmental Challenges of AI and AIoT Technologies   166
      3.7.4   Ethical, Societal, and Regulatory Challenges of AI and AIoT Technologies   168
      3.7.5   Challenges, Issues, and Considerations of Explainable AI (XAI) and Interpretable ML (IML)   170

3.8	EXISTING GAPS, THEIR IMPLICATIONS, AND THE CORRESPONDING FUTURE RESEARCH DIRECTIONS	173
3.9	CONCLUSION	174
	REFERENCES	177

CHAPTER 4 ▪ Artificial Intelligence of Things for Advancing Smart.Sustainable Cities: A Synthesized Case Study Analysis of City Brain, Digital Twin, Metabolism, and Platform ... 202

4.1	INTRODUCTION	203
4.2	RESEARCH METHODOLOGY	207
4.3	ARTIFICIAL INTELLIGENCE OF SMART SUSTAINABLE CITY THINGS: A NOVEL CONCEPTUAL FRAMEWORK FOR ENVIRONMENTAL URBAN PLANNING AND GOVERNANCE	210
4.4	RESULTS	217
	4.4.1 Case Study Analysis	218
	4.4.2 An Integrative Analysis	240
4.5	DISCUSSION	248
	4.5.1 Summary of Findings and Interpretation of Results	248
	4.5.2 Comparison with Existing Studies	249
	4.5.3 Challenges and Complexities Surrounding the Integration of City Brain, UDT, SUM, and Platform Urbanism	250
	4.5.4 Existing Gaps and Future Research Directions	253
4.6	CONCLUSION	254
	REFERENCES	256

CHAPTER 5 ▪ Artificial Intelligence of Sustainable Smart City Brain, Digital Twin, and Metabolism: Pioneering Data-Driven Environmental Management and Planning ... 268

5.1	INTRODUCTION	269
5.2	THEORETICAL BACKGROUND	275
	5.2.1 Sustainable Smart Cities	275

	5.2.2	Artificial Intelligence of Things	277
	5.2.3	Cyber-Physical Systems and Artificial Intelligence of Things	280
5.3	METHODOLOGY		284
5.4	RESULTS: ANALYSIS AND SYNTHESIS		287
	5.4.1	Functional and Architectural Dimensions	287
	5.4.2	Architectural Layers of Artificial Intelligence of Things Grounded in the Data Science Cycle	304
	5.4.3	The Dynamic Interplay of Artificial Intelligence of Things and Cyber-Physical Systems of Systems	307
	5.4.4	An Innovative Data-Driven Management and Planning Framework	312
5.5	DISCUSSION		320
	5.5.1	Summary of Findings and Interpretation of Results	320
	5.5.2	Comparative Analysis	321
	5.5.3	Benefits and Opportunities	322
	5.5.4	Implications for Research, Practice, and Policymaking	324
	5.5.5	Challenges, Barriers, and Limitations	325
	5.5.6	Suggestions for Future Research Directions	327
5.6	CONCLUSION		328
REFERENCES			330

CHAPTER 6 ▪ Artificial Intelligence of Things for Harmonizing Smarter Eco-City Brain, Metabolism, and Platform: An Innovative Framework for Data-Driven Environmental Governance — 343

6.1	INTRODUCTION	345
6.2	ESSENTIAL COMPONENTS OF THE PROPOSED INTEGRATED FRAMEWORK: INTERLINKAGES AND CONCEPTUAL DEFINITIONS	350
6.3	A SURVEY OF RELATED WORK	353
6.4	RESEARCH METHODOLOGY	356

6.5	RESULTS: ANALYSIS AND SYNTHESIS		360
	6.5.1	Artificial Intelligence of Things	360
	6.5.2	City Brain as a Large-Scale System of Artificial Intelligence of Things	364
	6.5.3	Smart Urban Metabolism	369
	6.5.4	Platform Urbanism	373
	6.5.5	Roles of Emerging Data-Driven Governance Systems in Enhancing Environmental Governance Processes and Strategies: A Theoretical and Empirical Analysis	380
	6.5.6	A Pioneering Framework for Data-Driven Environmental Governance in Smarter Eco-Cities	390
6.6	DISCUSSION		395
	6.6.1	Interpretation of Results	396
	6.6.2	Comparative Analysis	396
	6.6.3	Methodological Evaluation and Limitations	397
	6.6.4	Implications for Research, Practice, and Policymaking	398
	6.6.5	Suggestions for Future Research Directions	399
6.7	CONCLUSION		401
	REFERENCES		403

CHAPTER 7 ▪ Artificial Intelligence and Its Generative Power for Advancing the Sustainable Smart City Digital Twin: A Novel Framework for Data-Driven Environmental Planning and Design   416

7.1	INTRODUCTION		417
7.2	RELATED WORK		422
7.3	METHODOLOGY		423
7.4	CONCEPTUAL BACKGROUND		425
	7.4.1	Sustainable Smart Cities and Their Planning and Design Processes	425
	7.4.2	Artificial Intelligence	427

	7.4.3	Generative Artificial Intelligence, Large Language Models, and Pre-Trained Foundation Models	431
	7.4.4	Urban Digital Twin and Its Synergistic Integration with Artificial Intelligence	436
7.5	APPLICATIONS OF ARTIFICIAL INTELLIGENCE, GENERATIVE ARTIFICIAL INTELLIGENCE, AND URBAN DIGITAL TWIN IN URBAN PLANNING AND DESIGN		437
	7.5.1	Artificial Intelligence and Its Five Subfields	437
	7.5.2	Urban Digital Twin	449
	7.5.3	Generative Artificial Intelligence for Environmental Sustainability and Climate Change	456
	7.5.4	Generative Artificial Intelligence for Urban Planning and Design	460
7.6	BLUE CITY PROJECT CASE STUDY: GENERATIVE SPATIAL ARTIFICIAL INTELLIGENCE FOR LAUSANNE CITY DIGITAL TWIN		463
7.7	A SYNTHESIZED FRAMEWORK FOR GENAI-DRIVEN SUSTAINABLE SMART CITY PLANNING AND DESIGN		466
7.8	DISCUSSION		469
	7.8.1	Interpretation of Findings	470
	7.8.2	Comparative Analysis	470
	7.8.3	Implication for Research, Practice, and Policymaking	471
	7.8.4	Limitations and Mitigation Strategies	473
	7.8.5	Key Challenges and Risks of Generative Artificial Intelligence	474
	7.8.6	Suggestions for Future Research Directions	477
7.9	CONCLUSION		478
REFERENCES			479

CHAPTER 8 ■ Transforming Smarter Eco-Cities with Artificial Intelligence: Harnessing the Synergies of Circular Economy, Metabolic Circularity, and Tripartite Sustainability — 498

8.1 INTRODUCTION — 499
8.2 CONCEPTUAL DEFINITIONS AND DISCUSSIONS — 504
 8.2.1 Urban Circularity and Circular Economy: Commonalities and Differences — 504
 8.2.2 Circularity, Circular Economy, and Smart Urban Metabolism — 507
 8.2.3 Smart Urban Metabolic Circularity and Tripartite Sustainability — 510
 8.2.4 Smart Sustainable Cities: A Tripartite Sustainability Perspective — 512
 8.2.5 From Eco-Cities to Smarter Eco-Cities: An Environmental and Technological Trajectory Perspective — 517
8.3 RESEARCH DESIGN — 519
8.4 RESULTS — 521
 8.4.1 Artificial Intelligence and Artificial Intelligence of Things for Metabolic and Circular Futures in Smarter Eco-Cities — 522
 8.4.2 Smart Eco-Cities and Symbiotic Cities as Prominent Approaches to Environmentally Smart Sustainable Urbanism — 530
 8.4.3 The Role of Artificial Intelligence of Things in Advancing Metabolic Circularity in Smarter Eco-Cities: Integrating Eco-Cities, Smart Cities, and Circular Cities — 534
 8.4.4 Advanced Artificial Intelligence of Things Solutions for Waste Management in Smarter Eco-Cities — 537
 8.4.5 Artificial Intelligence and Artificial Intelligence of Things Solutions for Circular Economy in Smarter Eco-Cities — 540

	8.4.6	The 4R Framework and Artificial Intelligence: Their Contribution to Circular Economy and Tripartite Sustainability in Smarter Eco-Cities	546
	8.4.7	Exploring the Interrelationships among Circular Economy, SDGs, and Smart Cities	550
	8.4.8	Smarter Eco-Cities as an Emerging Approach to Data-Driven Smart Sustainable Cities	553
8.5	A NOVEL FRAMEWORK FOR INTEGRATING CIRCULAR ECONOMY, METABOLIC CIRCULARITY, AND TRIPARTITE SUSTAINABILITY THROUGH ARTIFICIAL INTELLIGENEC OF THINGS IN SMARTER ECO-CITIES		562
8.6	DISCUSSION		566
	8.6.1	Key Findings and Their Interpretation	567
	8.6.2	Circular Economy and Artificial Intelligence of Things in Smarter Eco-Cities	567
	8.6.3	Artificial Intelligence of Things in Smarter Eco-Cities	572
8.7	CONCLUSION		578
REFERENCES			580

INDEX, 595

Preface

RECENT ADVANCEMENTS IN ARTIFICIAL intelligence (AI) and AI of things (AIoT) are revolutionizing urban development, ushering in a new era of highly efficient, resilient, environmentally conscious, and technologically sophisticated urban environments. These technologies are transforming the way cities operate, function, evolve, and tackle complex environmental challenges. By facilitating and enhancing the integration, coordination, and effectiveness of urban systems, AI and AIoT unlock the tremendous potential to advance environmental sustainability goals. These technologies open up new pathways for urban innovation and transition, redefining how cities manage resources and respond to ecological pressures. Specifically, AI and AIoT contribute significantly to improved energy management, resource conservation, pollution reduction, infrastructure resilience, ecosystem monitoring, and disaster response—all of which are crucial for effective climate change mitigation and adaptation strategies.

Furthermore, AI and AIoT facilitate more dynamic and responsive urban processes and practices by leveraging interconnected devices, real-time monitoring, advanced data analytics, predictive modeling, and intelligent decision-making. This approach aligns closely with sustainable development goals, fostering a synergistic integration of data-driven management, planning, and governance systems in smarter eco-cities and sustainable smart cities, thus improving their efficiency, resilience, and overall sustainability. This integration encourages collaboration among urban planners, policymakers, technology experts, and community actors, thereby streamlining urban management and planning processes. Such collaboration is critical for addressing complex environmental challenges and achieving sustainability outcomes within the framework of urban governance through its structures, processes, and institutions.

The urban environment, encompassing the physical, social, and economic aspects of city life, is currently undergoing a significant shift.

Propelled by rapid advancements in computational technologies and collaborative models, and driven by a growing recognition of the urgent need to accelerate transitions toward environmental sustainability, this shift involves major changes in the built environment as well as social, economic, and ecological systems. It is manifested in revolutionizing decision-making processes by harnessing data-driven insights and fostering innovative solutions to effectively address the pressing challenges of modern urbanization, including resource scarcity, ecological degradation, and climate change.

Therefore, cities globally are responding to these challenges by exploring new approaches to sustainable development, focusing on strategies that ensure the long-term sustainability and resilience of urban environments. Such progress hinges on collaborative efforts among governments, institutions, businesses, communities, and citizens to formulate more effective policies, ensuring long-term ecological balance. By embracing these integrated and holistic approaches, cities can be better equipped to address the multifaceted challenges of modern urbanization, thereby fostering a sustainable future. In this context, the integration of AI and AIoT technologies into urban management, planning, and governance is proving to be a promising avenue for driving impactful and positive changes in cities. These technologies provide unprecedented capabilities to integrate urban systems, coordinate urban domains, enhance resource efficiency, lower environmental impacts, and improve the quality of life for citizens.

The book, *Artificial Intelligence of Things for Smarter Eco-Cities: Pioneering the Environmental Synergies of Urban Brain, Digital Twin, Metabolic Circularity, and Platform,* explores the transformative impact of advanced technologies, particularly AI and AIoT, on sustainable urban development. It highlights their pivotal role in facilitating the synergistic and collaborative integration of UB, UDT, and SUM as forms of cyber-physical systems of systems (CPSoS), as well as platform urbanism, within the dynamic landscapes of smarter eco-cities and sustainable smart cities. These data-driven management and planning systems signify a major shift toward platform urbanism, where digital platforms are designed to create interconnected, intelligent urban ecosystems. These systems facilitate various urban functions and services, thereby enhancing urban life and resource management through real-time data integration and analysis. This approach enables a cohesive framework that supports dynamic, responsive, and participatory modes of urban governance.

The book details how UB enhances real-time urban intelligence, UDT advances urban modeling and simulation, and SUM improves urban metabolic analysis, all within the framework of platform urbanism. Through a focus on these systems, the book underscores the untapped potential of effectively harnessing data-driven management, planning, and governance systems to enhance environmental sustainability outcomes and advance sustainable development goals. It represents an extensive body of research aimed at both understanding and applying AI and AIoT technologies in environmentally sustainable urban development, demonstrating their capacity to drive innovative and holistic solutions in emerging smarter eco-cities and sustainable smart cities.

The book emphasizes the critical role of environmental synergies in the context of smarter eco-cities and sustainable smart cities, showcasing the sophisticated incorporation of AI and AIoT technologies into UB, UDT, SUM, and the broader framework of platform urbanism. These synergies highlight the cooperative interactions among various data-driven urban systems, yielding environmental benefits that exceed the sum of their individual contributions. By advancing strategies for environmental sustainability and climate change mitigation and adaptation, the book delineates a new frontier in the application and integration of AI and AIoT technologies in emerging data-driven urban management, planning, and governance systems. This endeavor is pioneering in its approach to harmonizing the functioning of these advanced systems through these cutting-edge technologies. It positions the book at the forefront of technological and ecological innovation in urban development. The exploration of synergistically integrating UB, UDT, SUM, and platform urbanism reveals how AI and AIoT can be harnessed to enhance the efficiency, sustainability, resilience, and livability of urban environments.

The book offers comprehensive theoretical analyses, synthesizes rich evidence, and presents use cases and practical case studies alongside interdisciplinary insights. It seeks to unravel the complexities of the multifaceted research topic at hand and introduces several pioneering frameworks that integrate UB, UDT, SUM, and platform urbanism for practical implementation. These integrated approaches are poised to significantly transform how smarter eco-cities and sustainable smart cities are environmentally managed, planned, and governed. The book emphasizes the importance of interdisciplinary perspectives and highlights the holistic approach required to effectively develop, implement, and integrate computational technologies and collaborative models.

The book equips readers with a thorough understanding of the opportunities and challenges inherent in building smarter eco-cities and sustainable smart cities through innovative technologies and models. As they progress through the chapters, readers will gain deep insights into the transformative potential of AI, AIoT, and CPSoS in reshaping urban environments, advancing environmental sustainability, improving citizen well-being, and driving urban innovation. Additionally, readers will gain insights into the pioneering nature of AI- and AIoT-driven sustainable urban development, which provides a comprehensive and actionable framework for cities to evolve in response to future demands and pressures while ensuring environmental sustainability, maintaining technological advancement, and fostering resilient and adaptive urban ecosystems.

I would like to extend my sincere gratitude to all the contributors who have indirectly shared their expertise and insights. Their commitment and passion for advancing sustainable urban development have been crucial in shaping the content of this volume and instrumental in bringing this project to fruition. While this book adopts a groundbreaking approach, it is their invaluable input that has provided the foundational insights discussed throughout and the inspiration behind them. This has enriched this work, establishing it as a seminal reference and valuable resource for researchers, practitioners, policymakers, and anyone interested in the dynamic interplay of technology, innovation, and sustainability in the era of digital revolution.

In closing, I envision this book as a catalyst for spurring groundbreaking research endeavors, inspiring fresh ideas, stimulating practical implementations, sparking meaningful discussions, and shaping effective policy-making, all paving the way for a more sustainable future for our cities. By sharing knowledge and best practices, the goal is to collaboratively create urban environments that transcend mere intelligence to become vibrant hubs of resilience, inclusivity, and environmental stewardship, ultimately fostering prosperity and well-being for all.

Thank you for joining me on this exploratory journey into the transformative era of AI and AIoT. Together, we have embarked on an exploration of how these cutting-edge technologies can revolutionize urban management, planning, and governance, particularly in the context of environmental sustainability. I hope this journey has sparked your curiosity and provided valuable insights into the future of sustainable urban development.

Simon Elias Bibri
Lausanne

Acknowledgments

This book is the culmination of rich learning experiences and meaningful intellectual pursuits, made possible by the many individuals I have had the privilege to meet or collaborate with. I owe a debt of gratitude to those who have directly or indirectly, knowingly or unknowingly, contributed to my intellectual enrichment and growth. Their support has been the bedrock of my academic journey, fueling my passion for scientific research, fostering confidence in my intellectual capabilities, and ultimately inspiring me to pursue the path of an academic author.

I extend my heartfelt thanks to those who have indirectly contributed to this book, making it a truly enjoyable and rewarding intellectual endeavor. First and foremost, I am profoundly grateful to Professor Jeffrey Huang for his unwavering support, remarkable openness, and genuine kindness. His steadfast encouragement and the autonomy he entrusted me with have been pivotal in allowing me to explore new ideas and pursue bold initiatives, fostering my professional and intellectual growth in ways I had never anticipated. This trust granted me the liberty to make decisions, explore new horizons, and engage in ambitious endeavors that might not have been possible under more constrained circumstances. As a professor, I hold him in the highest regard for his integrity and exemplary human qualities.

I would also like to express my sincere gratitude to Professor John Krogstie for his unwavering support throughout my academic journey. His insightful guidance and constant encouragement have greatly enriched my academic experience and contributed to my professional development. His dedication to creating a supportive and intellectually stimulating environment during my PhD journey has been instrumental in shaping my scholarly pursuits and achievements.

Additionally, I would like to acknowledge the many scholars whose work and insights have made this book possible. Their dedication to advancing sustainable urban development and their innovative contributions to AI and AIoT technologies have been critical in shaping the content of this book.

Lastly, but certainly not least, I would like to extend my deepest gratitude to my sister, whose profound generosity has been the driving force behind my professional endeavors in more recent years. Her selflessness has provided me with the strength and determination to overcome challenges and confidently pursue my aspirations and goals.

Author

Simon Elias Bibri, PhD, is a senior research scientist and project coordinator at the Swiss Federal Institute of Technology Lausanne (EPFL), Institute of Computer and Communication Sciences (IINFCOM), School of Architecture, Civil and Environmental Engineering (ENAC), and Media and Design Laboratory (LDM). He has a diverse professional background, having served as head of a computer department, software engineer, information technology (IT) business engineer, project manager, green information and communications technology (ICT) and environmental sustainability strategist, research associate, assistant professor, and consultant. Currently, he is an expert in Artificial Intelligence and Applied Machine Learning for Climate Action for the United Nations Framework Convention on Climate Change (UNFCCC), an international expert in Sustainable Cities for the United Nations Industrial Development Organization (UNIDO), and an expert for Focus Group for Smart Sustainable Cities, Environmental Sustainability, and Digital Twin for the United Nations Agency of International Telecommunication Union (ITU). He also holds the position of Editor-in-Chief of the *International Journal of Environmental Studies* at Taylor & Francis Group.

Dr. Bibri's academic journey exemplifies both interdisciplinarity and transdisciplinarity, traversing and bridging the realms of science, technology, and society. This multifaceted background underscores his ability to navigate and contribute to various fields, thereby enriching the intersectional understanding and collaborative knowledge within these critical areas. His intellectual background encompasses computer science, computer engineering, systems science, innovation science, environmental science, social sciences, and humanities. He holds a bachelor's degree in computer engineering and an advanced management program (AMP) degree. Notably, he earned multiple master's degrees at esteemed Swedish universities, namely, Lund University, West University, Blekinge Institute of Technology, Stockholm University, Malmö University, and Mid-Sweden University. Additionally, he earned a PhD in computer science at the Norwegian University of Science and Technology (NTNU), specializing in computational technology and urban informatics, with a research focus on data-driven smart sustainable cities.

Dr. Bibri is a prolific author with numerous highly cited journal articles, six authored books, two edited books, and five co-edited books to his credit. His work has been cited over 9,500 times, resulting in an h-index of 47. His impact is further underscored by his recognition by Stanford and Elsevier as one of the top 2% of scientists globally for five consecutive years. He has also been listed by ScholarGPS in 2022 as the #1 highly ranked scholar in sustainable cities, #4 in big data, and #8 in sustainability in the world.

Dr. Bibri's research interests and expertise cover a broad spectrum of areas at the intersection of science and technology, urban development and planning, and strategic sustainable development. These areas include smart cities, sustainable cities, smarter eco-cities, sustainable smart cities, urban artificial intelligence, urban artificial intelligence of things, urban digital twin, smart urban metabolism, cyber-physical systems of systems, Internet of City Things, platform urbanism, environmental planning and governance, environmental sustainability, climate change mitigation and adaptation, smart sustainable energy, smart sustainable transportation, smart sustainable waste management, sustainability transitions, and technological innovation systems.

CHAPTER 1

Transformative Integration of Artificial Intelligence of Things and Cyber-Physical Systems of Systems

Advancing Smarter Eco-Cities and Sustainable Smart Cities

Abstract

This chapter sets the stage for an in-depth exploration of cutting-edge technologies and innovative models aimed at reshaping the landscape of environmental management, planning, and governance in smarter eco-cities and sustainable smart cities. It offers a comprehensive introduction to the book, covering background, justification, aim, objectives, motivation, significance, structure and content, and readership. It initiates with an overview of the intricate interplay of City Brain, urban digital twin (UDT), smart urban metabolism (SUM), and platform urbanism in the context

DOI: 10.1201/9781003536420-1

of artificial intelligence (AI) of things (AIoT) and cyber-physical systems (CPSs) of systems (CPSoS). Notably, there exists no work to date that has comprehensively investigated the synergistic and collaborative integration of these data-driven systems in advancing environmental sustainability goals in the rapidly evolving landscapes of smarter eco-cities and sustainable smart cities. The key topics are introduced in this inaugural chapter and further developed in the subsequent chapters. Specifically, each of these chapters will explore these technologies and models in depth, providing thorough analysis, evidence synthesis, and real-world case studies to demonstrate their complementary capabilities and untapped synergies to advance sustainable urban development. Emphasis will be placed on interdisciplinary perspectives to highlight the holistic approach necessary for the effective development, implementation, and integration of these technologies and models. Readers will be introduced to the pioneering nature of these emerging data-driven management, planning, and governance systems, particularly in how they can be effectively utilized and synergistically harnessed to accelerate the needed transition to environmental sustainability. They will be equipped with the knowledge and tools to understand the transformative potential of AI and AIoT in shaping the cities of tomorrow and leverage these integrated systems for tackling environmental challenges in the dynamic realms of smarter eco-cities and sustainable smart cities.

1.1 INTRODUCTION AND BACKGROUND

In an era increasingly marked by the rapid pace of urbanization and the increasing complexity of ecological degradation, the imperative for advancing sustainable urban development has become more pressing than ever before. In response to this unprecedented level of urgency, cities across the globe are actively seeking ways to address the daunting realities of rapid population growth, dwindling resources, and the looming specter of climate change. At the forefront of addressing these formidable challenges and wicked problems is the integration of cutting-edge technologies and innovative models. This integration has emerged as a beacon of hope, offering transformative solutions to navigate the complexities of urban management, planning, and governance to accelerate the needed transition to environmental sustainability. Central to this integration are both the recent advancements in artificial intelligence (AI) and the Internet of Things (IoT) and their groundbreaking convergence—known as AIoT.

The power of AI and its synergistic integration with IoT are driving the dawn of a new era in urban transformation across various domains of emerging smarter cities (Alahi et al. 2023; Bibri et al. 2023a, 2024a; Hoang 2024; Hoang et al. 2024; Jagatheesaperumal et al. 2024; Kuguoglu et al. 2021; Mishra and Singh 2023) and smarter eco-cities (Bibri et al. 2023a, 2024b). Most of these domains relate to environmental sustainability, highlighting the critical role of AI, IoT, and AIoT technologies in achieving green, resilient, and efficient urban environments. These domains encompass energy efficiency, transportation management, waste management, water management, pollution control, biodiversity conservation, ecosystem health, and climate change mitigation and adaptation (e.g., Bibri 2020; Bibri et al. 2023b; Bibri et al. 2024a; Chen et al. 2023; Mishra 2023; Seng et al. 2022; Ullah et al. 2020; Zhang and Toa 2021). Several recent studies have also addressed the significant potential of AIoT in advancing Sustainable Development Goals (SDGs) related to these domains and sustainable urban development (Mishra 2023; Thamik et al. 2024). Moreover, AI and AIoT play a crucial role in climate change mitigation and adaptation by enabling precise monitoring, efficient resource management, and predictive analytics to reduce emissions and enhance resilience (e.g., El Himer et al. 2022; Dheeraj et al. 2020; Jain et al. 2023; Leal Filho et al. 2022; Popescu et al. 2024; Rane et al. 2024; Rolnick et al. 2022; Shaamala et al. 2024; Tomazzoli et al. 2020; Samadi 2022).

AI and AIoT are marking the rise of smarter eco-cities and sustainable smart cities at the nexus of technological innovation and environmental responsibility. As urbanization accelerates globally, these two concepts have emerged as innovative solutions to address environmental and urban challenges. Both city models aim to create sustainable and livable environments by integrating advanced technologies and sustainable practices. Understanding their definitions, commonalities, and differences provides insight into how these urban strategies can shape the future of urban development. Smarter eco-cities are urban developments designed to integrate ecological principles with advanced technologies, especially AI and AIoT to create a sustainable, efficient, and livable environment. These cities prioritize environmental sustainability by using green infrastructure, renewable energy, and conservation methods. The goal is to minimize the ecological footprint while enhancing the quality of life for residents. Sustainable smart cities focus on utilizing advanced technologies to enhance the efficiency of urban services and infrastructure, improve the quality of life for residents, and ensure sustainability. These cities incorporate smart technologies such as IoT, big data analytics, and AI to manage resources efficiently, reduce waste, and optimize urban operations.

In view of the above, smarter eco-cities and sustainable smart cities both strive to leverage emerging data-driven technologies for advancing environmental sustainability goals, yet they approach these goals from slightly different angles. Smarter eco-cities prioritize environmental sustainability as their core objective, aiming to minimize environmental impact, conserve resources, promote biodiversity, enhance resilience, and mitigate climate change risks. In smarter eco-cities, AI and AIoT play a crucial role in environmental monitoring, and real-time data collected from IoT sensors and devices on environmental parameters are analyzed to understand environmental conditions, patterns, and trends. This data-driven approach informs decision-making processes for city planners and policymakers, allowing them to implement effective environmental strategies, policies, and interventions. Additionally, community engagement is prioritized to raise awareness about environmental issues, with AIoT-driven data used to involve citizens in sustainability initiatives.

On the other hand, sustainable smart cities encompass not only environmental sustainability but also social and economic aspects. They aim for a holistic approach, balancing environmental goals with other urban development priorities (Efthymiou and Egleton 2023; Mishra and Singh 2023; Zaidi et al. 2023).These cities heavily rely on AIoT and cyber-physical systems (CPSs) to manage various urban systems, including energy, transportation, water, and waste, which are interconnected to achieve overall sustainability goals (Alahi et al. 2023; Bibri et al. 2024a; Mylonas et al. 2021; Juma and Shaalan 2020; Singh et al. 2023a). Efficient resource management is a key focus of sustainable smart cities, achieved through technologies like energy-efficient buildings, smart grids, and optimized transportation systems. Data-driven approaches facilitated by AIoT and CPS contribute to more effective policymaking and governance. By analyzing real-time data, city authorities can make informed decisions to address environmental challenges and ensure sustainable development.

Enhancing the quality of life is a central goal for both smarter eco-cities and sustainable smart cities, making them integral parts of the future of urban development. Both aim to provide clean air, green spaces, efficient public services, and a healthy living environment. In terms of design approach, smarter eco-cities often incorporate biophilic design principles, promoting a close connection with nature through extensive green spaces, urban forests, and sustainable landscaping. Sustainable smart cities, however, focus on the integration of smart technologies into the urban fabric, such as smart grids, intelligent transportation systems, and automated building management systems. When it comes to technology utilization, smarter eco-cities use technology to support environmental goals. Sustainable smart cities, on the other hand, utilize a broader range of smart technologies for various urban functions, including governance, public safety, transportation, and health care.

The relationship between smarter eco-cities and sustainable smart cities is one of mutual dependence and technological synergy. While smarter eco-cities primarily focus on environmental sustainability, sustainable smart cities integrate social and economic considerations into their approach. They share complementary goals, with smarter eco-cities benefiting from the holistic approach of sustainable smart cities. Smarter eco-cities can integrate social and economic factors into their environmental initiatives, creating a holistic approach to urban sustainability. Meanwhile, sustainable smart cities can draw inspiration from smarter eco-cities' emphasis on nature-based solutions and environmental conservation, utilizing advanced technologies to enhance these efforts. By combining ecological design principles with smart technology, both types of cities can achieve more effective and comprehensive solutions for urban challenges.

Both classes of cities benefit from advancements in AI, AIoT, and CPSs of systems (CPSoS). As such, they contribute to data-driven environmental management, planning, and governance through the integration of these advanced technologies. This collaboration leads to the creation of more livable, resilient, and sustainable urban environments that address the complex challenges of modern urbanization, resource depletion, ecological degradation, and climate change.

AIoT serves as a driving catalyst and new frontier for emerging data-driven management, planning, and governance solutions for smarter eco-cities and sustainable smart cities. Prominent among these emerging AIoT-powered solutions are City Brain (Kierans et al. 2023; Liu et al. 2022; Xu et al. 2024), urban digital twin (UDT) (Beckett 2022; Bibri et al. 2024a; Zayed et al. 2023), smart urban metabolism (SUM) (e.g., Bibri et al. 2024b; Ghosh and Sengupta 2023; Peponi et al. 2022), and platform urbanism (Haveri and Anttiroiko 2023; Repette et al. 2021; ITU 2022). These solutions not only enable more efficient and effective management and planning of urban systems but also have the potential to advance environmentally sustainable urban development goals as a form of governance by goals. Their transformative potential lies in their ability to integrate data from various sources and domains, allowing for holistic and proactive decision-making that considers environmental impacts and fosters resilience in the face of changing conditions and climate change threats. Through the integration of these emerging AI or AIoT-driven solutions, both smarter eco-cities and sustainable smart cities can work toward achieving long-term environmental sustainability goals while enhancing the quality of life for their citizens.

The concepts of smart eco-cities and smart cities have evolved beyond mere environmental strategies and technological advancements, respectively, to encompass an integrated approach to urban development, emphasizing sustainability, resilience, citizen well-being, and innovation. Leveraging AI and AIoT technologies, these urban environments are evolving into smarter eco-cities and (environmentally) sustainable smart cities (e.g., Bibri et al. 2023a, b, 2024a, b; Bibri and Huang 2024; Gourisaria et al. 2023; Efthymiou and Egleton 2023; Mishra and Singh 2023; Zaidi et al. 2023; Singh et al. 2023; Matei and Cocosatu 2024). In these interconnected ecosystems, data-driven insights drive decision-making, optimize resource allocation, alleviate pollution levels, mitigate climate change risks, and improve the quality of life. In these dynamic landscapes, innovative initiatives, notably City Brain, UDT, SUM, and platform urbanism, have become at the forefront in recent years (Beckett 2022; Bibri et al. 2024a, b; Bibri and Huang 2024; D'Amico et al. 2021; Kierans et al. 2023; Weil et al. 2022). They leverage the capabilities of AI and AIoT to create urban environments that not only enhance but also seamlessly integrate urban management, planning, and governance processes.

In addition, the concept of CPS further emphasizes the interconnectedness of City Brain, UDT, SUM, and platform, highlighting their role in creating holistic urban ecosystems where technological innovation and environmental sustainability converge. CPS refers to interconnected systems of digital and physical components that collaborate to achieve specific goals in urban contexts, such as optimizing urban operations and services, enhancing resource allocation, improving transportation efficiency, mitigating air and noise pollution, and streamlining waste management processes. These systems integrate computing, communication, and control systems with physical processes to increase efficiency, effectiveness, and reliability. They seamlessly link smart city infrastructure with IoT and AI technologies (Sing et al. 2023a, b).

AI is used in CPS to analyze real-time data from sensors and actuators (IoT), enabling intelligent decision-making and adaptive control of physical processes. This integration enhances system performance, optimizes resource use, and improves the efficiency of operations (Sharma and Sharma 2022; Radanliev et al. 2020). The symbiotic relationship between AIoT and CPS forms a powerful framework that revolutionizes the operation and integration of data-driven management and planning systems in smarter eco-cities and sustainable smart cities. The integration of AIoT and CPS has amplified the transformative capabilities of City Brain, UDT, and SUM in managing and planning these urban environments. Singh et al. (2023a) demonstrate the significant potential of CPS in improving efficiency, environmental management and planning, and decision-making in smart city systems.

At its core, AIoT integrates AI's computational and analytical capabilities with the ubiquitous network of interconnected IoT sensors and devices. This integration transforms urban landscapes into sophisticated, interconnected ecosystems, where urban systems and platforms not only collect and exchange data but also possess the autonomy to analyze, learn, perform complex task, make decisions, and adapt to changing environments (Seng et al. 2022; Shi et al. 2020; Zhang and Tao 2021). AIoT empowers real-time data analytics, predictive modeling, and adaptive responses, thereby optimizing the overall performance and efficiency of IoT applications in smarter eco-cities and sustainable smart cities (Bibri et al. 2023a, 2024a, b; Bibri and Huang 2024). Most of its functionalities align with the core principles of CPS in terms of sensing, computation, communication, networking, and storage.

AIoT plays a key role in advancing CPS and CPSoS. AIoT enhances the capabilities of CPS and CPSoS by introducing intelligent decision-making and autonomous functionalities. AIoT amplifies each technical component of CPS and CPSoS, driving more efficient, adaptive, and secure operations, including sensing, communication network, network security, computation and control, actuation, software integration, and cybersecurity (e.g., Alowaidi et al. 2023; Gürkaş et al. 2023; Rajawat et al. 2022; Sharma and Sharma 2022), including in the context of smart cities (Juma and Shaalan 2020; Khan et al. 2021; Singh et al. 2023a, 2023b) and sustainable smart cities (Bibri and Huang 2024).

In the sensing component, AIoT enhances the data collection process by enabling smart sensors. These sensors gather raw data and incorporate AI algorithms to preprocess and analyze data at the edge. This local processing reduces the volume of data that needs to be transmitted, thus decreasing latency and bandwidth usage. AI-enabled sensors can also identify patterns and anomalies in real time, leading to more proactive and predictive maintenance and operations. Concerning the communication network, AIoT optimizes it by using AI techniques to manage and route data traffic efficiently. AI algorithms can predict network congestion, optimize routing paths, and allocate bandwidth dynamically based on the current demand and priority of data. This intelligent management ensures reliable and low-latency communication, which is critical for real-time CPS and CPSoS applications. Additionally, AI can enhance network security by detecting and mitigating cyber threats and anomalies in network behavior. The computation and control component benefits significantly from AIoT through advanced data analytics and machine learning (ML) models. AI processes the vast amounts of data collected by sensors to extract meaningful insights, enabling more accurate and efficient decision-making. AI algorithms can predict future states, optimize control strategies, and even perform autonomous control without human intervention. This leads to improved system performance, energy efficiency, and adaptability to changing conditions.

Regarding actuation, AIoT empowers actuators with intelligent control algorithms that enable more precise and adaptive responses. AI models can predict the optimal actuation required based on real-time data and context, ensuring more accurate control actions. This intelligent actuation enhances the overall reliability and efficiency of CPS and CPSoS. Furthermore, AIoT enhances the integration of software in CPS and CPSoS by providing advanced middleware solutions that facilitate seamless interoperability and data flow. AI-driven middleware can manage and orchestrate the interactions between sensors, communication networks, computation units, and actuators more efficiently. Moreover, AI-powered user interfaces offer predictive analytics and decision support tools, making it easier for operators to monitor and control CPS and CPSoS effectively. These intelligent interfaces can provide actionable insights and automate routine tasks, reducing the cognitive load on human operators. In terms of cybersecurity, AIoT plays a crucial role by implementing intelligent security systems that can detect, analyze, and respond to threats in real time. AI algorithms can identify unusual patterns of behavior, flag potential security breaches, and automatically initiate countermeasures to mitigate risks. This proactive approach to cybersecurity enhances the resilience of CPS and CPSoS against various cyber threats, ensuring the continuous and safe operation of critical systems.

Overall, AIoT significantly enhances the functionality and performance of CPS and CPSoS by embedding intelligence across all components. From smart sensing and efficient communication to advanced computation, intelligent actuation, seamless software integration, and robust cybersecurity, AIoT transforms CPS into more efficient, adaptive, and secure systems. As AIoT technologies continue to evolve, their impact on CPS will expand, leading to even greater levels of autonomy and intelligence in these systems.

Furthermore, AIoT underpins the functioning of City Brain, UDT, and SUM as forms of CPSoS, as well as platform urbanism, with the aim of advancing data-driven management, planning, and governance in smarter eco-cities and sustainable smart cities (Figure 1.1). CPS leverages IoT for deployment by integrating sensor data, connectivity, and real-time monitoring, with AI algorithms analyzing the collected data to enable intelligent decision-making and automation (Mohamed 2023) in smart cities (Singh et al. 2023a). The integration of AIoT amplifies IoT deployment by

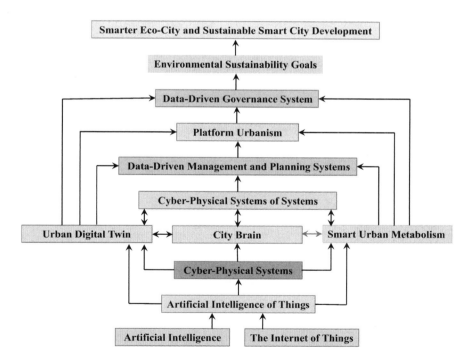

FIGURE 1.1 A framework for smarter eco-city and sustainable smart city development based on City Brain, UDT, SUM, and platform urbanism as AIoT-driven CPSoS.

infusing AI algorithms into the network of interconnected IoT devices as part of CPS and hence CPSoS, fostering advanced data analysis, real-time insights, intelligent decision-making, and autonomous control in sustainable smart cities as real-time urban environments (Bibri and Huang 2024).

City Brain, UDT, and SUM exemplify CPSoS tailored specifically for urban environments given their unique characteristics in terms of system size, complexity, and requirements. As such, they represent integrated systems of integrated systems that combine computational and physical components, tightly coupled and interacting with each other using sensors, computing elements, communication networks, and actuators to monitor, analyze, and control physical processes in real time.

City Brain, as an AIoT-driven platform, integrates data from various sources, including IoT sensors, cameras, and other urban infrastructure, to monitor and manage urban systems in real time. It analyzes these data to make informed decisions and optimize resource allocation, traffic flow, emergency response, and other aspects of urban management (Kierans et al. 2023; Liu et al. 2022; Zhang et al. 2019). In the context of CPSoS, it serves as a central intelligence hub that orchestrates the functioning of various urban systems, such as transportation, energy, waste, and public services. It connects with the physical infrastructure of the city through sensors and actuators, creating a feedback loop between the digital and physical worlds. By leveraging AI and IoT technologies, it contributes to the creation of a holistic urban ecosystem where cyber and physical components interact seamlessly to enhance efficiency, resilience, circularity, and sustainability (Kierans et al. 2023; Liu et al. 2022). Overall, City Brain exemplifies the integration of AIoT in CPSoS to create smarter and more responsive urban environments, where data-driven decision-making and automation play a crucial role in improving the quality of life for citizens.

UDT is a dynamic virtual replica of a city that mirrors its physical, spatial, and functional aspects in a digital realm. Integrating real-time data generated by IoT sensors and devices, with advanced AI models and algorithms, UDT monitors, models, simulates, and visualizes urban processes, infrastructures, systems, and dynamics. UDT relates to CPSoS through its role in creating a three-dimensional (3D) representation of physical urban environments, allowing for real-time monitoring, analysis, and optimization of various urban systems. Virtual simulation algorithms, DL architectures, and cyber-physical cognitive systems orchestrate the development of interconnected smart cities (Nica et al. 2023) toward sustainability. Introducing the concept of DTs as a cutting-edge paradigm for

CPS, Somma et al. (2023) highlight the versatility of DTs across various domains, with a particular focus on UDT, which aims to virtually replicate urban assets such as buildings, transportation infrastructure, energy grids, and waste management facilities. The authors demonstrate UDT applicability to real-world CPS. Agostinelli et al. (2020) apply DT and AI as a form of CPS to improve building energy management. The authors demonstrate the potential of DT and AI models in optimizing energy systems and promoting sustainability in residential areas. They underscore the importance of leveraging CPS technologies to address the challenges of urban energy management and achieve energy efficiency targets.

In urban management and planning, UDT in smart cities exemplifies CPSoS given their unique characteristics and complexities (Bibri et and Huang 2024; Mylonas et al. 2021). It serves as complex systems of systems, structured hierarchically with interconnected DT instances that process and share vast amounts of data governed by diverse access and usage policies (Somma et al. 2023). Bibri et al. (2024a) explore the interactions between AI, AIoT, UDT, data-driven planning, and environmental sustainability in sustainable smart cities. The authors delve into the complex ecosystem of these cities, uncovering nuanced dynamics and untapped synergies. Their findings provide valuable insights into the transformative potential of integrating AI, AIoT, and UDT technologies for sustainable urban development. However, the bidirectional interactivity between physical systems and their digital counterparts may often necessitate human intervention (Shahat et al. 2021; Weil et al. 2023). Overall, UDT, as a manifestation of CPSoS, enables cities to achieve a more comprehensive understanding of their urban environments and enhance decision-making processes related to urban management and planning. This integration ultimately leads to enhanced sustainability, resilience, and the quality of life in cities.

SUM integrates computational modeling with the physical dynamics of urban metabolism. It is an AI- or AIoT-powered model that aims to monitor, analyze, and optimize the flow and exchange of energy, materials, resources, and information in cities in real time (Bibri et al. 2024b; Ghosh and Sengupta 2023; Peponi et al. 2022). Accordingly, SUM relates to CPSoS through its focus on understanding and optimizing resource flows in urban environments. SUM applies principles from metabolic theory and industrial ecology to cities, treating them as living organisms with inputs, outputs, and internal processes (D'Amico 2020; Shahrokni et al. 2015). Exemplifying CPSoS by its very nature, it enables cities to gain insights into the dynamics of resource flows such as energy, water, and waste in urban systems. These insights empower urban planners and policymakers to identify inefficiencies, prioritize interventions, and optimize resource utilization. The integration of SUM and CPS into a form of CPSoS facilitates a holistic approach to urban management and planning, where decisions are informed by a comprehensive understanding of the urban metabolism. By optimizing resource flows and promoting circular economy principles, SUM contributes to the development of more sustainable and resilient cities.

In light of the above, City Brain, UDT, and SUM, augmented by AIoT, fit into the CPSoS framework through the integration of digital and physical components, real-time monitoring and control, interconnectedness and interactions, and dynamic and adaptive capabilities pertaining to systems of systems. The interplay of AIoT, CPS, City Brain, UDT, and SUM is crucial, as these components are intricately interconnected and mutually reinforce one another in shaping smart and intelligent urban environments. The role of AIoT emerges as a pivotal force, seamlessly integrated into the core functionalities of City Brain, UDT, and SUM as CPSoS. These data-driven management and planning systems set the stage for the transformative potential of AIoT to reshape smarter eco-cities and sustainable smart cities thanks to its integration with CPS.

Furthermore, City Brain, UDT, and SUM are not only instrumental in data-driven urban management and planning but also significantly contribute to data-driven governance processes. They empower policymakers and government officials to make informed decisions and formulate evidence-based policies that address complex urban challenges (Bibri et al. 2024a, b; D'Amico et al. 2020; Deng et al. 2021; Kierans et al. 2023; Liu et al. 2022; Shahrokni et al. 2015; Weil et al. 2022; Xu et al. 2024; Zhu 2021). Their collaborative functions facilitate coordination among stakeholders, including government agencies, businesses, communities, and citizens, to address environmental issues and implement effective measures. Through data-driven insights into key environmental factors, they inform environmental governance strategies. Together, they enable advanced modes of governance through data-driven insights, decision-making processes, stakeholder collaboration, and citizen engagement, ultimately leading to more responsive and resilient urban governance frameworks.

Platform urbanism entails using digital platforms and ecosystems to facilitate collaboration, innovation, and citizen participation in urban governance and development. These platforms typically leverage digital technologies to connect various stakeholders, including government agencies, businesses, communities, and citizens, to co-create and co-manage urban services, resources, and infrastructure. Research has examined the impact of platform urbanism on various aspects of urban life and governance (Caprotti et al. 2022; Haveri and Anttiroiko 2023; Katmada Katsavounidou and Kakderi 2023; Komninos and Kakderi 2019). The emergence of the city-as-a-platform concept (e.g., Repette et al. 2021; Komninos and Kakderi 2019) highlights the emerging paradigm of data-driven governance in the context of environmentally sustainable urban development (Bibri et al. 2024b).

Urban platforms represent a governance approach that embraces multi-stakeholder innovation in the discourse surrounding smart cities and sustainable smart cities (e.g., Haveri and Anttiroiko 2023; Ansell and Gash 2018, 2019; Noori et al. 2020; ITU 2022). Platform urbanism enables more inclusive and participatory decision-making processes in urban governance by providing channels for citizens to contribute ideas and engage with government agencies and other stakeholders. It also allows for the collection, analysis, and visualization of data from various sources to inform decision-making and policy formulation. By leveraging data analytics and visualization tools, platform urbanism enables government agencies and other stakeholders to analyze trends, identify patterns, and evaluate the impact of policies and interventions on urban systems. This data-driven approach helps optimize resource allocation, improve service delivery, and address urban challenges more effectively. Overall, platform urbanism complements data-driven governance by providing the infrastructure and tools necessary to collect, analyze, and act upon data in urban governance processes, ultimately leading to more responsive, adaptive, and accountable governance in cities.

In the context of CPSoS, the relationship among City Brain, UDT, SUM, and platform urbanism is intricate and symbiotic. City Brain serves as the central nervous system, leveraging data-driven insights, IoT, and AI algorithms to optimize urban operations and decision-making processes. UDT complements City Brain by providing a digital replica of the city, enabling real-time monitoring, simulation, and analysis of various urban systems and processes. SUM plays a key role in understanding and managing the city's resource flows, energy consumption, and environmental impacts, thereby informing sustainable decision-making. Platform urbanism facilitates the integration and coordination of diverse urban stakeholders, enabling collaborative governance and innovation.

City platforms, while not classified as CPSoS themselves, are integral components of a broader CPSoS framework in urban management, planning, and governance. They serve as digital ecosystems that integrate diverse data sources, technologies, and services to support urban operations, planning decisions, governance processes, and stakeholder participation. Interacting with other CPSoS elements like the sensors, actuators, and control systems associated with City Brain, UDT, and SUM, city platforms form interconnected networks that manage and optimize urban functions. Thus, while not CPSoS in isolation, city platforms play a crucial role within the broader CPSoS framework. They contribute to a holistic approach to urban management, planning, and governance that leverages advanced technologies, data analytics, and collaborative processes to create more efficient, sustainable, and resilient cities.

In addition, the integration of City Brain, UDT, SUM, and platform urbanism is poised to catalyze the emergence of Artificial Intelligence of Smarter Eco-City or Sustainable Smart City CPSoS as part of the Internet of Everything (IoE). Through their synergistic collaboration, these advanced technologies and models enable comprehensive data-driven management, planning, and governance solutions that optimize resource utilization, enhance resilience, and improve citizen well-being in urban environments. As interconnected components within the IoE ecosystem, they create a dynamic and adaptive urban infrastructure that responds intelligently to the needs of citizens, businesses, and the environment, thereby paving the way for the realization of smarter, more sustainable cities of the future.

Masserov and Masserov (2022) envision a future of a highly interconnected "Internet of Everything on a Smart Cyber-Physical Earth," with IoT and CPS driving a new smart revolution. The authors highlight the current challenges pertaining to the processing of vast amounts of data with current computing power and suggest that AI and data science offer solutions. They argue that IoT combined with AI could bring significant advancements, not only in cost-saving but also in enhancing human life. In a similar vein, Singh et al. (2023b) explore the role of intelligent systems and IoE in developing smart cities. The authors explain the complexities of CPS and smart city infrastructure, underscoring how AI and ML are crucial in enhancing the efficiency, connectivity, and responsiveness of these urban environments. They cover key areas such as the physical layer design of smart city infrastructure, smart sensors and actuators, and the broader applications and challenges of IoE in smart cities.

These transformations manifest in smarter eco-cities and sustainable smart cities through their brain, DT, and metabolism as AIoT-powered data-driven management and planning systems, which encapsulate the principles of CPS and represent CPSoS. Platform urbanism leverages digital platforms to create interconnected and data-driven urban ecosystems, integrating these AIoT and CPS-driven systems into a cohesive framework that supports dynamic and responsive governance.

Lastly, the integration of AIoT with CPSoS has the potential to revolutionize data-driven management, planning, and governance in the context of smarter eco-cities and sustainable smart cities in the field of environmental sustainability and climate change. AIoT enables the seamless integration of data from various sensors, devices, and systems deployed throughout these cities, generating vast amounts of real-time data on environmental factors, such as air quality, water quality, transportation efficiency, traffic flow, energy consumption, waste management, and climatic conditions. CPSoS facilitates the interoperability and coordination of diverse urban systems powered by AIoT, thereby allowing for efficient data collection, aggregation, and analysis across various urban domains.

Together, AIoT and CPSoS empower city managers, planners, authorities, and policymakers in smarter eco-cities and sustainable smart cities to make informed decisions based on real-time data insights. With these advanced technologies, these cities can implement comprehensive environmental monitoring systems that continuously collect and analyze data on environmental conditions and changes. This enables proactive management of environmental risks and rapid response to emerging issues through data-driven decisions in management, planning, and governance processes. This leads to increased efficiency, reduced resource consumption, and minimized environmental impact. Overall, the seamless integration of data-driven environmental management, planning, and governance facilitated by AIoT and CPSoS provides a holistic solution for creating future smarter eco-cities and sustainable smart cities. By leveraging these cutting-edge technologies, these cities can optimize resource utilization, improve environmental quality, engage citizens, promote well-being, enhance resilience, and build robust urban ecosystems.

1.2 RESEARCH PROBLEM

The research problem tackled in this book revolves around the crucial need to comprehend and harness the synergistic and collaborative interplay of

City Brain, UDT, SUM, and platform urbanism as data-driven management, planning, and governance systems in the dynamic landscapes of smarter eco-cities and sustainable smart cities. The objective is to enhance their environmental performance and strengthen their role in advancing environmental sustainability goals.

1.3 RESEARCH GAP

The research gap identified in this context underscores the lack of comprehensive exploration of the integrated application, impact, and theoretical underpinnings of these AI or AIoT-driven solutions and CPSoS on advancing environmental sustainability goals in the realms of smarter eco-cities and sustainable smart cities.

While City Brain, UDT, SUM, and city platform as individual data-driven systems have been studied in isolation, there exists a significant gap in research focusing on their synergistic interplay and collective potential to drive sustainable urban development. This gap highlights the need for a more holistic approach to studying these advanced systems and their dynamic interaction and mutual influence, particularly considering the rapidly evolving nature of both technological advancements and urban environments. Despite the growing body of research and practical endeavors in the individual domains, there is a notable absence of thorough investigations into the intricate dynamics and subtleties of these emerging domains and their contributions to fostering environmental sustainability in emerging urban landscapes. Therefore, addressing this research gap is essential for advancing our understanding of smarter eco-cities and sustainable smart cities and maximizing their potential to tackle pressing environmental challenges. Overall, this book stands as the pioneering effort in addressing this research gap, making it the first of its kind in this thorough investigation.

1.4 RESEARCH AIM AND OBJECTIVES

To address the notable gap in the current landscape of research and practice, this book embarks on a pioneering exploration into the interconnected synergy of City Brain, UDT, SUM, and platform urbanism within the framework of AIoT and CPSoS. It aims to examine how these advanced technologies and models can be synergistically harnessed to holistically advance data-driven environmental urban management, planning, and governance systems, thereby propelling the development of smarter eco-cities and sustainable smart cities. Embarking into this uncharted territory, the book endeavors to address a series of fundamental research questions and objectives outlined in the subsequent chapters. Through comprehensive theoretical analyses, evidence synthesis, case studies, and interdisciplinary insights, the book is set to unravel the complexities of this multifaceted research topic. In doing so, it introduces several pioneering frameworks and models that integrate City Brain, UDT, SUM, and platform urbanism for practical implementation. These integrated approaches are poised to drastically change how smarter cities and smart sustainable cities can be environmentally managed, planned, and governed.

1.5 RESEARCH MOTIVATION

The impetus for this book arises from the urgent need to confront the multifaceted challenges facing urban environments, such as rapid urbanization, resource depletion, ecological degradation, and climate change. This calls for the implementation of innovative strategies, approaches, and practices for enhancing environmental urban sustainability. The overarching goal is to offer insights and solutions for developing urban landscapes that are more efficient, resilient, sustainable, and centered around the well-being of citizens.

1.6 RESEARCH SIGNIFICANCE

The significance of this book lies in its innovative approach to unveiling the complex interplay between AIoT, CPSoS, and data-driven environmental urban management, planning, and governance systems. By adopting a holistic perspective, the book provides invaluable insights into the transformative power of integrating AIoT and CPSoS in shaping the cities of the future. Bridging the gap between theory and practice, its interdisciplinary framework equips urban planners, policymakers, researchers, and practitioners with actionable insights for promoting sustainability, resilience, and equitable development in urban environments. As cities continue to evolve and grapple with unprecedented challenges, this book serves as a roadmap for harnessing the untapped potential of emerging data-driven technologies and models to create more livable, inclusive, technologically advanced, and environmentally conscious urban landscapes. Ultimately, the research endeavor aims to advance knowledge in the field of AIoT-driven sustainable urban development and guide policy and decision-making processes. Policymakers and urban planners can develop more effective strategies to address urgent environmental challenges and foster sustainable urban development by understanding the synergies and collective potential of these technologies and models.

1.7 THE STRUCTURE AND CONTENT OF THE BOOK

This introductory chapter sets the stage for an exploration into transformative technologies and innovative models aimed at reshaping urban environmental management, planning, and governance in smarter eco-cities and sustainable smart cities. It provides a comprehensive overview of the book's scope, beginning with the intricate interplay of City Brain, UDT, SUM, and platform urbanism in the contexts of AIoT and CPSoS. Emphasizing the pioneering nature of these integrated systems, the chapter highlights their potential to accelerate the transition toward environmental sustainability. Each subsequent chapter will investigate these technologies and models, offering thorough analysis, evidence synthesis, and real-world case studies to illustrate their synergistic capabilities and their role in advancing sustainable urban development.

In more detail, this book will delve into the pioneering frameworks and models for data-driven environmental management, planning, and governance in smarter eco-cities and sustainable smart cities. This involves a thorough examination of their theoretical foundations, design principles, implementation strategies, empirical evidence, emerging trends, and anticipated challenges. These insights offer a comprehensive understanding of their relevance and significance in advancing research and practice in sustainable urban development. Through rigorous analysis, the book aims to elucidate underlying concepts and theories, scrutinize practical applications, highlight benefits and opportunities, and discuss implications for sustainable urban development. Its structure ensures a thorough exploration of key concepts, technologies, and methodologies intersecting AI, AIoT, CPSoS, UDT, SUM, platform urbanism, and data-driven urban management, planning, and governance in the context of environmental sustainability.

The chapters of the book cover various facets, starting with an overview of the theoretical and practical foundations that underpin smarter eco-cities and sustainable smart cities. Subsequently, they investigate the integration of AI and AIoT with their brain, DT, metabolism, and platform solutions. Finally, they discuss the role of data-driven approaches in enhancing their management, planning, and governance processes and practices, with the aim of achieving the goals of environmentally sustainable urban development. Case studies and real-world examples are utilized to illustrate the implementation and impact of emerging computational technologies and collaborative models in creating smarter, more sustainable cities. Through this structured approach, the book aims to provide readers with a comprehensive understanding of the opportunities and challenges inherent in leveraging AI and AIoT for sustainable urban development. The detailed structure of the remainder of the book is as follows:

Chapter 2—Smarter Eco-City Management, Planning, and Governance in the Era of Artificial Intelligence of Things: Theoretical and Conceptual Frameworks. This chapter aims to develop conceptual and theoretical frameworks for smarter eco-city management, planning, and governance, anchored in AI and AIoT. In pursuit of this aim, it undertakes a comprehensive analysis of the foundational underpinnings of smarter eco-cities, emphasizing the incorporation of AI and AIoT systems into urban management, planning, and governance practices. Highlighting the interdisciplinary and transdisciplinary nature of smarter eco-cities, this chapter underscores the significance of knowledge integration and fusion from diverse disciplines to address complex environmental challenges. The symbiotic relationship among the identified concepts, theories, technologies, and models plays a crucial role in shaping the future of sustainable urban development. Shedding light on these underpinnings and their intricate interconnections is key to laying the groundwork for a coherent framework for smarter eco-cities that facilitate the practical exploration and implementation of AI- and AIoT-driven urban initiatives. This chapter provides valuable insights and essential guidance for urban planners, policymakers, and scholars navigating the landscape of smarter eco-cities to understand their multifaceted dimensions and nuanced dynamics.

Chapter 3—Artificial Intelligence and Artificial Intelligence of Things Solutions for Smarter Eco-Cities: Advancing Environmental Sustainability and Climate Change Strategies. This chapter offers a comprehensive review of the dynamic landscape of smarter eco-cities, with a specific focus on the transformative role of emerging applied AI and AIoT solutions in advancing environmental sustainability and climate change strategies. Additionally, it identifies and analyzes potential technological complexities, ethical concerns, and the socioeconomic implications of AI and AIoT-driven urban transformations. Moreover, it derives a novel framework for smarter eco-city development based on synthesized evidence. This chapter demonstrates how emerging AI and AIoT applications can significantly contribute to enhancing environmental sustainability outcomes and advancing climate change strategies in the context of smarter eco-cities. The findings underscore the substantial potential of emerging AI and AIoT technologies in advancing the development of smarter eco-cities. AI and AIoT solutions offer benefits and opportunities for optimizing resource usage, enhancing infrastructure efficiency, monitoring environmental parameters, reducing carbon footprints, improving air and water quality, and fostering resilience to climate change. However, despite the significant opportunities of AI and AIoT to advance the development of smarter eco-cities, their implementation is fraught with formidable challenges that need to be addressed and overcome. The valuable insights derived from this study empower policymakers, practitioners, and researchers to navigate the complexities surrounding the integration of data-driven scientific urbanism and smart eco-urbanism. City stakeholders glean essential knowledge for informed decision-making, effective strategy implementation, and environmental policy development through the analysis of real-world implementations and practical use cases.

Chapter 4—Artificial Intelligence of Things for Advancing Smart Sustainable Cities: A Synthesized Case Study Analysis of City Brain, Digital Twin, Metabolism, and Platform. This chapter investigates the transformative potential and intricate interplay of City Brain, UDT, SUM, and platform urbanism—as prominent urban planning and governance systems—within the framework of "Artificial Intelligence of Smart Sustainable City Things." This investigation is guided by four specific objectives. First, it analyzes and delineates the roles played by City Brain, UDT, SUM, and platform urbanism in advancing data-driven environmental urban planning and governance. Second, it explores how the distinct functions of these systems can be seamlessly integrated. Third, it examines the contribution of AIoT functionalities and capabilities to the integrated operation of these systems. Finally, it identifies the key anticipated challenges and obstacles in the integration of these systems' functions and proposes strategies to address them. This chapter adopts a research design integrating case study analysis and literature review, allowing for a thorough and nuanced exploration of real-world implementations and their foundational underpinnings to enhance the depth and breadth of the study. The findings extend beyond theoretical enrichment, providing invaluable insights that empower urban planners, policymakers, and practitioners to adopt integrated and holistic solutions. By exploring the potentials, synergies, and opportunities demonstrated by the real-world implementation of City

Brain, UDT, SUM, and platform urbanism, city stakeholders acquire the essential knowledge for making informed decisions, including implementing effective strategies and designing policies that prioritize the collective intelligence of emerging data-driven planning and governance systems, thereby advancing sustainable urban development.

Chapter 5—Artificial Intelligence of Sustainable Smart City Brain, Digital Twin, and Metabolism: Pioneering Data-Driven Management and Planning. This chapter introduces a pioneering framework—named Artificial Intelligence of Sustainable Smart City Things—that synergistically interconnects urban brain (UB), UDT, and SUM as forms of CPSoS. Methodologically, it adopts an integrated approach that combines a thorough literature review with in-depth case studies. Through a detailed analysis and synthesis of a large body of knowledge, supported by empirical insights, this study establishes a solid foundation that elucidates the functionalities and architectures of UB, UDT, SUM, and AIoT and their synergistic interplay in the context of CPSoS. Furthermore, it sparks a discourse on the substantive effects of AIoT on the trajectory of environmentally sustainable urban development, covering opportunities, benefits, and implications, as well as challenges, barriers, limitations, and potential avenues for future research. Its primary contribution lies in the development of a pioneering framework for environmentally sustainable urban development, offering invaluable insights for researchers, practitioners, and policymakers alike. This framework serves as a guiding roadmap for spurring groundbreaking research endeavors, inspiring practical implementations, and informing policymaking decisions in the realm of emerging AIoT-driven, environmentally sustainable urban development.

Chapter 6—Artificial Intelligence of Things for Harmonizing Smarter Eco-City Brain, Metabolism, and Platform: An Innovative Framework for Data-Driven Environmental Governance. This chapter aims to explore the linchpin potential of AIoT in seamlessly integrating City Brain, SUM, and platform urbanism as data-driven governance systems to advance environmental governance in smarter eco-cities. Specifically, it introduces a pioneering framework that effectively leverages the synergies among these AIoT-powered governance systems to enhance environmental sustainability practices in smarter eco-cities. In developing the framework, the study employs configurative and aggregative synthesis approaches through an extensive literature review and in-depth case study analysis of publications spanning from 2018 to 2023. The study identifies key factors driving the co-evolution of AI and IoT into AIoT and specifies technical components constituting the architecture of AIoT in smarter eco-cities. A comparative analysis reveals commonalities and differences among City Brain, SUM, and platform urbanism within the frameworks of AIoT and environmental governance. These data-driven systems collectively contribute to environmental governance in smarter eco-cities by leveraging real-time data analytics, predictive modeling, and stakeholder engagement. The proposed framework underscores the importance of data-driven decision-making, optimization of resource management, reduction in environmental impact, collaboration among stakeholders, engagement of citizens, and formulation of evidence-based policies. The findings unveil that the synergistic and collaborative integration of City Brain, SUM, and platform urbanism through AIoT presents promising opportunities and prospects for advancing environmental governance in smarter eco-cities. The framework not only charts a strategic trajectory for stimulating research endeavors but also holds significant potential for practical application and informed policymaking in the realm of environmental urban governance. However, ongoing critical discussions and refinements remain imperative to address the identified challenges, ensuring the framework's robustness, ethical soundness, and applicability across diverse urban contexts.

Chapter 7—Artificial Intelligence and Its Generative Power for Advancing Sustainable Smart City Digital Twin: A Novel Framework for Data-Driven Planning and Design. This chapter explores the transformative potential of AI and its five subset models—ML, deep learning (DL), computer vision (CV), natural language processing (NLP), and generative AI (GenAI)—in reshaping sustainable smart city planning and design through UDT. To guide this investigation, four objectives are formulated: (1) Analyze and synthesize an extensive body of literature on AI and its five subset models, alongside UDT, with a specific focus on their foundational underpinnings and practical applications in the context of sustainable smart city planning and design; (2) investigate a case study of Lausanne City on Generative Spatial AI (GSAI), a specialized branch of GenAI, tailored for UDT; (3) develop a novel framework for GenAI-driven sustainable smart city planning and design; and (4) engage in a comprehensive discussion encompassing a comparative analysis, implications, limitations, and recommendations for future research avenues. The findings highlight the pivotal role of AI and its five subsets in enhancing data-driven planning and design processes through UDT. They also reveal the untapped potential of GenAI and GSAI in propelling the advancement of sustainable smart cities. These insights offer valuable guidance for researchers, practitioners, and policymakers, catalyzing groundbreaking research endeavors, facilitating practical implementations, and informing policy decisions in the realm of emerging GenAI-driven sustainable urban development.

Chapter 8—Transforming Smarter Eco-Cities with Artificial Intelligence of Things: Harnessing the Synergies between Metabolic Circularity and Tripartite Sustainability. This chapter delves into the transformative role of emerging AI and AIoT in harnessing the circular economy, metabolic circularity, and tripartite sustainability frameworks to advance the development of smarter eco-cities toward achieving SDGs. Findings indicate that AI and AIoT can significantly enhance data-driven decision-making, enabling smarter eco-cities to align with these goals. AI and AIoT technologies facilitate real-time data collection and analysis, supporting the circular economy's principles of resource regeneration and waste minimization. Additionally, they bolster urban metabolic circularity by optimizing the flow, reuse, and closed-loop systems of materials and energy within smarter eco-cities. Furthermore, these technologies promote tripartite sustainability by balancing environmental, economic, and social dimensions, ensuring a holistic approach to urban development. The contributions of this study lie in providing a comprehensive examination of AI and AIoT's capabilities in fostering sustainable urban development and providing actionable insights and valuable guidance for researchers, practitioners, and policymakers. The implications are profound, suggesting that the strategic deployment of AI and AIoT technologies can lead to more sustainable, resilient, and efficient urban environments. This work underscores the pivotal role of these advanced technologies in achieving urban sustainability goals and provides a framework for future research endeavors and practical implementations in the realm of smarter eco-cities.

1.8 THE NEED FOR THE BOOK

In the face of mounting environmental concerns, rapid urbanization, and technological advancements, there is a pressing demand for this book to address several pressing issues and bridge the gap between conventional urban practices and data-driven applied solutions. The following points outline the key areas where this book aims to make significant contributions:

Emerging technological integration: There is a growing need for practical guidance on how to harmonize advanced technologies and innovative models such as AI, IoT, AIoT, CPSoS, UDT, SUM, and city platforms in the realms of smarter eco-cities and sustainable smart cities.

Lack of system integration: Many of these cities struggle with fragmented systems for managing various aspects of urban life and planning and governing urban environments. This book advocates for the integration of different data-driven systems for enhancing these urban practices, providing readers with a holistic framework for sustainable urban development.

Environmental sustainability: AI- and AIoT-driven systems can significantly optimize resource usage, reduce waste, improve transportation infrastructure, preserve biodiversity, minimize environmental impact, and mitigate climate change in modern cities.

Environmental challenges: Climate change, pollution, and resource depletion pose significant challenges to urban environments. The book will help readers learn how to monitor, analyze, and mitigate these environmental issues effectively by harnessing the synergistic potential of AIoT and leveraging the complementary capabilities of CPSoS.

Urban data overload: Urban environments generate massive amounts of data, but making sense of it can be overwhelming. The book offers strategies for handling this data deluge, turning it into actionable insights for better decision-making in environmental management, planning, and governance. By integrating AIoT and CPSoS, readers will discover how these technologies enable the seamless assimilation and analysis of diverse data streams, facilitating integrated decision-making processes that are essential for addressing complex urban challenges effectively.

Interdisciplinary approach: This comprehensive book merges expertise from urban planning and design, urban governance, urban management, environmental science, computer science, and big data science and analytics for sustainable urban development. By exploring the integration of innovative models and frameworks, the book provides invaluable insights into the potential of AIoT and CPSOS to transform urban environments into smarter, more sustainable spaces that prioritize efficiency, resilience, and inclusivity.

1.9 TARGET AUDIENCE AND READERSHIP

This book welcomes a wide range of readers to embark on an illuminating journey through the dynamic terrain of emerging data-driven environmental management, planning, and governance systems in the context of smarter eco-cities and sustainable smart cities. As readers navigate through the chapters, they will gain profound insights into the transformative potential of AIoT-driven CPSoS in reshaping urban environments and fostering sustainability, resilience, citizen well-being, and innovation. This journey will underscore the importance of interdisciplinary and transdisciplinary perspectives, emphasizing the holistic approach essential for effectively developing, implementing, and integrating these technologies and models. Readers will also be introduced to the pioneering nature of emerging AIoT-driven sustainable urban development practices.

The book offers a unique opportunity for urban planners, policymakers, researchers, academics, professionals, and students in various fields to delve into the synergistic and collaborative integration of City Brain, UDT, SUM, and platform urbanism, all within the broader framework of environmental sustainability.

For urban planners and policymakers, the book serves as a comprehensive guide to understanding the innovative frameworks and models driving sustainable urban development. By critically engaging with the theoretical analyses, evidenced syntheses, case studies, and interdisciplinary insights presented, they can enhance their decision-making processes and develop strategies for more effective data-driven environmental urban management, planning, and governance. This equips them with the knowledge and tools necessary to navigate the complexities of modern urban environments and address pressing environmental challenges.

Researchers and academics will find the book to be a valuable resource for exploring new avenues of inquiry, experimenting with integrated data-driven solutions, and gaining a deeper understanding of the complexities surrounding AIoT-driven sustainable urban development practices. Through rigorous analysis and synthesis of existing literature, they can contribute to the advancement of knowledge in the field and identify opportunities for further research. This facilitates their engagement with novel concepts and cutting-edge methodologies to foster innovation and collaboration in the pursuit of sustainable urban development goals.

Master and PhD students pursuing degrees in urban AI or AIoT, urban computing, smart urbanism, eco-urbanism, urban planning, urban design, urban governance, environmental science, city modeling and simulation, sustainable development engineering, and related fields will benefit from the book's comprehensive coverage of key concepts, innovative approaches, and emerging trends. By studying the real-world case studies and theoretical frameworks presented, they can deepen their understanding of the challenges and opportunities facing urban environments and develop the critical thinking skills necessary to address them. This exposure to interdisciplinary perspectives and practical applications prepares them for leadership roles in shaping the future of sustainable urban development.

Professionals working in fields related to urban development, such as Information and Communications Technology (ICT) experts, computer scientists, urban scientists, city designers, city engineers, city managers, and urban sustainability consultants, will find practical insights and innovative solutions in this book. These insights can inform their practice and enhance their ability to create sustainable and resilient urban landscapes by integrating innovative data-driven approaches effectively. By leveraging the knowledge and strategies presented in the book, they can adapt their approaches to meet the evolving needs of cities while prioritizing environmental sustainability and resilience.

Whether readers have extensive knowledge in the field or are beginners seeking foundational understanding, the book offers a versatile approach to engagement. Readers with prior knowledge in the field can leverage their existing expertise to deepen their understanding of innovative solutions and emerging trends, while beginners can use the book for an in-depth examination of the complex landscape of AIoT-driven sustainable urban development. Regardless of their level of expertise, readers are encouraged to approach each chapter with curiosity and an open mind and a willingness to explore new ideas and perspectives.

Knowledgeable readers and beginners alike can approach the book in a manner that suits their needs and preferences. For those well-versed in the field, diving into the entire book at once can provide a comprehensive overview of the interconnected concepts and technologies shaping sustainable urban development. They may choose to explore specific chapters in more detail based on their interests and areas of expertise. On the other hand, beginners may benefit from starting with an overview of the book's key themes and concepts before delving into individual chapters.

Taking the time to reflect on each chapter's content and implications can enhance their understanding and facilitate a deeper engagement with the material over time. Whether reading the book cover to cover or exploring it gradually, both knowledgeable readers and beginners can extract valuable insights and practical knowledge to inform their understanding and practice in the realm of sustainable urban development.

1.10 THE FIVE KEY FEATURES OF THE BOOK

1. Interdisciplinary expertise: The book provides a comprehensive exploration of AIoT-driven sustainable urban development, drawing from diverse fields and integrating and fusing different disciplines. The inclusion of interdisciplinary perspectives ensures that readers gain a holistic understanding of sustainable urban development. The book offers a nuanced insight into the challenges and opportunities facing smarter eco-cities and sustainable smart cities by drawing on these perspectives.

2. Cutting-edge insights: The book presents the latest advancements in key concepts, models, and technologies shaping smarter eco-cities and sustainable smart cities. With a focus on innovative approaches and solutions, it offers actionable strategies and enhances practices for fostering resilient, efficient, inclusive, and environmentally sustainable urban environments.

3. In-depth exploration: The book offers an in-depth examination of emerging key concepts, models, and technologies, supported by extensive literature reviews and real-world case studies that illustrate their implementation and impact in various domains. It reflects a well-rounded approach to addressing the key aspects of sustainable smart urban development. The topics addressed are crucial for understanding the complexities of modern urban ecosystems.

4. Practical relevance: In addition to theoretical rigor and enrichment, the content offers practical recommendations, strategies, and interventions for policymakers, urban planners, researchers, and practitioners, ensuring its applicability in real-world scenarios. The emphasis on real-world examples throughout the chapters enhances the practical relevance of the content. Readers will benefit from learning how the emerging computational technologies and innovative models are

being implemented in actual urban contexts to address environmental challenges and drive sustainable development.

5. Holistic approach: By bridging theory and practice, the book equips readers with the tools and knowledge needed to address complex urban challenges comprehensively and foster sustainable development effectively. Through its integration of diverse perspectives and practical insights, it empowers stakeholders to implement holistic solutions that prioritize environmental sustainability in urban contexts.

1.11 AUTHOR'S CONTRIBUTION TO THE FIELD

The author, recognized as one of the leading scholars in the field, brings many years of experience and a deep understanding of various interconnected disciplines to the table. He has a rich multidisciplinary background encompassing computer science, computer engineering, system science, environmental science, innovation science, urban planning and design, urban governance, strategic sustainable development, and data-driven technologies. Leveraging his interdisciplinary and transdisciplinary expertise, he is uniquely equipped to offer nuanced insights and innovative perspectives on the complex challenges facing modern urban environments and the opportunities that can be realized to mitigate or overcome these challenges. His work is dedicated to leveraging cutting-edge technologies and innovative approaches to create more sustainable, resilient, livable, and environmentally conscious urban environments.

The author's extensive research contributions, spanning numerous publications and projects, have earned him international acclaim and positioned him as one of the thought leaders in the field. He has been listed by ScholarGPS as the #1 highly ranked scholar in sustainable cities, #4 in big data, and #8 in sustainability in the world in 2022. He has also been recognized as one of the top 2% scientists in the world for five consecutive years by Stanford and Elsevier.

He has been recognized for his impactful contribution to advancing knowledge and informing policies. His research has been focused on advancing SDGs and tackling contemporary challenges through the integration of emerging computational technologies and advanced models. Through active engagement with various international organizations, such as the United Nations Framework Convention on Climate Change (UNFCCC), the United Nations Industrial Development

Organization (UNIDO), and the United Nations Agency of International Telecommunication Union (ITU), he contributes expertise and knowledge to global initiatives addressing pressing environmental challenges and promoting sustainable development on a global scale. His involvement in international policy-related initiatives and projects provides valuable insights into the practical application of academic research in real-world settings. He can use his policy experience to offer practical recommendations for policymakers, urban planners, and other stakeholders involved in the development and implementation of sustainable smart urban initiatives.

REFERENCES

Agostinelli, S., Cumo, F., Guidi, G., & Tomazzoli, C. (2021). Cyber-physical systems improving building energy management: Digital twin and artificial intelligence. *Energies*, *14*(8), 2338.

Alahi, M. E. E., Sukkuea, A., Tina, F. W., Nag, A., Kurdthongmee, W., Suwannarat, K., & Mukhopadhyay, S. C. (2023). Integration of IoT-enabled technologies and artificial intelligence (AI) for smart city scenario: Recent advancements and future trends. *Sensors*, *23*(11), 5206.

Alowaidi, M., Sharma, S. K., AlEnizi, A., & Bhardwaj, S. (2023). Integrating artificial intelligence in cyber security for cyber-physical systems. *Electronic Research Archive*, *31*(4), 1876–1896. https://doi.org/10.3934/era.2023097.

Ansell, C., & Gash, A. (2018). Collaborative platforms as a governance strategy. *Journal of Public Administration Research and Theory*, *28*(1), 16–32.

Ansell, C., & Miura, S. (2019). Can the power of platforms be harnessed for governance? Public Administration. https://doi.org/10.1111/padm.12636.

Anttiroiko, A.-V. (2023). Smart circular cities: Governing the relationality, spatiality, and digitality in the promotion of circular economy in an urban region. *Sustainability*, *15*(17), 12680. https://doi.org/10.3390/su151712680.

Beckett, S. (2022). Smart city digital twins, 3D modeling and visualization tools, and spatial cognition algorithms in artificial intelligence-based urban design and planning. *Geopolitics, History, and International Relations*, *14*(1), 123–138. https://www.jstor.org/stable/48679657

Bibri, S. E. (2020). Data-driven environmental solutions for smart sustainable cities: Strategies and pathways for energy efficiency and pollution reduction. *Euro-Mediterranean Journal for Environmental Integration,* *5*(66). https://doi.org/10.1007/s41207-020-00211-w

Bibri, S. E., Alexandre, A., Sharifi, A., & Krogstie, J. (2023a). Environmentally sustainable smart cities and their converging AI, IoT, and big data technologies and solutions: An integrated approach to an extensive literature review. *Energy Informatics*, 6, 9.

Bibri, S. E., Huang, J., Jagatheesaperumal, S. K., & Krogstie, J. (2024a). The synergistic interplay of artificial intelligence and digital twin in environmentally planning sustainable smart cities: A comprehensive systematic review. *Environmental Science and Ecotechnology, 20*, 100433. https://doi.org/10.1016/j.ese.2024.100433.

Bibri, S. E., Huang, J., & Krogstie, J. (2024b). Artificial intelligence of things for synergizing smarter eco-city brain, metabolism, and platform: Pioneering data-driven environmental governance. *Sustainable Cities and Society*, 105516. https://doi.org/10.1016/j.scs.2024.105516.

Bibri, S. E., & Huang, J. (2025). Artificial intelligence of sustainable smart city brain and digital twin : A pioneering framework for advancing environmental sustainability. *Environmental Technology and Innovation*, (in press).

Bibri, S. E., Krogstie, J., Kaboli, A., & Alahi, A. (2023b). Smarter eco-cities and their leading-edge artificial intelligence of things solutions for environmental sustainability: A comprehensive systemic review. *Environmental Science and Ecotechnology, 19*. https://doi.org/10.1016/j.ese.2023.100330.

Caprotti, F., Chang, I.-C. C., & Joss, S. (2022). Beyond the smart city: A typology of platform urbanism. *Urban Transformations, 4*(4). https://doi.org/10.1186/s42854-022-00033-9.

Chen, L., Chen, Z., Zhang, Y., et al. (2023). Artificial intelligence-based solutions for climate change: A review. *Environmental Chemistry Letters*. https://doi.org/10.1007/s10311-023-01617-y

D'Amico, G., Arbolino, R., Shi, L., Yigitcanlar, T., & Ioppolo, G. (2021). Digital technologies for urban metabolism efficiency: Lessons from urban agenda partnership on circular economy. *Sustainability, 13*(11), 6043. https://doi.org/10.3390/su13116043.

D'Amico, G., Taddeo, R., Shi, L., Yigitcanlar, T., & Ioppolo, G. (2020). Ecological indicators of smart urban metabolism: A review of the literature on international standards. *Ecological Indicators, 118*, 106808.

Deng, T., Zhang, K., & Shen, Z.-J. (2021). A systematic review of a digital twin city: A new pattern of urban governance toward smart cities. *Journal of Management Science and Engineering, 6*, 125–134. https://doi.org/10.1016/j.jmse.2021.03.003.

Dheeraj, A., Nigam, S., Begam, S., Naha, S., Jayachitra Devi, S., Chaurasia, H. S., Kumar, D., Ritika, Soam, S. K., Srinivasa Rao, N., Alka, A., Sreekanth Kumar, V. V., & Mukhopadhyay, S. C. (2020). Role of artificial intelligence (AI) and internet of things (IoT) in mitigating climate change. In: Ch. Srinivasa, R. T. Srinivas, R. V. S. Rao, N. Srinivasa Rao, S. Senthil Vinayagam, & P. Krishnan (Eds.), *Climate Change and Indian Agriculture: Challenges and Adaptation Strategies* (pp. 325–358). ICAR-National Academy of Agricultural Research Management.

Efthymiou, I.-P., & Egleton, T. E. (2023). Artificial intelligence for sustainable smart cities. In: B. K. Mishra (Ed.), *Handbook of Research on Applications of AI, Digital Twin, and Internet of Things for Sustainable Development* (pp. 1–11). IGI Global.

El Himer, S., Ouaissa, M., Ouaissa, M., & Boulouard, Z. (2022). Artificial intelligence of things (AIoT) for renewable energies systems. In: S. El Himer, M. Ouaissa, A. A. A. Emhemed, M. Ouaissa, & Z. Boulouard (Eds.), *Artificial Intelligence of Things for Smart Green Energy Management: Studies in Systems, Decision and Control* (vol. 446). Springer. https://doi.org/10.1007/978-3-031-04851-7_1.

Ghosh, R., & Sengupta, D. (2023). Smart urban metabolism: A big-data and machine learning perspective. In: R. Bhadouria, S. Tripathi, P. Singh, P. K. Joshi, & R. Singh (Eds.), *Urban Metabolism and Climate Change*. Springer. https://doi.org/10.1007/978-3-031-29422-8_16.

Gourisaria, M. K., Jee, G., Harshvardhan, G., Konar, D., & Singh, P. K. (2022). Artificially intelligent and sustainable smart cities. In: P. K. Singh, M. Paprzycki, M. Essaaidi, & S. Rahimi (Eds.), *Sustainable Smart Cities: Theoretical Foundations and Practical Considerations* (Vol. 942, pp. 277–294). Springer.

Gürkaş Aydın, Z., & Kazanç, M. (2023). Using artificial intelligence in the security of cyber physical systems. *Alphanumeric Journal*, *11*(2), 193–206. https://doi.org/10.17093/alphanumeric.1404181.

Haveri, A., & Anttiroiko, A.-V. (2023). Urban platforms as a mode of governance. *International Review of Administrative Sciences*, *89*(1), 3–20. https://doi.org/10.1177/00208523211005855.

Hoang, T. T., Nguyen, T. N., Nawara, D., & Kashef, R. (2024). Connecting the indispensable roles of IoT and artificial intelligence in smart cities: A survey. *Journal of Information and Intelligence*, *2*(3), 261–285. https://doi.org/10.1016/j.jiixd.2024.01.003.

Hoang, T. V. (2024). Impact of integrated artificial intelligence and internet of things technologies on smart city transformation. *Journal of Technical Education Science*, *19*(special Issue 1), 64–73. https://doi.org/10.54644/jte.2024.1532.

ITU. (2022). The United for Smart Sustainable Cities (U4SSC). The International Telecommunication Union, https://u4ssc.itu.int (accessed 10 Dec 2022).

Jagatheesaperumal, S. K., Bibri, S. E., Huang, J Ganesan, S., & Jeyaraman, P. (2024). Artificial intelligence of things for smart cities: Advanced solutions for enhancing transportation safety. *Computers, Environment and Urban Systems*, *4*, 10. https://doi.org/10.1007/s43762-024-00120-6.

Juma, M., & Shaalan, K. (2020). Cyberphysical systems in the smart city: Challenges and future trends for strategic research. In: A. E. Hassanien, & A. Darwish (Eds.), *Intelligent Data-Centric Systems: Swarm Intelligence for Resource Management in Internet of Things* (pp. 65–85). Academic Press. https://doi.org/10.1016/B978-0-12-818287-1.00008-5.

Katmada, A., Katsavounidou, G., & Kakderi, C. (2023). Platform urbanism for sustainability. In: N. A. Streitz, & S. Konomi (Eds.), *Distributed, Ambient and Pervasive Interactions. HCII 2023*, Lecture Notes in Computer Science (vol. 14037). Springer. https://doi.org/10.1007/978-3-031-34609-5_3.

Khan, F., Kumar, R. L., Kadry, S., Nam, Y., & Meqdad, M. N. (2021). Cyber physical systems: A smart city perspective. *International Journal of Electrical and Computer Engineering (IJECE)*, *11*(4), 3609–3616. https://doi.org/10.11591/ijece.v11i4.pp3609-3616.

Kierans, G., Jüngling, S. & Schütz, D. (2023). Society 5.0 2023. *EPiC Series in Computing*, *93*, 82–96.

Komninos, N., & Kakderi, C. (Eds.). (2019). *Smart Cities in the Post-algorithmic Era: Integrating Technologies, Platforms and Governance*. Edward Elgar.

Koumetio, T. S. C., Diop, E. B., Azmi, R., & Chenal, J. (2023). Artificial intelligence based methods for smart and sustainable urban planning: A systematic survey. *Archives of Computational Methods in Engineering*, *30*(5), 1421–1438. https://doi.org/10.1007/s11831-022-09844-2.

Kuguoglu, B. K., van der Voort, H., & Janssen, M. (2021). The giant leap for smart cities: Scaling up smart city artificial intelligence of things (AIoT) initiatives. *Sustainability*, *13*(21), 12295. https://doi.org/10.3390/su132112295.

Leal Filho, W., Wall, T., Mucova, S. A. R., Nagy, G. J., Balogun, A.-L., Luetz, J. M., Ng, A. W., Kovaleva, M., Azam, F. M. S., & Alves, F. (2022). Deploying artificial intelligence for climate change adaptation. *Technological Forecasting and Social Change*, *180*, 121662.

Liu, W., Mei, Y., Ma, Y., Wang, W., Hu, F., & Xu, D. (2022). City brain: A new model of urban governance. In: M. Li, G. Bohács, A. Huang, D. Chang, & X. Shang (Eds.), *IEIS 2021*. Lecture Notes in Operations Research. Springer. https://doi.org/10.1007/978-981-16-8660-3_12.

Masserov, D. A., & Masserov, D. D. (2022). Applying artificial intelligence to the internet of things. *Russian Journal of Resources, Conservation and Recycling*, *9*(2). https://doi.org/10.15862/05ITOR222. https://resources.today/PDF/05ITOR222.pdf (in Russian).

Matei, A., & Cocoşatu, M. (2024). Artificial internet of things, sensor-based digital twin urban computing vision algorithms, and blockchain cloud networks in sustainable smart city administration. *Sustainability*, *16*(16), 6749. https://doi.org/10.3390/su16166749

Mishra, B. K. (Ed.). (2023). *Handbook of Research on Applications of AI, Digital Twin, and Internet of Things for Sustainable Development*. IGI Global. https://doi.org/10.4018/978-1-6684-6821-0.

Mishra, P., & Singh, G. (2023). Artificial intelligence for sustainable smart cities. In: *Sustainable Smart Cities*. Springer. https://doi.org/10.1007/978-3-031-33354-5_6.

Mohamed, K. S. (2023). Deep learning for IoT "Artificial Intelligence of Things (AIoT)". In: *Deep Learning-Powered Technologies. Synthesis Lectures on Engineering, Science, and Technology*. Springer. https://doi.org/10.1007/978-3-031-35737-4_3.

Mylonas, G., Kalogeras, A., Kalogeras, G., Anagnostopoulos, C., Alexakos, C., & Muñoz, L. (2021). Digital twins from smart manufacturing to smart cities: A survey. *IEEE Access*, *9*, 143222–143249.

Nica, E., Popescu, G. H., Poliak, M., Kliestik, T., & Sabie, O.-M. (2023). Digital twin simulation tools, spatial cognition algorithms, and multi-sensor fusion technology in sustainable urban governance networks. *Mathematics, 11*(9), 1981. https://doi.org/10.3390/math11091981.

Noori, N., Hoppe, T., & de Jong, M. (2020). Classifying pathways for smart city development: Comparing design, governance and implementation in Amsterdam, Barcelona, Dubai, and Abu Dhabi. *Sustainability, 12*, 4030.

Peponi, A., Morgado, P., & Kumble, P. (2022). Life cycle thinking and machine learning for urban metabolism assessment and prediction. *Sustainable Cities and Society, 80*, 103754. https://doi.org/10.1016/j.scs.2022.103754.

Popescu, S. M., Mansoor, S., Wani, O. A., Kumar, S. S., Sharma, V., Sharma, A., Arya, V. M., Kirkham, M. B., Hou, D., Bolan, N., & Chung, Y. S. (2024). Artificial intelligence and IoT driven technologies for environmental pollution monitoring and management. *Frontiers in Environmental Science, 12*, 1336088. https://doi.org/10.3389/fenvs.2024.1336088.

Radanliev, P., Roure, D. C. D., Van Kleek, M., Santos, O., & Ani, U. D. (2020). Artificial intelligence in cyber physical systems. *SSRN*. doi:10.2139/ssrn.3692592. https://ssrn.com/abstract=3692592

Rajawat, A. S., Bedi, P., Goyal, S. B., Shaw, R. N., & Ghosh, A. (2022). Reliability analysis in cyber-physical system using deep learning for smart cities industrial IoT network node. In: V. Piuri, R. N. Shaw, A. Ghosh, & R. Islam (Eds.), *AI and IoT for Smart City Applications (Studies in Computational Intelligence* (vol. 1002). Springer. https://doi.org/10.1007/978-981-16-7498-3_10.

Rane, N., Choudhary, S., & Rane, J. (2024). Artificial Intelligence and machine learning in renewable and sustainable energy strategies: A critical review and future perspectives. *SSRN*. https://doi.org/10.2139/ssrn.4838761.

Repette, P., Sabatini-Marques, J., Yigitcanlar, T., Sell, D., & Costa, E. (2021). The evolution of city-as-a-platform: Smart urban development governance with collective knowledge-based platform urbanism. *Land, 10*, 33. https://doi.org/10.3390/land10010033.

Rolnick, D., Donti, P. L., Kaack, L. H., Kochanski, K., Lacoste, A., Sankaran, K., Ross, A. S., Milojevic-Dupont, N., Jaques, N., & Waldman-Brown, A. (2022). Tackling climate change with machine learning. *ACM Computing Surveys (CSUR), 55*(2), 1–96.

Samadi, S. (2022). The convergence of AI, IoT, and big data for advancing flood analytics research. *Frontiers in Water, 4*, 786040. https://doi.org/10.3389/frwa.2022.786040.

Seng, K. P., Ang, L. M., & Ngharamike, E. (2022). Artificial intelligence Internet of Things: A new paradigm of distributed sensor networks. *International Journal of Distributed Sensor Networks*. https://doi.org/10.1177/15501477211062835.

Shaamala, A., Yigitcanlar, T., Nili, A., & Nyandega, D. (2024). Algorithmic green infrastructure optimization: Review of artificial intelligence driven approaches for tackling climate change. *Sustainable Cities and Society, 101*, 105182. https://doi.org/10.1016/j.scs.2024.105182.

Shahat, E., Hyun, C. T., & Yeom, C. (2021). City digital twin potentials: A review and research agenda. *Sustainability*, *13*, 3386. https://doi.org/10.3390/su13063386.

Shahrokni, H., Årman, L., Lazarevic, D., Nilsson, A., & Brandt, N. (2015). Implementing smart urban metabolism in the Stockholm Royal Seaport: Smart city SRS. *Journal of Industrial Ecology*, *19*(5), 917–929.

Sharma, R., & Sharma, N. (2022). Applications of artificial intelligence in cyber-physical systems. *Cyber Physical Systems*, 1–14. https://doi.org/10.1201/9781003202752-1.

Shi, F., Ning, H., Huangfu, W., Zhang, F., Wei, D., Hong, T., & Daneshmand, M. (2020). Recent progress on the convergence of the internet of things and artificial intelligence. *IEEE Network*, *34*(5), 8–15.

Singh, K. D., Singh, P., Chhabra, R., Kaur, G., Bansal, A., & Tripathi, V. (2023a). Cyber-physical systems for smart city applications: A comparative study. In: 2023 International Conference on Advancement in Computation & Computer Technologies (InCACCT) (pp. 871–876). Gharuan, India. https://doi.org/10.1109/InCACCT57535.2023.10141719.

Singh, P.K., Paprzycki, M., Essaaidi, M., Rahimi, S. (eds) (2023) Sustainable Smart Cities. Studies in Computational Intelligence, vol 942. Springer, Cham. https://doi.org/10.1007/978-3-031-08815-5_14

Singh, T., Solanki, A., Sharma, S. K., & Hachimi, H. (2023b). Smart sensors and actuators for Internet of Everything based smart cities: Application, challenges, opportunities, and future trends. *Intelligent Systems for IoE Based Smart Cities* (vol. 1, p. 61). https://doi.org/10.2174/9789815124965123010006.

Somma, A., De Benedictis, A., Zappatore, M., Martella, C., Martella, A., & Longo, A. (2023, December). Digital twin space: The integration of digital twins and data spaces. In: 2023 IEEE International Conference on Big Data (BigData) (pp. 4017–4025). IEEE.

Thamik, H., Cabrera, J. D. F., & Wu, J. (2024). The digital paradigm: Unraveling the impact of artificial intelligence and internet of things on achieving sustainable development goals. In S. Misra, K. Siakas, & G. Lampropoulos (Eds.), *Artificial Intelligence of Things for Achieving Sustainable Development Goals (Lecture Notes on Data Engineering and Communications Technologies, Vol. 192)*. Springer, Cham. https://doi.org/10.1007/978-3-031-53433-1_2

Tomazzoli, C., Scannapieco, S., & Cristani, M. (2020). Internet of Things and artificial intelligence enable energy efficiency. *Journal of Ambient Intelligence and Humanized Computing*, *14*, 4933–4954. https://doi.org/10.1007/s12652-020-02151-3.

Ullah, Z., Al-Turjman, F., Mostarda, L., & Gagliardi, R. (2020). Applications of artificial intelligence and machine learning in smart cities. *Computer Communications*, *154*, pp.313–323.

Weil, C., Bibri, S. E., Longchamp, R., Golay, F., & Alahi, A. (2023). Urban digital twin challenges: A systematic review and perspectives for sustainable smart cities. *Sustainable Cities and Society*, *99*, 104862.

Xu, Y., Cugurullo, F., Zhang, H., Gaio, A., & Zhang, W. (2024). The emergence of artificial intelligence in anticipatory urban governance: Multi-scalar evidence of China's transition to city brains. *Journal of Urban Technology*. https://doi.org/10.1080/10630732.2023.2292823.

Zaidi, A., Ajibade, S.-S. M., Musa, M., & Bekun, F. V. (2023). New insights into the research landscape on the application of artificial intelligence in sustainable smart cities: A bibliometric mapping and network analysis approach. *International Journal of Energy Economics and Policy*, 4, 287.

Zayed, S. M., Attiya, G. M., El-Sayed, A., & Hemdan, E. E.-D. (2023). A review study on digital twins with artificial intelligence and internet of things: Concepts, opportunities, challenges, tools and future scope. *Multimedia Tools and Applications*, 82, 47081–47107. https://doi.org/10.1007/s11042-023-15611-7.

Zhang, J., & Tao, D. (2021). Empowering things with intelligence: A survey of the progress, challenges, and opportunities in artificial intelligence of things. *IEEE Internet of Things Journal*, 8(10), 7789–7817. https://doi.org/10.1109/JIOT.2020.3039359.

Zhang, J., Hua, X. S., Huang, J., Shen, X., Chen, J., Zhou, Q., Fu, Z., & Zhao, Y. (2019). City brain: Practice of large-scale artificial intelligence in the real world. *IET Smart Cities*, 1, 28–37.

Zhu, W. (2021). Artificial Intelligence and urban governance: Risk conflict and strategy choice. *Open Journal of Social Sciences*, 9, 250–261. https://doi.org/10.4236/jss.2021.94019.

CHAPTER 2

Smarter Eco-City Management, Planning, and Governance in the Era of Artificial Intelligence of Things

Theoretical and Conceptual Frameworks

Abstract

The rapid pace of urbanization and the escalating challenges of environmental degradation have propelled the concept of smarter eco-cities to the forefront of sustainable urban development. These urban environments leverage cutting-edge technologies, especially artificial intelligence (AI) and AI of things (AIoT), to bolster environmental sustainability and resilience. This is facilitated through the integration of innovative models powered by AIoT, notably City Brain, urban digital twin (UDT), smart urban metabolism (SUM), and platform urbanism, which

DOI: 10.1201/9781003536420-2

harness data-driven approaches to optimize urban management, planning, and governance processes. However, despite the burgeoning interest in smarter eco-cities, the current scholarly landscape lacks a cohesive framework that integrates the diverse array of the underlying concepts, theories, technologies, and models essential for unraveling the complexities of this urban ecosystem. Further, the intricate interplay of City Brain, UDT, SUM, and platform urbanism highlights the need for a structured approach to understanding the dynamic landscape of smarter eco-cities. Therefore, this chapter aims to develop conceptual and theoretical frameworks for smarter eco-city management, planning, and governance, anchored in AI and AIoT. In pursuit of this aim, it undertakes a comprehensive analysis of the foundational underpinnings of smarter eco-cities, emphasizing the incorporation of AI and AIoT systems into urban management, planning, and governance practices. Highlighting the interdisciplinary and transdisciplinary nature of smarter eco-cities, this chapter underscores the significance of knowledge integration and fusion from diverse disciplines to address complex environmental challenges. The symbiotic relationship among the identified concepts, theories, technologies, and models plays a crucial role in shaping the future of sustainable urban development. Shedding light on these underpinnings and their intricate interconnections is key to laying the groundwork for a coherent framework for smarter eco-cities that facilitate the practical exploration and implementation of AI- and AIoT-driven urban initiatives. This chapter provides valuable insights and essential guidance for urban planners, policymakers, and scholars navigating the landscape of smarter eco-cities to understand their multifaceted dimensions and nuanced dynamics.

2.1 INTRODUCTION

The increasing challenges posed by rapid urbanization, resource depletion, and environmental degradation have elevated the urgency of sustainable urban development, emerging as a paramount priority for policymakers, urban planners, and researchers worldwide. In recent years, the notion of smarter eco-cities, which integrates data-driven smart urbanism and eco-urbanism models, has risen to prominence as a key pursuit in achieving sustainable urban development. These urban environments are positioned at the crossroads of urban planning and governance, technological integration and innovation, and sustainable transformation and transition. They seek to harness the power of cutting-edge technologies, particularly artificial intelligence (AI), the Internet of Things (IoT), and AI of things (AIoT), for the benefit of the environment and the well-being of their citizens. A comprehensive understanding of smarter eco-cities and their multifaceted dimensions is crucial for realizing their significant potential in advancing environmental sustainability goals.

At the forefront of this transformation in urban development are numerous interconnected concepts, theories, technologies, and models that underlie the notion of smarter eco-cities. Environmental planning and governance, supported by innovative solutions, form the bedrock upon which these urban landscapes are built, guiding decisions pertaining to land use, infrastructure development, natural resource management, climate change mitigation and resilience, and stakeholder engagement and collaboration (Bibri et al. 2024a, b). These decisions are increasingly informed by advancements in AI and AIoT, particularly machine learning (ML) and deep learning (DL), in smart cities (Alahakoon et al. 2020; Bibri et al. 2023a; Gourisaria et al. 2023; Efthymiou and Egleton 2023; Mishra 2023; Szpilko et al. 2023; Zaidi et al. 2023; Ullah et al. 2020, 2023) and smarter eco-cities (Bibri et al. 2023b). The integration of smart cities into the broader framework of smarter eco-cities further enhances the sophistication and effectiveness of these AI-driven approaches, fostering a more comprehensive and interconnected urban ecosystem poised to address the multifaceted challenges of sustainable urban development. Moreover, AI is increasingly transforming the landscape of urban planning (e.g., Bibri et al. 2024b; Bibri and Huang 2024; Koumetio et al. 2023; Marasinghe et al. 2024; Sanchez 2023b; Son et al. 2022) and urban governance (Ajuriaguerra Escudero and Abdiu 2022; Bibri et al. 2024a; Cugurullo et al. 2024; Nishant 2020; Samuel et al. 2023; Yigitcanlar et al. 2024). These transformations have unlocked new possibilities for data-driven decision-making, predictive modeling, and adaptive and collaborative approaches aimed at advancing environmental sustainability and mitigating climate change impacts in the context of smart eco-cities.

Furthermore, the groundbreaking convergence of AI and IoT—AIoT—has taken a center stage in pioneering innovative solutions for smart cities and smarter eco-cities, shaping their development practices and strategies, particularly in the field of environmental sustainability (e.g., Alahi et al. 2023; Bibri et al. 2023a, b; Bibri 2024a, b; El Himer et al. 2022;

Jagatheesaperumal et al. 2024; Mishra 2023; Seng et al. 2022; Thamik et al. 2024; Zhang and Toa 2021). This convergence is instrumental in driving data-driven management, planning, and governance processes in these interconnected urban landscapes. In this context, there exists a symbiotic relationship among these processes, all working in concert to achieve sustainable and resilient urban development. Data-driven urban management involves the collection, analysis, and utilization of vast amounts of urban data to optimize the functioning of city systems and infrastructure. By harnessing advanced technologies, city officials can gain valuable insights into urban dynamics and make informed decisions to enhance efficiency and resource allocation. Data-driven urban planning complements data-driven urban management by using data-driven insights to inform the development of urban infrastructure, land-use policies, and spatial planning strategies. By analyzing data on population growth, demographic trends, economic activity, environmental conditions, and spatial organizations, urban planners can design more sustainable and resilient urban environments. This approach enables cities to optimize the use of limited resources, minimize environmental impact, and improve the overall quality of life for residents (Koumetio et al. 2023; Son et al. 2022).

Meanwhile, data-driven urban governance involves the use of data and technology to enhance decision-making processes, policy formulation, and citizen engagement in urban governance (Bibri et al. 2024a). City officials can leverage data analytics and AI algorithms to identify emerging issues, assess policy outcomes, and solicit feedback from citizens. This participatory approach to governance fosters transparency, accountability, and trust between government institutions and the public, which is key to more effective and responsive urban governance. Worth noting is that addressing urban governance challenges can significantly contribute to overcoming urban planning challenges (e.g. Asadzadeh et al. 2023; Kramers et al. 2016; Stead 2021).

The synergy of data-driven urban management, planning, and governance in smarter eco-cities enables them to address complex urban challenges in a holistic and integrated manner. This interconnected approach harnesses the power of emerging AIoT-powered systems such as City Brain (e.g., Bibri 2024a; Bibri and Hunag 2024; Kierans, Jüngling, and Schütz 2023; Liu et al. 2022), urban digital twin (UDT) (e.g., Almusaed and Yitmen 2023; Beckett 2022; Bibri et al. 2024b; Zvarikova et al. 2022), smart urban metabolism (SUM) (e.g., Bibri et al. 2024a; Ghosh and Sengupta 2023), and platform urbanism (e.g., Bibri 2024a; Caprotti et al. 2022; ITU 2022; Katmada et al. 2023). These innovative systems not only exemplify synergistic opportunities but also hold high potential to drive the advancement of smarter eco-city development.

Leveraging the complementary capabilities of City Brain, UDT, SUM, and platform urbanism can effectively bridge the gap between data-driven environmental urban management, planning, and governance processes—through the unifying power of advanced technologies and collaborative models. The resulting synergistic interplay allows for data-driven, cooperative, coherent, and adaptive strategies to address the interconnected challenges of environmental management, planning, and governance in emerging smarter eco-cities. Moreover, smarter eco-cities epitomize a dynamic blend of academic rigor and practical innovation, thriving on interdisciplinary and transdisciplinary collaboration. They are founded on the seamless integration of diverse fields, as exemplified by concepts such as City Brain, UDT, SUM, and platform urbanism. These concepts transcend traditional boundaries, converging to shape a comprehensive approach to sustainable urban development. The interaction among these domains mirrors the complexities of urban ecosystems, with AIoT serving as a pivotal and unifying force.

Despite the burgeoning interest in smarter eco-cities, the current scholarly landscape lacks a cohesive framework that integrates the diverse array of underlying concepts, theories, technologies, and models essential for unraveling the complexity of this urban ecosystem. Further, the intricate interplay of City Brain, UDT, SUM, and platform urbanism highlights the need for a structured approach to elucidating the dynamic landscape of smarter eco-cities. The foundational principles of these data-driven management, planning, and governance systems, along with their interrelationships, require clarification to establish a cohesive framework for advancing environmental sustainability in smarter eco-cities and understanding the nuanced spectrum of their underlying dimensions. Addressing these gaps is crucial for unleashing the full potential of AI and AIoT technologies in advancing the sustainability agenda and effectively addressing urgent environmental challenges.

Against the backdrop outlined above, this chapter aims to explore the theoretical and conceptual underpinnings of smarter eco-cities, weaving together a diverse array of concepts, theories, technologies, and models. In pursuit of this overarching aim, four specific objectives are formulated as follows:

1. Examine the multifaceted foundations of smarter eco-cities in terms of their management, planning, and governance practices and the underlying computational technologies and models.

2. Elucidate the intricate interconnections and synergies among the core foundational elements of smarter eco-cities.

3. Formulate a novel conceptual framework for the effective management, planning, and governance of smarter eco-cities through the integration of AI and AIoT technologies and the associated models.

4. Establish a foundation to pave the way for the practical application of the identified technologies and models in smarter eco-city initiatives, laying the groundwork for their effective integration and implementation.

The significance of this chapter in the burgeoning field of smarter eco-cities lies in its comprehensive analysis of both conceptual frameworks and technological underpinnings. By systematically examining key concepts, theories, technologies, and models, this chapter contributes to the advancement of both theoretical understanding and technological application in the realm of intelligent and resilient cities. Furthermore, it sets the stage for the practical exploration and implementation of AIoT-driven approaches in environmental management, planning, and governance in the dynamic context of smarter eco-cities. These cities represent not merely theoretical constructs, but strategic pathways toward a sustainable urban future, and this chapter plays a pivotal role in realizing that vision.

This chapter is structured as follows: Section 2.2 describes and discusses the concepts and theories underlying smarter eco-cities. Section 2.3 is dedicated to unraveling the interconnections and synergies among these conceptual and theoretical foundations, emphasizing their roles in creating a holistic approach to sustainable urban development. Section 2.4 focuses on the practical applications that stem from the understanding of these foundations, setting the stage for exploring smarter eco-cities in practice. Section 2.5 provides a discussion on the integrated and holistic nature of smarter eco-cities. This chapter concludes, in Section 2.6, by summarizing the key takeaways, contributions, and implications.

2.2 THEORETICAL AND CONCEPTUAL FOUNDATIONS AND FRAMEWORKS

This section delves into the fundamental pillars that form the bedrock of smarter eco-cities as the primary focus of the book. This exploration entails introducing and explaining the core theoretical and conceptual foundations that underpin and define this emerging paradigm of urbanism. The primary aim is to establish a robust foundation, enabling readers to navigate the complexities, dynamics, and nuances associated with the phenomenon of smarter eco-cities.

The conceptual and theoretical frameworks serve to organize and structure the research in this book concerning smarter eco-cities. Against the backdrop of this chapter, the conceptual framework outlines the central themes, concepts, ideas, technologies, and relationships that are central to and will be explored throughout the book chapters. It provides readers with a roadmap for understanding the book's content and serves as a foundation for the discussion that follows. The theoretical framework, however, establishes the broader theoretical perspectives and frameworks that inform the book's approach and analysis and guides the research process and the interpretation of findings. It offers a coherent explanation of the phenomenon of smarter eco-cities based on the existing body of knowledge. It also helps situate the book in the larger scholarly discourse and provides readers with insight into the theoretical underpinnings of the book's arguments and conclusions. The conceptual framework is developed based on the theoretical framework in terms of existing theories, models, and empirical evidence, including new insights. Therefore, it is meant to be tailored to the specific research context and objectives of the book. Overall, the conceptual framework lays out the key concepts and themes, while the theoretical framework establishes the broader theoretical context and perspective of the book. Together, they help orient readers to the book's content and scholarly context. Accordingly, these interrelated frameworks are both essential components in the context of this research endeavor.

2.2.1 Smarter Eco-Cities

In recent years, the concept of smarter eco-cities has gained traction as a holistic approach to addressing the complex challenges of rapid urbanization, resource depletion, and environmental degradation. These cities leverage the latest technologies and solutions of smart cities to optimize their resource utilization and enhance their environmental performance. The evolving landscape of urban development has witnessed the convergence of several concepts, namely smart cities, eco-cities, smart eco-cities, and smarter eco-cities, underpinned by the integration of data-driven technologies and solutions and environmental technologies and strategies to advance environmentally sustainable urban development goals (Figure 2.1). This integrated framework envisions the coexistence of technology, environmental stewardship, and resilient urban ecosystems. Its value lies in facilitating a deep understanding of the foundational underpinnings of emerging smarter eco-cities.

The concept of smart cities has garnered substantial attention as a promising solution to address challenges related to sustainability, resource management, and urbanization. Various attempts have been made to define smart cities, resulting in diverse interpretations and developmental directions (e.g., Singh and Singla 2021; Toli and Murtagh 2020). The definition of smart cities has evolved significantly over the past two decades, transitioning from a technology-centric approach, encompassing infrastructures, architectures, platforms, systems, applications, and models, to a people-centric perspective that includes stakeholders, citizens, knowledge, services, and relevant data. Broadly, a smart city integrates cutting-edge technologies and data-driven approaches to optimize urban operations, strengthen resilience, enhance the quality of life, and promote sustainability (Mishra et al. 2022; Sharifi et al. 2024).

The working definition adopted in this chapter describes a smart city as an urban environment that leverages data-driven technologies and their applied solutions to conserve resources, minimize environmental impact, and enhance overall ecological well-being. Indeed, the tendency of smart city initiatives to prioritize economic benefits over environmental

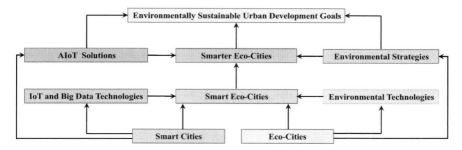

FIGURE 2.1 Smarter eco-cities and their underlying urban paradigms, technologies, solutions, and systems.

considerations (Ahvenniemi et al. 2017; Evans 2019; Toli and Murtagh 2020) has underscored the need for a greater emphasis on research and practical initiatives directed toward the environmental goals of sustainable development (Al-Dabbagh et al. 2022; Saravanan and Sakthinathan 2021; Shruti et al. 2022; Tripathi et al. 2022). This transition toward environmentally conscious smart cities marks a significant departure from conventional smart city paradigms and has garnered considerable attention in recent years. In this evolving landscape, sustainable smart cities are increasingly embracing and harnessing AI models and techniques to confront complex environmental and climate change challenges (Bibri et al. 2023b; Gourisaria et al. 2023; Efthymiou and Egleton 2023; Zaidi et al. 2023), establishing AI solutions as a pivotal driver in shaping the trajectory of their planning and governance practices.

The emphasis on environmental sustainability in the evolving landscape of smart cities seamlessly aligns with the core principles of smart eco-cities, which are fundamentally centered around environmental stewardship and sustainability. As smart cities evolve to prioritize ecological considerations alongside technological advancements, they inherently begin to resemble the ethos of smart eco-cities in terms of planning and governance. Both paradigms share a common goal of creating urban environments that not only optimize efficiency and improve quality of life but also prioritize environmental protection and resilience. This convergence signifies a natural progression toward holistic urban development models that place environmental sustainability at their forefront, bridging the gap between conventional smart city initiatives and the concept of smart eco-cities.

The concept of eco-cities, initially coined by Register in 1987, centers on the idea of urban environmental systems in which resource input and waste output are minimized. Eco-cities function as organisms that monitor the flow of resources and the management of waste (Bai 2007), evolving over time into diverse strategies and solutions to address and overcome environmental challenges. As a set of various types of models, they can be founded on the principles of urban ecology (Roseland 1997) or an amalgamation of sustainable urbanism and smart urbanism (Bibri 2021a). With respect to the latter, among these models, the prominent one is smart eco-cities, which combine environmental technologies and strategies with data-driven technologies and solutions to enhance their environmental performance. Smart eco-cities utilize primarily IoT and big data technologies and solutions of smart cities to achieve environmental sustainability goals (Bibri 2020; Bibri and Krogstie 2020a, b; Caprotti et al. 2017; Späth 2017; Tarek 2023).

Taking a step further, smarter eco-cities leverage and integrate AI and AIoT technologies and solutions with environmental technologies and strategies to maximize the performance of their sustainable systems. They also integrate these systems with smart systems, recognizing the synergies in their operations to achieve combined effects that exceed the sum of their separate effects on environmental sustainability. They place a strong emphasis on environmental sustainability and embrace innovative approaches to energy efficiency, sustainable transportation, waste management, water resources conservation, environmental monitoring, and climate change adaptation and resilience (see Bibri et al. 2023b for a systematic review). By incorporating emerging AI and AIoT technologies, they pursue a more holistic approach to urban development, striving to create more sustainable, resilient, environmentally friendly, and technologically advanced urban environments. Overall, the distinctive features of smarter eco-cities include the following:

- Innovative potential of AI and AIoT technologies in advancing environmental sustainability.
- Elevated sustainability outcomes through the application of AI and AIoT solutions.
- The synergistic integration of smart city systems and eco-city systems.
- Optimized performance and efficiency of smart eco-city systems.
- Enhanced urban planning and governance processes.

In summary, as smart eco-cities pivot their focus toward attaining environmental sustainability primarily through IoT and big data technologies, the evolution toward smarter eco-cities involves leveraging emerging AI and AIoT technologies to enhance environmental monitoring, optimize resource management, and implement innovative solutions for sustainable development. By doing so, they not only adhere to the core principles of technology- and data-driven urban development that are central to smart cities but also enhance their capacity to address environmental challenges more effectively and comprehensively.

2.2.2 Urban Governance, Planning, and Management: An Environmental Perspective

2.2.2.1 Urban Governance

Smarter eco-cities represent an advanced model of sustainability transitions, and hence, they are deeply intertwined with environmental governance. Urban governance refers to the set of structures, processes, and mechanisms through which urban areas are managed and regulated. It entails the decisions, policies, and actions taken by various stakeholders, including government authorities, institutions, community organizations, and citizens, to plan, manage, and finance urban areas. Accordingly, it "involves a continuous process of negotiation and contestation over the allocation of social and material resources and political power" (Avis 2023), "where conflicting or diverse interests may be accommodated, and cooperative action can be taken" (UN-Habitat 2023). The relationship between governance and eco-cities and smart eco-cities is self-reinforcing, with the former playing a central role in shaping the planning and development of the latter (Deng et al. 2021; Flynn et al. 2016; Joss 2011; Caprotti et al. 2017; Späth and Rohracher 2011; Späth et al. 2017) by formulating and executing policy (Pierre and Peters 2005) or multi-layered political authority (Bridge and Perreault 2009).

Urban policy—plans, laws, regulations, and actions that are adopted by local governments, mediated by civic institutions, and carried out by city agents—is crucial in aligning and mobilizing urban stakeholders. This process requires huge coordination and steering efforts, negotiations and disagreements, and countless decisions based on periodic and anecdotal evidence. This involves meta-governance, the regulation of self-regulation, where public authorities take on the role of meta-governor (i.e., spider in the web) of governance networks (Sehested 2009) based on values, norms, and principals (Kooiman and Jentoft 2009). As a comprehensive and inclusive approach, meta-governance is associated with "situationally appropriate governance frameworks" (Meuleman and Niestroy 2015).

Network governance is the most applied type of governance in eco-cities and smart eco-cities in European countries, such as Denmark, the Netherlands, Sweden, Finland, and Switzerland. Network governance is advocated by many political scientists in consensus-style democracies (Meuleman and Niestroy 2015). According to Sörensen and Torfing (2005), network governance consists of five key components: "relatively stable horizontal articulations of interdependent, but operationally autonomous actors; who interact through negotiations; which takes place within a regulative, normative, cognitive, and imaginary framework; that is self-regulating within limits set by external agencies; and which contributes to the production of a public purpose as an expression of plans, policies, and regulatory frameworks." However, while the networked mode of governance has the potential to increase legitimacy, empower citizens, and enable stakeholder buy-in (e.g. Sørensen and Torfing 2005), it is questionable as to how it influences democratic concerns, such as accountability, representation, and transparency (Nyseth 2008; Sørensen 2006), as well as its effectiveness in driving sustainable development (e.g., Connelly and Richardson 2008).

2.2.2.2 Environmental Governance
The focus of this chapter and the book is on environmental sustainability. In this light, environmental governance, which overlaps with smarter eco-city governance, denotes the system of processes, mechanisms, structures, and institutions through which decisions are made and actions are taken to manage and protect the environment. This system enables political actors to wield influence on environmental actions and outcomes (de Wit 2020; Lemos and Agrawal 2006; Vatn 2016). It involves the formulation and implementation of policies, regulations, and initiatives aimed at addressing environmental challenges and promoting sustainability). It encompasses a wide range of activities, including environmental planning, monitoring, evaluation, laws and regulations, enforcement, compliance, and stakeholder engagement.

The processes of environmental governance involve the identification of environmental issues, the development of policies and strategies to address them, and the coordination of efforts among various stakeholders, including government agencies, non-governmental organizations (NGOs), businesses, and communities. Accordingly, they entail the methodologies, protocols, processes, and procedures employed to develop, execute, and assess decisions within organizational frameworks (e.g., de Wit 2020; Vatn 2016; van Assche et al. 2020). They include data collection and analysis, public participation, consultation, collaboration, and coordination to ensure that decisions reflect the diverse interests and perspectives of stakeholders with respect to policy formulation and implementation.

Mechanisms in environmental governance constitute an array of tools and tactics employed for executing environmental policies and achieving desired ecological outcomes. This toolkit includes legal frameworks, policy instruments, agreements, partnerships, incentives, disincentives, monitoring and reporting, public participation processes, and technology-driven innovations (Azizi, Biermann and Kim 2019; Bibri et al. 2024a; de Wit 2020; Fung 2006; Gupta and van der Heijden 2019; Nishant et al. 2020). These mechanisms provide practical avenues through which decisions are made, actions are taken, and outcomes are realized in the context of environmental governance.

The structures of environmental governance refer to the overarching frameworks, organizations, hierarchies, and systems established to manage and regulate environmental issues (Gorris et al. 2019; Koch et al. 2021; Kramers et al. 2016). They encompass the legal, policy, and institutional frameworks that define how decisions are made, authority is distributed, and responsibilities are allocated in regard to guiding decision-making and resource and environmental management. They typically include government agencies responsible for environmental protection and management at various levels, such as national ministries or departments of environment, regional environmental authorities, and local environmental agencies. They also involve collaborations among government agencies, international organizations, NGOs, and other stakeholders, collectively addressing environmental challenges. They may involve the establishment of regulatory bodies, advisory councils, and other institutions tasked with overseeing environmental policies and programs. Overall, these structures serve as the foundational basis for managing environmental issues across various jurisdictions.

Institutions play a critical role in environmental governance by providing the frameworks and mechanisms for decision-making, coordination, and implementation. They include legislative bodies responsible for enacting environmental laws and regulations, executive agencies responsible for implementing and enforcing environmental policies, and judicial bodies responsible for adjudicating disputes and enforcing compliance with environmental laws (Azizi, Biermann and Kim 2019; de Wit 2020; Green and Hadden 2021; Vatn 2016).

Furthermore, effectively implementing environmental governance strategies is essential for attaining environmental sustainability goals. These strategies, often integrated into comprehensive frameworks, cover sustainable resource management, pollution control, biodiversity conservation, climate change mitigation and adaptation, circular economy initiatives, green infrastructure development, stakeholder engagement, policy coherence, science-based policies, monitoring and reporting, and capacity building (e.g., Bibri et al. 2023b; Bodin 2017; Beunen and Patterson 2019; de Wit 2020; Evans 2011; Nishant et al. 2020; Vatn 2016).

Overall, environmental governance is essential for promoting sustainable development and ensuring the long-term health and integrity of the environment. By establishing effective processes, mechanisms, structures, institutions, and strategies, environmental governance helps to address environmental challenges, protect natural resources, and promote the well-being of present and future generations.

2.2.2.3 Urban Planning

Urban planning involves designing, regulating, and managing land use and built environments to create livable, resilient, and inclusive cities. It involves a diverse set of theories and practices aimed at directing the spatial growth of cities and enriching their livability, ultimately striving to improve human well-being and promote environmentally conscious urban development. In the context of smarter eco-cities, urban planning plays a pivotal role in orchestrating the development and evolution of urban areas with a focus on ecological awareness, efficiency, and resilience. This involves addressing key challenges related to resource management, energy conservation, waste management, pollution control, ecosystem preservation, and climate change mitigation. This relates to environmental planning, which entails "a comprehensive framework of methodologies, plans, and strategies that revolve around the evaluation, management, and optimization of natural resources and ecosystems" (Bibri et al. 2024a).

Generally, urban planning involves intricate interactions both vertically, across different policy levels, and horizontally, between various policy sectors and public and private interests (Janin Rivolin 2012). These interactions occur within the framework of legally established objectives, tools, and processes, influencing existing urban developments and guiding future trajectories (Janin Rivolin 2012). Indeed, urban planning constitutes a multifaceted and systematic endeavor, encompassing a diverse array of activities, tasks, methods, and structures involving (Asadzadeh et al. 2023; Berisha et al. 2020; Bibri 2021a, b; Cotella and Janin Rivolin 2010; Hamel et al. 2021; Sharifi et al. 2024; Schmitt and Wiechmann 2018; Stead 2021):

- Developing plans and designs for the layout and organization of urban spaces, including streets, buildings, parks, and public amenities, to create functional, attractive, and sustainable environments.
- Establishing regulations, policies, and zoning ordinances to govern land use, building standards, and development activities, ensuring

compliance with legal requirements, and promoting orderly growth and development.

- Coordinating and overseeing the implementation of urban plans and projects, including infrastructure development, public service delivery, and environmental management, to achieve desired urban outcomes and community needs.

- Planning and managing interconnected urban infrastructure systems, such as transportation networks, water supply and sanitation systems, energy grids, telecommunications networks, and waste management facilities, to support urban functions and services efficiently.

- Conducting research, data analysis, and impact assessments to understand current urban conditions, identify trends and challenges, evaluate potential development scenarios, and assess the social, economic, and environmental implications of proposed projects and policies.

- Facilitating public participation and engagement processes to involve various stakeholders, including residents, businesses, community organizations, and government agencies, in decision-making processes and fostering collaboration, transparency, and accountability in urban planning and development.

- Integrating principles of environmental sustainability, resilience, and social equity into urban planning practices to minimize environmental impacts, enhance resource efficiency, promote equitable access to services and opportunities, and address social and spatial inequalities within urban areas.

- Exploring innovative approaches, technologies, and strategies to address emerging urban challenges, improve urban governance and management, and enhance the quality of life for urban residents, while fostering resilience in the face of global trends and disruptions.

- Anticipating and responding to dynamic changes and uncertainties, such as demographic shifts, economic fluctuations, technological advancements, and climate change impacts, by adopting flexible, adaptive, and resilient planning approaches that can accommodate evolving urban needs and conditions over time.

- Collaborating with neighboring jurisdictions, regional authorities, and international organizations to coordinate planning efforts, address cross-boundary challenges, and promote integrated and sustainable development across different scales and levels of governance.

In light of the above, urban planning operates across various conceptual domains, each serving distinct but interconnected purposes (Bibri 2021b), including:

- Strategic planning: Establishing long-term goals and visions for urban development, guiding the overall direction and growth of cities.
- Sustainable planning: Integrating environmental, social, and economic considerations to promote environmental sustainability and resilience in urban areas.
- Spatial planning: Allocating land use and determining the spatial organization of urban environments to optimize functionality, accessibility, and quality of life.
- Land-use planning: Deciding how land should be used and developed, considering factors such as zoning regulations, density, and compatibility of land uses.
- Transportation planning: Designing transportation systems and infrastructure to facilitate efficient movement of people and goods while reducing congestion, pollution, and reliance on fossil fuels.
- Local and regional planning: Coordinating planning efforts at local and regional levels to ensure consistency and compatibility between neighboring jurisdictions and to address shared challenges.
- Infrastructure planning: Planning and provision of essential urban infrastructure such as water supply, sanitation, energy, and telecommunications to support urban functions and activities.
- Environmental Planning: Integrating environmental considerations into urban development decisions to mitigate environmental impacts, preserve natural resources, and enhance ecological resilience.

These domains of urban planning operate across diverse scales, ranging from the micro-level of individual districts and neighborhoods to the macro-level of entire urban regions. Overall, urban planning encompasses a wide range of activities and responsibilities aimed at creating livable, sustainable, functional, and resilient urban environments that meet the diverse needs and aspirations of present and future generations. Urban planners strive to shape cities that are inclusive, equitable, prosperous, and environmentally sustainable by means of integrating various disciplines, engaging stakeholders, and adopting innovative approaches.

2.2.2.4 Urban Management

Urban management involves the day-to-day operation, administration, uptake, and maintenance activities necessary to ensure the efficient functioning of urban areas. City operation is about how cities run or are managed on a day-to-day basis. This includes the practical, logistical, and administrative aspects of city management. City functioning entails the roles or purposes that different parts of the city serve. This encompasses how various systems and infrastructures fulfill their intended roles. Furthermore, urban management activities include transport management, traffic management, mobility management, energy management, waste management, water management, environmental monitoring, building management, public services, and emergency response. Urban management involves the administration of resources, infrastructure, services, and policies to meet the diverse needs of residents and communities while promoting sustainable development and enhancing the quality of life. Unlike urban planning, which deals with long-term visioning and strategic decision-making, urban management deals with the implementation and execution of plans and policies in urban areas. It involves scheduling repairs, inspections, and renovations to ensure that infrastructure remains safe and functional. It ensures that services are delivered efficiently, effectively, and equitably to meet the needs of diverse urban populations. Urban managers engage with residents, community organizations, businesses, and other stakeholders to gather feedback, address concerns, and promote civic participation in decision-making processes (Bibri 2021c; Bibri and Krgostie 2021; Noori et al. 2020). Overall, effective urban management requires a holistic and integrated approach that considers the diverse needs and interests of urban stakeholders while striving to achieve sustainability, resilience, and livability in cities.

2.2.3 Urban Governance, Planning, and Management: Linkages, Dynamics, and Nuances

Urban management, planning, and governance are interconnected facets that shape the development and functioning of cities. Together, these three processes and practices play a vital role in shaping the growth, resilience, and livability of cities, influencing everything from land use and transportation to environmental sustainability and social equity.

The relationship between urban planning and urban governance is deeply interconnected and intertwined. Urban planning serves as the foundation by offering plans and strategies for urban development, while urban governance is responsible for executing and overseeing the implementation of these plans and strategies, ensuring compliance with established policies. Governance represents a policy process aimed at formulating sectoral coordination and integration and guiding stakeholder inclusion in planning practices (Schmitt and Danielzyk 2018). Traditional governance and planning systems are recognized for their capacity to coordinate essential actors and entities throughout the planning process, as well as to integrate relevant discourses and paradigms into the urban development agenda (Bush and Doyon 2019; Rivolin 2012). In essence, urban planning is described as a "policy-driven mode of governance" (Schmitt and Wiechmann 2018, p. 25) or an activity of governance (Stead 2021) in the intricate and dynamic contexts shaped by social, cultural, economic, and political institutions (Rivolin 2012; Stead 2021). Consequently, it can be viewed as a social process and a value-based collective effort of diverse stakeholders aimed at improving the built environment (Nuissl and Heinrichs 2011). Overall, the synergy between urban planning and urban governance lies in their complementary roles in and synergistic effects on shaping and managing urban environments.

The integration of urban planning and governance systems holds significant implications for environmental sustainability. By coordinating planning efforts with governance structures, smarter eco-cities can ensure that environmental concerns are effectively addressed in decision-making processes and policy implementation. This integration enables the development of more holistic and coordinated approaches to environmental management, fostering the protection of natural resources, reduction in pollution, promotion of sustainable land-use practices, and mitigation of climate change impacts. Moreover, it enhances accountability and transparency in environmental decision-making, thereby fostering greater public participation and stakeholder engagement in sustainability efforts. Overall, integrated urban planning and governance systems are essential for advancing environmental sustainability in cities and creating healthier, more resilient urban environments for current and future generations.

In smarter eco-cities, the synergy between urban management and urban planning is paramount for achieving the environmental objectives of sustainable development. This collaboration involves integrated decision-making processes where environmental considerations are incorporated into land-use plans and infrastructure designs. Together, these urban practices prioritize compact, mixed-use developments, green infrastructure integration, and the preservation of natural habitats to enhance biodiversity and ecosystem services. Data-driven approaches support their efforts to leverage advanced technologies to assess environmental conditions and monitor emerging trends. Moreover, community engagement plays a vital role, with stakeholders actively participating in decision-making processes to shape policies and projects that reflect local priorities and values. Ultimately, this integration fosters resilient, sustainable, and livable urban environments that prioritize environmental health and well-being.

In recent years, the landscape of environmental urban management, planning, and governance processes has been significantly influenced by the convergence of AI, IoT, and big data, collectively known as AIoT. This convergence has brought about groundbreaking advancements, revolutionizing how smart cities and smarter eco-cities collect, analyze, and harness data to inform decision-making and improve sustainability efforts (e.g., Ajuriaguerra Escudero and Abdiu 2022; Bibri and Hunag 2024, Nishant et al. 2020; Kamrowska-Załuska et al. 2021; Koumetio et al. 2023; Samuel et al. 2023; Sanchez 2023a). AIoT has demonstrated substantial potential in optimizing resource management, enhancing energy efficiency, streamlining waste management, improving transportation systems, conserving biodiversity, reducing environmental impacts, and strengthening climate change mitigation and resilience (Bibri et al. 2023b; Chen et al. 2023; El Himer et al. 2022; Leal Filho et al. 2022; Samadi 2022; Seng et al. 2022; Zhang and Tao 2021). Furthermore, the integration of AIoT holds profound implications for the convergence of environmental urban management, planning, and governance in the context of smarter eco-cities. By harnessing AIoT technologies, these cities can enhance their ability to monitor environmental conditions in real time, collect vast amounts of data on various aspects of urban life, and analyze these data to inform decision-making processes. Overall, AIoT offers a transformative framework for integrating environmental considerations into urban management, planning, and governance practices, fostering the development of smarter and more resilient eco-cities.

2.2.4 Environmental Sustainability and Climate Change

Environmental sustainability refers to the responsible and balanced utilization of natural resources and ecosystems to meet present needs without compromising the ability of future generations to meet their own needs. Morelli (2011) defines it as "meeting the resource and services needs of current and future generations without compromising the health of the ecosystems." It aims to create a harmonious relationship between human activities and the natural environment, thereby strengthening resilience, maximizing ecological value, and improving well-being. It involves sustainable transportation, energy conservation, renewable energy, sustainable materials, efficient land use, water resource conservation, circular economy, pollution control, and climate change mitigation (Bibri et al. 2023b; Nishant et al. 2020). In the context of smarter eco-cities, environmental sustainability emphasizes adopting and implementing environmentally oriented practices, policies, strategies, approaches, and technologies to minimize the negative impacts of urbanization on the environment concerning resource depletion and ecological degradation. As such, commitment to environmental sustainability serves as a guiding force, steering smarter eco-city development toward a more ecologically responsible and harmonious future. Integrating current technological advancements, especially AI and AIoT, into the management, planning, and governance systems of smarter eco-cities amplifies the potential for attaining environmental sustainability goals

A review study conducted by Nishant et al. (2020) on the relationship between AI and environmental sustainability highlights the need for future research to embrace multilevel perspectives, system dynamic methodologies, design thinking applications, and a thorough exploration of economic value considerations. This approach aims to elucidate how AI can provide prompt solutions while ensuring they do not inadvertently pose long-term risks to environmental sustainability goals. Moreover, AI holds the capacity to address information asymmetries and human cognitive and emotional biases, which often hinder devising effective solutions for environmental sustainability (Cullen-Knox et al. 2017). By harnessing AI and AIoT, smart cities and smarter eco-cities can potentially accelerate their approach and transform their thinking in developing science-based solutions and policies for tackling environmental challenges (Bibri et al. 2023b, 2024a). However, achieving this transformation and formulating innovative policies necessitates a fundamental paradigm shift. Ecological degradation and climate disruption are exceedingly complex phenomena, involving numerous trade-offs that often reduce problems to oversimplified explanations, prompting decision-makers to default to game-theoretical interactions driven by biased solutions.

The relationship between environmental sustainability challenges, strategies, and goals is multifaceted and dynamic, reflecting the intricate interplay between human activities and the natural environment. Environmental sustainability challenges arise from various factors, including population growth, urbanization, industrialization, resource exploitation, pollution, and climate change. These challenges pose significant threats to ecosystems, biodiversity, natural resources, and human well-being. In response to these challenges, environmental sustainability strategies aim to mitigate adverse impacts on the environment while promoting long-term ecological balance and resilience. These strategies, as applied in sustainable smart cities, encompass a wide range of initiatives, technologies, and practices aimed at reducing carbon emissions, conserving natural resources, protecting ecosystems, promoting renewable energy, enhancing biodiversity, and fostering sustainable development (Almalki et al. 2023; Mishra et al. 2022; Saravanan and Sakthinathan 2021; Tripathi et al. 2022). These strategies have also been bolstered by AI and AIoT in both sustainable smart cities and smarter eco-cities to advance environmental sustainability goals, improve urban resilience, optimize resource management, and enhance the overall quality of life for residents (Alahakoon et al. 2020; Bibri et al. 2023b, 2024a, b; Gourisaria et al. 2023; Efthymiou and Egleton 2023; Szpilko et al. 2023; Zaidi et al. 2023; Ullah et al. 2020).

AI offers vast potential not only for enhancing environmental sustainability outcomes but also for addressing climate change challenges by advancing data-driven insights and solutions across these interconnected fields (Bibri et al. 2023b; Chen et al. 2023; Jain et al. 2023; Leal Filho et al. 2022; Sandalow et al. 2023). Although climate change is a key aspect of environmental sustainability, the latter encompasses a broader aim of maintaining and improving overall environmental health. This involves ensuring that human activities are conducted in ways that do not deplete natural resources or harm ecosystems, thereby preserving the environment for future generations. Environmental sustainability is closely aligned with climate mitigation efforts, as both strive to achieve long-term ecological balance and minimize human-induced environmental damage.

Climate change involves the long-term alterations in temperature, precipitation, weather patterns, and other elements of the Earth's climate system. These changes are predominantly driven by human activities, including the burning of fossil fuels and deforestation, which lead to increased concentrations of greenhouse gases (GHGs) in the atmosphere, thereby driving climate change. The accumulation of GHGs, such as carbon dioxide and methane, enhances the greenhouse effect, trapping more heat and causing global temperatures to rise. This warming triggers a cascade of environmental impacts, including more frequent and severe weather events, rising sea levels, and disruptions to ecosystems and biodiversity. Addressing the root causes of climate change is essential for mitigating its effects and ensuring a sustainable future for the planet.

Tackling climate change is crucial because it threatens ecosystem stability, undermines sustainable development efforts, and endangers human health, livelihoods, and global economies. Thus, initiatives to mitigate climate change—such as reducing GHG emissions—and adapt to its inevitable impacts are essential components of broader environmental sustainability goals. Strategies to combat climate change focus on reducing GHG emissions and increasing the resilience of communities and ecosystems to climate impacts. Integrating AI and AIoT technologies enhances our understanding of climate challenges and facilitates more effective and timely interventions. AI and AIoT play a crucial role in climate change mitigation and adaptation by enabling precise monitoring, efficient resource management, and predictive analytics to reduce emissions and enhance resilience in urban environments (Bibri et al. 2023b; Chen et al. 2023; El Himer et al. 2022; Dheeraj et al. 2020; Jain et al. 2023; Leal Filho et al. 2022; Popescu et al. 2024; Rane et al. 2024; Rolnick et al. 2022; Tomazzoli et al. 2020; Samadi 2022). AI models, including in the context of AIoT, can process and analyze large, complex datasets, uncovering patterns and insights that were previously unattainable or difficult for humans to discern. This capability is vital for understanding climate dynamics, predicting future climate scenarios, and developing innovative solutions for both mitigation and adaptation efforts.

The overarching goal of environmental sustainability is to achieve a harmonious and sustainable relationship between human societies and the natural world. Thus, effective environmental sustainability strategies align with broader sustainability frameworks, such as the United

Nations Sustainable Development Goals (SDGs), which provide a comprehensive roadmap for addressing interconnected environmental, social, and economic challenges. Several SDGs directly relate to environmental sustainability, reflecting the interconnectedness between environmental protection, social development, and economic prosperity. Some of the key SDGs that address environmental sustainability in the context of smarter eco-cities and sustainable smart cities include SDG 6: Clean Water and Sanitation, SDG 7: Affordable and Clean Energy, SDG 11: Sustainable Cities and Communities, SDG 12: Responsible Consumption and Production, SDG 13: Climate Action, SDG 14: Life Below Water, and SDG 15: Life on Land. These SDGs highlight the importance of integrating environmental sustainability into broader development agendas, recognizing the critical role that environmental protection plays in achieving sustainable development outcomes. However, achieving environmental sustainability requires concerted efforts from governments, businesses, civil society organizations, and individuals as part of environmental governance structures and processes (Aziz et al. 2019; Green and Hadden 2021; de Wit 2020). It involves collective action, collaboration, innovation, and investment in sustainable technologies and practices. Moreover, fostering environmental literacy, raising awareness, and promoting behavior change are essential components of successful sustainability initiatives.

2.2.5 Artificial Intelligence

AI stands at the forefront of transformative technologies, reshaping cities and societies worldwide. With its ability to analyze vast amounts of data, learn from patterns, and make decisions and predictions, AI is revolutionizing how we approach various tasks across diverse domains. It is driving innovation and efficiency, transforming the way we work, live, and interact with technology. AI is the scientific and engineering discipline dedicated to developing computers or machines capable of performing complex tasks that typically require human intelligence. It is a broad field that aims to simulate human cognitive processes and functions, such as perception, learning, reasoning, problem-solving, language understanding and generation, and planning (Bibri 2015). It employs a variety of algorithms that are trained on extensive datasets to extract patterns from data, make decisions, draw predictions, recommend actions, and navigate unfamiliar scenarios, continually improving their performance by learning from experience with minimal or no human intervention. This capability for autonomous improvement over time, through exposure to new data, is termed ML. AI systems and machines, equipped to perceive their surroundings and independently execute tasks, hold the potential of improving their self-learning, adaptation, and evolution abilities. They can "learn from experience, adjust to new inputs, and perform human-like tasks" (Duan, Edwards and Dwivedi 2019, p. 63) to "interpret external data correctly, to learn from such data, and to use those learnings to achieve specific goals through flexible adaptation" (Kaplan and Haenlein 2019, p. 17). Alahakoon et al. (2020) introduce a novel suite of self-building AI solutions, characterized by their ability to adapt, self-structure, self-configure, and self-learn, facilitating dynamic data processing.

In the dynamic landscape of AI, a triad of fundamental components plays an integral role in shaping the capabilities and functionality of AI systems. These components encompass AI models, techniques, and algorithms, each serving a distinct yet interwoven purpose within the AI framework as applied to smart cities and smarter eco-cities. While AI models provide the overarching structure for specific tasks, AI techniques enhance performance through complex methods. At the same time, AI algorithms dictate the execution of tasks. Together, these components form the bedrock of AI in terms of facilitating data processing, learning, and task performance. Understanding their synergistic interplay and integrative nature is ke to navigating the multifaceted and dynamic field of AI. Furthermore, as AI becomes increasingly accessible and is applied in various domains, including smarter eco-cities and sustainable smart cities, there can be significant confusion surrounding the common terminology used. AI models, techniques, and algorithms are sometimes used interchangeably, thereby the relevance of describing and differentiating between these essential components.

AI models are high-level structures or architectures that define the objectives and tasks an AI system can perform. They are the foundation upon which AI capabilities are built. They serve as a framework for learning patterns from data, acting as the output of the algorithm and functioning like a program that can be executed on data to make predictions and decisions. These models often consist of layers, nodes, or elements that process data in a specific way. AI models include ML models, such as decision trees and support vector machines (SVMs), which are used for classification and regression tasks. DL employs models like convolutional neural networks (CNNs) for image recognition and recurrent neural networks (RNNs) for sequential data, such as time series or language processing. Natural language processing (NLP) uses models like BERT or GPT to understand and generate human language. Similarly, computer vision (CV) applies models like you only look once (YOLO) for object detection in images or videos. Fuzzy logic (FL) uses Mamdani and Sugeno fuzzy inference systems in control systems, decision-making processes, and expert systems. These models handle situations where the boundaries between categories are not clear. Natural computing (NC) uses models such as ant colony pptimization (ACO) and genetic algorithms (GAs) to mimic the behavior of biological systems such as the evolution of species (in GAs) or the foraging behavior of ants (in ACO) to solve optimization problems. Other NC models include swarm intelligence, cellular automata, deoxyribonucleic acid (DNA) computing, and quantum computing. Evolutionary computing (EC) applies models like GAs, particle swarm pptimization (PSO), and differential evolution (DE) to optimize tasks that are challenging for traditional methods, such as complex problem-solving or ML optimization. These models are inspired by natural processes like evolution, swarm behavior, and mutation, and are particularly useful for solving optimization problems in dynamic or complex environments where traditional algorithms may struggle to find optimal solutions. All these models are specialized to perform tasks in their respective domains, enabling advances in automation, data interpretation, and decision-making. They are designed to capture patterns, relationships, or knowledge from data and can be used for various applications, including environmental sustainability and climate change (Chen et al. 2023; Jain et al. 2023; Kaack et al. 2022; Leal Filho et al. 2022; Rolnick et al. 2022), including in the context of smart cities and smarter eco-cities (Bibri et al. 2023b; Nishant et al. 2020).

For example, NLP models are employed in environmental sustainability domains to analyze textual data from various sources to understand public sentiments, opinions, and trends related to environmental issues. CV models are utilized to interpret visual information by processing satellite imagery, aerial photographs, and other visual data to monitor land use alterations, track deforestation and forest degradation, and monitor wildfires and floods. FL models are designed to handle imprecise or uncertain data in environmental modeling and decision-making processes, allowing for the representation of vague concepts and fuzzy boundaries. NC models are used to simulate complex systems in ecology, such as modeling ecosystems, species interactions, and the effects of climate change on biodiversity. Lastly, EC models are employed to optimize solutions for environmental sustainability problems, such as optimizing resource allocation, energy efficiency, and waste management strategies.

AI is a vast field encompassing various subfields or domains, each focused on different aspects of intelligent systems (see Chapter 7 for further details). In addition to ML and its subset DL, NLP is centered on enabling computers to understand, interpret, and generate human language. CV allows machines to analyze and make decisions based on visual inputs. FL deals with handling uncertain or imprecise information, often used in systems that require approximate reasoning. NC draws inspiration from natural phenomena, utilizing models like neural networks and evolutionary algorithms to solve complex problems. EC, a subset of NC, focuses on optimization algorithms that mimic biological evolution processes, such as mutation and selection, to find optimal solutions in dynamic environments. Together, these subfields drive innovation in AI, addressing a wide range of real-world challenges.

AI techniques are methodologies or approaches used in AI models to enhance their performance. Techniques are usually part of the model's design and guide how it processes data. For example, transfer learning (TL) is a technique often used in neural network (NN) models to improve training efficiency and performance. In transfer learning, a model that has been pre-trained on one task (usually with a large dataset) is fine-tuned or adapted to perform a different but related task (see Chapter 7). This allows the model to leverage previously learned features, reducing the amount of data and training time required for the new task, while often improving performance. Another example is reinforcement learning, one of the ML techniques, which centers on optimizing a long-term cumulative reward through interactions with the environment. Reinforcement learning involves training models through trial and error, where an agent learns to make decisions by receiving feedback from the environment, making it particularly useful for applications requiring sequential decision-making. It can be divided into two categories: model-based and model-free. Model-based reinforcement learning utilizes function approximation for sample efficiency, while model-free reinforcement learning, including Monte Carlo (MC) and temporal difference (TD) approaches (e.g., Q-learning), addresses reinforcement learning problems without explicit models (Ullah et al. 2020). ML techniques also include supervised learning and unsupervised learning. In supervised learning, models are trained using labeled data, making them well-suited for tasks like classification and regression, where the desired outcomes are known. In contrast, unsupervised learning does not depend on labeled data and is used to uncover patterns or structures within datasets, such as in clustering or anomaly detection. Each paradigm offers distinct methods for addressing complex challenges, enabling the creation of robust and adaptable ML applications.

AI algorithms are step-by-step procedures or sets of rules that dictate how a specific task is executed. They operate on datasets to recognize patterns and rules, allowing them to create models that can then make predictions and decisions. They are often more granular and can be used across different models and techniques. They define the logic and mathematical operations that drive AI systems. For example, gradient descent is an algorithm used to optimize model parameters during training. It can be applied to various models and techniques to improve their performance. Another example is the reinforcement learning algorithm's active entity called an agent, which aims to achieve the specified objective by following an optimal policy, a sequence of actions maximizing overall long-term rewards. Agents must strike a balance between exploiting known actions and exploring new ones to enhance rewards (Ullah et al. 2020). Additionally, k-means clustering is an unsupervised learning algorithm that partitions a dataset into k distinct clusters, where each data point belongs to the cluster with the nearest mean, widely used in pattern recognition and data segmentation. Another example is the backpropagation algorithm, which is used in training neural networks by adjusting weights through the minimization of the error rate across iterations.

In summary, AI models provide the overarching structure for specific tasks, techniques are methods used in models to improve their performance, and algorithms are detailed procedures or rules that drive specific operations within AI systems. These three components work together to enable AI systems to process data, learn from them, and perform tasks effectively. AI empowers machines to exhibit intelligent behaviors through the analysis of their surrounding environment, the pursuit of defined goals, and the generation of rational decisions and conclusions.

2.2.6 The Internet of Things

IoT represents the interconnected network of physical objects embedded with sensing, processing, communication, and actuating capabilities, enabling data exchange with other devices over the Internet or other networks. It encompasses a wide range of things, from everyday objects to complex systems and infrastructures. Within the IoT ecosystem, there are web-enabled smart devices (e.g., smartphones, computers, appliances, and wearables) facilitating human-to-machine interaction, as well as embedded systems (e.g., sensors, processors, edge/fog nodes, and actuators) functioning without direct human involvement. An increasing number of IoT devices are continually connecting across the globe, contributing substantial volumes of data to analytical systems. It is estimated that the daily global data generation has reached 2.5 quintillion bytes, with a further surge anticipated to reach 463 exabytes by 2025 (Ghosh and Sengupta 2023).

Edge, fog, and cloud computing represent three closely interrelated paradigms that play a fundamental role in both IoT and AIoT ecosystems in smart cities and smarter eco-cities in the field of environmental sustainability (Bibri and Krogstie 2020a; Seng et al. 2023; Zhang and Tao 2021). Table 2.1 outlines the distinctive features of these concepts.

These computing paradigms collaborate to create a comprehensive IoT or AIoT ecosystem that efficiently manages data processing and analysis across various levels of the network infrastructure. IoT and AIoT have proven to drive innovation, improve decision-making, and create new opportunities in various applied domains of smarter eco-city management, planning, and governance, especially in the context of environmental sustainability (Bibri et al. 2023b, 2024a).

2.2.7 Artificial Intelligence of Things

The integration of AI with IoT, known as AIoT, has effectively addressed many challenges faced by traditional IoT systems. By incorporating AI algorithms and techniques into IoT devices and networks, AIoT enables devices to process and analyze data locally, reducing latency and bandwidth requirements while enhancing real-time responsiveness (Seng et al. 2022; Zhang and Tao 2021). This local processing capability enhances real-time decision-making, making IoT systems more responsive and efficient. Additionally, AIoT enables predictive analytics and proactive maintenance, allowing for the detection of potential issues before they escalate. AIoT has also revolutionized data handling capabilities by efficiently managing the colossal amounts of data generated by IoT devices (Shi et al. 2020). With AI algorithms integrated into IoT systems, AIoT enables intelligent data processing and analysis at the edge, reducing the need to transmit large volumes of raw data to centralized servers. This edge computing approach minimizes latency, conserves bandwidth, and enhances privacy and security by processing sensitive data locally (Fragkos et al. 2020; Song et al. 2020; Yang et al. 2020a, b).

Furthermore, AIoT leverages advanced analytics and ML to extract valuable insights from massive datasets, enabling predictive modeling,

TABLE 2.1 Distinction between Edge, Fog, and Cloud Computing

Computing Models	Description	Common Application Scenarios
Edge computing	Edge computing is a decentralized approach where data processing occurs in close proximity to data sources, often within the device or a nearby gateway. Its primary objective is to minimize latency and enable real-time responsiveness by executing computations locally. Edge computing excels in applications that demand quick decision-making and low-latency interactions, such as autonomous vehicles	Real-time decision support, low-latency applications
Fog computing	Fog computing extends edge computing by establishing a hierarchical architecture with multiple edge devices and gateways. Fog nodes are strategically positioned near data sources to conduct intermediate processing, data filtering, and preliminary analytics. This approach optimizes bandwidth usage and enhances system performance, making it ideal for scenarios involving distributed data sources and resource-constrained devices	Distributed data processing, resource-efficient computing
Cloud computing	Cloud computing relies on centralized remote servers for data storage, management, and processing. It offers substantial computational resources and storage capacities, making it well-suited for complex data analytics, ML, and large-scale processing. Cloud computing enables data access and analysis from virtually anywhere with an internet connection, making it the preferred choice for applications that require significant computation and storage capabilities	Complex data analytics, ML/DL large-scale processing

Source: Adapted from Bibri et al. (2023b).

anomaly detection, and optimization of operations. By combining AI and IoT technologies, AIoT has transformed data handling, making it more efficient, scalable, and responsive to the needs of diverse applications. Parihar et al. (2023) underscore the significance of the integration of IoT and AI within the AIoT framework, highlighting its advantages while examining its evolution, architecture, applications, and challenges. Al-Turjman et al. (2021) present a comprehensive exploration of AI–IoT, providing a holistic understanding by integrating both theoretical and practical perspectives of the AI paradigm for IoT. The work covers fundamentals, theoretical discussions, critical applications, and ML and DL algorithms. It addresses a range of AI–IoT applications across various domains, including environmental sustainability, climate change, smart applications, and NLP. Additionally, it covers applications in robotics, sustainability, smart grid management, security, privacy, and ethics, highlighting the broad and impactful scope of AI–IoT technologies. Overall, AIoT has transformed how IoT systems operate and perform, making them more intelligent, adaptive, and capable of addressing the complex demands of modern applications.

Concurrently, AI is experiencing a resurgence driven by the abundant and potent data flow generated by IoT sensors and devices. This resurgence is facilitated by increased computing storage capacity and the remarkable speed of real-time data processing. With the proliferation of IoT devices capturing vast amounts of data across various urban domains, AI algorithms can now leverage these data deluge to uncover valuable insights and patterns. Indeed, the escalating volume, variety, and velocity of data necessitates advanced data analytics models and innovative data visualization methods to extract and display valuable knowledge through applied machine intelligence. The interconnectedness of IoT and AI is facilitated by the colossal amount of data generated through the IoT infrastructure and analyzed by AI systems, contributing to the development of smarter and more sustainable urban systems. Moreover, the integration of ML and DL systems introduces edge intelligence to cloud-based IoT deployments, allowing AI algorithms to operate outside the constraints of the cloud and meet the high computing resource demands.

The symbiotic relationship among IoT, big data, and AI has led to improved insights and decision-making capabilities in various applications across diverse domains in smart cities and smarter eco-cities, mainly relying on ML and DL (Bibri et al. 2023a, b). AIoT-powered applications can analyze and respond to data streams in near real time, offering new opportunities for optimization, automation, and innovation across different urban domains. Alahi et al. (2023) explore the role of IoT in smart cities, encompassing its impact on city infrastructures, wireless communication technologies, AI algorithms, and their integration. Highlighting the potential of fifth-generation (5G) networks, the authors provide insights into creating sustainable urban environments. The exploration of IoT and AI integration offers a glimpse into the future of smart cities, emphasizing its positive influence on urban living and well-being. In their comprehensive review, Bibri et al. (2023b) examine the myriad applications of AIoT in smarter eco-cities, highlighting how this technological framework has significantly advanced various domains of environmental sustainability. The authors examine how AIoT integration enhances urban infrastructure, resource management, and environmental monitoring, leading to more efficient energy usage, waste management, and pollution reduction. Additionally, they discuss how AIoT-driven solutions enable the implementation of smart grids and intelligent transportation systems, contributing to the overall resilience and sustainability of urban ecosystems. Through their analysis, they shed light on the transformative potential of AIoT in shaping the future of smarter, more sustainable cities. Several other studies have explored and discussed the convergence of AI and IoT and its associated opportunities and challenges in the domains of environmental sustainability, climate change, smart cities, and beyond (Bibri et al. 2023a; Kuguoglu et al. 2021; Puri et al. 2019; Samadi 2022; Sleem and Elhenawy 2023).

In sum, the integration of AI and IoT, particularly through edge intelligence, represents a significant advancement in addressing the challenges faced by IoT deployments. With the ability to process data locally and in real time, AIoT systems offer enhanced efficiency, responsiveness, and decision-making capabilities across various domains, including smart cities, smarter eco-cities, and environmental sustainability. As research and development in this field continue to progress, AIoT is expected to play an increasingly pivotal role in shaping the future of interconnected systems and enabling transformative solutions for a wide range of applications.

2.2.8 Artificial Intelligence and Artificial Intelligence of Things: Commonalities and Differences

In the dynamic landscape of technological evolution, the distinctions between AI and AIoT play a key role in shaping the trajectory of applied data-driven solutions in emerging smarter eco-cities and sustainable smart cities. This section focuses on the convergence of AI and IoT, unraveling the underpinnings that fuel their synergistic interplay. Based on the previous discussion of these interconnected domains, it is important to discern the subtle differences and shared commonalities between AI and AIoT. This not only enriches their theoretical understanding but also lays the groundwork for practical applications that hold transformative potential in advancing data-driven environmental management, planning, and governance in smarter eco-cities and sustainable smart cities.

In the symbiotic relationship between AI and AIoT, certain commonalities weave through their intricate tapestry, including:

- Data utilization: Both AI and AIoT heavily rely on data. AI processes and analyzes data to make intelligent decisions, while AIoT involves the use of AI algorithms to interpret and act upon data collected primarily by IoT devices.
- Autonomy: Both AI and AIoT strive to enable systems to operate autonomously. AI seeks to create machines that can perform tasks without human intervention, and AIoT extends this autonomy to connected devices and systems in the IoT ecosystem.
- Learning capability: Another commonality between AI and AIoT is the emphasis on learning. AI involves ML/DL algorithms that enable systems to improve performance based on experience. In AIoT, learning capabilities are applied to enhance the adaptability and intelligence of IoT devices.

On the flip side, discernible differences distinguish AI and AIoT, including:

- Scope: AI is a broader concept that encompasses the development of intelligent machines capable of a wide range of tasks. AIoT, however, specifically focuses on enhancing IoT systems and devices with AI capabilities.
- Application: AI can be applied in various domains, and AIoT is particularly applied in the context of IoT, enhancing the intelligence and decision-making abilities of interconnected devices.
- Interconnectivity: The focal point of their divergence is encapsulated in the interconnectivity aspect, where AIoT emphasizes the integration of AI into interconnected devices within the IoT framework, specifically addressing the synergy between AI and IoT.

As the commonalities and differences between AI and AIoT are unraveled, a nuanced understanding emerges, delineating their roles in shaping the technological landscape. While their shared reliance on data and pursuit of autonomy create a foundation for synergy, the distinct scopes and applications of AI and AIoT carve or create unique trajectories.

2.2.9 Machine Learning

AI involves developing computers capable of processing, interpreting, and predicting outcomes based on datasets, using algorithms as governing rules. These algorithms are trained on extensive datasets to extract patterns from data, make decisions, draw predictions, recommend actions, and navigate unfamiliar scenarios, continually improving their performance by learning from experience with minimal or no human intervention. This capability for autonomous improvement over time, through exposure to new data, is termed ML. ML stands out as one of the most widely applied subsets of AI in the development of smarter eco-cities and smart sustainable cities, particularly in the context of environmental sustainability and climate change. ML techniques enable cities to analyze vast amounts of data, predict trends, and optimize resource management, leading to more efficient energy use, waste management, water management, and transportation systems (e.g., Bibri et al. 2023a, b; Gourisaria et al. 2023; Efthymiou and Egleton 2023; Fang et al. 2023; Mishra 2023; Nishant et al. 2020; Ullah et al. 2020).

ML plays a key role in deciphering patterns in data to create predictive models. It is a subset of AI that involves the development of algorithms, techniques, and models that enable computer systems to learn from data and improve their performance on specific tasks over time without being explicitly programmed. The core idea of ML is to provide computer systems with the ability to automatically learn from experiences and adapt to their surroundings, allowing them to identify patterns and make predictions and decisions. Numerous studies have thoroughly examined and discussed ML techniques, algorithms, architectures, opportunities, and challenges (e.g., Azevedo et al. 2024; Naeem et al. 2023; Sharma et al. 2021; Verma et al. 2024).

ML can be categorized into several techniques or methodologies, each with its own characteristics and applications and applied to different domains (e.g., Antonopoulos et al. 2020; Garcoa et al. 2019; Donti and Kolter 2021; Mosavi et al. 2019; Ullah et al. 2020; Wang et al. 2019; Zhang and Tao 2021). These include environmental sustainability and climate change in relation to both AI and AIoT solutions applied to smart cities and smarter eco-cities (e.g., Alahakoon et al. 2020; Bibri et al. 2023b, 2024a, b; Din et al. 2019; Fang et al. 2023; Gourisaria et al. 2023; Huntingford et al. 2019; Leal Filho et al. 2023; Nishant et al. 2020). The choice of ML subtype can vary significantly based on the nature of the learning project, the method of data input, and the desired learning outcome, as well as the specific problem and data availability for training and inference. Important to note is that ML encompasses a wide array of subtypes (Table 2.2) based on the aforementioned studies.

TABLE 2.2 A Comprehensive Overview of ML Subtypes, Techniques, and Common Tasks

ML Subtypes	Description	Methods and Algorithms	Common Tasks
Supervised learning	In supervised learning, the algorithm is trained on a labeled dataset, where input data are paired with the correct output. It is provided with known datasets and desired inputs and outputs and is gradually corrected until it reaches a high level of accuracy	Linear regression, support vector machines (SVMs), decision trees (DTs), random forest (RF), ANNs (e.g., multi-layer perceptron), generalized linear models (GLMs), Bayesian networks (BNs)	Image classification, object recognition, speech recognition, prediction
Unsupervised learning	Unsupervised learning operates autonomously or without specific guidance, relying solely on unlabeled and unclassified input data to train AI networks in uncovering hidden patterns, answers, and distributions. It can analyze and categorize data without predefined criteria	K-means clustering, hierarchical clustering, principal component analysis (PCA), t-distributed stochastic neighbor embedding (t-SNE), GANs, auto-encoder	Segmentation, clustering, image compression, anomaly detection, topic modeling

(*Continued*)

TABLE 2.2 (*Continued*) A Comprehensive Overview of ML Subtypes, Techniques, and Common Tasks

ML Subtypes	Description	Methods and Algorithms	Common Tasks
Semi-supervised learning	Semi-supervised learning combines supervised and unsupervised learning, using a small amount of labeled and a large amount of unlabeled data. It is useful when labeled data are limited and learns to label the unlabeled data	Self-training, o-training, multi-view learning, tri-training	Document classification with limited labeled data, image recognition with few labeled examples, speech recognition with sparse labels
Reinforcement learning	Reinforcement learning involves training an algorithm to make sequences of decisions. The algorithm interacts with an environment and receives feedback in the form of rewards or punishments. It employs a trial-and-error approach to achieve an optimal outcome, guided by predefined actions, parameters, and target values	Q-learning, deep Q network (DQN), policy gradient methods (PGMs), actor-critic model (ACM), MC tree search (MCTS)	Game playing, autonomous vehicle, robotic control, decision-making in dynamic environments
Self-supervised learning	Self-supervised learning is a form of unsupervised learning where the algorithm creates its own labels from the data, often by predicting parts of the data from other parts, creating "self-generated supervision"	Contrastive learning, bidirectional encoder representations from transformers (BERTs), simple contrastive learning (SimCLR), contrastive predictive coding (CPC)	Pre-training for NLP, image representations for CV, audio representations for speech recognition

(*Continued*)

TABLE 2.2 (*Continued*)   A Comprehensive Overview of ML Subtypes, Techniques, and Common Tasks

ML Subtypes	Description	Methods and Algorithms	Common Tasks
Online learning	Online learning updates the model continuously as new data become available, ideal for real-time applications	Online gradient descent (OGD), stochastic gradient descent (SGD), passive-aggressive algorithms, perceptron algorithm, adaptive learning rate models	Adaptive personalization in recommender systems, continuous monitoring in IoT, fraud detection in real time
Transfer learning	Transfer learning trains a model on one task and then uses it to enhance performance on a different but related task, leveraging knowledge gained from one domain to benefit another	Fine-tuning pre-trained models, domain adaptation, multi-task learning, knowledge distillation, progressive NNs (PNNs)	Image classification with pre-trained models, natural language understanding with transfer learning
Few-shot learning	Few-shot learning is a variation of supervised learning where the model is trained to make accurate predictions with very few examples, valuable in scenarios with limited labeled data	N-shot learning, meta-learning, siamese networks, prototypical networks, relation networks	Image recognition with limited examples, gesture recognition with sparse training data
Transduction learning	Transduction learning is a type of semi-supervised learning where the algorithm learns from both labeled and unlabeled data, making predictions on new, unlabeled data points	Transductive SVM, label propagation, self-training, co-training	Image segmentation with limited annotations, sentiment analysis with partially labeled data, enhancing pre-trained models with unlabeled data, prediction accuracy

(*Continued*)

TABLE 2.2 (Continued) A Comprehensive Overview of ML Subtypes, Techniques, and Common Tasks

ML Subtypes	Description	Methods and Algorithms	Common Tasks
Multitasking learning	Multitasking learning, or multi-task learning, trains a model to perform multiple related tasks simultaneously, leveraging shared knowledge and correlations among tasks to improve performance	• Multi-task NNs • Multi-head attention models • Taskonomy • Joint training • Modular networks	Urban climate modeling with simultaneous predictions, autonomous systems with multiple objectives, air quality prediction, energy demand forecasting
Federated learning	Federated learning is a decentralized approach where ML models are trained across multiple devices or servers holding local data, maintaining data privacy and security	Federated averaging, federated learning with differential privacy, federated transfer learning, secure aggregation	Collaborative traffic pattern analysis without sharing raw data, regional emission analysis without centralized data sharing, privacy-preserving analysis of transportation patterns and emission data
Swarm learning	Swarm learning is inspired by natural swarm intelligence, where individual agents (ML models) collaborate in a decentralized and self-organizing manner to learn from data. Applicable in distributed environments with limited agent communication and coordination	PSO, ant colony optimization (ACO), bee algorithm, grey wolf optimizer (GWO), artificial bee colony algorithm (ABCA)	Environmental monitoring, resource optimization in IoT networks, waste management, adaptive control systems in autonomous vehicles

By presenting these components side by side, Table 2.1 illustrates the versatile landscape of ML and provides a comprehensive understanding of ML components. The subtypes of ML offer a diverse set of methods and algorithms to address a wide range of problems, each tailored for specific use cases and applications.

However, a common trend in research and application narrows the focus to three key subtypes: supervised, unsupervised, and reinforcement learning. The predominant focus on these three subtypes stems from their foundational roles in addressing a wide range of real-world problems. These subtypes represent the foundational pillars of ML, with applications spanning a broad spectrum of domains. Their prominence reflects their versatility and effectiveness in solving complex challenges and their unique strengths and adaptability to a wide array of problems. This has led to concentrated research and application efforts.

2.2.10 Deep Learning, Machine Learning, and Generative AI as Subfields of Artificial Intelligence

DL is increasingly gaining prominence and is also gaining prominence in the development of smarter eco-cities and smart sustainable cities, particularly in the field of environmental sustainability and climate change. Its capabilities in processing complex data structures and identifying intricate patterns make DL instrumental in addressing sustainability challenges in urban ecosystems. With DL, cities can enhance their capacity for real-time monitoring, decision-making, and adaptive responses to environmental changes (e.g., Aqib et al. 2019; Dong et al. 2021; Han et al. 2021; Li 2021; Kim and Cho 2019; Sen et al. 2022; Shen et al. 2019; Son et al. 2022), thereby advancing the transition toward more resilient and sustainable urban ecosystems.

ML and DL are interconnected subfields in the broader field of AI. AI encompasses a wide spectrum of subsets that aim to create computer systems that can exhibit intelligent thinking and behavior. DL is a subset of ML and a branch of AI (Figure 2.2). It involves training artificial NNs (ANNs) with multiple layers to learn hierarchical representations of data for advanced data-driven tasks. DL algorithms can autonomously discover patterns and extract features from raw data, making them particularly well-suited for NLP and CV tasks. Hence, they operate without the need for human-designed rules. Numerous studies have extensively researched and discussed DL concepts, architectures, applications, challenges, and future directions (e.g., Alhijaj and Khudeyer 2023; Alom et al. 2019; Alzubaidi et al. 2021; Dong et al. 2021; Joshi et al. 2023; Khan et al. 2020; Mathew et al. 2021).

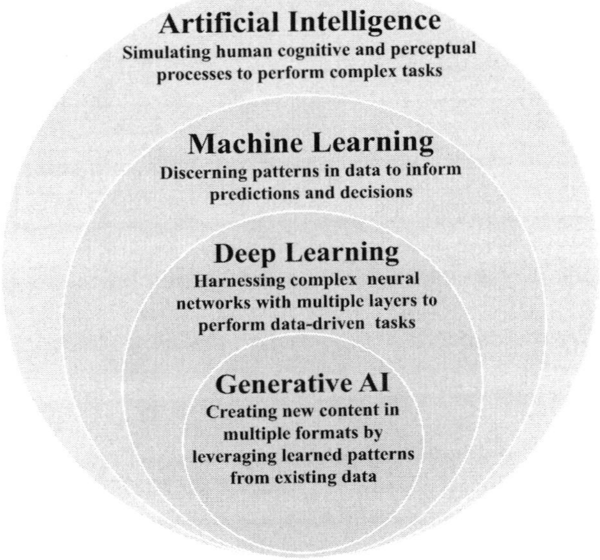

Artificial Intelligence
Simulating human cognitive and perceptual processes to perform complex tasks

Machine Learning
Discerning patterns in data to inform predictions and decisions

Deep Learning
Harnessing complex neural networks with multiple layers to perform data-driven tasks

Generative AI
Creating new content in multiple formats by leveraging learned patterns from existing data

FIGURE 2.2   Artificial Intelligence and its common and emeging subfieds.

Incorporating insights from the information processing patterns observed in the human brain, DL harnesses ANNs composed of layers of neurons, akin to algorithms. These layers contribute to a unique interpretation of the input data they receive. It relies on extensive datasets to establish mappings between input data and corresponding labels. DL specializes in creating and fine-tuning complex NNs for data analysis, predictive forecasting, speech recognition, object detection, and decision-making. Their success has enhanced performance standards and opened up novel opportunities in areas like CV, NLP, generative AI, and decision-making, making DL a highly sought-after and influential field. For example, convolutional NNs (CNNs) are particularly versatile in CV analytics, finding applications in tasks such as image recognition and object detection (Joshi et al. 2023). DL's remarkable capability to enable computers to learn from extensive datasets and tackle intricate tasks has led to its widespread adoption (Alhijaj and Khudeyer 2023; Alzubaidi et al. 2023).

DL differs from conventional ML in its ability to handle large and complex datasets with minimal feature engineering, thanks to its deep architecture. While both ML and DL involve algorithms learning from data, the key distinction lies in the complexity of data representations. DL excels in handling unstructured data, such as images, audio, and text, due to its ability to automatically discover hierarchical features. ML, in contrast, may struggle with such data and rely on feature engineering. ML systems rely on manually created feature extractors, which prove to be non-scalable when dealing with large datasets, while DL can address the limitations of shallow networks, enabling efficient training and the abstraction of hierarchical representations for complex, multi-dimensional training data (Shrestha and Mahmood 2019). In addition, Shrestha and Mahmood (2019) examine multiple optimization techniques for enhancing training accuracy and reducing training duration, delving into the mathematical foundations of recent deep network training algorithms, and discussing existing limitations, improvements, and practical applications. Among the most commonly applied deep architectural models are, as shown in Table 2.3, CNNs, recurrent neural networks (RNNs), generative adversarial networks (GANs), deep reinforcement learning (DRL), long short-term memory (LSTM), variational autoencoders (VAEs), transfer learning with pre-trained models, and capsule networks (CapsNets) (e.g., Alhijaj and Khudeyer 2023; Alzubaidi et al. 2021; Alom et al. 2019; Joshi et al. 2023; Mathew et al. 2021; Radford et al. 2021; Raffel et al. 2020; Shrestha and Mahmood 2019).

DL has emerged as a powerful tool with the potential to revolutionize the way we approach environmental sustainability and tackle the pressing challenges of climate change. As climate-related issues intensify, DL's capabilities in data analysis, prediction, and optimization are increasingly harnessed to drive innovations in fields ranging from renewable energy management and transportation planning to water resources conservation, ecological monitoring, and climate modeling (Almalaq and Zhang 2018; Puri et al. 2019; Cai et al. 2019; Han et al. 2021; Kim and Cho 2019; Shen et al. 2019; Shen 2018). In this context, DL offers the prospect of more accurate and efficient solutions for advancing sustainable development goals, making it a pivotal technology in the quest for sustainable, climate-resilient, and environmentally conscious cities.

GenAI has rapidly advanced in recent years, transforming various industries and sectors with its ability to create new content based on patterns in vast amounts of data (Castelli and Manzoni 2022; Ooi et al. 2023). GenAI models excel in learning from extensive datasets, enabling the generation of original content that replicates or mirrors the patterns and structures of the training data (e.g., Goodfellow et al. 2020; Kingma & Dhariwal, 2018). Notable GenAI models include GPT-3 (Generative Pre-trained Transformer 3) (Brown et al., 2020), BERT (Wolf et al. 2020), CLIP

TABLE 2.3 An Overview of DL Techniques, Tasks, and Applications

DL Technique/Algorithm	Description	Example Applications
CNNs	Specialized for image analysis, CNNs use convolutional layers to detect spatial hierarchies of features	Image recognition and classification, object detection, remote sensing, environmental monitoring, land-cover classification, species detection
RNNs	Suited for sequence data, RNNs maintain a memory of previous inputs, allowing them to capture temporal dependencies	Language translation, sentiment analysis, time-series analysis, climate prediction, air quality prediction, climate modeling
LSTM	A type of RNN designed to address the vanishing gradient problem, enabling better learning of long-term dependencies	Speech recognition, sentiment analysis, smart grid optimization, energy consumption prediction, demand response
GANs	Comprising a generator and discriminator, GANs create synthetic data by learning from real data distributions	Synthetic image generation, environmental simulation, anomaly detection, data augmentation
Autoencoders	Used for unsupervised learning, autoencoders aim to reconstruct input data and learn efficient representations	Data compression, anomaly detection
DRL	Merging DL with reinforcement learning, this approach enables agents to learn optimal actions in dynamic environments	Robotic control, autonomous vehicles, smart grid optimization, climate-informed resource management, autonomous environmental monitoring, eco-friendly transportation planning, renewable energy management
Transfer learning with pre-trained models	Leveraging knowledge from pre-trained models on large datasets to enhance performance on specific tasks	Fine-tuning a pre-trained image classification model for a new dataset, biodiversity monitoring, land-cover classification, climate pattern recognition, pollution source identification, deforestation monitoring
CapsNets	Proposed as an improvement to traditional NNs, CapsNets aim to handle hierarchical relationships more effectively	Object recognition in CV

(Contrastive Language–Image Pre-training) (Radford et al. 2021), and Generative Adversarial Networks (GANs) (Goodfellow 2020). The expansion of GenAI into urban domains is particularly significant, as it facilitates the creation of intelligent systems capable of simulating, optimizing, and predicting the behavior of cities, infrastructure, and environments (see Chapter 7). By leveraging large-scale datasets, GenAI is transforming how urban systems are modeled, planned, and managed, contributing to smarter, more sustainable cities, especially in the context of environmental sustainabiiity and climate change (see Chapter 3 and Chapter 7 for further detail).

GenAI models rely extensively on ML and DL algorithms, which form the foundation of GenAI's capabilities. ML and DL play essential roles in enabling GenAI to both analyze and understand the intricate patterns in large datasets, as well as generate highly sophisticated outputs that closely resemble real-world data. DL, in particular, leverages neural networks with multiple layers to learn hierarchical representations of data, making it especially powerful for tasks like image synthesis, language processing and production, and video generation. For example, GANs, a prominent GenAI framework, consist of two neural networks—a generator and a discriminator—that engage in a competitive process, resulting in the generation of realistic synthetic data. The generator creates new data samples, while the discriminator evaluates their authenticity (Goodfellow 2020). This adversarial training process, powered by DL, allows GANs to generate remarkably lifelike images, text, and even video. In addition, large-scale transformer models like GPT-3 and BERT utilize DL techniques, particularly attention mechanisms, to capture complex patterns in language, allowing them to generate highly coherent and contextually relevant text. These models are a direct result of advances in DL architectures, including the scaling of models to billions of parameters, enabling them to generate rich, human-like outputs across a wide range of applications. Therefore, the generative power of AI is fundamentally driven by advancements in ML and DL. These subdomains have unlocked new possibilities in creating novel content and revolutionizing digital transformation across multiple fields, including the advancement of digital twins (Xu et al. 2024) and urban planning and design in the context of sustainable smart cities (see Chapter 7).

As GenAI models continue to evolve, their application is expanding beyond traditional areas like language production and visual content creation. This shift is paralleled by the emergence of Pretrained Foundation

Models (PFMs) in specialized fields, where they offer powerful tools for tackling complex, domain-specific challenges. These models are now being harnessed for tasks in spatial, geographical, and diffusional modeling, enabling transformative applications in urban planning and environmental management (see Chapter 7 for a detailed discussion and illustrative examples). Spatial foundation models leverage large datasets of geospatial information to predict urban growth patterns, land-use changes, and infrastructure development, providing critical insights for urban planning and sustainable development. Geographical models, focusing on large-scale environmental and topographical data, help cities manage resources, such as water, energy, and land more effectively, optimizing urban resilience to natural hazards. Diffusional models are important in urban contexts for simulating the spread of pollutants, heat, and other substances through urban environments, informing strategies for air quality control, urban cooling, and public health interventions.

2.2.11 City Brain

City Brain exemplifies the application of AI in city management and how AI and IoT converge to transforming city operations and services. By processing real-time data from IoT-enabled sensors and devices integrated into various urban systems (e.g., transportation, energy, water, waste, and environment), City Brain provides valuable insights that enhance collaborative decision-making and automation processes. Within the framework systems, seeks to understand the technical and socioeconomic processes that drive these flows and how they impact urban sustainability, and has often been studied in a linear way—resources are consumed and waste is produced. Kennedy et al. (2007, p. 44) define UM as "the total sum of the technical and socioeconomic processes that occur in cities, resulting in growth, production of energy, and elimination of waste." UM plays a fundamental role in the development of the eco-city (Kennedy et al. 2011), an urban system "for which the inflows of materials and energy and the disposal of wastes do not exceed the capacity of its hinterlands" (Kennedy et al. 2007, p. 44). This concept emphasizes sustainable resource use and waste management to minimize environmental impact.

The comprehensive analysis of UM, which requires substantial data on urban metabolic flows (Kennedy et al. 2007; Weisz and Steinberger 2010), has led to the emergence of SUM. This advanced system collects and integrates colossal amounts of high spatial and temporal resolution data from various urban systems, allowing for real-time processing, analysis, and visualization. It involves the application of data-driven, technology-enhanced approaches to monitor, manage, and optimize resource flows, energy consumption, and waste production within urban areas. SUM represents a significant advancement in understanding how both smart eco-cities and smart cities consume resources, generate waste, and impact the environment (e.g., Bibri and Krogstie 2020a; D'Amico et al. 2021; Shahrokni et al. 2015). SUM builds upon UM by integrating advanced technologies and data analytics to optimize resource management and enhance urban sustainability and resilience. It enables cities to monitor and analyze their metabolic processes more effectively, leading to informed decision-making, targeted interventions, and policy formulation for resource efficiency, waste reduction, and environmental sustainability. The integration of AI and AIoT in SUM represents a significant advancement in environmental sustainability in the context of smarter eco-city and sustainable smart city management, planning, and governance. AI enhances its analytical capabilities for improving real-time monitoring and resource management (Ghosh and Engupta 2023; Peponi et al. 2022). AI algorithms analyze vast amounts of data collected from IoT devices embedded in urban infrastructure and systems to gain insights into urban metabolic processes in real time. They can predict future trends, identify inefficiencies, and suggest interventions to enhance resource efficiency and reduce environmental impact. Moreover, AIoT systems facilitate the automation of various processes, allowing smarter eco-cities and sustainable of AIoT, IoT acts as a digital nervous system, collecting and transmitting data, while AI functions as the brain, extracting meaningful knowledge from interconnected sensors and devices across this vast network. City Brain harnesses AI and IoT technologies for cognition, optimization, decision-making, prediction, and intervention across various urban domains (Zhang et al. 2019).

The construction and implementation of City Brain vary depending on the specific applied domain, incorporating a wide range of technical and functional components (e.g., Gao et al. 2021; Liu et al. 2018, 2021). These components are carefully integrated and customized to address the unique needs and challenges of each city, ensuring the effectiveness of the City Brain system in enhancing urban management, planning, and governance processes (e.g., Bibri et al. 2024a; Kierans et al. 2023; Liu et al. 2022; Wang et al. 2022). This adaptability allows City Brain to address a wide range of urban issues, ultimately leading to more sustainable and efficient urban environments.

City Brain, as an AIoT-driven large-scale system, significantly enhances urban management, planning, and governance processes in the context of smarter eco-cities and sustainable smart cities (Bibri et al. 2024a; Bibri and Huang 2024). In terms of urban management, City Brain enables real-time data collection and analysis from various IoT devices, allowing city officials to monitor and manage critical infrastructure, such as transportation systems, energy systems, waster systems, waste systems, utilities, and public services, more efficiently to mitigate environmental impacts. For urban planning, City Brain provides valuable insights derived from data analytics, helping city planners make informed decisions about land use, infrastructure development, and environmental conservation. In terms of governance, City Brain facilitates better resource allocation, improved management, enhanced intervention measures, stakeholder collaboration, and more effective responses to environmental challenges, contributing to the overall sustainability of the city. Through its advanced analytics capabilities, City Brain empowers city authorities to make informed decisions, streamline operations, predict scenarios, and create smarter, and more sustainable cities.

2.2.12 Smart Urban Metabolism

UM is a framework for describing and analyzing the dynamic flow of energy, materials, and resources within cities. It involves quantifying the inputs and outputs to understand the efficiency and sustainability of urban smart cities to respond quickly to changing conditions and achieve greater resilience. AI- and AIoT-driven insights aid in making well-informed decisions and implementing responsive policies to create more sustainable and environmentally conscious cities. Through the integration of AI and AIoT, SUM offers a holistic approach to managing urban metabolism, contributing to the creation of smarter, more resilient cities.

Important to note is that there is a large range of indicators to consider in SUM, which often makes it difficult to select the right ones for developing, monitoring, and evaluating the city's metabolic pathways (D'Amiro et al. 2020). This is a challenge for policymakers and planners as to what, how, and how often to measure, adding to how to consider the multiple dimensions of the indicators (UN-Habitat 2016). Nevertheless, research on sustainable smart cities contributes to the understanding of SUM (ITU 2016a). In this regard, the United Nations (UN), in collaboration with the International Telecommunication Union (ITU), has developed the United Nations Economic Commission for Europe–ITU (UNECE–ITU) "smart sustainable indicators" to assess the smartness and metabolic pathways of cities to comply with the SDGs (D'Amiro et al. 2020). These indicators include information about electricity supply and demand-response tracking, public transportation network, traffic monitoring, solid waste collection, water supply monitoring, wastewater sewer connections, and governance (ITU 2016a, b). Several international intergovernmental organizations also provide indicator platforms to support the transition toward SUM concerning the best practices in these areas (D'Amiro et al. 2020). Worth pointing out is that to capture and analyze huge datasets on environmental sustainability indicators requires advanced AI models.

In the context of this chapter, SUM, as an AIoT-driven model, plays a crucial role in enhancing urban management, planning, and governance processes in smarter eco-cities and sustainable smart cities. In urban management, it enables real-time monitoring and optimization of resource flows such as energy, water, and waste through IoT sensors and AI algorithms (Ghosh and Engupta 2023; Peponi et al. 2022). This allows city authorities to identify inefficiencies, reduce resource consumption, and improve overall urban sustainability. In urban planning, SUM provides insights into the dynamics of resource flows and their impact on the urban environment. This information helps urban planners design more resilient infrastructure, prioritize green spaces, and develop sustainable transportation networks. In governance, SUM supports evidence-based decision-making by providing data-driven insights into the city's resource use patterns and environmental performance. It facilitates the implementation of policies and strategies aimed at reducing carbon emissions, enhancing resource efficiency, and promoting circular economy principles, ultimately contributing to the long-term sustainability of the city (Bibri et al. 2024a).

2.2.13 Urban Digital Twin

The DT concept has recently gained prominence in the urban landscape, marking a significant advancement in the management, planning, and governance of urban systems. Van DerHorn and Mahadevan (2021) define DT as "a virtual representation of a physical system (and its associated environment and processes) that is updated through the exchange of information between the physical and virtual systems." UDT is a virtual replica or simulation of a city, encompassing its various components such as buildings, infrastructure, systems, and spatial and natural features. It integrates data from multiple sources to provide a dynamic representation of urban areas.

It is worth noting that there is no universally accepted definition or standardized blueprint for constructing DT or UDT, and this is yet to be established. Consequently, diverse definitions have been proposed, each emphasizing different aspects based on specific domains of application. These domains include governance, transport, recycling, energy, product design, construction, and health monitoring (Weil et al. 2023). Similarly, UDT demonstrates versatility across various applied domains, encompassing urban planning (Bibri et al. 2024b; Schrotter and Hürzeler 2020), infrastructure management (Ferre-Bigorra et al. 2022; Saeed et al. 2022), transportation management (Lv et al. 2021; Nochta et al. 2021), governance (Deng et al. 2021), water management (Pedersen et al. 2021), energy management (Austin et al. 2020; Francisco et al. 2020; Shen et al. 2021), land use (Bindajam et al. 2021), and disaster management (Fan et al. 2021; Ford and Wolf 2020).This versatility underscores the potential of UDT to transform smarter eco-city and sustainable smart city systems by providing comprehensive computational models that facilitate efficient and informed decision-making pertaining to urban management, planning, and governance (Bibri et al. 2024a; Bibri and Huang 2024). By creating virtual replicas of these urban environments, urban planners, policymakers, and stakeholders can test innovative solutions and derive data-driven insights for advancing the development of more efficient, resilient, and environmentally friendly urban spaces.

The advancement of UDT is intricately linked to the integration of AI and AIoT technologies. AI enables UDT to process vast amounts of data generated by IoT sensors, providing insights and predictive analytics crucial for sustainable urban development. Through AI and AIoT, UDT can identify patterns, optimize resource allocation, and simulate various scenarios to inform decision-making processes and respond to various urban conditions by identifying and implementing targeted intervention measures. Moreover, AI- and AIoT-driven UDT can autonomously adapt to dynamic urban environments, improving efficiency and responsiveness. AIoT further enhances UDT capabilities by enabling real-time data analysis, adaptive systems, and intelligent automation. This synergy empowers UDT to address complex urban challenges with greater effectiveness, ultimately contributing to the development of smarter, more sustainable cities.

AI- and AIoT-driven UDT capabilities have been evidenced by several studies conducted in recent years. Zvarikova et al. (2022) propose UDT algorithms leveraging three-dimensional (3D) spatiotemporal simulations, ML, and DL for accurate urban modeling and simulation.

Austin et al. (2020) integrate semantic knowledge representation and ML in their UDT architecture, demonstrating the tangible benefits of AI in smart city contexts. Similarly, Beckett (2022) highlights the considerable potential of AI-integrated UDT in bolstering urban planning and design strategies through the incorporation of 3D modeling, visualization tools, and spatial cognition algorithms. In a comprehensive systematic review, Bibri et al. (2024b) provide a thorough analysis and synthesis of a large body of literature to uncover the intricate interactions among AI, AIoT, UDT, data-driven planning, and environmental sustainability. The authors elucidate the nuanced dynamics and untapped synergies among these technologies, models, and domains in the complex ecosystem of sustainable smart cities. The integration of these technologies and models is revolutionizing data-driven planning strategies and practices, thereby supporting the achievement of environmental sustainability goals. The study identifies and synthesizes the theoretical and practical foundations that support the convergence of AI, AIoT, UDT, data-driven planning, and environmental sustainability into a comprehensive framework. It also investigates how the integration of AI and AIoT transforms data-driven planning to enhance the environmental performance of sustainable smart cities. Additionally, it explores how AI and AIoT can augment UDT capabilities to improve data-driven environmental planning processes. The integration of AI, AIoT, and UDT technologies significantly enhances data-driven environmental planning processes, providing innovative solutions to complex environmental challenges and promoting sustainable urban development. Beyond theoretical enrichment, the study offers valuable insights into the transformative potential of integrating AI, AIoT, and UDT technologies to advance sustainable urban development practices. Overall, these studies underscore the transformative potential of AI and AIoT in advancing UDT, thereby shaping sustainable, resilient, efficient, and environmentally conscious urban landscapes and profoundly impacting the trajectory of sustainable urban development.

However, digital models are not intended to replicate the original system with the same level of detail or to mirror every aspect of it, which inherently distinguishes the model from the original system (Batty 2018). Digital models serve as simplified representations that capture essential elements and dynamics of real systems while omitting non-essential or less relevant details. This selective representation allows for efficient analysis, simulation, and experimentation, facilitating understanding and decision-making processes in various domains, including urban planning, infrastructure management, urban governance, and environmental sustainability. Moreover, incorporating AI and AIoT technologies into digital models enables dynamic adaptation and predictive capabilities, enhancing their ability to simulate real-world systems with greater accuracy and responsiveness. By leveraging AI and AIoT, digital models can continuously learn from data streams, updating their representations in real time and offering more insightful and actionable insights for sustainable urban development.

2.2.14 Platform Urbanism

Platform urbanism, an emerging urban development paradigm driven by digital ecosystems, has emerged due to the intensification of platformization. This phenomenon, which has gained traction in recent years (Poell et al. 2019), is facilitated and supported by AIoT processes, namely "digital instrumentation, digital hyper-connectivity, datafication, and algorithmization" (Bibri et al. 2024a). It represents a shift toward interconnected digital platforms that enable various urban functions and services, reshaping the way cities are designed, experienced, and governed. Platformization refers to the trend toward digital platforms that enable interaction, collaboration, coordination, co-creation, and transactions among multiple stakeholders. This shift reflects a fundamental change in how services and information are accessed, shared, and utilized in various domains, including urban development and governance. Digital platforms create value, promote innovation, share knowledge, improve resource efficiency, and facilitate application development (Evans and Schmalensee 2016; Parker et al. 2016; Poell et al. 2019).

In the context of this chapter, platformization is closely linked to the principles of platform urbanism, which systematically reshape urban processes and practices by introducing digital platforms across various urban domains. These platforms, as reprogrammable digital infrastructures, enable diverse stakeholders to connect and interact for various purposes in different urban contexts. These digital arrangements and intermediations rely on "the systematic collection, algorithmic processing, and circulation of data" (Poell et al. 2019) to reorganize urban life.

Caprotti et al. (2024, p. 4) define platform urbanism as "urban development and urban life facilitated by a growing number of digitally enabled socio-technical assemblages that engender new forms of social, economic, and political intermediations." At the heart of these intermediations is the "digital mediation of cities" (Barns 2019) enabled by AI and AIoT technologies (Bibri et al. 2024a). These play a key role in these intermediations by enabling real-time data processing, advanced analytics, and decision-making capabilities. The wealth of these data enables platforms to offer tailored and data-driven services, optimize resource allocation, and improve the efficiency of urban activities and processes. The integration of AI and AIoT in platform urbanism enhances the effectiveness of intermediations by providing stakeholders with actionable insights and supporting collaborative decision-making. This leads to the creation of smarter, more sustainable cities.

Platform urbanism has significantly influenced urban governance by transforming the dynamics of decision-making, participation, and innovation in smarter eco-cities and sustainable smart cities. The rise of digital platforms has introduced new modes of engagement between governments, citizens, and private actors, enabling more inclusive and collaborative approaches to urban governance (e.g., Ansell and Gash 2018; Ansell and Miura 2019; Repette et al. 2021). In particular, through these platforms, citizens can actively participate in decision-making processes, provide feedback, and co-create solutions with local authorities and other stakeholders. By embracing digital platforms, urban governance has become more adaptive, transparent, and citizen-centric, marking a significant advancement from traditional modes of governance (Haveri and Anttiroiko 2023; Hodson et al. 2021; Repette et al. 2021).

2.3 THE INTERCONNECTED CONCEPTS SHAPING THE DEVELOPMENT OF SMARTER ECO-CITIES

In the quest for sustainable urban development, the notion of smarter eco-cities has risen as a beacon of innovation and environmental stewardship. Smarter eco-cities aspire to strike a harmonious balance between environmental sustainability, economic prosperity, and the well-being of their inhabitants. Key to achieving this vision is the convergence of various concepts, including environmental management, planning, and governance, as well as AI, AIoT, ML, DL, UDT, City Brain, SUM, and platform urbanism. These conceptual components are intricately interwoven, working in synergy to create urban environments that are not only smarter but also eco-friendlier, resilient, and adaptive to the evolving challenges of the modern urban landscape. Here, the focus is on the synergistic interplay of these concepts and the illumination of their roles in shaping the development of smarter eco-cities of the future.

Environmental management, planning, and governance processes, supported by City Brain, UDT, SUM, and platform urbanism as data-driven systems, are fundamental to the development of smarter eco-cities. These systems ensure that smarter eco-cities can effectively address environmental challenges and promote ecological resilience by leveraging advanced technologies and data-driven approaches—thanks to their enabling technologies.

The incorporation of AI and AIoT technologies with UDT is a cornerstone of smarter eco-cities. These technologies enable data-driven decision-making and the automation of various urban systems. They provide the backbone for real-time data collection and analysis, which is crucial for optimizing resource management and enhancing the efficiency of urban systems and services. ML and DL, as subsets of AI that are applied to AIoT, contribute to analyzing the vast amounts of data generated in smarter eco-cities, with UDT providing dynamic and up-to-date datasets for analysis and visualization. These AI models offer the capability to predict patterns, optimize resource usage, and provide valuable insights for informed decision-making in areas like energy management, transport management, traffic management, waste management, water management, and environmental monitoring. Moreover, they are at the core of City Brain, an urban management system that is driven by AI and AIoT and can be empowered by UDT. In this context, City Brain serves as the central nervous system of smarter eco-cities. It monitors and manages various aspects of urban life, supported with real-time data and analytics from UDT, making smarter eco-cities more efficient and sustainable.

The insights derived from UDT form the foundation of SUM's operations, which are dedicated to understanding and enhancing the resource and energy flow of smarter eco-cities. In this complex process, AI and data analytics, also empowered by UDT, assume a crucial role. They not only monitor but also optimize these critical resource flows, thus reducing waste and fostering sustainability and resilience. Moreover, the integration of AI and AIoT technologies contributes significantly to the efficient reuse and recycling of resources, effectively aligning with the circular economy's overarching objective of waste minimization. In essence, the dynamic insights from UDT facilitate the unification of SUM's resource optimization and circular economy principles, working in tandem to improve the development of smarter eco-cities.

Platform urbanism has become essential to environmental planning and governance in the context of smarter eco-cities. It leverages digital platforms as powerful tools to not only connect diverse urban stakeholders but also foster collaboration and innovation. One of the key functions of platform urbanism is the integration of data from multiple sources. These sources encompass a wide array of information, ranging from the data generated by IoT devices to insights produced by AI and analytics, all of which are created, managed, and analyzed by UDT. UDT, as a dynamic mirror of smarter eco-cities, continuously captures and updates data regarding their operations, infrastructure, and environmental conditions in real time. This wealth of data serves as a foundational resource for platform urbanism, providing a comprehensive and up-to-date understanding of the dynamics of smarter eco-cities. The ability to integrate data from diverse sources empowers platform urbanism to create a responsive and adaptive urban environment. By amalgamating data streams, platform urbanism can offer a holistic view of the city's current state and trends. This knowledge is of strategic value for urban planners and decision-makers as it enables them to make informed choices related to environmental governance.

Furthermore, platform urbanism facilitates enhanced communication and coordination among various stakeholders involved in environmental planning and governance. City authorities, local businesses, environmental organizations, and residents can engage in collaborative efforts, sharing insights and resources efficiently through digital platforms. This collaborative approach can lead to the development of more effective and sustainable environmental policies, which are essential for the governance of smarter eco-cities. In a nutshell, platform urbanism contributes to both urban planning and governance. It creates a platform for data sharing and innovation, empowering urban planners and policymakers in their efforts and enhancing overall governance.

Overall, the interconnected concepts form a multifaceted web, where technology, data, and sustainability principles synergize to create intelligently advanced and environmentally conscious cities. Effective environmental management, planning, and governance, empowered by AI- and AIoT-driven systems, serve as foundational pillars for the development of smarter eco-cities. This collective synergy yields urban environments that are adaptive and well-equipped to tackle the challenges of a constantly changing future.

2.4 A NOVEL CONCEPTUAL FRAMEWORK AND ITS ESSENTIAL COMPONENTS

The conceptual framework (Figure 2.3) presented in this section is derived based on insights from the synthesized studies in the previous section, which focused on understanding the integration of AI and AIoT technologies, City Brain, UDT, SUM, platform urbanism (governance perspective), and environmental sustainability in the context of smarter eco-cities. This framework aims to provide a comprehensive understanding of the interconnectedness of these components and their roles in promoting environmental sustainability in urban contexts. By synthesizing key concepts and theories from the literature, this framework offers a structured approach to addressing the complex challenges of sustainability in modern cities, grounded in empirical evidence and practical insights.

Rapid urbanization and environmental degradation pose significant challenges to environmental sustainability in smarter eco-cities. To address these challenges, a holistic approach integrating AI and AIoT technologies, operational management processes, urban planning systems, and

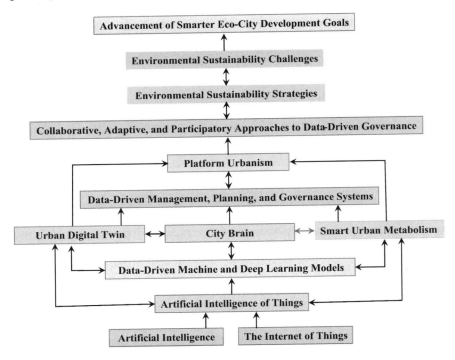

FIGURE 2.3 A novel conceptual framework for data-driven environmental management, planning, and governance in smarter eco-cities.

governance mechanisms is essential. City Brain, UDT, SUM, and platform urbanism represent emerging data-driven approaches to environmental urban management, planning, and governance. Each of these systems contributes uniquely to these practices, highlighting the importance of their integration.

AI and AIoT play crucial roles in data collection, analysis, and decision-making across urban management, planning, and governance domains. They process data from various urban systems, feeding into City Brain, UDT, SUM, and platform urbanism, which utilize the processed data to generate insights and recommendations for urban managers, planners, and decision-makers.

Specifically, City Brain serves as an AI-powered system collecting and analyzing real-time data to optimize urban operations and enhance public services to reduce environmental impact and hence inform urban planning and governance processes. UDT provides a virtual replica of the city, integrating real-time data and simulations to support urban planning, infrastructure management, and operational decision-making, as well as governance processes. SUM analyzes resource and energy flows in urban systems to promote sustainability and resilience. It contributes to environmental planning and governance by providing insights into patterns of consumption and waste generation. This information enables policymakers to develop targeted interventions and policies for promoting sustainability and resilience in smarter eco-cities. Platform urbanism serves as the overarching framework for environmental governance, providing the structures and processes for communication, collaboration, coordination, citizen engagement, and transparent decision-making. Together, these components offer a powerful toolkit for addressing the complex challenges of environmental sustainability in smarter eco-cities.

Furthermore, these components interact synergistically. City Brain integrates data from UDT to create a real-time understanding of the city's current state, allowing for dynamic decision-making and resource allocation. It also analyzes SUM data to identify patterns and trends in resource consumption, enabling the optimization of urban systems to minimize waste and improve efficiency. UDT provides a platform for simulating and visualizing SUM data, allowing urban planners to explore different scenarios and evaluate the impact of interventions on the city's metabolism. Platform urbanism enhances governance systems by providing tools for collaboration, information sharing, and citizen engagement, thereby strengthening environmental governance in smarter eco-cities.

Proposition 1: The integration of AI and IoT technologies, urban planning systems, governance mechanisms, and operational management processes enhances the environmental performance of smarter eco-cities.

Proposition 2: The synergy between City Brain, UDT, SUM, and platform urbanism creates a robust framework for achieving environmental sustainability goals in smarter eco-cities.

Proposition 3: Platform urbanism facilitates transparent and participatory governance, leading to more sustainable decision-making and implementation of environmental policies while fostering citizen participation in environmental governance initiatives.

This integrated conceptual framework emphasizes the interconnectedness of AI and AIoT technologies, urban planning systems, governance mechanisms, operational management processes, City Brain, UDT, SUM, and platform urbanism in promoting environmental sustainability in smarter eco-cities. Together, these components form a cohesive system that leverages data-driven approaches to address the complex challenges of environmental degradation, resource depletion, and urbanization. City Brain, UDT, and SUM provide essential tools for environmental management, planning, and governance, each contributing unique insights and capabilities. Platform urbanism fosters citizen-centric, environmentally conscious choices, participatory and adaptive governance approaches, in a holistic approach to sustainable urban development. By integrating these components, this framework offers a comprehensive solution for building smarter eco-cities that are resilient, resource-efficient, environmentally sustainable, and technologically advanced.

2.5 DISCUSSION

The exploration of the theoretical and conceptual underpinnings of smarter eco-cities presented in this chapter sheds light on the complex landscape of urban sustainability and technological integration. By weaving together a diverse array of concepts, theories, technologies, and models, this study aimed to provide a comprehensive understanding of the foundational elements that contribute to the development of smarter eco-cities. Through the fulfillment of four specific objectives, this discussion section delves into the significance and future directions arising from the findings.

The multifaceted examination of smarter eco-cities has revealed the intricate connections between management, planning, governance practices, and the underlying computational technologies and models. The integration of AI and AIoT technologies has emerged as a central theme, demonstrating their potential to enhance the efficiency and sustainability of urban systems through emerging data-driven systems. This synergy is of crucial importance for unraveling the complexities of smarter eco-cities. The significance of this study lies in providing readers with a structured understanding of a multifaceted topic that extends across various domains of knowledge.

By explicating each concept, readers gain a profound insight into the individual dimensions that constitute the complex ecosystem of smarter eco-cities. Each concept, with its distinct characteristics and contributions, serves as a piece of the puzzle. When these pieces are integrated, readers can observe how they interlock to create a comprehensive approach to sustainable urban development. This highlights the inherently interdisciplinary and transdisciplinary nature of smarter eco-cities. Discussing the interlinkages between the identified theories and theories allows readers to value how knowledge from various disciplines is integrated and fused to tackle complex environmental urban challenges. It also provides a roadmap for breaking down silos and fostering collaboration across disciplinary fields and applied domains and hence between scholars and practitioners. The practical application of the identified concepts in the real-world setting of smarter eco-cities is where the actual impact lies.

Understanding AIoT, City Brain, UDT, SUM, and platform urbanism is the first step toward their effective implementation in urban management, planning, and governance for advancing environmental sustainability in smarter eco-cities. Readers equipped with a grasp of all these concepts are better prepared to address real-world issues and work toward promoting environmental sustainability. In addition, smarter eco-cities are hubs of innovation and this foundation serves as a launchpad for innovative problem-solving. When readers understand the relationships among these concepts, they gain better understanding of how to contribute to innovative solutions in smarter eco-cities. These urban environments are envisioned as models for urban development that reduce resource consumption, mitigate environmental impact, and enhance the quality of life. A clear understanding of their multifaceted dimensions is the foundation for achieving these objectives. In essence, by clarifying the foundations and their interplay within the framework of smarter eco-cities, readers are empowered to navigate the complex and multifaceted landscape of smarter eco-cities. This foundation empowers readers to embark on a journey of practical exploration, using the knowledge gained to actively contribute to the development of these urban environments.

Furthermore, providing a solid foundation for the practical exploration of these concepts in the context of emerging smarter eco-cities is central to the objectives of this chapter. This serves as the bedrock upon which the entire book's journey is built. This chapter not only introduces the key concepts but also sets the stage for how they can be effectively applied in real-world scenarios across the subsequent chapters. The chapter acts as a bridge between theory and practice, ensuring that readers have the necessary conceptual and theoretical foundations to embark on the journey of this practical application. Consequently, readers will gain insights into how these concepts can be translated into tangible solutions for resource management, infrastructure development, and improved quality of life for urban residents. The ultimate goal is to empower readers to participate in bringing about urban transformation by means of spurring groundbreaking research endeavors.

Moving forward, future research should focus on refining and expanding the proposed conceptual framework to address emerging challenges and opportunities in smarter eco-city development. Furthermore, the identification of synergies among the core foundational elements of smarter eco-cities highlights the importance of holistic approaches to urban sustainability. Future studies could explore the dynamics of these interconnections in greater depth, considering the socioeconomic, environmental, and technological factors at play. Additionally, empirical research is needed to validate the effectiveness of the proposed framework in real-world settings, providing insights into its applicability and scalability.

Establishing a foundation for the practical implementation of identified technologies and models is crucial for advancing smarter eco-city initiatives. This involves not only technological integration but also addressing policy, regulatory, and governance challenges. Future research should investigate strategies for fostering collaboration among various stakeholders, ensuring inclusive decision-making processes, and promoting the equitable distribution of benefits across communities.

Overall, the findings presented in this chapter contribute to a deeper understanding of the theoretical and practical dimensions of smarter eco-cities. By addressing key objectives and delineating future research avenues, this study lays the groundwork for further advancements in sustainable urban development and the integration of cutting-edge technologies.

2.6 CONCLUSION

This chapter explored the theoretical and conceptual underpinnings of smarter eco-cities, weaving together a diverse array of concepts, theories, technologies, and models. It provided a detailed analysis of the blossoming field of smarter eco-cities, positioned at the intersection of urban management, planning, and governance; technological advancement; and urban transformation. It addressed the existing gap in the scholarly landscape by providing conceptual and theoretical frameworks that integrate diverse elements, thereby contributing to a more comprehensive understanding of smarter eco-cities and their pivotal role in advancing sustainable urban development.

This chapter proposed a systematic approach for integrating AI and AIoT systems into urban management, planning, and governance processes. Emphasizing the interdisciplinary and transdisciplinary nature of the field, it underscores the significance of integrating and fusing knowledge from diverse domains to address complex environmental challenges. The synergistic relationship among the identified concepts, technologies, and models plays a pivotal role in shaping the future of sustainable urban development. Elucidating these interconnected foundational elements is key to establishing a coherent framework for smarter eco-city development. This framework demonstrates the relevance and importance of interdisciplinary and transdisciplinary approaches for the practical implementation of data-driven management, planning, and governance systems in smarter eco-city initiatives based on City Brain, UDT, SUM, and platform urbanism. It serves as a guiding compass, providing invaluable insights for researchers, practitioners, and policymakers navigating the landscape of smarter eco-cities to comprehensively grasp their multifaceted underlying dimensions.

The implications of this research extend beyond academia to inform practice and policymaking. For researchers, this study serves as a call to action to further explore and refine the foundations of emerging smarter eco-cities, as well as conduct empirical studies to validate and operationalize the proposed framework in real-world settings. By advancing our understanding of the complex dynamics at play in smarter eco-city development, researchers can contribute to the creation of more efficient, technologically advanced, and environmentally conscious urban environments.

For practitioners, this chapter provides actionable insights for advanced management, planning, and governance systems in smarter eco-city initiatives. Practitioners can enhance the sustainability, resilience, and livability of urban environments by embracing AI and AIoT and leveraging their synergistic potential in advancing environmentally sustainable urban development.

For policymakers, this chapter underscores the importance of integrating AI and AIoT into sustainable urban development strategies and policies to address the multifaceted challenges of rapid urbanization, resource depletion, environmental degradation, and climate change. Policymakers can create the necessary conditions for the emergence of smarter eco-cities by fostering an enabling policy environment and investing in innovative management, planning, and governance models. Overall, the implications of this research extend far beyond the confines of academia, shaping the future of urban development in an increasingly complex and interconnected world.

REFERENCES

Ahvenniemi, H., Huovila, A., Pinto-Seppä, I., Airaksinen, M., & Valvontakonsultitoy, R. (2017). What are the differences between sustainable and smart cities? *Cities, 60*, 234–245. https://doi.org/10.10.1016/j.cities.2016.09.009.

Ajuriaguerra Escudero, M. A. & Abdiu, M. (2022). Artificial intelligence in European urban governance. In: J. Saura, & F. Debasa (Eds.), *Handbook of Research on Artificial Intelligence in Government Practices and Processes* (pp. 88–104). IGI Global. https://doi.org/10.10.4018/978-1-7998-9609-8.ch006

Alahakoon, D., Nawaratne, R., Xu, Y., De Silva, D., Sivarajah, U., & Gupta, B. (2020). Self-building artificial intelligence and machine learning to empower big data analytics in smart cities. *Information Systems Frontiers, 25*(1), 221–240. https://doi.org/10.1007/s10796-020-10056-x.

Alahi, M. E. E., Sukkuea, A., Tina, F. W., Nag, A., Kurdthongmee, W., Suwannarat, K., & Mukhopadhyay, S. C. (2023). Integration of IoT-enabled technologies and artificial intelligence (AI) for smart city scenario: Recent advancements and future trends. *Sensors, 23*(11), 5206.

Al-Dabbagh, R. H. (2022). Dubai, the sustainable, smart city. *Renewable Energy and Environmental Sustainability, 7*(3), 12.

Alhijaj, J. A., & Khudeyer, R. S. (2023). Techniques and applications for deep learning: A review. *Journal of Al-Qadisiyah for Computer Science and Mathematics, 15*(2), 114–126. https://doi.org/10.10.29304/jqcm.2023.15.2.1236

Almalaq, A., & Zhang, J. J. (2018). Evolutionary deep learning-based energy consumption prediction for buildings. *IEEE Access, 7*, 1520–1531.

Almalki, F. A., Alsamhi, S. H., Sahal, R., Hassan, J., Hawbani, A., Rajput, N. S., Saif, A., Morgan, J., & Breslin, J. (2021). Green IoT for eco-friendly and sustainable smart cities: future directions and opportunities. *Mobile Networks and Applications.* https://doi.org/10.1007/s11036-021-01790-w

Almusaed, A., & Yitmen, I. (2023). Architectural reply for smart building design concepts based on artificial intelligence simulation models and digital twins. *Sustainability*, 15(6), 4955.

Alom, M. Z., Taha, T. M., Yakopcic, C., Westberg, S., Sidike, P., Nasrin, M. S., Hasan, M., Van Essen, B. C., Awwal, A. A., & Asari, V. K. (2019). A state-of-the-art survey on deep learning theory and architectures. *Electronics*, 8(3):292.

Al-Turjman, F., Nayyar, A., Devi, A., & Shukla, P. K. (Eds.). (2021). *Intelligence of Things: AI-IoT Based Critical Applications and Innovations*. Springer. https://doi.org/10.10.1007/978-3-030-82800-4

Alzubaidi, L., Zhang, J., Humaidi, A. J., Al-Dujaili, A., Duan, Y., Al-Shamma, O., Santamaría, J., Fadhel, M. A., Al-Amidie, M., & Farhan, L. (2021). Review of deep learning: Concepts, CNN architectures, challenges, applications, future directions. *Journal of Big Data*, 8(1). https://doi.org/10.10.1186/s40537-021-00444-8.

Ansell, C., & Gash, A. (2018). Collaborative platforms as a governance strategy. *Journal of Public Administration Research and Theory*, 28(1), 16–32.

Ansell, C., & Miura, S. (2019). Can the power of platforms be harnessed for governance? *Public Administration*. https://doi.org/10.10.1111/padm.12636

Antonopoulos, I., Robu, V., Couraud, B., Kirli, D., Norbu, S., Kiprakis, A., Flynn, D., Elizondo González, S. I., & Wattam, S. (2020). Artificial intelligence and machine learning approaches to energy demand-side response: A systematic review. *Renewable and Sustainable Energy Reviews*, 130, 109899.

Aqib, M., Mehmood, R., Alzahrani, A., Katib, I., Albeshri, A., & Altowaijri, S. (2019). Rapid transit systems: Smarter urban planning using big data, in-memory computing, deep learning, and GPUs. *Sustainability*, 11(10), 2736. https://doi.org/10.10.3390/su11102736

Asadzadeh, A., Fekete, A., Khazai, B., Moghadas, M., Zebardast, E., Basirat, M., & Kötter, T. (2023). Capacitating urban governance and planning systems to drive transformative resilience. *Sustainable Cities and Society*, 96, 104637. https://doi.org/10.10.1016/j.scs.2023.104637

Austin, M., Delgoshaei, P., Coelho, M., & Heidarinejad, M. (2020). Architecting smart city digital twins: Combined semantic model and machine learning approach. *Journal of Management in Engineering*, 36(4), 04020026.

Avis, W. R. (2023). Sustainable Development Goals (Urban Governance Topic Guide). GSDRC, University of Birmingham. GSDRC. https://gsdrc.org/topic-guides/urban-governance/concepts-and-debates/what-is-urban-governance/ (accessed 17 March 2023).

Azevedo, B. F., Rocha, A. M. A. C., & Pereira, A. I. (2024). Hybrid approaches to optimization and machine learning methods: A systematic literature review. *Machine Learning*. https://doi.org/10.10.1007/s10994-023-06467-x

Azizi, D., Biermann, F., & Kim, R. E. (2019). Policy integration for sustainable development through multilateral environmental agreements. *Global Governance*, 25, 445–475. https://doi.org/10.10.1163/19426720-02503005

Bai, X. (2007) Industrial ecology and the global impacts of cities. *Journal of Industrial Ecology*, 11(2):1–6.

Barns, S. (2019). Negotiating the platform pivot: From participatory digital ecosystems to infrastructures of everyday life. *Geography Compass*, 13, e12464

Batty, M. (2018). Digital twins. *Environment and Planning B: Urban Analytics and City Science, 45*(5), 817–820. https://doi.org/10.10.1177/2399808318796416

Beckett, S. (2022). Smart city digital twins, 3D modeling and visualization tools, and spatial cognition algorithms in artificial intelligence-based urban design and planning. *Geopolitics, History, and International Relations, 14*(1), 123–138. https://www.jstor.org/stable/48679657

Berisha, E., Cotella, G., Janin Rivolin, U., & Solly, A. (2020). Spatial governance and planning systems and the public control of spatial development: A European typology. *European Planning Studies*, 1–20. https://doi.org/10.10.1080/09654313.2020.1726295

Beunen, R., & Patterson, J. J. (2019). Analyzing institutional change in environmental governance: Exploring the concept of 'institutional work'. *Journal of Environmental Planning and Management, 62*(1), 12–29. https://doi.org/10.1080/09640568.2016.1257423

Bibri, S. E. (2015). Ambient intelligence: A new computing paradigm and a vision of a next wave in ICT. In: *The Human Face of Ambient Intelligence (Atlantis Ambient and Pervasive Intelligence* (vol. 9). Atlantis Press. https://doi.org/10.10.2991/978-94-6239-130-7_2

Bibri, S. E. (2020). Data-driven environmental solutions for smart sustainable cities: Strategies and pathways for energy efficiency and pollution reduction. *Euro-Mediterranean Journal for Environmental Integration, 5*(66). https://doi.org/10.1007/s41207-020-00211-w

Bibri, S. E. (2021a). Data-driven smart eco-cities and sustainable integrated districts: A best-evidence synthesis approach to an extensive literature review. *European Journal of Futures Research, 9*, 1–43. https://doi.org/10.10.1186/s40309-021-00181-4.

Bibri, S. E. (2021b). Data-driven smart sustainable cities of the future: Urban computing and intelligence for strategic, short-term, and joined-up planning. *Computational Urban Science, 1*, 1–29.

Bibri, S. E. (2021c). The underlying components of data-driven smart sustainable cities of the future: A case study approach to an applied theoretical framework. *European Journal of Futures Research, 9*(13). https://doi.org/10.1186/s40309-021-00182-3

Bibri, S. E., & Huang, J. (2025). Artificial intelligence of sustainable smart city brain and digital twin : A pioneering framework for advancing environmental sustainability. *Environmental Technology and Innovation* (in press).

Bibri, S. E., & Krogstie, J. (2020a). Environmentally data-driven smart sustainable cities: Applied innovative solutions for energy efficiency, pollution reduction, and urban metabolism. *Energy Informatics, 3*, 29. https://doi.org/10.10.1186/s42162-020-00130-8.

Bibri, S. E., & Krogstie, J. (2020b). Data-driven smart sustainable cities of the future: A novel model of urbanism and its core dimensions, strategies, and solutions. *Journal of Futures Studies, 25*(2), 77–94. https://doi.org/10.6531/JFS.202012_25(2).0009

Bibri, S. E., & Krogstie, J. (2021). A novel model for data-driven smart sustainable cities of the future: A strategic roadmap to transformational change in the era of big data. *Future Cities and Environment*, *7*(1), 3. https://doi.org/10.10.5334/fce.116

Bibri, S. E., Alexandre, A., Sharifi, A., & Krogstie, J. (2023a). Environmentally sustainable smart cities and their converging AI, IoT, and big data technologies and solutions: An integrated approach to an extensive literature review. *Energy Informatics*, *6*, 9.

Bibri, S. E., Huang, J., & Krogstie, J. (2024a). Artificial intelligence of things for synergizing smarter eco-city brain, metabolism, and platform: Pioneering data-driven environmental governance. *Sustainable Cities and Society*, 105516. https://doi.org/10.10.1016/j.scs.2024.105516

Bibri, S. E., Huang, J., Jagatheesaperumal, S. K., & Krogstie, J. (2024b). The synergistic interplay of artificial intelligence and digital twin in environmentally planning sustainable smart cities: A comprehensive systematic review. *Environmental Science and Ecotechnology*, *20*, 100433. https://doi.org/10.10.1016/j.ese.2024.100433

Bibri, S. E., Krogstie, J., Kaboli, A., & Alahi, A. (2023b). Smarter eco-cities and their leading-edge solutions for environmental sustainability: A comprehensive systemic review. *Environmental Science and Ecotechnology*, *19*. https://doi.org/10.10.1016/j.ese.2023.100330

Biermann, F., Pattberg, P., Van Asselt, H., & Zelli, F. (2009). The fragmentation of global governance architectures: A framework for analysis. *Global Environmental Politics*, *9*(4), 14–40.

Bindajam, A. A., Mallick, J., Talukdar, S., Islam, A. R. M. T., & Alqadhi, S. (2021). Integration of artificial intelligence–based lulc mapping and prediction for estimating ecosystem services for urban sustainability: Past to future perspective. *Arabian Journal of Geosciences*, *14*, 1–23.

Bodin, Ö. (2017). Collaborative environmental governance: Achieving collective action in social-ecological systems. *Science*, *357*(6352), eaan1114. https://doi.org/10.1126/science.aan1114

Bridge, G., & Perreault, T. (2009). Environmental governance. In: N. Castree, D. Demeritt, D. Liverman, & B. Rhoads (Eds.), *A Companion to Environmental Geography* (pp. 475–497). Blackwell Publishing.

Brown, T., Mann, B., Ryder, N., Subbiah, M., Kaplan, J. D., Dhariwal, P., Neelakantan, A., Shyam, P., Sastry, G., Askell, A., Agarwal, S., Herbert-Voss, A., Krueger, G., Henighan, T., Child, R., Ramesh, A., Ziegler, D., Wu, J., Winter, C., Hesse, C., Chen, M., Sigler, E., Litwin, M., Gray, S., Chess, B., Clark, J., Berner, C., McCandlish, S., Radford, A., Sutskever, I., & Amodei, D. (2020). Language models are few-shot learners. *Proceedings of NeurIPS*, 1877–1901.

Bush, J., & Doyon, A. (2019). Building urban resilience with nature-based solutions: How can urban planning contribute? *Cities*, *95*, 102483. https://doi.org/10.10.1016/j.cities.2019.102483

Cai, Q., Abdel-Aty, M., Sun, Y., Lee, J., & Yuan, J. (2019). Applying a deep learning approach for transportation safety planning by using high-resolution transportation and land use data. *Transportation Research Part A, Policy and Practice, 127*, 71–85.

Caprotti, F., Cowley, R., Bailey, I., Joss, S., Sengers, F., Raven, R., Spaeth, P., Jolivet, E., Tan-Mullins, M., & Cheshmehzangi, A. (2017). *Smart Eco-City Development in Europe and China: Policy Directions*. University of Exeter (SMART-ECO Project.

Chen, L., Chen, Z., Zhang, Y.., Liu, Y., Osman, A. I., Farghali, M., Hua, J., Al-Fatesh, A., Ihara, I., Rooney, D. W., & Yap, P.-S. (2023). Artificial intelligence-based solutions for climate change: A review. *Environmental Chemistry Letters, 21*, 2525–2557. https://doi.org/10.1007/s10311-023-01556-7

Connelly, S., & Richardson, T. (2008). Effective deliberation in stakeholder engagement. In: A. Bonn, T. Allott, K. Hubacek, & J. Stewart (Eds.), *Drivers of Environmental Change in Uplands* (pp. 376–392). Routledge.

Cotella, G., & Janin Rivolin, U. (2010). Institutions, discourse and practices: Towards a multidimensional understanding of EU territorial governance. In: 24 AESOP Annual Conference, Finland, July 7–10, 2010 (pp. 7–10).

Cugurullo, F., Caprotti, F., Cook, M., Karvonen, A., McGuirk, P., & Marvin, S. (2024). *Artificial Intelligence and the City: Urbanistic Perspectives on AI*. Taylor & Francis, Routledge.

Cullen-Knox, C., Eccleston, R., Haward, M., Lester, E., & Vince, J. (2017). Contemporary challenges in environmental governance: Technology, governance and the social licence. *Environmental Policy and Governance, 27*(1), 3–13.

D'Amico, G., Arbolino, R., Shi, L., Yigitcanlar, T., & Ioppolo, G. (2021). Digital technologies for urban metabolism efficiency: Lessons from urban agenda partnership on circular economy. *Sustainability, 13*(11), 6043. https://doi.org/10.10.3390/su13116043.

D'Amico, G., Taddeo, R., Shi, L., Yigitcanlar, T., & Ioppolo, G. (2020). Ecological indicators of smart urban metabolism: A review of the literature on international standards. *Ecological Indicators, 118*, 106808.

de Wit, M. P. (2020). Environmental governance: Complexity and cooperation in the implementation of the SDGs. In: W. Leal Filho, A. Azul, L. Brandli, A. Lange Salvia, & T. Wall (Eds.), *Affordable and Clean Energy. Encyclopedia of the UN Sustainable Development Goals*. Springer. https://doi.org/10.10.1007/978-3-319-71057-0_25-1

Deng, T., Zhang, K., & Shen, Z.-J. (2021). A systematic review of a digital twin city: A new pattern of urban governance toward smart cities. *Journal of Management Science and Engineering, 6*, 125–134. https://doi.org/10.10.1016/j.jmse.2021.03.003.

Deng, W., Cheshmehzangi, A., Ma, Y., & Peng, Z. (2021). Promoting sustainability through governance of eco-city indicators: A multi-spatial perspective. *International Journal of Low-Carbon Technologies, 16*(1), 61–72.

Dheeraj, A., Nigam, S., Begam, S., Naha, S., Jayachitra Devi, S., Chaurasia, H. S., Kumar, D., Ritika, Soam, S. K., Srinivasa Rao, N., Alka, A., Sreekanth Kumar, V. V., & Mukhopadhyay, S. C. (2020). Role of artificial intelligence (AI) and internet of things (IoT) in mitigating climate change. In: Ch. Srinivasa, R. T. Srinivas, R. V. S. Rao, N. Srinivasa Rao, S. Senthil Vinayagam, P. Krishnan (Eds.), *Climate Change and Indian Agriculture: Challenges and Adaptation Strategies* (pp. 325–358). ICAR-National Academy of Agricultural Research Management.

Din, I. U., Guizani, M., Rodrigues, J. J. P. C., Hassan, S., & Korotaev, V. V. (2019). Machine learning in the internet of things: Designed techniques for smart cities. *Future Generation Computer Systems, 100,* 826–843.

Dong, S., Wang, P., & Abbas, K. (2021). A survey on deep learning and its applications. *Computer Science Review, 40.* https://doi.org/10.10.1016/j.cosrev.2021.100379.

Donti, P. L., & Kolter, J. Z. (2021). Machine learning for sustainable energy systems. *Annual Review of Environment and Resources, 46,* 719–747. https://doi.org/10.1146/annurev-environ-020220-061831

Duan, Y., Edwards, J. S., & Dwivedi, Y. K., (2019). Artificial intelligence for decision making in the era of big data – Evolution, challenges and research agenda. *International Journal of Information Management, 48,* 63–71.

Efthymiou, I.-P., & Egleton, T. E. (2023). Artificial intelligence for sustainable smart cities. In: B. K. Mishra (Ed.), *Handbook of Research on Applications of AI, Digital Twin, and Internet of Things for Sustainable Development* (pp. 1–11). IGI Global.

El Himer, S., Ouaissa, M., Ouaissa, M., & Boulouard, Z. (2022). Artificial intelligence of things (AIoT) for renewable energies systems. In: S. El Himer, M. Ouaissa, A. A. A. Emhemed, M. Ouaissa, Z. & Boulouard (Eds.), *Artificial Intelligence of Things for Smart Green Energy Management. Studies in Systems, Decision and Control* (vol. 446). Springer. https://doi.org/10.10.1007/978-3-031-04851-7_1

Evans, J. (2011). *Environmental Governance.* Routledge.

Evans, J., Karvonen, A., Luque-Ayala, A., Martin, C., McCormick, K., Raven, R., & Palgan, Y. V. (2019). Smart and sustainable cities? Pipedreams, practicalities and possibilities. *Local Environment, 24*(7), 557–564. https://doi.org/10.10.1080/13549839.2019.1624701.

Fan, C., Zhang, C., Yahja, A., & Mostafavi, A. (2021). Disaster city digital twin: A vision for integrating artificial and human intelligence for disaster management. *International Journal of Information Management, 56,* 102049.

Fang, B., Yu, J., Chen, Z., Osman, A. I., Farghali, M., Ihara, I., & Yap, P.-S. (2023). Artificial intelligence for waste management in smart cities: A review. *Environmental Chemistry Letters.* https://doi.org/10.10.1007/s10311-023-01604-3

Ferre-Bigorra, J., Casals, M., & Gangolells, M. (2022). The adoption of urban digital twins. *Cities, 131,* 103905. https://doi.org/10.10.1016/j.cities.2022.103905

Flynn, A., Yu, L., Feindt, P., & Chen, C. (2016). Eco-cities, governance and sustainable lifestyles: The case of the Sino-Singapore Tianjin Eco-City. *Habitat International*, 53, 78–86.

Ford, D. N., & Wolf, C. M. (2020). Smart cities with digital twin systems for disaster management. *Journal of Management in Engineering*, 36, 04020027.

Fragkos, G., Tsiropoulou, E. E., & Papavassiliou, S. (2020). Artificial intelligence enabled distributed edge computing for Internet of Things applications. In: Proceedings of the 2020 16th International Conference on Distributed Computing in Sensor Systems (DCOSS) (pp. 450–457). IEEE.

Francisco, A., Mohammadi, N., & Taylor, J. E. (2020). Smart city digital twin-enabled energy management: Toward real-time urban building energy benchmarking. *Journal of Management in Engineering*, 36, 04019045. https://doi.org/10.10.1061/(ASCE)ME.1943-5479.0000741

Fung, A. (2006). Varieties of participation in complex governance. *Public Administration Review*, 66(s1), 66–75.

García, C. G., Núñez Valdéz, E. R., García Díaz, V., Pelayo García-Bustelo, B. C., & Cueva Lovelle, J. M. (2019). A review of artificial intelligence in the internet of things. *International Journal of Interactive Multimedia and Artificial Intelligence*, 5(4), 9–20. https://doi.org/10.9781/ijimai.2018.03.004.

Ghisellini, P., Cialani, C., & Ulgiati, S. (2016). A review on circular economy: The expected transition to a balanced interplay of environmental and economic systems. *Journal of Cleaner Production*, 114, 11–32.

Ghosh, R., & Sengupta, D. (2023). Smart urban metabolism: A big-data and machine learning perspective. In: R. Bhadouria, S. Tripathi, P. Singh, P. K. Joshi, & R. Singh (Eds.), *Urban Metabolism and Climate Change*. Springer. https://doi.org/10.10.1007/978-3-031-29422-8_16.

Goodfellow, I., Pouget-Abadie, J., Mirza, M., Xu, B., Warde-Farley, D., Ozair, S., Courville, A., & Bengio, Y. (2020). Generative adversarial networks. *Communications of the ACM*, 63(11), 139–144. https://doi.org/10.1145/3422622

Gorris, M., Glaser, R., Idrus, A., & Yusuf, A. (2019). The role of social structure for governing natural resources in decentralized political systems: Insights from governing a fishery in Indonesia. *Public Administration*, 97(3), 654–670. https://doi.org/10.10.1111/padm.12586

Gourisaria, M. K., Jee, G., Harshvardhan, G. M., Konar, D., Singh, P. K. (2023). Artificially intelligent and sustainable smart cities. In: P. K. Singh, M. Paprzycki, M. Essaaidi, & S. Rahimi (Eds.), *Sustainable Smart Cities. Studies in Computational Intelligence* (vol. 942). Springer. https://doi.org/10.10.1007/978-3-031-08815-5_14

Green, J. F., & Hadden, J. (2021). How did environmental governance become complex? Understanding mutualism between environmental NGOs and International Organizations. *International Studies Review*, 23(4), 1–21.

Gupta, J., & van der Heijden, J. (2019). Introduction: The co-production of knowledge and governance in collaborative approaches to environmental management. *Current Opinion in Environmental Sustainability*, 39, 45–51.

Hamel, P., Hamann, M., Kuiper, J. J., Andersson, E., Arkema, K. K., Silver, J. M., & Guerry, A. D. (2021). Blending ecosystem service and resilience perspectives in planning of natural infrastructure: Lessons from the San Francisco Bay Area. *Frontiers in Environmental Science, 9*, 1–13. https://doi.org/10.10.3389/fenvs.2021.601136

Han, T., Muhammad, K., Hussain, T., Lloret, J., & Baik, S. W. (2021). An efficient deep learning framework for intelligent energy management in IoT networks. *IEEE Internet of Things Journal, 8*(5), 3170–3179. https://doi.org/10.10.1109/JIOT.2020.3013306.

Haveri, A., & Anttiroiko, A.-V. (2023). Urban platforms as a mode of governance. *International Review of Administrative Sciences, 89*(1), 3–20. https://doi.org/10.10.1177/00208523211005855

Hodson, M., Kasmire, J., McMeekin, A., Stehlin, J. G., & Ward, K. (2021). *Urban Platforms and the Future City. Transformations in Infrastructure, Governance, Knowledge and Everyday Life*. Routledge and Taylor and Francis.

Huntingford, C., Jeffers, E. S., Bonsall, M. B., Christensen, H. M., Lees, T., & Yang, H. (2019). Machine learning and artificial intelligence to aid climate change research and preparedness. *Environmental Research Letters, 14*(12), 124007.

International Telecommunication Union. (2016a). *ITU-T Y.4901/L.1601 – Key Performance Indicators Related to the Use of Information and Communication Technology in Smart Sustainable Cities*. ITU.

International Telecommunication Union. (2016b). *ITU-T Y.4902/L.1602 – Key Performance Indicators Related to the Sustainability Impacts of Information and Communication Technology in Smart Sustainable Cities*. ITU.

Jagatheesaperumal, S. K., Bibri, S. E., Huang, J Ganesan, S., & Jeyaraman, P. (2024). Artificial intelligence of things for smart cities: Advanced solutions for enhancing transportation safety. *Computational Urban Science, 4*, 10. https://doi.org/10.10.1007/s43762-024-00120-6

Jain, H., Dhupper, R., Shrivastava, A., Kumar, D., & Kumari, M. (2023). AI-enabled strategies for climate change adaptation: Protecting communities, infrastructure, and businesses from the impacts of climate change. *Computational Urban Science, 3*, Article 25. https://doi.org/10.1007/s43762-023-00041-6

Janin Rivolin, U. (2012). Planning systems as institutional technologies: A proposed conceptualization and the implications for comparison. *Planning Practice and Research, 27*(1), 63–85. https://doi.org/10.10.1080/02697459.2012.661181

Joshi, K., Kumar, V., Anandaram, H., Kumar, R., Gupta, A., & Krishna, K. H. (2023). A review approach on deep learning algorithms in computer vision. In: N. Mittal, A. Kant Pandit, M. Abouhawwash, & S. Mahajan (Eds.), *Intelligent Systems and Applications in Computer Vision* (1st ed., p. 15). CRC Press.

Joss, S. (2011). Eco-city governance: A case study of Treasure Island and Sonoma Mountain Village. *Journal of Environmental Policy & Planning, 13*(4), 331–348.

Kaack, L. H., Donti, P. L., Strubell, E., Kamiya, G., Creutzig, F., & Rolnick, D. (2022). Aligning artificial intelligence with climate change mitigation. *Nature Climate Change, 12*(6), 518-527.

Kamrowska-Załuska, D. (2021). Impact of AI-based tools and urban big data analytics on the design and planning of cities. *Land, 10,* 1209. https://doi.org/10.10.3390/land10111209

Kaplan, A., & Haenlein, M. (2019). Siri, Siri, in my hand: Who's the fairest in the land? On the interpretations, illustrations, and implications of artificial intelligence. *Business Horizons, 62*(1), 15–25. https://doi.org/10.10.1016/j.bushor.2018.08.004

Katmada, A., Katsavounidou, G., & Kakderi, C. (2023). Platform urbanism for sustainability. In: N. A. Streitz, & S. Konomi (Eds.), *Distributed, Ambient and Pervasive Interactions. HCII 2023,* Lecture Notes in Computer Science, vol 14037. Springer, Cham. https://doi.org/10.1007/978-3-031-34609-5_3

Kennedy, C., Cuddihy, J., & Engel-Yan, J. (2007). The changing metabolism of cities. *Journal of Industrial Ecology, 11*(2), 43–59.

Kennedy, C., Pincetl, S., & Bunje, P. (2011). The study of urban metabolism and its applications to urban planning and design. *Environmental Pollution, 159*(8–9), 1965–1973.

Khan, A., Sohail, A., Zahoora, U., & Qureshi, A. S. (2020). A survey of the recent architectures of deep convolutional neural networks. *Artificial Intelligence Review, 53*(8), 5455–5516. https://doi.org/10.10.1007/s10462-020-09825-6.

Kierans, G., Jüngling, S., & Schütz, D. (2023). Society 5.0 2023 (EPiC Series in Computing, vol. 93), pp. 82–96

Kim, J.-Y., & Cho, S.-B. (2019). Electric energy consumption prediction by deep learning with state explainable autoencoder. *Energies, 12*(4), 739.

Kingma, D. P., & Dhariwal, P. (2018). Glow: Generative flow with invertible 1x1 convolutions. arXiv preprint arXiv:1807.03039.

Kirchherr, J., Reike, D., & Hekkert, M. (2017). Conceptualizing the circular economy: An analysis of 114 definitions. *Resources, Conservation and Recycling, 127,* 221–232.

Koch, L., Gorris, P., & Pahl, W. C. (2021). 'Narratives, narration and social structure in environmental governance'. *Global Environmental Change, 69,* 102317. https://doi.org/10.10.1016/j.gloenvcha.2021.102317 (accessed 23 July 2021).

Kooiman, J., & Jentoft, S. (2009). Meta-governance: Values, norms and principles, and the making of hard choices. *Public Administration, 87*(4), 818–836.

Koumetio, T. S. C., Diop, E. B., Azmi, R., & Chenal, J. (2023). Artificial intelligence based methods for smart and sustainable urban planning: A systematic survey. *Archives of Computational Methods in Engineering, 30*(5), 1421–1438. https://doi.org/10.1007/s11831-022-09844-2

Kramers, A., Wangel, J., & Höjer, M. (2016). Governing the smart sustainable city: The case of the Stockholm Royal Seaport. *ICT for Sustainability, 46,* 99–108.

Kuguoglu, B. K., van der Voort, H., & Janssen, M. (2021). The giant leap for smart cities: Scaling up smart city artificial intelligence of things (AIoT) initiatives. *Sustainability, 13*(21):12295. https://doi.org/10.10.3390/su132112295.

Leal Filho, W., Wall, T., Mucova, S. A. R., Nagy, G. J., Balogun, A.-L., Luetz, J. M., Ng, A. W., Kovaleva, M., Azam, F. M. S., & Alves, F. (2022). Deploying artificial intelligence for climate change adaptation. *Technological Forecasting and Social Change, 180,* 121662.

Lemos, M. C., & Agrawal, A. (2006). Environmental governance. *Annual Review of Environment and Resources*, *31*, 297–325. https://doi.org/10.10.1146/annurev.energy.31.042605.135621

Levoso, A., Gasol, C. M., Martinez-Blanco, J., Durany, X. G., Lehmann, M., & Gaya, R. F. (2020). Methodological framework for the implementation of circular economy in urban systems. *Journal of Cleaner Production*, *248*, 119227.

Li, X. (2021). Examining the spatial distribution and temporal change of the green view index in New York City using Google Street View images and deep learning. *Environment and Planning B: Urban Analytics and City Science*, *48*(7), 2039–2054. https://doi.org/10.10.1177/2399808320962511

Liu, F., Liu, F. Y., & Shi, Y. (2018). City brain, a new architecture of smart city based on the internet brain. In: IEEE 22nd International Conference on Computer Supported Cooperative Work in Design, Nanjing, 9–11 May 2018, Article No. 8465164.

Liu, F., Ying, L., & Yunqin, Z. (2021). Discussion on the definition and construction principles of city brain. In: 2021 IEEE 2nd International Conference on Big Data, Artificial Intelligence and Internet of Things Engineering (ICBAIE), Nanchang, China. https://doi.org/10.10.1109/ICBAIE52039.2021.9390064.

Liu, W., Mei, Y., Ma, Y., Wang, W., Hu, F., & Xu, D. (2022). City brain: A new model of urban governance. In: M. Li, G. Bohác, A. Huan, D. Chang, & X. Shang (Eds.), *IEIS 2021*. Lecture Notes in Operations Research. Springer. https://doi.org/10.10.1007/978-981-16-8660-3_12

Lv, Z., Li, Y., Feng, H., & Lv, H. (2021). Deep learning for security in digital twins of cooperative intelligent transportation systems. *IEEE transactions on intelligent transportation systems*, *23*(9), 16666–16675.

Marasinghe, R., Yigitcanlar, T., Mayere, S., Washington, T., & Limb, M. (2024). Computer vision applications for urban planning: A systematic review of opportunities and constraints. *Sustainable Cities and Society*, *100*, 105047. https://doi.org/10.10.1016/j.scs.2023.105047.

Mathew, A., Amudha, P., & Sivakumari, S. (2021). Deep learning techniques: An overview. In: *Advances in Intelligent Systems and Computing* (pp. 599–608). https://doi.org/10.10.1007/978-981-15-3383-9_54.

Meuleman, L., & Niestroy, I. (2015). Common but differentiated governance: A metagovernance approach to make the SDGs Work. *Sustainability*, *7*(9), 12295–12321. https://doi.org/10.10.3390/su70912295.

Mishra, B. K. (Ed.). (2023). *Handbook of Research on Applications of AI, Digital Twin, and Internet of Things for Sustainable Development*. IGI Global https://doi.org/10.10.4018/978-1-6684-6821-0

Mishra, P., & Singh, G. (2023). Artificial intelligence for sustainable smart cities. In: *Sustainable Smart Cities*. Springer. https://doi.org/10.10.1007/978-3-031-33354-5_6

Mishra, R. K., Kumari, C. L., Chachra, S., Krishna, P. S. J., Dubey, A., & Singh, R. B. (Eds.). (2022). Smart cities for sustainable development. In: *Advances in Geographical and Environmental Sciences*. Springer.

Morelli, J. (2011). Environmental sustainability: A definition for environmental professionals. *Journal of Environmental Sustainability*, *1*(1), 2.

Mosavi, A., Salimi, M., Faizollahzadeh Ardabili, S., Rabczuk, T., Shamshirband, S., & Varkonyi-Koczy, A. R. (2019). State of the art of machine learning models in energy systems. *A Systematic Review, 12*(7), 1301.

Naeem, S., Ali, A., Anam, S., & Ahmed, M. M. (2023). An unsupervised machine learning algorithms: Comprehensive review. *International Journal of Computer and Digital Systems, 13*, 911–921.

Nishant, R., Kennedy, M., & Corbett, J. (2020). Artificial intelligence for sustainability: Challenges, opportunities, and a research agenda. *International Journal of Information Management, 53*, 102104.

Nochta, T., Wan, L., Schooling, J. M., & Parlikad, A. K. (2021). A socio-technical perspective on urban analytics: The case of city-scale digital twins. *Journal of Urban Technology, 28*, 263–287. https://doi.org/10.10.1080/10630732.2020.1798177

Noori, N., Hoppe, T., & de Jong, M. (2020). Classifying pathways for smart city development: comparing design, governance and implementation in Amsterdam, Barcelona, Dubai, and Abu Dhabi. *Sustainability, 12*, 4030.

Nuissl, H., & Heinrichs, D. (2011). Fresh wind or hot air – Does the governance discourse have something to offer to spatial planning? *Journal of Planning Education and Research, 31*(1), 47–59. https://doi.org/10.10.1177/0739456X10392354

Nyseth, T. (2008). Network governance in contested urban landscapes. *Planning Theory and Practice, 9*, 497–514. https://doi.org/10.10.1080/14649350802481488.

Ooi, K. B., Tan, G. W. H., Al-Emran, M., Al-Sharafi, M. A., Capatina, A., Chakraborty, A., Dwivedi, Y. K., Huang, T. L., Kar, A. K., Lee, V. H., Loh, X. M., Micu, A., Mikalef, P., Mogaji, E., Pandey, N., Raman, R., Rana, N. P., Sarker, P., Sharma, A., Teng, C. I., Wamba, S. F., & Wong, L. W. (2023). The potential of generative artificial intelligence across disciplines: Perspectives and future directions. *Journal of Computer Information Systems*, 1–32. https://doi.org/10.1080/08874417.2023.2261010

Parihar, V., Malik, A., Bhawna, Bhushan, B., & Chaganti, R. (2023). From smart devices to smarter systems: The evolution of artificial intelligence of things (AIoT) with characteristics, architecture, use cases, and challenges. In: B. Bhushan, A. K. Sangaiah, & T. N. Nguyen (Eds.), *AI Models for Blockchain-Based Intelligent Networks in IoT systems: Concepts, Methodologies, Tools, and Applications* (Vol. 6, pp. 1–28). Springer.

Pedersen, A. N., Borup, M., Brink-Kjær, A., Christiansen, L. E., & Mikkelsen, P. S. (2021). Living and prototyping digital twins for urban water systems: Towards multipurpose value creation using models and sensors. *Water, 13*, 592. https://doi.org/10.10.3390/w13050592

Peponi, A., Morgado, P., & Kumble, P. (2022). Life cycle thinking and machine learning for urban metabolism assessment and prediction. *Sustainable Cities and Society, 80*, 103754. https://doi.org/10.10.1016/j.scs.2022.103754

Pierre, J., & Peters, B. G. (2005). *Governing Complex Societies: Trajectories and Scenarios*. Palgrave Macmillan.

Poell, T., Nieborg, D., & Van Dijck, J. (2019). Platformisation. *Internet Policy Review*, 8, 1–13.

Popescu, S. M., Mansoor, S., Wani, O. A., Kumar, S. S., Sharma, V., Sharma, A., Arya, V. M., Kirkham, M. B., Hou, D., Bolan, N., & Chung, Y. S. (2024). Artificial intelligence and IoT driven technologies for environmental pollution monitoring and management. *Frontiers in Environmental Science*, 12, 1336088. https://doi.org/10.10.3389/fenvs.2024.1336088

Puri, V., Jha, S., Kumar, R., Priyadarshini, I., Son, L. H., Abdel-Basset, M., Elhoseny, M., & Long, H. V. (2019). A hybrid artificial intelligence and Internet of Things model for generation of renewable resource of energy. *IEEE Access*, 7, 111181–111191.

Radford, A., Kim, J. W., Hallacy, C., Sastry, G., Askell, A., Mishkin, P., Ramesh, A., Goh, G., Agarwal, S., Clark, J., Krueger, G., & Sutskever, I. (2021). Learning transferable visual models from natural language supervision.

Raffel, C., Shazeer, N., Roberts, A., Lee, K., Narang, S., Matena, M., Zhou, Y., & Li, W., Liu, P. J. (2020). Exploring the limits of transfer learning with a unified text-to-text transformer. *The Journal of Machine Learning Research*, 21(1):140, Pages 5485–5551

Rane, N., Choudhary, S., & Rane, J. (2024). Artificial Intelligence and machine learning in renewable and sustainable energy strategies: A critical review and future perspectives. *SSRN*. https://ssrn.com/abstract=4838761. https://doi.org/10.10.2139/ssrn.4838761

Repette, P., Sabatini-Marques, J., Yigitcanlar, T., Sell, D., & Costa, E. (2021). The evolution of city-as-a-platform: Smart urban development governance with collective knowledge-based platform urbanism. *Land*, 10, 33. https://doi.org/10.10.3390/land10010033.

Rivolin, U. J. (2012). Planning systems as institutional technologies: A proposed conceptualization and the implications for comparison. *Planning Practice and Research*, 27(1), 63–85. https://doi.org/10.10.1080/02697459.2012.661181.

Rolnick, D., Donti, P. L., Kaack, L. H., Kochanski, K., Lacoste, A., Sankaran, K., Ross, A. S., Milojevic-Dupont, N., Jaques, N., Waldman-Brown, A., Luccioni, A. S., Maharaj, T., Sherwin, E. D., Karthik Mukkavilli, S.., Körding, K. P., Gomes, C. P., Ng, A., Hassabis, D., Platt, J. C., Creutzig, F., Chayes, J., & Bengio, Y. (2022). Tackling climate change with machine learning. *ACM Computing Surveys (CSUR)*, 55(2), 1–96.

Roseland, M. (1997). Dimensions of the eco-city. *Cities*, 14(4), 197–202. doi:10.1016/s0264-2751(97)00003-6

Saeed, Z. O., Mancini, F., Glusac, T., & Izadpanahi, P. (2022). Future city, digital twinning and the urban realm: A systematic literature review. *Buildings*, 12, 685. https://doi.org/10.10.3390/buildings12050685

Samadi, S. (2022). The convergence of AI, IoT, and big data for advancing flood analytics research. *Frontiers in Water*, 4, 786040. https://doi.org/10.10.3389/frwa.2022.786040

Samuel, P., Jayashree, K., Babu, R., & Vijay, K. (2023). Artificial intelligence, machine learning, and IoT architecture to support smart governance. In: K. Saini, A. Mummoorthy, R. Chandrika, & N. Gowri Ganesh (Eds.), *AI, IoT, and Blockchain Breakthroughs in E-Governance* (pp. 95–113). IGI Global. https://doi.org/10.10.4018/978-1-6684-7697-0.ch007

Sanchez, T. W. (2023a). Planning on the verge of AI, or AI on the verge of planning. *Urban Science, 7*, 70. https://doi.org/10.10.3390/urbansci7030070

Sanchez, T. (2023b). *Planning with Artificial Intelligence*. American Planning Association. https://www.planning.org/publications/report/9270237/.

Sandalow, D., McCormick, C., Kucukelbir, A., Friedmann, J., Nagrani, T., Fan, Z., Halff, A., d'Aspremont, A., Glatt, R., Méndez Leal, E., Karl, K., & Ruane, A. (2023). Artificial intelligence for climate change mitigation roadmap (ICEF Innovation Roadmap Project, December 2023). Available at https://www.icef.go.jp/roadmap/.

Saravanan, K., & Sakthinathan, G. (2021). *Handbook of Green Engineering Technologies for Sustainable Smart Cities*. CRC Press.

Schmitt, P., & Danielzyk, R. (2018). Exploring the planning-governance nexus: Introduction to the special issue. *Disp, 54*(4), 16–20. https://doi.org/10.10.1080/02513625.2018.1562792.

Schmitt, P., & Wiechmann, T. (2018). Unpacking spatial planning as the governance of place: Extracting potentials for future advancements in planning research. *Disp, 54*(4), 21–33. https://doi.org/10.10.1080/02513625.2018.1562795

Schrotter, G., & Hürzeler, C. (2020). The digital twin of the city of Zurich for urban planning. *PFG, 88*, 99–112. https://doi.org/10.10.1007/s41064-020-00092-2

Sehested, K. (2009). Urban planners as network managers and metagovernors. *Planning Theory & Practice, 10*(2), 245–263.

Seng, K. P., Ang, L. M., & Ngharamike, E. (2022). Artificial intelligence internet of things: A new paradigm of distributed sensor networks. *International Journal of Distributed Sensor Networks*. https://doi.org/10.10.1177/15501477211062835

Shahrokni, H., Årman, L., Lazarevic, D., Nilsson, A., & Brandt, N. (2015). Implementing smart urban metabolism in the Stockholm Royal Seaport: Smart city SRS. *Journal of Industrial Ecology, 19*(5), 917–929.

Sharifi, A., Allam, Z., Bibri, S. E., & Khavarian-Garmsir, A. R. (2024). Smart cities and sustainable development goals (SDGs): A systematic literature review of co-benefits and trade-offs. *Cities, 146*, 104659. https://doi.org/10.10.1016/j.cities.2023.104659.

Sharma, N., Sharma, R., & Jindal, N. (2021). Machine learning and deep learning applications: A vision. *Global Transitions Proceedings, 2*(1), 24–28. https://doi.org/10.10.1016/j.gltp.2021.01.004

Shen, C. (2018). A transdisciplinary review of deep learning research and its relevance for water resources scientists. *Water Resources Research, 54*(11), 8558–8593. https://doi.org/10.10.1029/2018WR022643

Shen, J., Saini, P. K., & Zhang, X. (2021). Machine learning and artificial intelligence for digital twin to accelerate sustainability in positive energy districts. In: X. Zhang (Ed.), *Data-driven Analytics for Sustainable Buildings and Cities: From Theory to Application* (pp. 411–422). CRC Press.

Shen, R., Huang, A., Li, B., & Guo, J. (2019). Construction of a drought monitoring model using deep learning based on multi-source remote sensing data. *International Journal of Applied Earth Observation and Geoinformation, 79*, 48–57. https://doi.org/10.10.1016/j.jag.2019.03.006

Shi, F., Ning, H., Huangfu, W., Zhang, F., Wei, D., Hong, T., & Daneshmand, M. (2020). Recent progress on the convergence of the Internet of Things and artificial intelligence. *IEEE Network, 34*(5), 8–15.

Shrestha, A., & Mahmood, A. (2019). Review of deep learning algorithms and architectures. *IEEE Access: Practical Innovations, Open Solutions, 7*, 53040–53065.

Shruti, S., Singh, P. K., Ohri, A., & Singh, R. S. (2022). Development of environmental decision support system for sustainable smart cities in India. *Environmental Progress & Sustainable Energy, 41*(5), e13817.

Sihvonen, S., & Ritola, T. (2015). Conceptualizing ReX for aggregating end-of-life strategies in product development. *Procedia CIRP, 29*, 639–644.

Sleem, A., & Elhenawy, I. (2023). Survey of artificial intelligence of things for smart buildings: A closer outlook. *Journal of Intelligent Systems and Internet of Things, 8*(2), 63–71. https://doi.org/10.10.54216/JISIoT.080206

Son, T. H., Weedon, Z., Yigitcanlar, T., Sanchez, T., Corchado, J. M., & Mehmood, R. (2023). Algorithmic urban planning for smart and sustainable development: Systematic review of the literature. *Sustainable Cities and Society, 94*, 104562.

Song, H., Bai, J., Yi, Y., Wu, J., & Liu, L. (2020). Artificial intelligence enabled internet of things: Network architecture and spectrum access. *IEEE Computational Intelligence Magazine, 15*(1), 44–51. https://doi.org/10.10.1109/MCI.2019.2954643.

Sørensen, E. (2006). Metagovernance: The changing role of politicians in processes of democratic governance. *The American Review of Public Administration, 36*, 98–114. https://doi.org/10.10.1177/0275074005282584.

Sörensen, E., & Torfing, J. (2005). The democratic anchorage of network governance. *Scandinavian Political Studies, 28*, 195–218.

Späth, P., & Rohracher, H. (2011). The "eco-cities" Freiburg and Graz: The social dynamics of pioneering urban energy and climate governance. In: H. Bulkeley, V. Castan-Broto, M. Hodson, S. & Marvin (Eds.), *Cities and Low Carbon Transitions. Routledge Studies in Human Geography* (pp. 88–106). Routledge.

Späth, P., Hawxwell, T., John, R., Li, S., Löffler, E., Riener, V., & Utkarsh, S. (2017). *Smart-Eco Cities in Germany: Trends and City Profiles*. University of Exeter Press.

Stead, D. (2021). Conceptualizing the policy tools of spatial planning. *Journal of Planning Literature, 36*(3), 297–311. https://doi.org/10.10.1177/0885412221992283.

Suárez-Eiroa, B., Fernández, E., Méndez-Martínez, G., & Soto-Oñate, D. (2019). Operational principles of circular economy for sustainable development: Linking theory and practice. *Journal of Cleaner Production, 214*, 952–961.

Szpilko, D., Naharro, F. J., Lăzăroiu, G., Nica, E., & Gallegos, A. D. L. T. (2023). Artificial intelligence in the smart city: A literature review. *Engineering Management in Production and Services, 15*(4), 53–75. https://doi.org/10.10.2478/emj-2023-0028.

Tarek, S. (2023). Smart eco-cities conceptual framework to achieve UN-SDGs: A case study application in Egypt. *Civil Engineering and Architecture, 11*(3), 1383–1406. https://doi.org/10.10.13189/cea.2023.110322.

Thamik, H., Cabrera, J.D.F., & Wu, J. (2024). The Digital Paradigm: Unraveling the Impact of Artificial Intelligence and Internet of Things on Achieving Sustainable Development Goals. In S. Misra, K. Siakas, & G. Lampropoulos (Eds.), Artificial Intelligence of Things for Achieving Sustainable Development Goals (Lecture Notes on Data Engineering and Communications Technologies, Vol. 192). Springer, Cham. https://doi.org/10.1007/978-3-031-53433-1_2

Toli, A. M., & Murtagh, N. (2020). The concept of sustainability in smart city definitions. *Frontiers in Built Environment, 6*, 77. https://doi.org/10.3389/fbuil.2020.00077.

Tomazzoli, C., Scannapieco, S., & Cristani, M. (2020). Internet of Things and artificial intelligence enable energy efficiency. *Journal of Ambient Intelligence and Humanized Computing, 14*, 4933–4954. https://doi.org/10.10.1007/s12652-020-02151-3

Tripathi, S. L., Ganguli, S., Kumar, A., & Magradze, T. (2022). *Intelligent Green Technologies for Sustainable Smart Cities*. John Wiley & Sons.

Uçar, E., Le Dain, M.-A., & Joly, I. (2020). Digital technologies in circular economy transition: Evidence from case studies. *Procedia Cirp, 90*, 133–136.

Ullah, A., Anwar, S. M., Li, J., Nadeem, L., Mahmood, T., Rehman, A., & Saba, T. (2023). Smart cities: The role of Internet of Things and machine learning in realizing a data-centric smart environment. *Complex Intelligent Systems, 10*, 1607–1637.

Ullah, Z., Al-Turjman, F., Mostarda, L., & Gagliardi, R. (2020). Applications of artificial intelligence and machine learning in smart cities. *Computer Communications, 154*, pp.313–323.

UN-Habitat. (2016). *New Urban Agenda*. United Nations.

UN-Habitat. (2023). Urban Governance. https://unhabitat.org/topic/urban-governance (accessed 12 March 2023).

Van Assche, K., Beunen, R., Gruezmacher, M., & Duineveld, M. (2020). Rethinking strategy in environmental governance. *Journal of Environmental Policy & Planning, 22*(5), 695–708, https://doi.org/10.1080/1523908X.2020.1768834

VanDerHorn, E., & Mahadevan, S. (2021). Digital twin: Generalization, characterization and implementation. Decision Support Systems, 145, 113524. https://doi.org/10.1016/j.dss.2021.113524

Vatn, A. (2016). *Environmental Governance: Institutions, Policies and Actions*. Edward Elgar

Venkata Mohan, S., Amulya, K., & Annie Modestra, J. (2020). Urban biocycles: Closing metabolic loops for resilient and regenerative ecosystem: A perspective. *Bioresource Technology, 306*, 123098.

Verma, T., Kishore Kumar, B., Rajendar, J., & Kumara Swamy, B. (2024). A review on quantum machine learning. In: A. Kumar, & S. Mozar (Eds.), Proceedings of the 6th International Conference on Communications and Cyber Physical Engineering (ICCCE 2024), Lecture Notes in Electrical Engineering (vol. 1096). Springer. https://doi.org/10.10.1007/978-981-99-7137-4_39

Wang, L., Chen, X., Xia, Y., Jiang, L., Ye, J., Hou, T., Wang, L., Zhang, Y., Li, M., Li, Z., et al. (2022). Operational data-driven intelligent modelling and visualization system for real-world, on-road vehicle emissions—A case study in Hangzhou City, China. *Sustainability, 14*(9), 5434. https://doi.org/10.3390/su14095434

Wang, P., Yao, J., Wang, G., Hao, F., Shrestha, S., Xue, B., Xie, G., & Peng, Y. (2019). Exploring the application of artificial intelligence technology for identification of water pollution characteristics and tracing the source of water quality pollutants. *Science of the Total Environment, 693*, 133440. https://doi.org/10.10.1016/j.scitotenv.2019.07.246

Wolf, T., Debut, L., Sanh, V., Chaumond, J., Delangue, C., Moi, A., Cistac, P., Rault, T., Louf, R., Funtowicz, M., Davison, J., Shleifer, S., von Platen, P., Ma, C., Jernite, Y., Plu, J., Xu, C., Le Scao, T., Gugger, S., Drame, M., Lhoest, Q., & Rush, A. M. (2020). Transformers: State-of-the-art natural language processing. In Proceedings of the 2020 Conference on Empirical Methods in *Natural Language Processing: System Demonstrations* (pp. 38–45). Association for Computational Linguistics (ACL). doi:10.18653/V1/2020.EMNLP-DEMOS.6

Weil, C., Bibri, S., Longchamp, R., Golay, F., & Alahi, A. (2023). A systemic review of urban digital twin challenges and perspectives for sustainable smart cities. *Sustainability of Cities and Society, 99*, 104862. https://doi.org/10.10.1016/j.scs.2023.104862

Weisz, H., & Steinberger, J. K. (2010). Reducing energy and material flows in cities. *Current Opinion in Environmental Sustainability, 2*(3), 185–192.

Xu, H., Omitaomu, F., Sabri, S., Li, X., & Song, Y. (2024). Leveraging generative AI for smart city digital twins: A survey on the autonomous generation of data, scenarios, 3D city models, and urban designs. arXiv. https://doi.org/10.48550/arXiv.2405.19464

Yang, L., Chen, X., Perlaza, S. M. et al. (2020a). Special issue on artificial-intelligence-powered edge computing for Internet of Things. *IEEE IoT Journal, 7*(10), 9224–9226.

Yang, S., Xu, K., Cui, L., Ming, Z., Chen, Z., & Ming, Z. (2020b). EBI-PAI: Towards an efficient edge-based IoT platform for artificial intelligence. *IEEE IoT Journal, 8*, 9580–9593.

Yao, T., Huang, Z., & Zhao, W. (2020). Are smart cities more ecologically efficient? Evidence from China. *Sustainable Cities and Society, 60*, 102008.

Yigitcanlar, T., David, A., Li, W., Fookes, C., Bibri, S. E., & Ye, X. (2024). Unlocking artificial intelligence adoption in local governments: Best practice lessons from real-world implementations. *Smart Cities, 7*(4), 1576–1625. https://doi.org/10.3390/smartcities7040064

Yigitcanlar, T., Senadheera, S., Marasinghe, R., Bibri, S. E., Sanchez, T., Cugurullo, F., & Sieber, R. (2024b). *The Local Government and Artificial Intelligence Nexus: A Five-Decade Scientometric Analysis on Evolution, State-of-the-Art, and Emerging Trends*. https://doi.org/10.10.2139/ssrn.4739812.

Zaidi, A., Ajibade, S.-S. M., Musa, M., & Bekun, F. V. (2023). New insights into the research landscape on the application of artificial intelligence in sustainable smart cities: A bibliometric mapping and network analysis approach. *International Journal of Energy Economics and Policy, 4*, 287.

Zayed, S. M., Attiya, G. M., El-Sayed, A., & Hemdan, E. E.-D. (2023). A review study on digital twins with artificial intelligence and internet of things: Concepts, opportunities, challenges, tools and future scope. *Multimedia Tools and Applications, 82*, 47081–47107. https://doi.org/10.1007/s11042-023-15611-7.

Zhang, J., & Tao, D. (2021). Empowering things with intelligence: A survey of the progress, challenges, and opportunities in artificial intelligence of things. *IEEE Internet of Things Journal, 8*(10), 7789–7817. https://doi.org/10.10.1109/JIOT.2020.3039359.

Zhang, J., Hua, X. S., Huang, J., Shen, X., Chen, J., Zhou, Q., Fu, Z., & Zhao, Y. (2019). City brain: Practice of large-scale artificial intelligence in the real world. *IET Smart Cities, 1*, 28–37.

Zvarikova, K., Horak, J., & Downs, S. (2022). Digital twin algorithms, smart city technologies, and 3d spatio-temporal simulations in virtual urban environments. *Geopolitics, History and International Relations, 14*(1), 139–154.

CHAPTER 3

Artificial Intelligence and Artificial Intelligence of Things Solutions for Smarter Eco-Cities

Advancing Environmental Sustainability and Climate Change Strategies

Abstract

The groundbreaking convergence of Artificial Intelligence (AI) and the Internet of Things (IoT)—known as AIoT—has offered transformative opportunities to enhance the environmental performance of smart cities and shape the development of smart eco-cities. The concept of smarter eco-cities, characterized by the seamless integration of technological advancements, data-driven approaches, and environmental strategies, represents an emerging

DOI: 10.1201/9781003536420-3

paradigm of urbanism. However, there is a notable gap in research where environmental sustainability and climate change have been addressed separately or in relation to smart cities in the realm of AI and AIoT adoption, but less so in the context of emerging smarter eco-cities. There is a lack of focus on how these technologies can be tailored to the unique challenges and opportunities of smarter eco-cities and contribute to their development. Therefore, this chapter offers a comprehensive review of the dynamic landscape of smarter eco-cities, with a specific focus on the transformative role of emerging applied AI and AIoT solutions in advancing environmental sustainability and climate change strategies. Additionally, it identifies and analyzes potential technological complexities, ethical concerns, and the socioeconomic implications of AI and AIoT-driven urban transformations. Moreover, it derives a novel framework for smarter eco-city development based on synthesized evidence. This chapter demonstrates how emerging AI and AIoT applications can significantly contribute to enhancing environmental sustainability outcomes and advancing climate change strategies in the context of smarter eco-cities. The findings underscore the substantial potential of emerging AI and AIoT technologies in advancing the development of smarter eco-cities. AI and AIoT solutions offer benefits and opportunities for optimizing resource usage, enhancing infrastructure efficiency, monitoring environmental parameters, reducing carbon footprints, improving air and water quality, and fostering resilience to climate change. However, despite the significant opportunities of AI and AIoT to advance the development of smarter eco-cities, their implementation is fraught with formidable challenges that need to be addressed and overcome. The valuable insights derived from this study empower policymakers, practitioners, and researchers to navigate the complexities surrounding the integration of data-driven scientific urbanism and smart eco-urbanism. City stakeholders glean essential knowledge for informed decision-making, effective strategy implementation, and environmental policy development through the analysis of real-world implementations and practical use cases.

3.1 INTRODUCTION

The compounding challenges arising from rapid urbanization, ecological degradation, and climate change have prompted a transformative shift in urban development. Rapid urbanization has led to increased pressure on resources and infrastructure, exacerbating issues such as pollution, energy consumption, and waste generation. Meanwhile, ecological degradation and climate change have heightened the urgency for smart cities to adopt sustainable practices and mitigate their environmental footprint. In response, smart cities are turning to innovative technologies to address these challenges effectively. Indeed, the rapid advancement of data-driven technologies in smart cities has significantly improved their environmental performance across various domains. These advancements have particularly accelerated recently, resulting from the wide adoption of Artificial Intelligence (AI) and the Internet of Things (IoT) and the impactful contribution of their groundbreaking convergence in smart cities (Alahi 2023; Bibri et al. 2023a; Cugurullo et al. 2024; Hoang 2024). This adoption has enabled smart cities to optimize resource management, reduce environmental impacts, create environmentally conscious urban environments, and mitigate and adapt to climate change (Alam et al. 2023; Bibri et al. 2023a, b; Jagatheesaperumal et al. 2023, 2024; Mishra 2023; Nishant et al. 2020; Shaamala et al. 2024; Ullah et al. 2020). AI and AIoT solutions facilitate the creation of smart sustainable urban environments by enabling real-time decision-making and adaptive management and planning strategies.

The synergistic interplay between AI and IoT represents a transformative shift in urban development, offering cities unprecedented opportunities to address pressing environmental challenges. By combining AI's analytical capabilities with the IoT's sensor-based data collection, AIoT systems can extract actionable insights from vast amounts of urban data in real-time. These insights empower cities to make informed decisions, implement targeted interventions, and optimize operational management and planning across various domains. Furthermore, the convergence of AI and IoT enables smart cities to proactively address environmental challenges, such as air and water pollution, waste management, energy consumption, and climate change mitigation. Thamik et al. (2024) discuss the applications of AIoT for Sustainable Development Goals (SDGs), exploring how it can be utilized to achieve specific SDGs related to environmental protection, energy efficiency, renewable energy, water management, sustainable urban development, and communities, among others. Through predictive analytics and optimization algorithms, AIoT systems can anticipate demand fluctuations, identify inefficiencies, and dynamically adjust resource distribution to meet urban needs while minimizing environmental impact. Therefore, AI and AIoT stand as a driving force reshaping both technological and urban landscapes, playing a crucial role in advancing both smart cities (Alahi et al. 2023; Alam et al. 2022; Hoang 2024; Szpilko et al. 2023; Zhang and Toa 2021) and smart eco-cities (Bibri et al. 2023b) toward sustainable smart cities (Gourisaria et al. 2023; Iris-Panagiota and Egleton 2023; Mishra and Singh 2023; Zaidi et al. 2023) and smarter eco-cities (Bibri et al. 2023b, 2024a).

Smart cities and eco-cities are interconnected through their shared goal of environmental sustainability, utilizing advanced technologies to address resource management and climate change challenges. While eco-cities focus on creating urban environments that prioritize ecological balance, green spaces, renewable energy, and sustainable practices to reduce environmental impact (Bibri 2021a; Bibri and Krogstie 2020a; Späth and Rohracher 2011), smart cities enhance these efforts with data-driven solutions to optimize urban infrastructure, improve environmental monitoring, and manage resources more efficiently. Together, they create sustainable urban environments that are efficient, resilient, and adaptive to environmental challenge, and needs. In short, this pertains to smart eco-cities, which involve the integration of environmental technologies and data-driven technologies, including IoT and big data analytics, to achieve different objectives of sustainability (Bibri 2021b; Caprotti et al. 2017, 2020; Späth et al. 2017; Tarek 2023).

The evolution of smart eco-cities into smarter eco-cities has been primarily accelerated by the recent advancements made in smart cities in terms of leveraging AI and AIoT technologies to enhance environmental sustainability and climate change mitigation. AI algorithms analyze vast amounts of data collected from sensors and IoT devices deployed throughout these urban environments to make real-time decisions regarding energy management, waste management, water management, transportation management, environmental monitoring, and climate mitigation. Concerning the latter, AI's emerging applications span both climate change mitigation and adaptation efforts. In climate mitigation in urban environments, AI-driven technologies can improve energy management by optimizing grid operational efficiency, forecasting renewable energy outputs, and managing energy demand. As a result, smart eco-cities are becoming increasingly adept at balancing technological innovation with ecological responsibility, paving the way for a smarter, greener future thanks to AI and AIoT. These urban environments not only leverage cutting-edge technologies for enhanced connectivity and intelligence but also prioritize environmental efficiency and resilience in the face of the mounting challenges posed by resource depletion and climate change.

The impact of AI on urban systems has been consistently expanding, with its computational and analytical capabilities experiencing exponential growth to accommodate the increasing influx of data from diverse sources facilitated by IoT. The potential of IoT lies in its capacity to facilitate real-time data analysis through advanced AI models and algorithms (Mastorakis et al. 2020; Parihar et al. 2023; Seng et al. 2022), a synergy poised to catalyze environmentally sustainable urban development (Alahi et al. 2023; Bibri et al. 2023a; Masoumi and van Genderen 2023; Mishra 2023). AIoT enables seamless communication and coordination among urban systems, leading to more efficient and sustainable operations. Moreover, AI and AIoT play a crucial role in climate change mitigation and adaptation by enabling real-time monitoring, efficient resource management, and predictive analytics to reduce emissions and enhance resilience to environmental changes (Chen et al. 2023; El Himer et al. 2022; Dheeraj et al. 2020; Leal Filho et al. 2022; Popescu et al. 2024; Rane et al. 2024; Tomazzoli et al. 2020; Samadi 2022). Overall, AIoT is on the verge of major transformations and innovations poised to take place across many domains, promising improvements in resource efficiency, energy consumption reduction, waste reduction, transportation enhancement, biodiversity conservation, and environmental protection.

The increasing integration of AI and AIoT technologies into the planning and management systems of smart cities and sustainable smart cities (Bibri et al. 2024a, b; Bibri and Huang 2024; Kamrowska-Załuska 2021; Marasinghe et al. 2024; Son et al. 2023) is poised to reshape the development of smarter eco-cities. This integration fosters a more interconnected and resilient urban ecosystem, where environmental sustainability is at the forefront of urban development strategies and technological innovation. Consequently, smarter eco-cities will emerge as a model of environmental innovation and urban transformation that prioritize sustainability, resilience, and the quality of life.

However, there is a notable gap in research where environmental sustainability and climate change have been addressed separately or often in relation to smart cities in the realm of AI and AIoT adoption, but less so in the context of emerging smarter eco-cities. While existing studies have explored the role of AI and AIoT in addressing various areas of environmental sustainability and climate change in smart cities, there is a lack of focus on how these technologies can be tailored to the unique challenges and opportunities of smarter eco-cities and specifically contribute to their development. This requires a more comprehensive approach that integrates technological advancements with sustainable urban development strategies. This gap highlights the need for further research to explore how AI and AIoT can be leveraged to create and optimize eco-friendly urban environments, ultimately propelling the development of smarter eco-cities. Especially, the focus of current smart eco-cities extends beyond climate change to encompass renewable energy, resource conservation, transport efficiency, and biodiversity protection (Bibri 2022; Kramers et al. 2016; Caprotti et al. 2016, 2017; Späth et al. 2017). These strategies and principles form the driving forces behind the evolution of smart eco-cities.

To address the identified gap and build upon and expand on the recent work of Bibri et al. (2023b), this study provides a comprehensive review of the evolving landscape of smarter eco-cities. It specifically focuses on the transformative role of emerging applied AI and AIoT solutions in advancing environmental sustainability and climate change strategies. Additionally, it identifies and analyzes potential technological complexities, ethical concerns, and the socioeconomic implications of AI and AIoT-driven urban transformations. Understanding both the opportunities and challenges of applied AI and AIoT solutions is pivotal for informed decision-making and effective policymaking in the pursuit of more sustainable, resilient, and intelligent urban futures. To achieve the overarching aim, the following research questions are formulated:

RQ1: How can emerging AI and AIoT technologies be utilized in the advancement of smarter eco-cities?

RQ2: How do AI and AIoT technologies contribute to promoting environmentally sustainable urban development practices, and what potential benefits and opportunities do they provide in the context of smarter eco-cities?

RQ3: How can the integration of emerging applied AI and AIoT solutions within the framework of smarter eco-cities contribute to achieving environmental sustainability goals and supporting effective climate change migration strategies?

RQ4: What challenges and obstacles are encountered in the implementation of AI and AIoT solutions for the development of smarter eco-cities?

In essence, the study highlights the potential of AI and AIoT technologies in shaping the future of sustainable urban development, in addition to providing a roadmap for advancing the discourse on smarter eco-cities and facilitating interdisciplinary collaborations. The insights derived from this comprehensive review not only inform researchers and practitioners in the field but also guide policymakers in making informed decisions regarding the adoption and implementation of AI and AIoT technologies in smart sustainable urban development. By highlighting the solutions, opportunities, benefits, and challenges in the field of smarter eco-cities, this comprehensive systemic review further facilitates the advancement of research, policy, and practice in pursuing more sustainable and technologically advanced urban environments. Overall, the study offers a nuanced understanding of the transformative landscape of smarter eco-cities, driven by AI and AIoT technologies. By examining their applied solutions, contributions, and challenges, this research aspires to provide a foundation for policymakers, urban planners, and researchers to navigate the complex interplay of technology, the environment, and society in the pursuit of sustainability transitions.

This chapter is structured as follows: Section 3.2 presents a survey of related work. Section 3.3 describes and outlines the research methodology. Section 3.4 presents the results of the literature review. Section 3.5 illustrates and describes a novel conceptual framework derived from insights distilled from the reviewed studies. Section 3.6 provides a detailed discussion. Section 3.7 identifies the existing gaps, their implications, and the corresponding future research directions. This chapter concludes, in Section 3.8, with a summary of key findings, contributions, implications, limitations, and some final thoughts.

3.2 CONCEPTUAL DEFINITIONS AND DISCUSSIONS

While this study involves various key concepts addressed in more detail in Chapter 2, the primary focus is on AIoT and its five pillars due to their high relevance and significance in environmental sustainability and climate change in the context smarter eco-cities.

3.2.1 Artificial Intelligence of Things and Its Pillars

AIoT is the convergence of AI and IoT, where AI techniques are integrated into IoT devices and systems to enhance their capabilities. In the context of AI, its subfields such as Machine Learning (ML), Deep Learning (DL), Computer Vision (CV), and Natural Language Processing (NLP) (see Chapter 7 for a detailed account and discussion) play crucial roles in AIoT applications in smart cities and smarter eco-cities (Alahi et al. 2023; Bibri et al. 2023b; Bibri 2024b; Rane et al. 2024; Tomazzoli et al. 2020; Seng et al., 2022; Zhang and Tao 2021). In these applications, DL focuses on training neural networks with multiple layers to extract high-level features from data, ML algorithms enable systems to learn from data and make predictions or decisions without explicit programming, and CV enables machines to interpret and understand the visual world. The application of AIoT varies across different domains, from manufacturing to transportation and urban development. In the domain of smarter eco-cities, AIoT plays a pivotal role in optimizing urban systems for environmental sustainability and resilience. Koffka (2023) explores the groundbreaking convergence of AI and IoT, reshaping the technological landscape. This fusion has led to interconnected objects and systems imbued with intelligence, autonomy, and remarkable capabilities across various domains, including smart sustainable cities. This work provides a comprehensive examination of AIoT, covering its fundamental principles, technological foundations, real-world applications, and associated challenges. It delves into the origins of AI and IoT, detailing IoT technologies and architectures, as well as AI fundamentals such as ML and DL. By showcasing real-world AIoT applications, it demonstrates the tangible impact of this emerging technology. Additionally, critical concerns such as security, privacy, interoperability, and ethics are addressed, emphasizing the importance of a responsible AIoT ecosystem.

Depending on the applied domain, an AIoT system (Figure 3.1) can be structured around five pillars: sensing, perceiving, learning, and visualizing, culminating in decision-making (Bibri et al. 2023b, 2024a). These components are essential in harnessing the power of the data collected by IoT devices to make informed decisions and drive positive environmental outcomes.

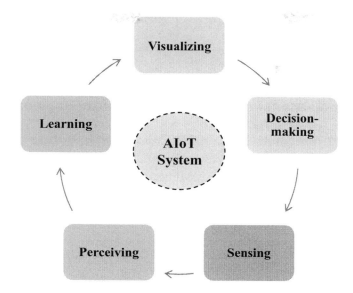

FIGURE 3.1 Five pillars of artificial intelligence of things system. Adapted from Bibri et al. (2024a).

1. Sensing: At the core of the AIoT system are various sensors deployed throughout the city to collect data on environmental parameters. These sensors can be integrated into different urban infrastructure components, including buildings, roads, public transport, and green spaces. They generate vast amounts of data that are then processed and analyzed using AI techniques, including DL and ML, to perceive patterns and trends in the data.

2. Perceiving: Once data are collected from sensors, the AIoT system processes and analyzes this information to derive meaningful insights about the city's environment. Perceiving involves using ML algorithms to detect patterns and trends in the data. For instance, the system may identify correlations between air pollution levels and traffic congestion or detect abnormal water usage patterns indicative of leaks in the water supply network.

3. Learning: The AIoT system continuously learns from the data it collects and the insights it derives from it to continuously improve its understanding of the urban environment and evaluate the effectiveness of interventions aimed at achieving environmental sustainability goals. ML algorithms are typically employed to adapt and

improve over time, enabling the system to better understand the complex dynamics of the urban environment. Learning allows the system to refine its models and predictions, leading to more accurate and actionable information for decision-making.

4. Visualizing: This pillar involves visualizing the insights derived from the data and using them to inform decision-making processes. Visualization tools, such as dashboards and interactive maps, present the analyzed data in a user-friendly format for city officials, urban planners, and residents, providing actionable insights, facilitating informed decision-making to create smarter, more environmentally sustainable cities.

5. Decision-making: Decision-making is facilitated by providing actionable recommendations based on the evaluated data and insights. For instance, city authorities may use the information to implement policies for reducing emissions, optimizing energy usage, or improving public transportation routes.

As an example, an AIoT system in a smarter eco-city could integrate air quality sensors across the urban area. These sensors continuously collect data on pollutants. The system processes this data, identifying patterns and trends in air quality. Through ML, it learns to predict air quality fluctuations based on factors like traffic congestion and weather conditions. It visualizes the data on public dashboards. City officials can then make informed decisions about implementing policies to mitigate pollution and improve the overall environmental quality of the city. Overall, the integration of AIoT technologies in smarter cities represents a significant step towards creating more sustainable and resilient urban environments.

3.2.2 Environmental Sustainability and Climate Change

Environmental sustainability refers to the responsible interaction with the environment to avoid depletion or degradation of natural resources and allow for long-term environmental quality. It encompasses a wide range of practices and principles aimed at maintaining and improving the health of the planet, such as optimizing energy efficiency, conserving water, reducing waste, protecting biodiversity, and minimizing pollution. It focuses on creating a balance where human needs are met without compromising the ability of future generations to meet their own needs. Climate change refers to the long-term alteration of temperature and weather patterns across the globe, primarily caused by human activities, such as burning fossil fuels, deforestation, and industrial processes that release Greenhouse Gases (GHG) into the atmosphere. These activities contribute to global warming, which in turn leads to various environmental impacts, such as sea-level rise, increased frequency and intensity of extreme weather events, and disruptions to ecosystems and biodiversity. While climate change can occur naturally, current trends are strongly driven by human activities, especially the emission of GHG like carbon dioxide (CO_2).

In the context of environmental sustainability and climate change, energy concerns have emerged as a major global issue in contemporary society, with intensive energy consumption standing as the foremost contributor to global warming. Energy utilization and supply issues are intricately linked to various environmental challenges, including air pollution, GHG emissions, radioactive emissions, deforestation, ozone layer depletion, acid rain, water and land use, and wildlife loss (Salim et al. 2018). Moreover, the current energy supply impacts the economic and social sectors, leading to a range of pressing socioeconomic challenges. The steady expansion of the global economy and the burgeoning population have led to an exponential surge in energy demand (Osman et al. 2022; Yang et al. 2023). This growing demand exacerbates climate change by escalating GHG emissions, underscoring the urgent need for sustainable energy solutions and enhanced energy efficiency to mitigate its impacts.

Implementing effective and comprehensive energy management strategies is crucial to transitioning to a carbon-neutral society and addressing the environmental challenges posed by climate change. This entails prioritizing sustainable energy sources and promoting environmental conservation practices worldwide. In reference to AI-driven approaches to tackling climate change, AI primarily improves air quality, enhances biodiversity and ecosystem security, increases energy efficiency, promotes public health, mitigates heat island effects, and manages water resources in the context of green infrastructure (Shaamala et al. 2024).

AI and AIoT can significantly contribute to environmental sustainability and climate change mitigation and adaptation (Chen et al. 2023; Dheeraj et al. 2020; Jain et al. 2023; Leal Filho et al. 2022; Popescu et al. 2024; Rane et al. 2024). The latter involve reducing GHG emissions, enhancing resilience to climate-related hazards, and transitioning towards sustainable practices and technologies. As AI subfields, ML, DL, CV, and NLP collectively offer advanced data analysis models, enhanced predictive modeling, and sophisticated decision-making processes that are essential for addressing complex climate and environmental challenges. AI is set to transform climate action and environmental outcomes through these technical capabilities, which demonstrate AI's role in processing and analyzing vast datasets to produce actionable insights, ultimately driving efficient and effective responses to these challenges. Mitigation focuses on reducing GHG emissions and addressing the root causes of climate change to stabilize the climate system. This aligns closely with broader environmental sustainability goals of conserving resources, reducing pollution, and promoting sustainable practices to minimize environmental impact. On the other hand, adaptation involves adjusting to the impacts of climate change that are already occurring or inevitable in the future. While adaptation strategies can contribute to environmental sustainability by enhancing resilience and reducing vulnerability to climate impacts, their primary focus is on responding to changing environmental conditions rather than preventing or reducing those changes.

For example, AI can assist in predicting and managing climate risks, advancing the accuracy of early warning systems for natural disasters such as floods, droughts, and wildfires. AI-powered models can process and analyze extensive datasets from sources such as satellite imagery, IoT sensors, observational data, and climate model outputs to detect patterns and predict the impacts of climate change. These data-driven insights are critical for boosting disaster response strategies, protecting vulnerable communities, and allocating resources more effectively, thereby informing policy-making and environmental planning.

3.3 A SURVEY OF RELATED WORK

The section provides a comprehensive survey of the relevant review studies conducted in the fields of environmental sustainability, climate change, and smart cities in relation to AI and AIoT technologies and solutions. The objective of providing the current research landscape is to highlight the aims, objectives, findings, contributions, and prevailing trends of these studies for comparative and justification purposes. The examination of the existing body of knowledge is intended to pinpoint gaps and potential avenues for the exploration of smarter eco-cities as an emerging paradigm of urbanism. This examination establishes the groundwork for this comprehensive review, empowering us to synthesize and analyze the findings of multiple theoretical, empirical, and review studies in a structured and cohesive manner.

3.3.1 Artificial Intelligence for Environmental Sustainability and Climate Change

In recent years, the literature on AI and AIoT applications for environmental sustainability and climate change has seen notable growth, expanding across various domains and disciplines. Numerous reviews have addressed the role of AI or AIoT in enhancing and advancing different facets of environmental sustainability and climate change.

The review studies on energy conservation and renewable energy (e.g., Akhter et al. 2019; Alsadi 2018; Das et al. 2018; Khan et al. 2018; Jha et al. 2017; Mhlanga 2023; Mosavi et al. 2019; Rane et al. 2024; Su et al. 2022) significantly enhance our understanding of diverse methods and strategies aimed at achieving energy efficiency and promoting renewable energy sources in various contexts. They underscore the pivotal role of technological advancements, the formulation of effective policy frameworks, and the necessity for behavioral changes to realize sustainable energy practices. In water resources conservation, the review work (e.g., Bertone, Burford and Hamilton 2018; Oyebode and Stretch 2019; Shen 2018; Tyralis et al. 2019; Mahardhika and Putriani 2023) makes noteworthy contributions to the existing knowledge of water management practices. It explores efficient water use, strategies for water conservation, and climate change impact on water resources. It accentuates the imperative for integrated water resource management and the adoption of sustainable water use practices to combat water scarcity and ensure the long-term sustainability of water resources. In examining waste management, the review studies (e.g., Fang et al. 2023; Nasir, Azir and Al-Talib 2023; Mounadel et al. 2023) underscore the substantial impact of AI on waste management practices, spanning from improved waste sorting to optimized collection routes, predictive maintenance, and decision support. These technological advancements have the potential to enhance resource efficiency, minimize environmental impact, and foster sustainable practices. Concerning sustainable transportation, the review studies (e.g., Abduljabbar et al. 2019; Liyanage et al. 2019; Nti et al. 2022) shed light on various facets, including electric vehicles, intelligent transportation systems, and multimodal transportation options. These explorations underscore the potential of sustainable transportation solutions to curtail CO_2 emissions, improve air quality, and enhance urban mobility.

In the area of climate change mitigation and adaption, the review studies (Leal Filho et al. 2022; Chan et al. 2023; Raseman et al. 2017; Shaamala et al. 2024; Yang and Liu 2020) deepen our understanding of the challenges and opportunities inherent in addressing the impacts of climate change. Exploring mitigation measures, adaptation strategies, and policy frameworks, these studies chart a course to reduce GHG emissions and fortify resilience against climate change. Lastly, the review studies on biodiversity and ecosystem services (e.g., Ochoa and Urbina-Cardona 2017; Salcedo-Sanz et al. 2016; Toivonen et al. 2019; Nti et al. 2022) provide insights into the importance of biodiversity conservation and the role of ecosystem services in sustaining human well-being. They emphasize the need for conservation measures, habitat restoration, and the integration of ecosystem services into decision-making processes for sustainable development.

AI plays a critical role in biodiversity conservation for climate mitigation and adaptation. It enhances carbon sequestration and protecting vital ecosystems, as well as monitors ecosystems to identify high-carbon storage areas for conservation (Chen et al. 2023). It has also the ability to develop innovative carbon storage methods, such as creating promising materials for sustainable CO_2 management (Zhang et al. 2022). While biodiversity conservation is associated with mitigation when it involves preserving ecosystems that store carbon (e.g., wetlands, forests, grasslands), it has also elements of adaptation such as maintaining ecosystem resilience. AI is being used to protect biodiversity by monitoring ecosystem changes, predicting climate impacts on species and tracking endangered ones, and analyzing environmental changes. It helps manage ecosystem health, enabling timely interventions to maintain resilience, and supports habitat restoration by identifying areas most vulnerable to climate shifts. Additionally, AI assists in planning adaptive measures for species migration and adjusting conservation priorities based on real-time data.

Collectively, these review studies offer invaluable insights into diverse facets of environmental sustainability and climate change, enriching the knowledge base within their respective fields. They underscore the significance of holistic approaches, interdisciplinary integration, and the consideration of technological and policy dimensions to effectively address environmental challenges and advance sustainable development practices.

3.3.2 Artificial Intelligence and the Internet of Things for Smart Cities

Several review studies have explored the interconnection between AI, IoT, and Big Data in smart cities, approaching the topic from various perspectives and in various domains beyond environmental sustainability. A set of relevant review studies is presented in Table 3.1.

TABLE 3.1 Relevant Review Studies on Artificial Intelligence and the Internet of Things and Their Integration in Smart Cities

Citations	Focus	Main Contributions
Bibri et al. (2023b)	Research trends and driving factors of environmentally sustainable smart cities, with a focus on the convergence of AI, IoT, and Big Data technologies.	Highlighting the rapid growth of environmentally sustainable smart cities due to large-scale digital transformation, decarbonization, and technological advancement, as well as the need to address the ethical risks and environmental costs associated with AI, IoT, and Big Data.
Szpilko et al. (2023).	The influence of AI in smart cities in terms of its benefits in enhancing efficiency and quality of life, as well as the challenges it poses regarding data security and privacy due to the widespread use of IoT technologies.	Systematically classifying scientific research related to AI in smart cities, providing insights into urban mobility innovations, highlighting persistent security concerns, advocating for standardized regulations in energy management and sustainability practices, emphasizing the need for further investigation into AI's applications to address ethical and societal implications in various sectors.
Herath and Mittal (2022)	The use of AI in enhancing comfort and cost-effectiveness, covering healthcare, education, environment, waste management, agriculture, mobility, transportation, risk management, and security.	Identification of key sectors where AI adoption in smart cities is most prominent and highlighting the significant impact of AI algorithms such as ANN, RNN/LSTM, CNN/R-CNN, DNN, and SVM/LS-SVM across various domains in smart cities.
Dadashzadeh and Zarezadeh (2022)	The role of AI and ML in the evolution of smart cities and how AI can contribute to achieving new standards of urbanization, optimal energy consumption, environmental sustainability, and economic development in smart cities.	Presenting the applications of ML techniques in various aspects of smart cities, including, energy grids, public lighting, natural resources management, water management, environmental monitoring, waste management, transport, mobility, and logistics.

(Continued)

TABLE 3.1 (*Continued*)   Relevant Review Studies on Artificial Intelligence and the Internet of Things and Their Integration in Smart Cities

Citations	Focus	Main Contributions
Navarathna and Malagi (2018)	Role of AI in smart city analysis, with a focus on AI techniques in analyzing data generated by smart city systems.	Providing insights into the application of AI techniques in analyzing smart city data, and highlighting AI's potential to enhance decision-making, resource allocation, and overall efficiency and sustainability in smart cities.
Ullah et al. (2020)	The role of AI, ML, and DL in the evolution of smart cities.	Highlighting the efficient use of AI, ML, and DL in the design of optimal policies and their applications in transportation, energy efficiency, cybersecurity, and healthcare systems.
Dominiković et al. (2021)	The role of AI in the advancement of smart cities.	Emphasizing the evolving role of AI in supporting intelligent systems in smart cities and offering insights into current research on AI in smart cities in terms of applications and challenges.

The findings of these studies offer valuable insights for scholars, practitioners, and policymakers involved in developing data-driven technology solutions and implementing environmental policies for smart cities. They contribute to understanding the challenges, opportunities, and ethical considerations associated with the integration of AI, IoT, and Big Data technologies in smart city initiatives.

The preceding review studies examining the intersections of AI and AIoT with environmental sustainability, climate change, and smart cities have established a strong framework for understanding how to create environmentally sustainable and technologically sophisticated urban environments. Nonetheless, in light of the rapidly evolving landscape of AI and AIoT, there arises a compelling need to thoroughly explore the specific confluence of these cutting-edge technologies in the established model of smart eco-cities, unraveling the spectrum of its intricate dimensions. The flourishing area of AI and AIoT introduces novel avenues for addressing the complex challenges posed by ecological degradation and climate disruption in urban ecosystems.

However, no review study has thoroughly explored the interconnections and synergies among environmental sustainability, climate change, and smart cities and their intersection with smarter eco-cities in the context of AI and AIoT technologies. In other words, the current landscape of research on smarter eco-cities lacks a comprehensive examination of the interplay of emerging data-driven technologies, green technologies, and environmental strategies.

Undertaking a comprehensive review dedicated to the examination of the burgeoning landscape of smarter eco-cities, with a specific focus on the transformative role played by AI and AIoT applications in addressing environmental sustainability and climate change challenges in smart cities, is intended to bridge the existing gaps between these applied solutions and the extant body of research on these three domains, all contextualized in the distinct paradigm of smarter eco-cities. This comprehensive review seeks to offer novel insights into the dynamic realm of smart eco-urbanism, enhancing the understanding of its diverse domains by synthesizing and analyzing a wide array of studies from various domains and disciplines.

3.4 MATERIALS AND METHODS

This comprehensive review focuses on the dynamic landscape of smarter eco-cities, with a specific emphasis on the transformative role played by AI and AIoT applications in tackling environmental sustainability and climate change challenges. To unravel the intricacies of these applied technology solutions and understand their contributions to advancing the various aspects of environmental sustainability in the context of smarter eco-cities, a systematic approach was employed to address the three questions guiding the review study.

The Preferred Reporting Items for Systematic Reviews and Meta-analyses (PRISMA) approach was adopted (Figure 3.2). Web of Science (WoS) and Scopus were selected as academic databases given their broad coverage of high-quality, peer-reviewed studies related to the topic on focus. Keywords such as "smart eco-cities," "smart cities," "environmental sustainability," "climate change," "artificial intelligence," "artificial intelligence of things," "machine learning," and "deep learning" were systematically combined to retrieve pertinent studies.

The search was limited to the timeframe of the last 8 years to ensure relevance and accuracy, yielding 453 documents in total. The selected timeframe of 2017–2024 encompasses the key facets of the subject of smarter eco-cities, particularly in relation to AI and AIoT technologies and their

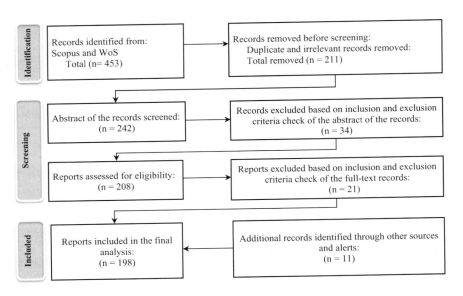

FIGURE 3.2 The PRISMA flowchart for literature search and selection. Adapted from Page et al. (2021).

role in advancing environmental sustainability, climate change, and smart cities.

The inclusion criteria filtered studies based on their relevance, reliability, language, publication date, and publication type (article, conference paper, or book chapter), providing definitive primary information. Exclusion criteria were applied to remove studies unrelated to the focal topics and irrelevant to the research aim and questions. Based on these predefined inclusion/exclusion criteria, the selection process involved initial screening based on titles and abstracts, followed by a detailed full-text review for eligibility.

In light of the above, the selection process involved initial screening based on titles and abstracts, followed by a detailed full-text review for eligibility. The search query retrieved a total of 453 records from two databases (Scopus=289, WoS=164). After removing duplicates and irrelevant documents, 179 records were eliminated. Subsequently, titles and keywords were scrutinized, leading to the exclusion of 32 more records. The remaining 242 records underwent abstract screening against the inclusion and exclusion criteria, resulting in the removal of 34 records. The full-text screening of the remaining 208 records led to the exclusion of an additional 21 records. Ultimately, this process yielded a final selection of 187 publications. Additionally, 11 extra records were included through other sources and alerts, bringing the total number of records included in the final analysis to 198. Throughout the process, critical appraisal was conducted to assess the quality of the selected studies.

Information relevant to the research questions was systematically extracted, including details on AI and AIoT applications for environmental sustainability and climate change, contributions to sustainable urban development practices, linkages between smart cities and smart eco-cities, potential benefits, encountered challenges, and knowledge gaps. A qualitative synthesis framework encompassing configurative, aggregative, and narrative approaches was employed to organize and categorize the extracted data, facilitating a nuanced understanding of the emergent themes. To ensure the credibility and reliability of the selected studies, a quality assessment was conducted. Peer-reviewed articles and studies from reputable sources were given precedence, and an evaluation was applied to ascertain the reliability and robustness of the methodologies employed in each study by assessing the strengths and weaknesses of the research approaches.

The collected data were analyzed and synthesized to provide a comprehensive overview of the current state, emerging trends, commonalities, differences, opportunities, and challenges in the realm of smarter eco-cities, with a focus on AI and AIoT applications. The findings of the synthesized studies were amalgamated using specified conceptual and descriptive categories. The categorization evolved as more precise themes were identified and refined through actions such as combination, separation, and discarding. From these identified categories, themes were organized to provide fresh interpretations extending beyond the findings of the synthesized studies. The findings of the review study were presented, offering valuable insights for researchers, policymakers, and practitioners engaged in the field of environmentally sustainable urban development.

3.5 RESULTS: AI AND AIoT APPLICATIONS FOR ENVIRONMENTAL SUSTAINABILITY, CLIMATE CHANGE, AND SMART CITIES

In presenting the results, all relevant data pertaining to the evolving interdisciplinary domain of smarter eco-cities are integrated. This integration spans a comprehensive range of theoretical, empirical, and practical evidence, addressing diverse facets of smarter eco-cities and emphasizing their synergies to increase the benefits of environmental sustainability and address its challenges.

In recent years, the utilization and integration of AI, IoT, and Big Data have experienced a remarkable surge and risen to prominence, particularly in the realms of environmental sustainability, climate change mitigation, and the development of smart cities. The period spanning from 2017 to 2024 has witnessed unprecedented growth in the application of these transformative technologies, driven by a confluence of influential factors. The widespread adoption of IoT and Big Data technologies in smart cities for advancing environmental sustainability during 2016-2020 has laid a robust foundation (e.g., Bibri 2019, 2020; Hashem et al. 2016; Perera et al. 2017; Rathore et al. 2016; Sinaeepourfard et al. 2016). This has paved the way for the seamless integration of AI into the IoT infrastructure across diverse domains of smart cities. It also marked the materialization of smart eco-cities (e.g., Caprotti et al. 2016, 2017). Moreover, the accelerated digital transformation of smart cities and smart eco-cities, further accelerated by COVID-19, has been a driving force behind the surge of data-driven technologies. Furthermore, imperatives such as decarbonization and the pressing need for sustainable development practices have underscored the urgency of deploying innovative technological solutions to advance environmental sustainability. The rapid advancements of data-driven technologies, coupled with their integration, have created synergies that have reshaped the landscape of environmental management and planning in smart cities. All these factors contribute to setting the stage for exploring the dynamic and evolving intersection of AI and IoT in the contexts of environmental sustainability, climate action, and the development of smart cities towards this direction, emphasizing the ongoing transformative impact of technological advancements on emerging smarter eco-cities and their core systems and domains.

3.5.1 The First Period: 2017–2019

3.5.1.1 Empirical Research on the Relationship between AI Environmental Sustainability

The advancement of emerging smarter eco-cities has been notably propelled by the outcome of several empirical studies on environmental sustainability during the first period (Table 3.2). These studies have made significant contributions by employing AI and ML models and techniques to address and overcome environmental sustainability challenges.

TABLE 3.2 Overview of Artificial Intelligence and Machine Learning Techniques in the Applied Domains of Environmental Sustainability

Areas	AI and ML Techniques	Tasks and Applications	Citations
Water resources conservation	ANN, SVM, FL, ANFIS, LR, KNN, ANN, DT, GA, DSS	Stream flow prediction and water quality analysis; Water chemistry analysis and water quality assessment; Hydro-meteorological forecasting and leak detection	Demirci et al. (2019); Poul et al. (2019); Al-Mukhtar and Al-Yaseen (2019); Najafzadeh and Ghaemi (2019); Calp (2019); Ferreira et al. (2019); Rojek and Studzinski (2019).
Energy conservation and renewable energy	ANN, FL, SVM, DT, ES, EC, BN, ANFIS, DNN	Energy operation, production, distribution, maintenance, and planning	González et al. (2018); Guiqiang et al. (2018); Merizalde et al. (2019); Puri et al. (2019); Sellak et al. (2017)
Sustainable transportation	ANN, DT, Time Series, NC, FL, SVM, LR, DSS	Traffic forecasting; Transport management	Alrukaibi et al. (2019); Zhou et al. (2019); Abdelkafi and Täuscher (2016); Aymen and Mahmoudi (2019); Jones and Leibowicz (2019); Solaymani (2019)
Biodiversity conservation and ecosystem services	FL, GA, ARIES, BN	Biodiversity protection and ecosystem service assessment	Rodriguez-Soto et al. (2017); Sharps et al. (2017); Toivonen et al. (2019); Willcock et al. (2018); Zhang et al. (2017)

The research conducted on the conservation of water resources yields invaluable insights that can significantly shape effective water resources management strategies. These findings play a crucial role in advancing the cause of water conservation and fostering sustainable water use within the framework of emerging smart eco-cities. The integration of AI techniques and algorithms in energy conservation and renewable energy studies has been instrumental in enhancing energy efficiency and promoting the adoption of renewable energy sources in the context of smarter eco-cities. This innovative approach has not only contributed to the reduction of energy consumption but has also facilitated the seamless integration of sustainable energy practices.

In the domain of sustainable transportation, studies have provided essential knowledge for the enhancement of transportation planning and management. These insights contribute to the creation of more efficient and environmentally sustainable mobility solutions in the context of emerging smarter eco-cities. The research in this domain serves as a catalyst for the development of intelligent transportation systems that promote eco-friendly modes of travel, ultimately fostering a greener and more sustainable urban environment.

Additionally, investigations into biodiversity conservation and ecosystem services have deepened the understanding of the intricate relationships between biodiversity, ecosystem services, and the overall sustainability of emerging smarter eco-cities. The knowledge derived from these studies informs decision-makers on how to harmonize urban development with the preservation of natural ecosystems. By recognizing the importance of biodiversity and ecosystem services, smarter eco-cities can develop strategies that balance ecological conservation with urban growth.

In essence, the collective findings and insights from these diverse studies contribute to evidence-based decision-making processes. They serve as a foundation for informed policy development and implementation, promoting sustainable practices in the ongoing development and evolution of smart eco-cities towards smarter eco-cities. This multidisciplinary approach underscores the significance of holistic research in creating urban environments that are not only technologically advanced but also environmentally conscious and sustainable.

3.5.1.2 Theoretical and Literature Research on the Relationship between AI and Environmental Sustainability

A diverse array of theoretical and literature review studies has significantly contributed to the discourse surrounding the development of smarter eco-cities, spanning multiple domains (Table 3.3).

The exploration of water resource conservation has yielded valuable insights, offering effective strategies for the management of water resources. In the realm of energy conservation and renewable energy, dedicated studies have made substantial contributions by optimizing energy operations, promoting energy efficiency, and advocating the integration of renewable energy technologies. The research on sustainable transportation has provided nuanced insights into critical aspects such as traffic management, transportation planning, and the enhancement of overall efficiency and sustainability in transportation systems. The studies on biodiversity and

TABLE 3.3 An Overview of Artificial Intelligence and Machine Learning Techniques in Theoretical and Review Studies on Environmental Sustainability

Areas	AI and ML Techniques	Citations
Water resources conservation	ANN, SVM, FL, GA, ANFIS	Mehr et al. (2018); Oyebode and Stretch (2019); Sahoo et al. (2019); Valizadeh et al. (2017)
Energy conservation and renewable energy	LR, ANN, SVM, NC, EC	Akhter et al. (2019); Alsadi and Khatib (2018); Das et al. (2018); Dawoud, Lin, and Okba (2018); Khan et al. (2018); Youssef, El-Telbany, and Zekry (2017); Wang and Srinivasan (2017)
Sustainable transportation	ANN, DT, NC, SVM, FL, time series models	Abduljabbar et al. (2019); Jiang and Zhang (2019); Liyanage et al. (2019)
Biodiversity and ecosystem services	SVM, ANN, GA, FL, ARIES	Kalmykov and Kalmykov (2015); Salcedo-Sanz et al. (2016); Ochoa and Urbina-Cardona (2017)

ecosystems significantly enrich our understanding of biodiversity dynamics and contribute essential knowledge for the sustainable stewardship of ecosystems. Collectively, these theoretical and literature reviews play a pivotal role in propelling the evolution of smart eco-cities towards smarter eco-cities. Their cumulative outcomes not only provide valuable insights but also lay the foundation for evidence-based approaches, empowering the development of more sustainable, resilient, and smart eco-cities.

3.5.1.3 Empirical Research on the Relationship between AI and Climate Change

In the empirical domain of AI research on climate change, specific emphasis has been placed on scenario analysis, disaster management and resilience, marine resources management, and mitigation and adaption. This targeted focus has resulted in substantial contributions that can drive the advancement of smarter eco-cities. Table 3.4 outlines the pivotal contributions of AI, ML, and DL techniques in addressing key challenges and advancing knowledge in the field of climate change. It sheds light on the rich tapestry of studies harnessing these technologies to foster sustainable practices and enhance our understanding of environmental dynamics.

The research conducted on scenario analysis has significantly contributed to our understanding of the impacts, vulnerabilities, and adaptation strategies related to climate change. These insights extend to unraveling the interdependencies and interactions among various facets of environmental

TABLE 3.4 An Overview of Artificial Intelligence and Machine Learning Techniques, Tasks, and Applications in Climate Change Research

Domains	AI Techniques	Tasks and Applications	Citations
Scenario analysis	EC, FL, ANN, BN, SVM, GA, neuro-fuzzy, GANs, CNNs, RNNs, NLP	GHG emissions, natural disasters, ocean and gryosphere, atmospheric forecasting, temperature increase projections, sea-level rise, biodiversity impact assessment, extreme weather events.	Atsalakis (2016); Bowes et al. (2019); Ise and Oba (2019); Jeon et al. (2018); Mokhtarzad et al. (2017); Piasecki et al. (2018); Shen et al. (2019); Skiba et al. (2017); Raji et al. (2019); Ullah et al. (2018); Woo (2019); Yin et al. (2017)
Disaster management	NLP, CV, AI-driven GIS, DT, RF, SVM, DSS	Predictive modeling, damage assessment, environmental hazard detection, sentiment analysis, image analysis, object recognition, spatial analysis, mapping, visualization, resource allocation optimization.	Anbarasan et al. (2020); Arachie et al. (2020); Choubin et al. (2019); Lee and Chien (2020); Ji et al. (2019); Salluri et al. (2020); Sun et al. (2020)
Marine resources management	GA, ML-based Species Distribution Models (SDMs), NLP, Time series models, CNNs, RNNs, Autonomous Underwater Vehicles (AUVs), ROVs, DSS	Water pollution control, water quality monitoring, pollution prevention, acidification mitigation, habitat and species protection.	Lu et al. (2019); Mahmood et al. (2017); Nunes et al. (2020); Villon et al. (2018); Wang et al. (2019)

(Continued)

TABLE 3.4 (*Continued*)   An Overview of Artificial Intelligence and Machine Learning Techniques, Tasks, and Applications in Climate Change Research

Domains	AI Techniques	Tasks and Applications	Citations
Mitigation	ML, DL, NLP, GA, Neural Networks (CNNs, RNNs)	Carbon footprint reduction, energy optimization, emission prediction, renewable energy management.	Youssef et al. (2017); Jha et al. (2017); Das et al. (2018); Liu et al. (2019); Nishant et al. (2020); Puri et al. (2019); Wang et al. (2019); Woo (2019)
Adaptation	ML, DL, CNNs, RNNs, NLP, Time Series Forecasting, SDMs, AUVs	Climate change impact assessment, crop yield prediction, water resource management, disaster risk reduction, biodiversity monitoring.	Bowes et al. (2019); Crane-Droesch (2018); Jeon et al. (2018); Kiba et al. (2017); Mokhtarzad et al. (2017); Ullah et al. (2018); Yeung et al. (2020)

sustainability. Furthermore, scenario analysis has played a crucial role in managing GHG emissions, fostering the development of effective strategies to reduce carbon footprints. This body of research serves as a foundation for informed decision-making in the pursuit of sustainable and eco-friendly urban development. In the domain of natural disaster analysis, key insights have significantly contributed to fortifying communities against the impacts of disasters. These contributions extend beyond theoretical insights to practical applications, enabling effective preparedness, response, and recovery measures in the face of natural calamities. Turning attention to marine resources management, empirical studies have yielded new perspectives on addressing diverse challenges and enhancing the sustainable stewardship of marine resources. The outcomes of these studies are instrumental in guiding policies and practices for the responsible use and conservation of marine ecosystems.

The research conducted on climate change mitigation and adaptation has significantly enhanced the strategies to tackle climate challenges. Mitigation research has utilized AI techniques to optimize energy consumption, reduce carbon emissions, and improve renewable energy management. On the adaptation front, AI models have been crucial in assessing climate impacts, predicting agricultural yields, managing water resources, and enhancing disaster risk reduction. These contributions are vital for developing robust approaches to both mitigate the effects of climate change and adapt to its unavoidable impacts.

In summary, these empirical studies stand as invaluable sources of insights and advancements in understanding and addressing climate change challenges in the context of emerging smarter eco-cities. The knowledge generated from these investigations informs academic discourse and provides practical guidance for policymakers, urban planners, and stakeholders working towards the realization of sustainable and resilient urban environments.

3.5.1.4 Research on the Relationship between AI and AIoT and Smart Cities

In the evolving landscape of smart cities, AI and AIoT research has undergone distinct phases, each contributing to a deeper understanding of the interplay between technology and urban development in the realm of environmental sustainability. Table 3.5 encapsulates the thematic progression, showcasing the diverse applications of AI and AIoT in addressing environmental challenges and shaping the development of smarter eco-cities.

This overview illuminates the transformative journey of AI and AIoT in environmental sustainability, underscoring their impact on the development of smarter eco-cities across different research phases.

TABLE 3.5 An Overview of Artificial Intelligence and Artificial Intelligence of Things Research on Environmental Sustainability and Climate Change in Smart Cities

AI and AIoT Research Focus in Smart Cities	Citations
Environmental sustainability	Ampatzidi et al. (2017); Aymen and Mahmoudi (2019); Mathur and Modani (2016); Navarathna and Malagi (2018); Sonetti et al. (2018); Vázquez-Canteli et al. (2019)
Climate change mitigation in urban planning and development	Barnes et al. (2019); Balogun et al. (2020); Huntingford et al. (2019); Nishant et al. (2020)
AI-powered IoT architecture	Muhammad et al. (2019); Yu et al. (2018)
Cognitive AI and knowledge-based AI	Aymen and Mahmoudi (2019); Guo et al. (2018); Park et al. (2019)
AIoT for energy management and optimization and self-driving vehicles	Antonopoulos et al. (2019); Zhang and Letaief (2019); Tomazzoli et al. (2020)

In addition, Ashqar et al. (2017) employed ML algorithms, including RF, Least-Squares Boosting (LSBoost), and Partial Least-Squares Regression (PLSR), to model bike availability at San Francisco Bay Area Bike Share stations. While univariate models (RF and LSBoost) outperform the multivariate PLSR model in predictive accuracy, the latter proves effective in networks with spatially correlated stations. Significantly, the research identifies station neighbors and an optimal 15-minute prediction horizon as influential factors. This comparative analysis contributes insights into the trade-offs between univariate and multivariate approaches, offering practical recommendations for improving predictive precision in bike-sharing systems.

3.5.2 The Second Period: 2020–2024

Throughout the second period from 2020 to 2024, there was a notable surge in empirical, theoretical, and literature research on AI and AIoT, spanning the domains of environmental sustainability, climate change, and smart cities. This accelerated growth in research activity has significantly deepened and broadened our understanding of these interconnected fields and has yielded substantial contributions to the evolving landscape of emerging smarter eco-cities. The expanding body of work not only enhances our knowledge base but also propels the development of strategies and solutions to effectively tackle and surmount the complex environmental challenges faced by contemporary urban environments.

3.5.2.1 Research on the Relationship between AI and Environmental Sustainability

While the primary domains of AI research in environmental sustainability maintained their significance, there was a slight decline in academic attention compared to the preceding period. Notably, a shift occurred in academic interests in COVID-19 during this period, transitioning from an emphasis on addressing environmental sustainability challenges to a heightened focus on the extensive deployment and application of digital technologies within the framework of smart cities (Bibri et al. 2023b). Despite this shift, the main areas of environmental sustainability retained their appeal for AI research, as outlined in Table 3.6.

As mentioned in Table 3.7, studies across various domains, including energy conservation and renewable energy, air quality, sustainable transportation, clean water security, and biodiversity conservation, have made substantial contributions to the advancement of smarter eco-cities

TABLE 3.6 An Overview of Artificial Intelligence Models, Techniques, and Applications in Environmental Sustainability Research

Domain	AI Models and Techniques	Tasks and Applications	Citations
Energy conservation and renewable energy	ML, DL, CNNs, RNNs, GA, ANN, PSO, SVM, FL NLP	Energy consumption prediction, renewable energy forecasting, energy efficiency optimization, demand-side management, grid stability control, smart grid management, energy load forecasting, renewable resource optimization, energy storage management, power generation scheduling	Al-Othman et al. (2022); Boza and Evgeniou (2021); Jin et al. (2022); Liu et al. (2021); Liu et al. (2022); Olabi et al. (2023), Qin et al. (2024); Shin et al. (2021); Zhao et al. (2024)
Air quality and environmental pollution monitoring and management	ML, DL, CNNs, SVM, Random Forest, RNNs, KNN, GA	Air pollution level prediction, real-time air quality forecasting, satellite imagery analysis, pollution level classification, feature selection and prediction, time-series air quality forecasting, pollution source identification, long-term air quality prediction, pollution control system optimization	Awasth et al. (2024); Delanoë et al. (2023); Neo et al. (2024); Popescu et al. (2024); Sowmya and Ragiphani (2022); Schürholz et al. (2020); Schultz (2022)
Sustainable Transportation	ML, DL, GA, EC, ANN, Spatial DNA, reinforcement learning ANN, FL, SVM, PSO, KNN, DT, CNNs, NLP	Energy planning, transportation connectivity, energy consumption analysis in household transportation, urban traffic management and surveillance, transport network capacity assessment, jobs-housing and commuting corridors balance optimization, emission reduction optimization, fleet management for public and private transport, Mobility-as-a-Service (MaaS) integration, autonomous vehicle coordination, electric vehicle (EV) charging infrastructure planning, predictive maintenance for transport infrastructure, demand-responsive transportation systems.	Amiri et al. (2021); Aschwanden (2021); Fatemidokht et al. (2021); Liao et al. (2021); Nikitas et al. (2020); Nishant et al. (2020); Nti et al. (2022); Olayode et al. (2020); Othman (2022); Saleh et al. (2022); Puri et al. (2020); Silva et al. (2022); Tyagi and Aswathy (2021)

(Continued)

TABLE 3.6 (*Continued*) An Overview of Artificial Intelligence Models, Techniques, and Applications in Environmental Sustainability Research

Domain	AI Models and Techniques	Tasks and Applications	Citations
Clean water security and conservation	ML, DL, ANN, CNNs, RNNs, GA, PSO, NLP, DSS	Water quality management, water supply quantity optimization, water control, water treatment, sanitation, leak detection and prevention, water conservation strategies, groundwater management, water demand forecasting.	Chang et al. (2023); Mahardhika1 and Putriani (2023); Martínez-Santos and Renard (2020); Rashid and Kumari (2023); Mahardhika1 and Putriani (2023); Zhu et al. (2022).
Biodiversity conservation	ML, DL, FL, SVM, CNNs, RNNs, GA, NLP, ARIES, AI-driven GIS, AI-powered drones	Natural capital improvement and protection, ecosystem health preservation, habitat and species protection, ecosystem restoration, biodiversity conservation, marine pollution prevention, wildlife monitoring and tracking, invasive species detection and control, ecosystem services valuation, climate change impact prediction on ecosystems, land use and deforestation monitoring, genetic diversity analysis	Debus et al. (2024); Dominguez et al. (2022); Granata et al. (2020); Jahani and Rayegani (2020); Liu et al. (2023); Nti et al. (2022); Santangeli et al. (2020); Silvestro et al. (2023)

through the application of diverse AI, ML, and DL models and algorithms. These collective efforts showcase the integral role of AI in addressing critical environmental challenges and promoting the development of more sustainable and intelligent eco-cities.

Furthermore, recent advancements in AI have significantly enhanced marine conservation efforts. Nunes et al. (2020) demonstrated how AI-aided automated image analysis can expedite coral reef conservation by providing rapid and accurate assessments of coral health. Ward et al. (2022) reviewed the broader application of AI in conserving marine life, emphasizing the role of AI in biodiversity and ecosystem management, which enhances the effectiveness of conservation strategies. Isabelle and Westerlund (2022) categorized various AI-based opportunities in

TABLE 3.7 Review Studies on Artificial Intelligence and the Internet of Things and Their Convergence

Citations	Focus	Main Contributions
Wang et al. (2023)	The application of graph neural networks (GNNs) for data acquisition in AIoT and the challenges of handling noisy and adversarial data generated by heterogeneous sensors in massive IoT deployments.	Presentation of the latest advancements in using GNNs for the horizontal task of data acquisition in AIoT, proposal of a unified GNN pipeline based on the encoder-decoder paradigm; systematic categorization and summarization of emerging technologies that address challenges in AIoT data acquisition, particularly in handling noisy and adversarial data; and outlining future research directions for GNN-based AIoT data acquisition, highlighting potential advancements and areas for improvement.
Mohamed (2020)	The integration of AI with IoT to manage the vast amount of data generated by IoT devices and the fundamentals of IoT and AI, algorithms used in AI, challenges faced when combining AI with IoT, and applications of AI systems in IoT.	Emphasizing the importance of self-optimizing networks and software-defined networks in AIoT systems, IoT applications and the relationship between AI and IoT, the role of AI in IoT applications, and enhancing IoT performance using ML and DL techniques.
Mukhopadhyay et al. (2021)	Sensors in IoT systems and their integration with AI.	Highlighting the importance of efficient, intelligent, and connected sensors for smart decision-making and collaborative communication, as well as the integration of advanced AI into IoT to enable sensors to detect performance degradation and identify patterns, among others.
Shi et al. (2020)	The convergence of IoT and AI under AIoT.	Comparing knowledge-enabled AI and data-driven AI, emphasizing their advantages and disadvantages and highlighting the recent progress in AI integration across the IoT architecture with respect to sensing, network, and application layers.

(Continued)

TABLE 3.7 (*Continued*) Review Studies on Artificial Intelligence and the Internet of Things and Their Convergence

Citations	Focus	Main Contributions
Zhang and Tao (2021)	The potential of AIoT to empower IoT systems	Demonstrating the potential of AI, especially DL techniques, in enhancing IoT speed, intelligence, and safety; shedding light on the AIoT architecture concerning cloud computing, fog computing, and edge computing; and highlighting AIoT applications, challenges, and opportunities.
Mastorakis et al. (2020)	A broader perspective on the convergence of AI and IoT	Elucidating AI methods in IoT in regard to research trends, industry needs, and practical implementations, as well as balancing theoretical foundations with real-world applications through case studies and best practices.

wildlife, ocean, and land conservation, highlighting the potential of AI to improve conservation outcomes across different environments. Şeyma (2023) focused on the use of AI for the identification and management of Marine Protected Areas (MPAs), showcasing how AI technologies can streamline the establishment and monitoring of MPAs to protect marine ecosystems.

Bakker's (2022) article "Smart Oceans: Artificial Intelligence and Marine Protected Area Governance" discusses the innovative use of AI-enabled Mobile Marine Protected Areas (MMPAs), which have dynamic boundaries that shift in response to the movements of endangered species. This approach leverages various digital tools, including ML, CV, and environmental sensors, to enable real-time, adaptive management of ocean resources, aimed at enhancing biodiversity conservation. Samaei and Hassanabad (2024) examine the vital role of the interplay between seas, marine industries, and AI in promoting sustainable development. Their study highlights how AI can enhance the efficiency and sustainability of marine operations. By integrating AI technologies, the marine sector can better manage resources, reduce environmental impacts, and support the broader goals of sustainable development.

3.5.2.2 Research on the Relationship between AI and Climate Change
Throughout the second period, a surge in research endeavors reflected heightened scholarly attention on the intersection of climate change with AI and AIoT. The rapid digital transformation catalyzed and expedited by the "outbreak" and impacts of COVID-19 played a pivotal role in bolstering climate action and tackling environmental challenges (Balogun et al. 2020; Falk et al. 2020). Consequently, there was a substantial uptick in the adoption of AI and AIoT technologies in climate change mitigation and adaptation (Chen et al. 2023; Leaf Filho et al. 2022; Nti et al. 2022; Samadi et al. 2022; Shaamala et al. 2024).

In response to the escalating challenges posed by climate change, innovative technologies have emerged as crucial tools in addressing urgent environmental concerns. Here, the focus is on climate change mitigation and adaptation, focusing on the transformative roles played by AI and AIoT in reshaping sustainable urban development. Cities, significant contributors to GHG emissions, are at the forefront of urgent climate action. A comprehensive review conducted by Bibri et al. (2023b) underscores AI's potential to mitigate climate change by integrating knowledge, design strategies, and innovative technologies across urban systems and domains, including transportation, energy, water, and waste management. Ivanova et al. (2023) contribute to this narrative by acknowledging the growing influence of AI across diverse domains. They predict its continuous expansion and emphasize the prevalence of narrow AI in technical fields. Their focus on applying narrow AI to investigate the impact of climate change on transport infrastructure showcases the controllability and capabilities of AI, particularly in studying climate change effects on transportation systems.

Building on the themes explored in the first period, disaster resilience and management have garnered increased attention in theoretical and empirical research on AI applications for climate change. Studies by Buis (2020), Khalilpourazari and Pasandideh (2021), Raza et al. (2020), and Rolnick et al. (2022) have delved into early warning systems, resilience planning, and simulation and prediction. Additionally, numerous studies have focused on disaster management, including research by Abid et al. (2021), Arora et al. (2023), Demertzis et al. (2021), Fan et al. (2021), Kankanamge et al. (2021), Kamal and Bhaumik (2022), Saleem and Mehrotra (2022), Tan et al. (2021), Velev and Zlateva (2023), and Zhang et al. (2023). These studies have employed a variety of AI, ML, and DL techniques, as well as the IoT, for predictive modeling, damage assessment, environmental hazard detection, sentiment analysis, image analysis, object recognition, spatial analysis, mapping, and visualization.

Leal Filho et al.'s (2022) systematic review offers a comprehensive overview of AI, ML, and DL studies conducted between 2020 and 2022, encompassing a wide range of climate change adaptation areas. Their review highlights several key themes and applications, including:

- Water utilization management: Utilizing AI for efficient water resource management, optimizing usage, and ensuring sustainable supply.
- Disaster response and management: Implementing AI-driven solutions for effective disaster response, including early warning systems, real-time monitoring, and resource allocation.
- Wildfire-related applications: Employing AI for evacuation planning, wildfire prediction, prevention strategies, and susceptibility mapping to mitigate the impact of wildfires.
- Flood prediction and protection: Developing AI models to predict flood events, enhance preparedness, and design protective infrastructure.
- Deforestation and forest degradation: Using AI to detect and monitor deforestation activities, assess forest health, and implement conservation strategies.
- Drought tolerance determination: Applying AI to identify drought-tolerant crops and optimize agricultural practices under water-scarce conditions.
- Carbon emissions forecasting: Leveraging AI to predict carbon emissions, helping in the formulation of mitigation strategies, and tracking progress towards emission reduction targets.
- Energy demand prediction: Utilizing AI to forecast energy demand, facilitating better energy management and the integration of renewable energy sources.
- Wind generation prediction: Employing AI techniques to predict wind patterns and optimize wind energy generation.

- Sub-grid processes modeling: Enhancing climate models by incorporating AI-driven sub-grid process simulations to improve accuracy and reliability.
- Adaptation policy: Informing and shaping adaptation policies through AI-driven insights, enabling policymakers to make data-driven decisions for effective climate adaptation.

These themes reflect the diverse and innovative applications of AI, ML, and DL in addressing the multifaceted challenges of climate change adaptation. By synthesizing recent advancements, Leal Filho et al.'s (2022) review underscores the potential of these technologies to enhance resilience and support sustainable development in the face of climate change.

In a recent comprehensive review study by Chen et al. (2023) on state-of-the-art research and applications of AI in mitigating the adverse effects of climate change, the authors focus on several key areas, including energy efficiency, carbon sequestration and storage, weather and renewable energy forecasting, grid management, building design, transportation, precision agriculture, industrial processes, reducing deforestation, and the development of resilient cities. The findings underscore the role of AI-enhanced energy efficiency in significantly reducing climate change impacts, with smart manufacturing potentially lowering energy consumption, waste, and carbon emissions by 30%–50%. Specifically, AI can reduce energy use in buildings by up to 50%. The integration of AI with smart grids can optimize power system efficiency, cutting electricity bills by 10%–20%. Intelligent transportation systems were shown to reduce CO_2 emissions by approximately 60%. Furthermore, AI applications in natural resource management and resilient city design are pivotal in promoting sustainability.

The integration of AI with carbon capture technologies presents a promising pathway for mitigating climate change by enhancing carbon emission reduction efforts. By leveraging advanced AI techniques, such as ML and DL, these technologies can improve the efficiency of capturing and storing CO_2, as well as optimize operational processes, leading to greater overall resource efficiency. Furthermore, AI-driven systems can facilitate real-time monitoring and predictive maintenance, ensuring that carbon capture infrastructures operate at peak efficiency, thereby reducing emissions at critical points in industrial and energy sectors. When applied to smarter eco-cities, AI's role in carbon capture is particularly valuable, as it can be integrated into urban systems to reduce the carbon footprint of cities by managing emissions from transportation, energy consumption, and industrial processes. In this context, AI-powered carbon capture technologies become an essential component in achieving urban sustainability, contributing to smarter eco-city designs that prioritize both environmental sustainability and climate resilience.

Manikandan et al. (2024) highlight how AI can significantly improve the management and operation of carbon capture systems, making them more effective in reducing industrial emissions and enhancing resource efficiency. By analyzing large datasets from carbon capture plants, AI can optimize system performance and identify patterns at a scale beyond current capabilities. AI-powered sensors and monitoring systems can detect operational failures early, enabling timely interventions. Furthermore, AI supports generative design in the discovery of new carbon-absorbing materials, such as metal-organic frameworks and polymers, which are crucial in industrial carbon capture. AI also enhances reservoir simulations and optimizes CO_2 injection systems for storage and enhanced oil recovery. The study seeks to optimize injection processes while minimizing CO2 emissions by applying AI algorithms to real-time data on reservoir geology and production. Finally, the integration of AI with renewable-based carbon capture, supported by AI-driven smart grid systems, can further improve the overall efficiency of carbon capture methods.

Priya et al. (2023) build on this by reviewing the integration of AI with carbon capture techniques, highlighting the importance of AI in addressing carbon emissions. The study discusses the role of ML, DL, and hybrid techniques in optimizing carbon capture processes. It also covers AI tools, frameworks, and mathematical models used in this domain, and explores the landscape of AI-related carbon capture patents. It highlight the role of AI in facilitating generative design and modeling to create new materials that enhance carbon capture efficiency. The review emphasizes the potential of AI to advance carbon capture technologies, aiding in the fight against climate change and supporting the achievement of SDG 13 – climate action.

In a broader context, Al-Sakkari et al. (2024) assess the potential of AI to optimize carbon capture, utilization, and sequestration (CCUS) systems, which are crucial for decarbonizing the energy and industrial sectors. The study reviews existing CCUS technologies and identifies AI-driven approaches, including ML, for modeling, simulation, and optimization of CCUS operations. It highlights advanced AI methods like deep reinforcement learning and generative DL for optimizing materials selection and plant design, aiming to reduce CCUS system costs. The study also advocates for integrating life cycle assessment (LCA) with AI to enhance the overall CCUS value chain.

Delanoë et al. (2023) address the challenges and trade-offs of using AI to reduce CO_2 emissions. They propose a method to assess both the positive and negative impacts of AI models, considering the emissions from AI model training and use. Their findings indicate that, while AI can significantly reduce CO_2 emissions, the balance between positive and negative impacts varies depending on scale. When scaled up, AI's contributions to emission reductions outweigh the associated carbon footprint, complementing the optimistic view presented by the other studies.

Overall, the synthesis of these studies underscores AI's transformative potential in optimizing carbon capture technologies, discovering new materials, and advancing CCUS systems. While the environmental benefits of AI in carbon capture are significant, careful consideration of AI's own carbon footprint is essential for maximizing its overall impact in climate change mitigation.

Several AI applications in climate change align with smart city and smart eco-city initiatives and contribute to achieving the environmentally-focused SDGs, including SDG11, SDG13, SDG 7, and SDG 9. AI technologies enhance our understanding of climate dynamics by processing and analyzing vast amounts of data, offering more advanced computational, analytical, and organizational capabilities and enabling the identification of patterns and trends that might be imperceptible to human analysis (Walsh et al. 2020; Nishant et al. 2020; Shaamala et al. 2024). In particular, AI emerges as a pivotal change agent, enabling carbon neutrality in various urban domains (Stein 2020). These collective efforts underscore the integral role of AI in addressing critical environmental challenges and climate risks and advancing the development of smarter eco-cities.

Moreover, AI tools are revolutionizing agriculture by enabling precision farming, optimizing planting strategies and irrigation systems, predicting crop yields under varying climate conditions, and managing water resources more efficiently. Precision agriculture harnesses cutting-edge technologies, including advanced sensors and predictive analytics, to gather real-time data on key environmental factors, labor costs, and availability, ultimately improving agricultural productivity and decision-making (Raj et al. 2021). By incorporating AI, precision agriculture not only enhances crop yields but also promotes sustainability by reducing the environmental impact of farming practices (Ampatzidis et al., 2020; Reddy et al. 2022; Wei et al. 2020). Furthermore, when combined with genome analysis and gene-editing techniques, AI-driven precision agriculture enables the development of crops tailored to specific land conditions, optimizing plant growth and overall production efficiency (Joseph et al. 2021).

Overall, AI demonstrates immense potential to enhance and accelerate global efforts toward sustainable climate practices. Through advanced data analysis, predictive modeling, and real-time monitoring, AI empowers decision-makers to implement more effective strategies for both climate mitigation and adaptation, driving more informed and impactful actions for a sustainable future.

3.5.2.3 Research on the Role of Generative Artificial Intelligence in Environmental Sustainability and Climate Change

GenAI has rapidly advanced in recent years, transforming various domains, including environmental sustainability and climate change initiatives. In the context of smarter eco-cities, GenAI plays a key role in optimizing urban planning by analyzing vast datasets to predict environmental impacts and resource consumption. By applying models that simulate complex ecosystems, GenAI can contribute to the creation of more sustainable urban environments that balance development with ecological preservation (see Chapter 7). This technology helps smarter eco-cities to mitigate the effects of climate change and adapt to it by designing resilient infrastructures that incorporate nature-based solutions.

GenAI models utilize massive datasets to generate new content—such as text, images, audio, or code—based on learned patterns. These models excel at understanding and replicating complex structures and relationships within data, enabling them to produce original outputs that mirror the patterns present in their training data (e.g., Goodfellow et al. 2020; Kingma & Dhariwal 2018). Unlike traditional AI systems, which operate using predefined rules or predictions based solely on historical data, GenAI models learn from large datasets and generate entirely new outputs by identifying intricate patterns and dependencies. Pretrained Foundation Models (PFMs) (e.g., Bommasani et al. 2022; Jakubik et al. 2023; Janowicz 2023; Zhou et al. 2023) illustrate this versatility. PFMs are trained on vast datasets and are capable of handling a broad range of tasks across multiple domains. Within this category, Large Language Models (LLMs) (e.g., Brown et al. 2020; Goodfellow et al. 2020; Radford et al. 2021; Wolf et al. 2020) are specifically designed to understand and generate human language, making them essential for NLP applications.

GenAI's influence has also significantly expanded into urban domains, where it enables the creation of intelligent systems that can simulate, optimize, and predict the behavior of cities, infrastructure, and environments (see Chapter 7). By leveraging large datasets, GenAI is transforming how urban systems are modeled and managed, driving the development of smarter, more sustainable cities. This is particularly relevant and important in addressing the challenges of environmental sustainability and climate change (see Chapter 7). GenAI models offer promising solutions for overcoming these challenges by enhancing their adaptability to the complexities of environmental systems. These models can analyze vast datasets and simulate intricate environmental processes, enabling more accurate predictions of climate impacts and helping design effective mitigation strategies. Their flexibility allows them to model various scenarios in areas such as ecosystem management, resource optimization, and urban planning, supporting efforts to create sustainable, resilient solutions in response to the evolving challenges of climate change.

GenAI is increasingly being applied in environmental and climate-related contexts, transforming both research and practical applications. Butler and Lupton (2024) provide a comprehensive review that explores how GenAI and LLMs are being utilized in agriculture, sustainability, biodiversity conservation, and climate change. Their work underscores GenAI's transformative potential in driving innovation in ecosystem and environmental management. Building on this, Richards et al. (2024) examine how GenAI can automate and scale nature-based solutions, such as ecosystem services reporting, biodiversity-friendly design, and landscape scenario visualization. While their findings emphasize GenAI's ability to scale these solutions, they also highlight risks such as bias, misinformation, and high energy consumption. To address these challenges, the authors advocate for integrated ethical and social research to ensure the responsible use of GenAI technologies.

In biodiversity conservation, Hirn et al. (2022) apply GenAI models, specifically Generative Adversarial Networks (GANs) and Variational Autoencoders (VAEs), to investigate species coexistence patterns in ecological communities—an area crucial for biodiversity conservation but difficult to study using traditional methods due to the complexity of species interactions. GANs effectively simulated realistic species compositions and preferences for different soil types, while VAEs achieved over

99% accuracy in their predictions. Notably, the study reveals that more complex species interactions tend to reduce the positive effects typically observed in simpler relationships, providing deeper insights into species coexistence. By analyzing these complex interactions, the study demonstrates how GenAI models can surpass traditional methods, particularly in understanding high-order species dynamics. This innovative approach highlights the potential of GenAI to advance ecological research and offer new ways to study biodiversity in complex ecosystems.

In climate resilience, Paramesha et al. (2024) explore how integrating GenAI models, particularly ChatGPT, can enhance resilience across multiple domains, with a special focus on climate resilience. Their study shows that ChatGPT improves communication, supports policy development, and facilitates real-time data analysis—essential components for effective climate preparedness and urban resilience. Moreover, ChatGPT bolsters social and community resilience by fostering communication and providing mental health support during crises.

In the context of disaster preparedness, McCormack and Grierson (2024) focus on using GenAI for immersive, narrative-driven simulations that help users experience the effects of climate disasters. These simulations enhance preparedness by allowing stakeholders to better plan and respond to climate events through realistic, experiential learning tools, highlighting GenAI's value in disaster preparedness.

In relation to urban transportation systems, Tupayachi et al. (2024) examine the use of AI-driven LLMs to generate scientific ontologies for optimizing intermodal freight transportation. Through the ChatGPT API and NLP techniques, the authors develop an integrated workflow for developing scenario-based ontologies. The study produces knowledge graphs that improved data and metadata modeling, streamlined the integration of complex datasets, and facilitated the coupling of multi-domain simulation models. A case study was used to validate the methodology, demonstrating its effectiveness in enhancing decision-making processes within complex urban freight systems.

In conclusion, GenAI has demonstrated immense potential across various domains, including environmental sustainability, climate change adaptation, and urban resilience. It provides powerful tools for optimizing ecosystems, urban planning, and disaster preparedness by leveraging advanced models like LLMs, GANs, and VAEs. However, as the technology advances, addressing challenges such as bias and ethical concerns will be crucial to ensuring its responsible and effective implementation.

3.5.2.4 The Convergence of AI and IoT in Environmental Sustainability and Climate Change

The integration of AI and IoT within the framework of AIoT plays an innovative role in advancing environmental sustainability goals and mitigating climate change. In this context, IoT devices, equipped with sensors and actuators, play a crucial role in collecting real-time data on a wide range of environmental parameters and climate change conditions across urban environments. This extensive network of data acquisition forms the bedrock for comprehensive environmental management and monitoring, providing a wealth of data for analysis and decision-making. IoT generates substantial amounts of data, which necessitate the utilization of AI for analysis and decision-making for diverse practical applications in smart cities, environmental sustainability, and climate change.

In the rapidly evolving landscape of technology, the convergence of AI and IoT has sparked transformative possibilities, giving rise to the burgeoning field of AIoT. As this technological synergy gains prominence, coupled with the focus of this study, it is highly relevant to review key existing literature to understand the multifaceted dimensions and applications of AIoT (Table 3.7).

In addition, Parihar et al. (2023) explore the convergence of AI and IoT into AIoT. The relevance of this convergence lies in its transformative potential, where smart devices are integrated into smarter systems, offering enhanced capabilities in various domains. By exploring characteristics, architectures, and use cases, the authors provide insights into how AIoT can revolutionize existing paradigms, particularly in the context of sustainable operations and resilient infrastructure. This exploration underscores the importance of harnessing AIoT technologies for building intelligent networks in IoT systems.

These review studies collectively contribute valuable insights into the advancements, benefits, opportunities, challenges, and applications of AIoT, laying the groundwork for further research, development, and implementation in this dynamic and promising technological domain. Through a comparative analysis of these studies, a nuanced understanding of the current state of knowledge in the realm of AIoT is provided, supported by other studies on the application of AIoT in environmental sustainability and climate change (e.g., Bibri et al. 2023a, b; Bibri and Hunag 2024; Parihar et al. 2023; Rane et al. 2024; Tomazzoli et al. 2020; Seng et al. 2022; Szpilko et al. 2023; Zhang and Tao 2021).

Facilitating the seamless flow of data, IoT enables efficient data transmission to centralized platforms, such as cloud computing. This ensures that data collected from distributed sources is readily available for processing and analysis. Connectivity protocols employed by IoT guarantee not only the efficiency of data transfer but also the security of sensitive environmental information.

The voluminous data generated by IoT is effectively handled through sophisticated AI models, including ML and DL (Zhang and Tao 2021). These models process the data to identify patterns, anomalies, and trends. The result is a wealth of insights into environmental conditions, empowering decision-makers with a comprehensive understanding of the dynamics at play. One of the key strengths of AIoT as an integrated system is its ability to employ predictive analytics. AI-driven predictive models utilize both historical and real-time data collected by IoT to forecast environmental changes. This capability facilitates proactive decision-making and timely responses to emerging environmental challenges.

IoT's synergy with AI goes beyond analysis to automation and control. This integration allows for the autonomous control of various systems and processes. For example, in the realm of energy management, smart grids can dynamically adjust energy distribution based on real-time demand and the availability of renewable energy sources (El Himer et al. 2020), thereby promoting energy efficiency and conservation. Furthermore, resource optimization is a notable outcome of the collaboration between IoT and AI. The sophisticated ML and DL techniques and algorithms optimize resource allocation and utilization, minimizing waste and maximizing efficiency in critical areas such as energy consumption, water usage, and transportation. Additionally, the remote monitoring capabilities of IoT-enabled devices reduce the need for physical interventions, minimizing the human impact on sensitive ecosystems. This aspect aligns with the broader goal of fostering sustainable practices and minimizing environmental footprints. Overall, IoT's technical capabilities within the AIoT framework offer a holistic approach to environmental sustainability. From data collection to analysis and automation, this integrated system enables informed decision-making and environmentally conscious practices across diverse domains, contributing to a more sustainable and resilient future.

In the domain of climate change mitigation and adaptation, IoT as a core element of AIoT unfolds a multifaceted landscape of technical and practical contributions (Table 3.8) based on several studies (e.g., Anbarasan et al. 2020; Bibri et al. 2023b; Chen et al. 2023; Dheeraj et al. 2020; El Himer

TABLE 3.8 Key Technical and Practical Aspects of Artificial Intelligence of Things in Climate Change Mitigation and Adaptation

Technical and Practical Aspects	Description
Data sensing and collection	• Continuous monitoring of relevant environmental parameters by IoT devices with embedded sensors. • Utilization of collected data to gain comprehensive insights into climate change patterns and trends.
Network connectivity	• Ensuring seamless data transmission and communication between IoT devices and central repositories. • Implementation of robust communication protocols, including 5G/6G, to facilitate rapid and efficient data sharing.
Data analysis	• Processing of extensive datasets from IoT sensors by AI algorithms to identify patterns and trends related to climate change. • Application of ML/DL models for predictive analytics, anticipating changes and evaluating potential impacts.
Remote sensing technology	• Deployment of IoT-connected satellites and drones with AI-enabled remote sensing technology for critical environmental data.
Real-time decision support	• Provision of real-time climate data and actionable insights to decision-makers by IoT systems. • Facilitation of adaptive strategies and informed policy development based on real-time information.
Weather monitoring and early warning systems	• IoT-connected weather and climate monitoring stations, in conjunction with AI-driven forecasting models. • Development of early warning systems to enhance preparedness for extreme weather events and natural disasters.
Real-time energy consumption tracking	• Energy consumption monitoring by IoT devices in real time. • Analysis of energy data by AI algorithms to optimize energy use, identify conservation opportunities, and promote renewables.

(Continued)

TABLE 3.8 (*Continued*)   Key Technical and Practical Aspects of Artificial Intelligence of Things in Climate Change Mitigation and Adaptation

Technical and Practical Aspects	Description
Smart infrastructure and urban planning	• Integration of IoT-enabled infrastructure, such as smart buildings, to support resilient urban planning. • Monitoring of structural integrity and climate-related risks through sensors, enabling adaptive responses.
Ecosystem monitoring	• Tracking changes in flora and fauna behavior and health by IoT sensors in ecosystems. • Utilization of AI algorithms to assess the impact of climate change on biodiversity and guide conservation efforts.

et al. 2022; Leal Filho et al. 2022; Lee and Chien 2020; Popescu et al. 2024; Rane et al. 2024; Swarna and Bhaumik 2022; Tomazzoli et al. 2020; Samadi 2022). IoT's synergy with AI plays a key role in addressing the intricate challenges posed by climate change.

The technical and practical capabilities of IoT within the AIoT framework emerge as indispensable tools in climate change mitigation and adaptation endeavors. Their role in fostering data-driven decision-making, optimizing resources, and building resilience proves crucial in navigating the complex landscape of climate-related challenges.

As two examples for illustrative purposes, AI and IoT technologies have shown significant potential in both climate change mitigation and disaster management.

Dheeraj et al. (2020) explore the role of AI and IoT technologies in mitigating climate change by creating environmentally friendly and high-performing systems. The authors demonstrate the potential of AI and IoT-based technologies in managing the impacts of climate change by optimizing resource utilization and reducing human interference. By integrating IoT and AI, data collected from field sensors is analyzed to monitor various environmental factors such as soil moisture, weather conditions, fertilization levels, and irrigation systems. The results indicate that this integration helps increase crop production, leading to higher incomes for farmers and demonstrating the potential of AI and IoT to address climate change effectively.

Popescu et al. (2024) explored the integration of AI, sensors, and IoT in monitoring environmental pollution, including air pollutants, water contaminants, and soil toxins. AI-driven sensor systems present promising solutions for detecting and responding to environmental risks, addressing growing concerns about the impact of hazardous substances on ecosystems and human health. The study emphasizes the significant potential of ML methods in environmental science, highlighting the complexities of predicting and tracking pollution changes due to the environment's dynamic nature and the challenges of data sharing. It stresses the importance of developing advanced monitoring systems that can efficiently detect, analyze, and mitigate environmental hazards, ultimately contributing to improved human well-being and ecosystem protection.

Swarna and Bhaumik (2022) explore the integration of AI IoT devices to enhance the prevention, response, and recovery phases of disaster management. The study focuses on developing a platform that combines multiple AI components, IoT devices, and data sources into a unified system to improve disaster management practices. The study resulted in the creation of an integrative AI platform designed to handle real-time data collection and analysis through IoT devices. This platform utilizes a microservices architecture and an easy-to-use dataflow graph-based programming model, which enhances reusability and allows for the flexible integration of different AI components and data sources. Two use cases in disaster prevention were highlighted, demonstrating the platform's capability to implement predictive monitoring and efficient response strategies. The study emphasized the platform's potential to streamline the development process, reduce engineering efforts, and increase the practical applicability of AI in diverse disaster management scenarios.

Overall, the integration of AI and IoT technologies presents transformative opportunities in both mitigating climate change impacts and enhancing disaster management. These technologies not only optimize resource utilization and agricultural productivity but also improve disaster preparedness and response through real-time data analysis and predictive monitoring. The ongoing advancements in AIoT continue to promise innovative solutions, reinforcing the need for further research and practical applications to address global climate and disaster challenges effectively.

3.5.2.5 Research on the Relationship between AI and AIoT and Smart City Systems and Domains

In contrast to the first period, there has been a significant proliferation and widespread adoption of AI and AIoT technologies, particularly drawing increased research attention due to their applications in smart cities. This trend indicates a notable shift, signifying the integration of AI and AIoT solutions originally developed for distinct areas of environmental sustainability and climate change. The evolution suggests a convergence wherein smart cities are increasingly incorporating and leveraging the advancements in AI and AIoT, which were previously conceptualized for addressing diverse challenges in the domains of environmental sustainability and climate change.

AI and AIoT applications play a pivotal role in various domains of smart cities, particularly in relation to environmental sustainability. Puri et al. (2020) explore the transformative impact of integrating IoT and AI technologies into urban transportation, specifically focusing on smart bicycle sharing systems in major cities. Equipping bicycles with advanced monitoring devices, including gas sensors, wireless modules, GPS, cameras, and fall detection sensors, enhances efficiency and safety. The authors introduced an analytical framework using ANN and supported SVM for data analysis, finding that ANN outperforms SVM in terms of accuracy. The proposed system demonstrates potential for optimizing urban transportation, offering precise data collection and transmission. This work contributes useful insights to advancing smart transportation, emphasizing the role of intelligent bicycle sharing schemes in promoting greener and more efficient city mobility in both developed and developing areas.

Herath and Mittal (2022) explore the utilization of AI in smart cities across various sectors, including environment, waste management, agriculture, mobility, transportation, and risk management. The aim is to identify the impact of AI in different domains and how it contributes to making cities smarter and more efficient. Their analysis reveals that mobility and energy sectors have the most significant influence on AI adoption in smart cities. AI algorithms such as ANN, RNN/LSTM, CNN/R-CNN, DNN, and SVM/LS-SVM have been found to have a high impact on various aspects of smart cities.

Yigitcanlar et al. (2020) conducted a systematic literature review, identifying the key contributions of AI and ML to environmental sustainability. These contributions include optimizing energy production and consumption, climate change risks forecasting, natural environment changes monitoring, and transport systems operation. AI applications extend to the management of city transport systems, encompassing shared autonomous mobility-on-demand, autonomous cities, and autonomous vehicles (Cugurullo 2020; Acheampong and Cugurullo 2019; Cugurullo et al. 2021; Hawkins and Nurul 2019). Zhang and Tao (2021) synthesized studies highlighting AIoT applications using DL in smart transportation, addressing areas such as traffic management, logistics, and in-car driver behavior monitoring, as well as in smart grids.

Han et al. (2020) proposed a DL-based framework for load forecasting in energy management, connecting the IoT network to smart grids for effective energy demand and supply activities. El Himer et al. (2022) explore the role of AIoT in creating novel opportunities in distributed energy resources, with a specific emphasis on renewable energy sources like solar and wind. Puri et al. (2020) introduce an AIoT system that harnesses energy through various sensors, capturing stress from human body weight, heat generated by movement, and sunlight. Their study showcases the system's precision in predicting power output from renewable resources, demonstrating its potential for accurate and efficient energy generation forecasting. Sleem and Elhenawy (2023) emphasize AIoT's contribution to smart building development, reducing energy consumption, improving occupant comfort and productivity, and enhancing safety and security.

The study by Seng et al. (2022) investigates the applications of AIoT in various sectors, including energy and smart grids, industry and smart buildings, vehicles and smart transportation, as well as robotics and CV. Their comprehensive analysis reveals the extensive use of AIoT across these domains, showcasing its potential in shaping the future of smart cities. The results of their investigation contribute to a deeper understanding of how AIoT technologies are being implemented to improve efficiency and functionality in urban environments. Additionally, the study highlights the importance of AIoT in addressing key challenges faced by smart cities, such as energy management, infrastructure optimization, and transportation systems. Overall, their research provides valuable insights into the role of AIoT in advancing smart city development and lays the groundwork for further exploration in this field.

Alahi et al. (2023) investigate how the integration of AI and IoT can enhance sustainability, productivity, and the quality of life in smart cities. The authors outline the characteristics of smart cities and discuss IoT architecture. They explore different AI algorithms applicable in smart city scenarios and discuss the potential contributions of integrating IoT and AI, particularly with 5G networks, in advancing urban environments. Their study contributes by emphasizing the opportunities created by combining IoT and AI in smart cities. They highlight the potential to significantly improve urban living conditions while promoting sustainability and productivity.

The comprehensive review study conducted by Bibri et al. (2023b) demonstrates a significant evolution in the integration of AI and AIoT technologies in environmental sustainability in the context of smart cities. This integration aimed to address key environmental challenges, such as climate change, air pollution, and resource management. The deployment of AI and AIoT focused on optimizing various urban systems. These technologies played a crucial role in the accelerated digital transformation of smart cities, offering solutions for resource efficiency, emissions reduction, and enhanced decision-making. The amalgamation of smart and sustainable city concepts gained momentum, giving rise to environmentally sustainable smart cities. Noteworthy applications of AI included energy conservation, sustainable transportation, water resource management, climate change scenario analysis, and waste reduction. In waste management, AI and AIoT have facilitated the development of advanced waste collection systems.

Fang et al. (2023) provide a comprehensive review, covering waste-to-energy, smart bins, waste-sorting robots, waste generation models, and more. The benefits include reduced transportation distance, improved waste pyrolysis, carbon emission estimation, and energy conversion. Nasir and Aziz Al-Talib (2023) discussed challenges in waste classification, highlighting the potential of AI and image processing techniques. Mounaded et al. (2023) focused on AI techniques in Municipal Solid Waste (MSW) management, using ANN for various related problems.

Addressing the limitations and intricacies associated with conventional methods, particularly in assessing the status and waste levels in bins, has propelled the development and adoption of advanced techniques. These encompass Particle Swarm Optimization (PSO) techniques (e.g., Qiao et al. 2020), Artificial Neural Networks (ANN) (e.g., Vu et al. 2019), and the Backtracking Search Algorithm (BSA) (e.g., Akhtar et al. 2017) for optimizing waste collection processes. Waste vehicle routing has also seen the application of Genetic Algorithms (GA) and nearest neighborhood search algorithms (e.g., Mi et al. 2017). Despite these advancements, precision remains a challenge, and there is a notable need for techniques that address cost and emission concerns while considering factors such as bin capacity, waste weight, collection frequency, vehicle capacity and maintenance, and trip rate (Hannan et al. 2020). Novel approaches are essential to overcome existing drawbacks and enhance the efficiency of waste management systems.

While several studies have explored the concept of Smart Urban Metabolism (SUM), which incorporates IoT and Big Data technologies to address the challenges faced by growing smart cities and smart eco-cities (Bibri and Krogstie 2020b; D'Amico et al. 2020, 2021; Shahrokni et al. 2015), new approaches are increasingly integrating AI and ML models and techniques into SUM for better efficiency. This is to address the increasing complexity of understanding, analyzing, and monitoring the flows of energy, materials, and resources in urban areas. Ghosh and Sengupta (2023) emphasize the role of AI, ML, and Big Data technologies in advancing SUM. By employing the self-learning approach to ML models and predictive analytics, these technologies facilitate the effective handling and analysis of multidimensional datasets, providing novel insights. This work proposes that SUM, empowered by these advanced technologies, can serve as an effective approach for urban planners to identify and address complex issues in the flow of energy and materials in urban environments.

Bibri et al. (2024a) focus on integrating AI and AIoT into SUM to propel the development of smarter eco-cities. This fusion empowers SUM to enhance sustainability, resilience, and governance effectiveness by facilitating data-driven decision-making and targeted interventions. According to the authors, the pivotal contributions of AI and AIoT include the development of predictive models and optimization algorithms, transforming urban metabolic processes for more resource-efficient and sustainable smart eco-city development. This work highlights the key role of these advanced technologies in realizing real-time city functions and informed decision-making, emphasizing their significance in the successful implementation of environmental policies for smarter eco-cities.

Peponi et al. (2022) propose a novel, evidence-based methodology to understand and manage these complexities, aiming to bolster the resilience of urban processes under the concept of smart and regenerative urban metabolism. By integrating Life Cycle Thinking and ML, they assess the metabolic processes of the city's urban core using multidimensional indicators, including urban ecosystem service dynamics. They developed and trained a multilayer perceptron (MLP) network to identify key metabolic drivers and predict changes by 2025. The model's performance was validated through prediction error standard deviations and training graphs. Results indicate significant drivers of urban metabolic changes include employment and unemployment rates (17%), energy systems (10%), and various factors such as waste management, demography, cultural assets, and air pollution (7%), among others. This research framework serves as a knowledge-based tool to support policies for sustainable and resilient urban development.

The evolution and pervasive integration of AI and AIoT technologies in the context of smart cities have significantly influenced the trajectory of smart eco-cities. The incorporation of these advanced technologies has enabled smart eco-cities to enhance their environmental sustainability initiatives and further optimize resource management and reduce environmental impacts. The application of AI and AIoT solutions in smart eco-cities has led to more efficient energy consumption, improved waste management systems, and enhanced monitoring of ecological parameters. This paradigm shift has facilitated the emergence of innovative models of urban development, fostering a holistic approach that integrates technology, sustainability, and environmental consciousness. Smart eco-cities are now positioned to leverage the data-driven insights provided by AI and AIoT to create smarter, more resilient urban environments that prioritize eco-friendly practices and contribute to a sustainable and harmonious coexistence with the environment.

3.5.2.6 The Environmental Impacts of AI in Autonomous and Shared Autonomous Electric Vehicles: Opportunities and Risks

AI is at the forefront of innovation in the transportation sector, particularly in the development of autonomous vehicles (AVs), shared autonomous electric vehicles (SAEVs), and smart city infrastructure. These technologies promise to revolutionize mobility by reducing accidents, improving fuel efficiency, and alleviating traffic congestion, all while potentially delivering significant environmental benefits. In smart cities, AI can further optimize traffic management, reduce urban emissions, and create more efficient public transportation systems, contributing to overall sustainability. However, alongside these advantages, there are also emerging concerns about the broader environmental impacts of AVs, SAEVs, and AI-integrated smart cities. This synthesis of studies explores both the opportunities and risks associated with AI-enabled transportation systems, focusing on their contributions to reducing emissions, improving energy efficiency, and mitigating climate change, while also addressing the potential unintended consequences such as increased noise, land use changes, and operational challenges.

Tyagi and Aswathy (2021) and Othman (2022) highlight AI's central role in enabling AVs to reduce accidents, enhance mobility, and improve fuel efficiency, potentially lowering emissions and optimizing traffic patterns. Othman (2022) also provides a broad overview of AVs' benefits, including reduced energy consumption and emissions, while linking AVs with the concept of vehicle sharing to amplify these effects.

The intersection of AVs with vehicle sharing, particularly SAEVs, is a recurring theme. Ahmed et al. (2023) and Jones and Leibowicz (2019) discuss how SAEVs can alleviate urban congestion, reduce greenhouse gas (GHG) emissions, and make efficient use of urban spaces. They emphasize that SAEVs, especially when integrated with renewable energy sources and advanced charging techniques, offer substantial environmental and economic advantages over privately owned vehicles (POVs). Jones and Leibowicz (2019) also suggest that widespread adoption of SAEVs could have a larger impact on reducing vehicle travel emissions than carbon taxation in the short-to-medium term. Liao et al. (2021) further elaborate on the economic and environmental benefits of Vehicle-to-Grid (V2G) services, demonstrating how V2G can reduce operating costs and emissions for SAEVs by leveraging grid integration.

However, some studies caution against optimistic projections. Saleh et al. (2022) note potential risks related to SAEV adoption, such as increased mileage and fleet turnover, which may offset environmental benefits. Similarly, Ahmed et al. (2023) find that SAEVs tend to have a higher environmental footprint than privately owned electric vehicles (POEVs), mainly due to increased power consumption and deadheading (empty trips). Both studies propose that the implementation of circular economy practices and optimized charging can mitigate these impacts. In particular, Ahmed et al. (2023) show that such practices can significantly reduce the Global Warming Potential (GWP), water footprint, and energy demand of SAEVs, with even more substantial reductions if clean energy sources are utilized.

The broader environmental implications of Connected and Autonomous Vehicles (CAVs) and electric vehicles are explored by Kopelias et al. (2020) and Silva et al. (2022). These studies focus on various factors influencing the environmental impacts of AVs, including vehicle and road network design, user behaviors, and externalities such as noise and land use changes. Both studies underscore that while AVs hold potential for reducing emissions and improving traffic flow, they may also introduce unintended consequences, such as increased noise pollution and adverse changes in land use that impact soil and water quality. Silva et al. (2022) also emphasize the need for more research on the effects of AVs on noise and light pollution. Lastly, Othman (2022) discusses legal and regulatory challenges, noting that existing frameworks are not yet fully equipped to address the rapid advancements in AV technology. The study calls for more detailed research to fully understand and harness the potential environmental benefits of AVs.

In synthesizing these studies, it becomes evident that while AI-enabled AVs and SAEVs offer environmental advantages—particularly through innovations like V2G services, optimized charging, and circular economy practices—they also present risks that need to be carefully managed. On the positive side, Liao et al. (2021) and Ahmed et al. (2023) highlight the importance of advanced practices in reducing emissions and costs. However, Saleh et al. (2022) and Silva et al. (2022) raise concerns about the unintended environmental impacts, such as increased mileage, noise pollution, and changes in land use, suggesting that a balanced and informed approach is crucial for maximizing the sustainability potential of AVs and SAEVs.

3.6 A NOVEL CONCEPTUAL FRAMEWORK FOR SMARTER ECO-CITY DEVELOPMENT

The framework illustrated in Figure 3.3 is derived based on the insights distilled from the reviewed studies, outlining the intricate relationship among smart cities, smarter eco-cities, and the applied solutions of AI and AIoT technologies, environmental strategies, environmental challenges, and environmental goals.

Smarter eco-cities represent a new approach to urban development, leveraging advanced technologies to create more sustainable, resilient, and environmentally conscious urban environments. At the core of the framework are emerging applied AI and AIoT solutions, which enable these cities to collect, analyze, and act upon data in real-time to optimize and enhance various aspects of urban life, particularly in relation to urban management and planning.

The relationship between smart cities; smarter eco-cities; emerging applied AI and AIoT solutions; and environmental strategies, challenges, and goals forms a complex yet interconnected framework crucial for addressing the challenges of ecological degradation, resource depletion, and climate change. This framework encompasses environmental sustainability goals, such as reducing carbon emissions and improving resource

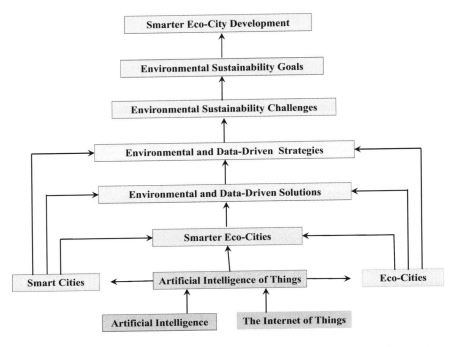

FIGURE 3.3 A novel framework for smarter eco-city development and its underlying technologies, solutions, strategies, challenges, and goals.

efficiency, which are essential for mitigating these challenges. Applied AI and AIoT solutions are central to achieving these goals in smarter eco-cities. These cities can monitor environmental indicators in real-time, predict and mitigate climate change impacts, and optimize resource management strategies by leveraging AI and AIoT. These technologies not only improve environmental outcomes but also contribute to the overall resilience of these cities in the face of climate change.

Moreover, the integration of AI and AIoT solutions enables smarter eco-cities to support climate change migration strategies. Through data-driven decision-making, these cities can identify vulnerable areas, develop adaptive measures, and enhance disaster preparedness. For example, AI algorithms can analyze historical weather data to predict and prepare for extreme weather events, while IoT sensors can monitor air quality to protect public health during environmental crises. Additionally, urban planning and infrastructure design can incorporate climate-resilient features to accommodate migration patterns and ensure the well-being of residents.

In smarter eco-cities, the relationship between environmental sustainability strategies, challenges, and goals is synergistic. AI and AIoT technologies enable the implementation of innovative sustainability strategies by providing the tools needed for efficient and effective management of urban environments, such as data-driven decision-making, resource optimization, predictive maintenance, and enhanced automation and efficiency. These technologies address significant environmental challenges. Ultimately, the goals of environmental sustainability in smarter eco-cities aim to create a balanced and sustainable urban ecosystem that leverages technology to enhance the quality of life, ensure resource efficiency, and mitigate the impacts of climate change. Moving forward, it is essential for policymakers, urban planners, and technologists to collaborate and innovate within this framework to create cities that are not only smarter but also environmentally sustainable and inclusive.

3.7 DISCUSSION

This chapter offers a comprehensive review of the current landscape of smarter eco-cities, with a specific focus on the transformative role of emerging applied AI and AIoT solutions in advancing environmental sustainability goals and mitigating climate change. The discussion will primarily focus on these aspects: summarizing the findings of the review, interpreting the results in light of the research questions, the role of AI and AIoT technologies in smart cities in shaping the dynamics of smarter eco-cities, as well as challenges, risks, and related mitigation strategies or remedies.

3.7.1 Summary and Interpretation of Results/Findings

The literature review reveals that emerging AI and AIoT technologies are playing a significant role in the development of smarter eco-cities. These technologies are being utilized across various sectors, including energy management, transportation, waste management, and environmental monitoring, to promote sustainability and mitigate the effects of climate change. Additionally, the review highlights the potential benefits and opportunities provided by AI and AIoT in improving urban infrastructure, enhancing resource efficiency, and reducing carbon emissions.

In response to the research questions posed in this chapter, the findings underscore the transformative potential of AI and AIoT technologies in reshaping sustainable urban development practices and addressing environmental challenges in smarter eco-cities. The review elucidates how these technologies contribute to enhancing the efficiency and effectiveness of urban systems, optimizing resource allocation, and enabling data-driven decision-making processes. However, it also acknowledges the challenges and complexities associated with the implementation of AI and AIoT solutions, including technical, ethical, and socioeconomic considerations. These include issues such as data privacy, algorithmic biases, and the potential exacerbation of socioeconomic inequalities. Moreover, the reliance on technological solutions may overlook the importance of human-centered approaches and community engagement in sustainable urban development.

3.7.2 Transformative Influence: Data-Driven Technologies of Smart Cities Shaping the Smart Eco-City Landscape and Dynamics

AI, IoT, and Big Data technologies are reshaping the landscape of smart cities and influencing the evolution of smart eco-cities towards greater environmental performance and efficiency. Since the mid-2010s, these data-driven technologies have progressively impacted eco-cities, integrating with their core domains to foster a more intelligent and sustainable approach under what has been known as smart eco-cities. As smart solutions advance and intertwine with sustainable technologies, they offer innovative strategies to address complex challenges in the realm of smart eco-cities. The continuous integration of AI, IoT, and Big Data with environmentally sustainable development practices is making smart eco-cities even smarter. In particular, the essence of AIoT lies in leveraging the synergies between smart city technologies and eco-city strategies, optimizing processes and operations, and enhancing functions and practices. This integration creates numerous opportunities to realize environmental sustainability goals. The ongoing progress in this advanced technology promises to provide increasingly sophisticated approaches, further augmenting the capabilities of existing smart eco-cities in their pursuit of environmental sustainability transitions in the defining context of smarter eco-cities.

Within the framework of AIoT, the integration of AI, ML, and DL into IoT systems and applications is influencing the landscape of smart eco-cities, offering innovative solutions for sustainable urban development. AI plays a crucial role in processing and interpreting vast datasets generated by IoT devices in smart eco-cities. AI models and algorithms, especially ML and DL, enable informed decision-making, contributing to predictive analytics for anticipating environmental changes and optimizing city operations based on both real-time and historical data patterns.

In the realm of environmental monitoring and management, the deployment of IoT devices with sensors captures real-time environmental data. AI processes these data for applications such as monitoring air quality, managing waste, and controlling pollution levels. ML models, capable of adapting and optimizing processes, further enhance these environmental management efforts, providing flexibility in response to changing conditions. DL aids in image recognition for monitoring ecosystems, identifying environmental hazards, and managing green spaces effectively.

AI contributes significantly to energy efficiency in smart eco-cities. Optimization of energy consumption in buildings, street lighting, and transportation systems is achieved through these algorithms. Additionally, ML models play a vital role in resource allocation, learning and adapting to patterns in resource usage for water management, waste reduction, and sustainable urban planning.

In the domain of urban transportation and mobility, AI-driven traffic management systems predict traffic patterns, optimize signal timings, and support intelligent transportation systems, resulting in reduced congestion, lower emissions, and improved overall transit efficiency. ML models analyze data from various sources to optimize public transportation routes, schedules, and capacity, contributing to a more sustainable and efficient public transit system. In the planning of smart infrastructure, DL is employed for structural health monitoring of buildings and infrastructure. This application aids in predicting maintenance needs, reducing the risk of failures, and ensuring the longevity of urban infrastructure. Overall, the symbiotic integration of AI, ML, and DL in smart eco-cities is transformative. These technologies collectively strive towards creating cities that are not only more resilient and efficient but also responsive to the needs of the environment.

However, the dynamics of smarter eco-cities must involve scrutiny of AI and AIoT technologies as pivotal elements instigating transformations across various domains. Investments in large-scale AI and AIoT digital ecosystems are poised to yield positive effects on smarter eco-cities, potentially fostering feedback mechanisms that, in turn, drive increased adoption of these ecosystems and stimulate additional future investments. In essence, AI and AIoT technologies are anticipated to thrive by offering innovative applications that address the pressing challenges of environmental sustainability and climate change. Consequently, a positive feedback loop may emerge wherein the successful implementation of these applications increases the likelihood of further adoption through network effects, learning, adaptation, and coordination (Bibri et al. 2023b). While it is expected that the relationship between outcomes, investments, and implementations will propel the transition of smart eco-cities towards environmental sustainability, asserting a strong causal link resulting from these interconnections requires precision and accuracy. This is particularly crucial given the intricate interplay of various internal and external factors. Therefore, comprehending the dynamics of smarter eco-cities remains a formidable and uncertain challenge, a complexity that will be further explored in the ensuing discussion.

3.7.3 Environmental Challenges of AI and AIoT Technologies

In the rapidly evolving landscape of urban development, the use and integration of AI and IoT technologies stands as a transformative force, weaving its influence into the fabric of smarter eco-cities. As these technologies seamlessly integrate into the various domains of these cities, it becomes imperative to not only celebrate their potential for groundbreaking innovations but also to critically scrutinize the environmental costs and risks they bring. This pertains to the multifaceted dimensions of the environmental implications inherent in the pervasive use and adoption of AI and AIoT technologies in the context of smarter eco-cities, unraveling the complexities surrounding their adoption in the pursuit of sustainability and heightened efficiency. To facilitate a nuanced exploration, Table 3.9 delineates the specific risks and costs associated with these technologies, juxtaposed with strategic remedies and solutions aimed at fostering a harmonious coexistence between technological advancement and environmental well-being (Almalki et al. 2021; Bibri et al. 2023b; Brevini 2020; Dauvergne 2020; Lannelongue et al. 2020; Kaplan and Haenlein 2020; Raman et al. 2024; Schwartz et al. 2020; Wang and Liao 2020).

In conclusion, in the dynamic context of smarter eco-cities, AI and AIoT technologies offer unparalleled opportunities and prospects for innovation and operational efficiency. However, it is crucial to acknowledge that their adoption carries inherent environmental costs and risks. From the heightened energy consumption associated with sophisticated algorithms to the challenges of managing e-waste and the potential for resource depletion, a scrutiny of these implications becomes indispensable. Striking a balance between technological advancement and sustainable practices is imperative to mitigate these risks and ensure a harmonious coexistence between technological progress and environmental preservation. Navigating this intricate landscape in smarter eco-cities demands the utmost commitment to the responsible and ethical deployment of AI and AIoT technologies,

TABLE 3.9 Environmental Risks of Artificial Intelligence and Artificial Intelligence of Things Technologies and Potential Remedies

Risks/Costs	Description	Remedies/Solutions
Energy consumption	The increased demand for computing power in AI models and the constant operation of IoT devices contribute to elevated energy consumption. The energy-intensive nature of training sophisticated AI algorithms and maintaining vast IoT networks can strain existing energy resources.	Optimize algorithms and models for energy efficiency. Utilize renewable energy sources for AI data centers and IoT device operations. Implement energy-efficient hardware and design principles.
E-waste generation	The rapid evolution of AI and IoT devices results in shorter product life cycles and frequent upgrades, leading to increased electronic waste generation. Disposal and management of e-waste pose environmental challenges, as it often contains hazardous materials that require careful handling and recycling.	Design products with longevity and upgradeability in mind. Promote recycling programs for AI and IoT devices. Encourage manufacturers to use environmentally friendly materials and facilitate responsible e-waste disposal.
Resource depletion	The production of AI hardware components and IoT devices relies on the extraction of rare minerals and metals. This can contribute to resource depletion and environmental degradation, particularly in regions where mining activities are prevalent. The extraction process often has ecological consequences, impacting biodiversity and ecosystems.	Promote sustainable and responsible mining practices. Encourage the use of recycled materials in manufacturing. Invest in research for alternative materials with less environmental impact.
Carbon emissions	The manufacturing processes of AI hardware and IoT devices, coupled with the energy-intensive data centers supporting AI operations, result in significant carbon emissions. The carbon footprint of AI and IoT technologies is a concern, considering the global push towards reducing greenhouse gas emissions and combating climate change.	Transition to renewable energy sources for manufacturing processes and data centers. Implement energy-efficient design standards for AI hardware. Offset carbon emissions through reforestation and other carbon sequestration initiatives.

marking a pivotal stride toward a future that is not only technologically advanced but inherently sustainable.

3.7.4 Ethical, Societal, and Regulatory Challenges of AI and AIoT Technologies

While AI and AIoT technologies usher in a new era of possibilities, it is still important to navigate the ethical, social, and regulatory challenges that accompany their increased use and widespread deployment. These challenges, as illustrated in Table 3.10, involve a spectrum of open issues and considerations (Ahmad et al. 2020; Bibri et al. 2023a, b, 2024a, b; Floridi et al. 2020; Greenfield 2018; Hoffmann 2019; Kaplan and Haenlein 2020; Kassens-Noor and Hintze 2020; Koffka 2023; Larsson and Heintz 2020; Paracha et al. 2024; Petit 2018; Raman et al. 2024; Vinuesa et al. 2020; Wazid et al. 2022; Yigitcanlar et al. 2020). Navigating these challenges is essential for ensuring the responsible and

TABLE 3.10 Ethical, Social, and Regulatory Challenges of Artificial Intelligence and Artificial Intelligence of Things Technologies

Challenges	Description
Privacy concerns	The extensive data collection in AI and AIoT systems raises concerns about individual privacy in smarter eco-cities. Continuous monitoring and analysis of personal and environmental data may lead to unauthorized access, surveillance, and the potential misuse of sensitive information.
Security issues	The integration of AI and IoT technologies introduces security challenges, encompassing the risk of cyber threats, data breaches, and the compromise of critical infrastructure in smarter eco-cities. Safeguarding against these security issues is imperative to maintain the integrity and functionality of interconnected systems.
Bias and fairness	AI algorithms may inadvertently or intentionally perpetuate biases, impacting decision-making processes in smarter eco-cities. These biases can manifest in various domains, including resource allocation, urban planning, and public services, potentially exacerbating existing societal inequalities.
Job displacement and inequality	The automation facilitated by AI and AIoT technologies has the potential to displace certain jobs, raising concerns about unemployment and contributing to social inequality in smarter eco-cities. Efforts are needed to address the societal impact of technological advancements and ensure inclusive economic growth.

(*Continued*)

TABLE 3.10 (*Continued*) Ethical, Social, and Regulatory Challenges of Artificial Intelligence and Artificial Intelligence of Things Technologies

Challenges	Description
Lack of transparency and accountability	The complex nature of AI algorithms and IoT systems may lead to a lack of transparency, making it challenging to understand and interpret the decision-making processes. This opacity can hinder accountability, raising questions about the responsibility and liability of AI and AIoT technologies in smarter eco-cities.
Access and inclusivity	The deployment of advanced technologies may create a digital divide, limiting access to benefits in smarter eco-cities. Ensuring inclusivity in the adoption of AI and AIoT technologies is crucial to prevent the exacerbation of social disparities and to guarantee that the benefits of technological advancements are accessible to all members of the community.
Regulatory frameworks	The rapidly evolving landscape of AI and AIoT technologies introduces regulatory challenges, as existing frameworks may struggle to keep pace. Establishing robust regulations that balance innovation with ethical considerations is essential for creating a cohesive regulatory environment in smarter eco-cities.
AI governance	The governance of AI and AIoT technologies poses challenges related to decision-making processes, ethical guidelines, and establishing responsible practices. Establishing effective AI governance frameworks is crucial to ensuring responsible and ethical deployment of AI and AIoT technologies in the realm of smarter eco-cities.

ethical use of AI and AIoT technologies in the complex landscape of smarter eco-cities.

While technological advancements, exemplified by AI and AIoT systems, hold the promise of bolstering the efficiency and effectiveness of urban processes and practices, it is essential to acknowledge that certain fundamental functions should not be delegated to these systems. The decision-making processes guided by civic and public values necessitate the nuanced judgment and ethical considerations inherent to human discernment. AI and AIoT systems can undoubtedly serve as invaluable tools, yet their role should be one of support and enhancement, complementing the endeavors of individuals and institutions rather than supplanting or overshadowing the significance of these foundational values. The delicate interplay between civic and public values and emerging technologies demands meticulous management to ensure they persist as the moral and ethical bedrock of our society and government.

It is safe to argue that the rapid adoption of AI and AIoT technologies, coupled with policy regulations and unforeseen events, harbors the potential to metamorphose vibrant metropolises into desolate ghost cities. The complete evolution of AI and AIoT may signify an erosion of moral and societal values, instigating concerns about the potential decline of humanity. Regardless, one of the most formidable challenges in urban transitions and digital transformations lies in the significant costs, risks, and uncertainties entwined with the integration of AI and AIoT into the realms of smart urbanism and eco-urbanism.

All in all, in the transformative journey toward smarter eco-cities, the ethical, social, and regulatory challenges posed by the implementation of AI and AIoT technologies across the core domains of these cities are both intricate and multifaceted. Addressing these challenges is key to cultivating a technological landscape that aligns with the core values of society. Striving for technological innovation in smarter eco-cities, a holistic approach that navigates these challenges ethically, socially, and inclusively will pave the way for a sustainable and socially equitable urban future.

3.7.5 Challenges, Issues, and Considerations of Explainable AI (XAI) and Interpretable ML (IML)

In the dynamic landscape of smarter eco-cities, the adoption of AI and ML systems brings unprecedented opportunities for innovation. However, the increased complexity of these systems necessitates a careful examination of the challenges, issues, and considerations related to XAI and IML, as outlined in Table 3.11. These challenges, issues, and considerations have been the focus of numerous studies in recent years (e.g., Ahmed et al. 2022; García-Magariño et al. 2019; Ghonge 2023; Javed et al. 2023; Kabir et al. 2021; Linardatos et al. 2021; Mayuri, Vasile and Indranath 2023; Thakker et al. 2020). Therefore, it is of crucial importance to discuss the complexities associated with XAI and IML systems in the context of smarter eco-cities in terms of the responsible and effective integration of AI and AIoT technologies into the systems and domains of these cities. Especially, XAI and IML both play crucial roles in the development of AI and AIoT applications with respect to well-informed and collaborative decision-making processes within the framework of smarter eco-cities.

Striving for technological advancements in smarter eco-cities, a comprehensive approach that navigates the challenges associated with XAI and IML will not only enhance the transparency, accountability, and

TABLE 3.11 Challenges and Considerations of Explainable Artificial Intelligence and Interpretable Machine Learning Systems

Challenges	Issues and Considerations
Complexity and opacity	The intricate nature of advanced AI and ML models often leads to a lack of transparency, making it challenging to comprehend the decision-making processes. This opacity poses a significant challenge in the context of smarter eco-cities, where the transparency of decision-making is crucial for ensuring accountability and fostering public trust.
Ethical implications	The ethical considerations surrounding XAI and IML systems involve the responsible use of data and the potential biases embedded in algorithms. In the context of smarter eco-cities, ethical implications become paramount as these technologies influence decisions related to resource allocation, energy usage, and environmental management.
Stakeholder engagement	In the development and deployment of XAI and IML systems, engaging with diverse stakeholders is crucial to ensure that the technology aligns with the needs and values of the community. In smarter eco-cities, where citizen participation is integral, fostering meaningful engagement becomes imperative for the success of these technologies.
Regulatory frameworks	The absence of comprehensive regulatory frameworks for XAI and IML poses challenges in ensuring compliance with ethical standards and legal requirements. In the context of smarter eco-cities, establishing robust regulatory frameworks becomes essential to navigate the ethical and legal complexities associated with these technologies.
Integration with human decision-makers	Harmonizing the outputs of XAI and IML systems with human decision-makers is a critical consideration. Ensuring that these technologies complement and augment human judgment rather than replace it is vital for the successful integration of XAI and IML in the decision-making processes of smarter eco-cities.

trustworthiness of AI and AIoT systems but also contribute to the ethical and sustainable development of urban environments. As a result, a recent surge in research has been dedicated to XAI and IML to address the various concerns raised by the deployment of AI and AIoT across diverse urban systems and domains. A notable contribution by Mayuri, Vasile, and Indranath (2023) presents a spectrum of applications in XAI and IML. The authors introduce various methodologies and techniques aimed at rendering AI and ML models more transparent and interpretable. Ghonge (2023) extends this exploration by addressing numerous case studies and practical use cases of XAI. Additionally, the author examines the impacts and challenges of integrating XAI into smart city systems, providing valuable insights into its diverse applications. Further enriching this landscape, Javid et al. (2023) undertake a comprehensive examination of XAI in the context of smart cities. Their focus spans across current and future developments, emerging trends, driving factors, notable use cases, as well as the challenges inherent in the integration of XAI in smart urban environments. The authors not only outline ongoing research projects but also shed light on standardization efforts, lessons learned from previous implementations, and the technical hurdles that researchers and practitioners may encounter in the pursuit of XAI and IML in smart city contexts. This multifaceted exploration contributes significantly to our understanding of the evolving landscape of XAI and IML in the domain of smart cities.

XAI and IML methodologies play an indispensable role in propelling the sustainable evolution of AI and AIoT solutions, fostering societal trust in the realms of both environmental and socioeconomic sustainability. These methodologies elucidate pivotal factors such as accuracy, fairness, openness, responsibility, and human-centeredness in decision-making processes empowered by AI and AIoT, effectively addressing pressing ethical and governance concerns. In the context of smarter eco-cities, where data-driven decision-making stands as a linchpin, the imperative arises for the development of models, techniques, and tools that are explainable and interpretable. This need is underscored by the intricate nature of smarter eco-city dynamics, where comprehensibility is paramount for fostering trust among stakeholders. Crucially, the collaborative efforts of AI and AIoT experts, environmental scientists, urban planners, and policy makers become necessary. This collaborative nexus ensures the meaningful integration of AI and AIoT technologies, creating a synergy that propels environmental sustainability initiatives and effectively addresses challenges posed by climate change in the dynamic landscape of smarter eco-cities. Looking forward, the trajectory of research endeavors assumes a significant role in realizing the transparent, effective, and ethically sound applications of XAI and IML methods in AI and AIoT solutions. This advancement is not merely a technological leap but a strategic imperative for enhancing environmental sustainability in smarter eco-cities, thereby ensuring equitable outcomes for all stakeholders involved in the transformative and transitional journey towards a more sustainable and intelligent urban future.

3.8 EXISTING GAPS, THEIR IMPLICATIONS, AND THE CORRESPONDING FUTURE RESEARCH DIRECTIONS

The rapidly evolving landscape of smarter eco-cities necessitates a thorough exploration of the transformative role played by emerging AI and AIoT applications in addressing pressing challenges related to environmental sustainability and climate change. This endeavor is driven by the overarching goal of not only understanding the current state of these applications in smarter eco-cities but also discerning and highlighting the existing gaps, their implications, and the corresponding future research directions at the nexus of AI and AIoT applications and the development of smarter eco-cities (Table 3.12).

By addressing these knowledge gaps, understanding their implications, and pursuing these research avenues, the field of smarter eco-cities can advance its implementation and impact. The identified gaps, their implications, and the corresponding research directions underscore the

TABLE 3.12   Existing Gaps, Implications, and Future Research Directions

Existing Gaps	Implications	Corresponding Future Research Directions
Interdisciplinary integration	The lack of interdisciplinary perspectives may hinder a holistic understanding of the societal impact and acceptance of these technologies in smarter eco-city development.	Investigate strategies for human-centric design in AI and AIoT applications, emphasizing user acceptance, inclusivity, and usability in the context of smarter eco-cities.
Ethical and societal implications	Neglecting ethical dimensions may lead to unintended consequences and hinder the societal acceptance of these technologies.	Develop governance models and policy frameworks that facilitate responsible and equitable deployment of AI and AIoT technologies in smarter eco-city development, addressing legal, regulatory, and ethical considerations.
Long-term environmental impact assessment	Without a comprehensive understanding of the environmental ramifications, the sustainability claims of these technologies remain uncertain.	Establish methodologies for quantifying the environmental and societal impact of AI and AIoT applications, including their influence on citizen engagement, health and well-being, and community dynamics.

(*Continued*)

TABLE 3.12 (Continued) Existing Gaps, Implications, and Future Research Directions

Existing Gaps	Implications	Corresponding Future Research Directions
Data privacy and security concerns	Neglecting robust data protection measures may result in breaches, compromising citizens' privacy and hindering the broader adoption of these technologies.	Explore methodologies for enhancing the explainability and transparency of AI and AIoT systems, addressing data privacy concerns, and fostering trust among citizens.
Socioeconomic disparities in technology access	Failure to address these disparities may exacerbate existing societal inequalities, posing challenges to the equitable development of smarter eco-cities.	Develop community engagement strategies to empower citizens, particularly those in marginalized communities, ensuring inclusivity and co-creation in technology access and decision-making.
Environmental justice considerations	Without a comprehensive understanding, there is a risk of unintentionally exacerbating environmental injustices in the development and implementation of smarter eco-city technologies.	Explore methodologies for assessing the resilience of AI and AIoT systems, especially in the context of environmental justice, to ensure the equitable distribution of benefits and mitigate potential risks.

multifaceted nature of developing more sustainable and technologically advanced urban environments. Delving into aspects such as data privacy, socioeconomic disparities, environmental justice considerations, and other critical aspects, this synthesis of knowledge paves the way for responsible and inclusive advancements. The tabulated account encapsulates a comprehensive overview, offering a roadmap for researchers, practitioners, and policymakers in the pursuit of smarter, more resilient, and more environmentally conscious eco-cities.

3.9 CONCLUSION

This chapter offered a comprehensive review of the current landscape of smarter eco-cities, with a specific focus on the transformative role of emerging applied AI and AIoT solutions in advancing environmental sustainability and climate change mitigation. It also identified and analyzed potential technological complexities, ethical concerns, and the socioeconomic implications of AI and AIoT-driven urban transformations. In doing so, it addressed a significant gap where environmental sustainability and climate change have been addressed separately or in relation to smart cities in the realm of AI and AIoT adoption, but less so in the context of emerging smarter eco-cities.

The aim of this chapter was to provide a nuanced understanding of the current state and the potential avenues for leveraging AI and AIoT technologies to foster sustainable urban development. In answering the research questions, not only were emerging applications uncovered but also the opportunities, benefits, challenges, and barriers associated with the integration of AI and AIoT in the context of smarter eco-cities were elucidated.

This chapter demonstrated how emerging AI and AIoT applications can contribute to enhancing environmental sustainability and mitigating climate change in the context of smarter eco-cities. The findings underscored the significant potential of emerging AI and AIoT technologies in advancing the development of smarter eco-cities. These cities can harness these technologies to promote environmentally sustainable urban development practices. AI and AIoT solutions offer opportunities for optimizing resource usage, enhancing infrastructure efficiency, monitoring environmental indicators, reducing carbon footprints, improving air and water quality, and strengthening resilience to climate change. Overall, the findings emphasize the transformative role that AI and AIoT technologies can play in advancing the sustainability agenda of cities and fostering the development of smarter, greener urban environments.

The investigation identified a spectrum of emerging AI and AIoT applications crucial for advancing smarter eco-cities. These solutions include energy conservation and renewable energy, sustainable transportation, traffic management, water resource management, waste management, biodiversity conservation, environmental monitoring and control, climate change adaptation and mitigation, and disaster resilience and management, as well as urban planning, all contributing to the creation of more responsive, efficient, and sustainable urban environments.

AI and AIoT technologies contribute significantly to promoting environmentally sustainable urban development practices in smarter eco-cities. From optimizing resource utilization and improving energy efficiency to fostering circular economy principles, these technologies offer a myriad of benefits and opportunities, enhancing the overall resilience and sustainability of urban ecosystems.

The integration of emerging applied AI and AIoT solutions within the framework of smarter eco-cities significantly contributes to achieving environmental sustainability goals and supporting effective climate change migration strategies. The conceptual framework highlights the crucial role of advanced technologies in addressing the challenges of urbanization, ecological degradation, and climate change. Smarter eco-cities can effectively monitor environmental indicators, optimize resource management, and enhance resilience to climate change impacts by integrating applied AI and AIoT solutions into their management and planning processes and practices.

The conceptual framework provides a roadmap for policymakers, urban planners, and technologists to develop holistic approaches to building sustainable and resilient urban environments. Furthermore, understanding the relationships among smarter eco-cities, AI and AIoT solutions, environmental sustainability goals, and climate change migration strategies is essential for guiding future research, policy development, and urban development initiatives aimed at creating cities that are both smart and environmentally sustainable.

While the potential benefits are vast, difficulties and barriers exist in the implementation of AI and AIoT applications for smarter eco-cities. In other words, despite the substantial potential of AI and AIoT in advancing the development of smarter eco-cities, their application is fraught with formidable challenges that need to be addressed and overcome. Environmental risks, data privacy, cybersecurity, public trust, socioeconomic disparities, lack of regulatory and governance frameworks, and environmental justice considerations present significant hurdles that need to be surmounted or mitigated to ensure ethical, equitable, and responsible deployment of AI and AIoT technologies.

The methodology employed in this literature review demonstrates comprehensiveness, encompassing a wide range of sources and providing a thorough analysis of the current state of research on AI and AIoT applications in smarter eco-cities. The systematic search strategies ensure the inclusion of relevant literature, and the rigorous criteria for analysis contribute to the reliability of the findings. However, a critical perspective highlights potential limitations of the methodology. While the review covers a broad spectrum of research, there may be biases in the selection of studies, such as a focus on English-language publications or a tendency to prioritize certain theoretical perspectives. Additionally, the exclusion of alternative viewpoints and grey literature could limit the scope and depth of the analysis. Future research should consider these limitations and strive for greater inclusivity and diversity in the literature review process.

This literature review contributes to the understanding of the dynamic interplay of AI, AIoT, and the development of smarter eco-cities. By synthesizing existing knowledge, emerging applications were highlighted, contributions to sustainable urban development were clarified, and the challenges inherent in the integration of AI and AIoT were identified together with strategies for mitigating them. The identification of existing gaps and future research avenues serves as a foundation for further exploration and advancements in this evolving field.

The implications of the findings extend beyond academia, offering valuable insights for policymakers, urban planners, technology developers, and researchers. Recognizing the potential benefits and navigating challenges responsibly is paramount to realizing the vision of smarter and more environmentally conscious eco-cities. This review sets the stage for informed decision-making, ethical considerations, and collaborative efforts to shape the future of urban living. The valuable insights derived from this study empower policymakers, practitioners, and researchers to navigate the complexities surrounding the integration of data-driven scientific urbanism and eco-urbanism. City stakeholders glean essential knowledge for informed decision-making, effective strategy implementation, and environmental policy development through the analysis of real-world implementations and practical use cases.

To sum up, the integration of AI and AIoT applications in the context of smarter eco-cities represents a transformative paradigm in urban development. While the exploration has provided a comprehensive overview, the evolving nature of data-driven technologies necessitates continuous examination and adaptation. The identified gaps and future research avenues outlined in this review offer a roadmap for future scholarly endeavors, aiming to drive the field forward. By navigating the complexities of creating more sustainable urban environments, the synergistic relationship among AI, AIoT, and smarter eco-cities emerges as a promising frontier with the potential to reshape the way we live, interact, and coexist in urban environments.

REFERENCES

Abdelkafi, N., & Täuscher, K. (2016). Business models for sustainability from a system dynamics perspective. *Organization & Environment, 29*(1), 74–96.

Abduljabbar, R. L., Dia, H., Liyanage, S., & Bagloee, S. A. (2019). Applications of artificial intelligence in transport: An overview. *Sustainability, 11*(1), 189. https://doi.org/10.3390/su11010189

Abid, S. K., Sulaiman, N., Chan, S. W., Nazir, U., Abid, M., Han, H., Ariza-Montes, A., & Vega-Muñoz, A. (2021). Toward an integrated disaster management approach: How artificial intelligence can boost disaster management. *Sustainability, 13*(22), 12560.

Acheampong, R. A., & Cugurullo, F. (2019). Capturing the behavioural determinants behind the adoption of autonomous vehicles: Conceptual frameworks and measurement models to predict public transport, sharing and ownership trends of self-driving cars. *Transportation Research Part F: Traffic Psychology and Behaviour, 62,* 349–375.

Ahmed, A. A., Nazzal, M. A., Darras, B. M., & Deiab, I. M. (2023). Global warming potential, water footprint, and energy demand of shared autonomous electric vehicles incorporating circular economy practices. *Sustainable Production and Consumption, 36,* 449–462. https://doi.org/10.1016/j.spc.2023.03.022

Ahmad, M. A., Teredesai, A., & Eckert, C. (2020). Fairness, accountability, transparency in AI at scale: Lessons from national programs. In: Proceedings of the 2020 Conference on Fairness, Accountability, and Transparency, Barcelona, Spain, 27–30 January 2020 (pp. 690–699).

Ahmed, I., Jeon, G., & Piccialli, F. (2022). From artificial intelligence to explainable artificial intelligence in industry 4.0: A survey on what, how and where. *IEEE Transactions on Industrial Informatics, 18,* 5031–5042.

Akhtar, M., Hannan, M. A., Begum, R. A., Basri, H., & Scavino, E. (2017). Backtracking search algorithm in CVRP models for efficient solid waste collection and route optimization. Waste Management, 61, 117–128.

Akhter, M. N., Mekhilef, S., Mokhlis, H., & Mohamed Shah, N. (2019). Review on forecasting of photovoltaic power generation based on machine learning and metaheuristic techniques. *13*(7), 1009–1023. https://doi.org/10.1049/iet-rpg.2018.5649

Alahakoon, D., Nawaratne, R., Xu, Y., De Silva, D., Sivarajah, U., & Gupta, B. (2020). Self-building artificial intelligence and machine learning to empower big data analytics in smart cities. *Information Systems Frontiers, 25*(1), 221–240. https://doi.org/10.1007/s10796-020-10056-x

Alahi, M. E. E., Sukkuea, A., Tina, F. W., Nag, A., Kurdthongmee, W., Suwannarat, K., & Mukhopadhyay, S. C. (2023). Integration of IoT-enabled technologies and artificial intelligence (AI) for smart city scenario: Recent advancements and future trends. *Sensors, 23*(11), 5206. https://doi.org/10.3390/s23115206

Alam, T., Gupta, R., Qamar, S., & Ullah, A. (2022). Recent applications of artificial intelligence for sustainable development in smart cities. In: M. Al-Emran, & K. Shaalan (Eds.), *Recent Innovations in Artificial Intelligence and Smart Applications: Studies in Computational Intelligence* (vol. 1061). Springer. https://doi.org/10.1007/978-3-031-14748-7_8

Aleh, M., Milovanoff, A., Posen, I. D., MacLean, H. L., & Hatzopoulou, M. (2022). Energy and greenhouse gas implications of shared automated electric vehicles. *Transportation Research Part D: Transport and Environment, 105,* 103233. https://doi.org/10.1016/j.trd.2022.103233

Almalki, F. A., Alsamhi, S. H., Sahal, R., Hassan, J., Hawbani, A., Rajput, N. S., Saif, A., Morgan, J., & Breslin, J. (2021). Green IoT for eco-friendly and sustainable smart cities: Future directions and opportunities. *Mobile Networks and Applications*. https://doi.org/10.1007/s11036-021-01790-w

Al-Mukhtar, M., & Al-Yaseen, F. (2019). Modeling water quality parameters using data-driven models, a case study abu-ziriq marsh in south of Iraq. 6(1), 24. https://doi.org/10.3390/hydrology6010024

Al-Othman, A., Tawalbeh, M., Martis, R., Dhou, S., Orhan, M., Qasim, M., & Ghani Olabi, A. (2022). Artificial intelligence and numerical models in hybrid renewable energy systems with fuel cells: Advances and prospects. *Energy Conversion and Management, 253*, 115154. https://doi.org/10.1016/j.enconman.2021.115154

Al-Sakkari, E. G., Ragab, A., Dagdougui, H., Boffito, D. C., & Amazouz, M. (2024). Carbon capture, utilization and sequestration systems design and operation optimization: *Assessment and perspectives of artificial intelligence opportunities. Science of the Total Environment, 917*, 170085. https://doi.org/10.1016/j.scitotenv.2024.170085

Ampatzidis, Y., Partel, V., & Costa, L. (2020). Agroview: cloud-based application to process, analyze and visualize UAV-collected data for precision agriculture applications utilizing artificial intelligence. *Computers and Electronics in Agriculture, 174*, 105457. https://doi.org/10.1016/j.compag.2020.105457

Alrukaibi, F., Alsaleh, R., & Sayed, T. (2019). Applying machine learning and statistical approaches for travel time estimation in partial network coverage. 11(14), 3822. https://doi.org/10.3390/su11143822

Alsadi, S., & Khatib, T. (2018). Photovoltaic power systems optimization research status: A review of criteria, constrains, models. *Techniques, and Software Tools, 8*(10), 1761. https://doi.org/10.3390/app8101761

Amiri, S. S., Mottahedi, S., Lee, E. R., & Hoque, S. (2021). Peeking inside the black-box: Explainable machine learning applied to household transportation energy consumption. *Computers, Environment and Urban Systems, 88*, 101647

Ampatzidis, Y., De Bellis, L., & Luvisi, A. (2017). iPathology: *Robotic Applications and Management of Plants and Plant Diseases*, 9(6), 1010. https://doi.org/10.3390/su9061010

Anbarasan, M., Muthu, B., Sivaparthipan, C., Sundarasekar, R., Kadry, S., Krishnamoorthy, S., & Dasel, A. A. (2020). Detection of flood disaster system based on IoT, big data and convolutional deep neural network. *Computers & Communications, 150*, 150–157. https://doi.org/10.1016/j.comcom.2019.10.011

Antonopoulos, I., Robu, V., Couraud, B., Kirli, D., Norbu, S., Kiprakis, A., Flynn, D., Elizondo-Gonzalez, S., & Wattam, S. (2020). Artificial intelligence and machine learning approaches to energy demand-side response: A systematic review. *Renewable and Sustainable Energy Reviews, 130*, 109899. https://doi.org/10.1016/j.rser.2020.109899.

Arachie, C., Gaur, M., Anzaroot, S., Groves, W., Zhang, K., & Jaimes, A. (2020, April). Unsupervised detection of sub-events in large scale disasters. Proceedings of the AAAI Conference on Artificial Intelligence, 34(01), 354–361. https://doi.org/10.1609/aaai.v34i01.5370

Arora, S., Kumar, S., & Kumar, S. (2023). Artificial intelligence in disaster management: A survey. In: M. Saraswat, C. Chowdhury, C. Kumar Mandal, & A. H. Gandomi (Eds.), Proceedings of International Conference on Data Science and Applications. Lecture Notes in Networks and Systems (vol. 552). Springer. https://doi.org/10.1007/978-981-19-6634-7_56

Aschwanden, G., Wijnands, J. S., Thompson, J., Nice, K. A., Zhao, H. F., & Stevenson, M. (2021). Learning to walk: Modeling transportation mode choice distribution through neural networks. *Environment and Planning B: Urban Analytics and City Science, 48*, 186–199.

Ashqar, H. I., Elhenawy, M., Almannaa, M. H., Ghanem, A., Rakha, H. A., & House, L. (2017, June). Modeling bike availability in a bike-sharing system using machine learning. In: Proceedings of the 2017 5th IEEE International Conference on Models and Technologies for Intelligent Transportation Systems (MT-ITS) (pp. 374–378). IEEE. 10.1109/MTITS.2017.8005700

Atsalakis, G. S. (2016). Using computational intelligence to forecast carbon prices *Applied Soft Computing, 43*, 107–116.

Awasthi, A., Pattnayak, K. C., Dhiman, G., & Tiwari, P. R. (Eds.). (2024). Artificial intelligence for air quality monitoring and prediction (1st ed.). CRC Press. https://doi.org/10.1201/9781032683805

Aymen, F., Mahmoudi, C., & Sbita, L. (2019). A novel energy optimization approach for electrical vehicles in a smart city. *Energies, 12*(5):929. https://doi.org/10.20944/preprints201901.0214.v1.

Bakker, K. (2022). Smart oceans: Artificial intelligence and marine protected area governance. *Earth System Governance, 13*, 100141. https://doi.org/10.1016/j.esg.2022.100141

Balogun, A. L., Marks, D., Sharma, R., Shekhar, H., Balmes, C., Maheng, D., Arshad, A., & Salehi, P. (2020). Assessing the potentials of digitalisation as a tool for climate change adaptation and sustainable development in urban centres. *Sustainable Cities and Society, 53*, 101888

Barnes, E. A., Hurrell, J. W., Ebert-Uphoff, I., Anderson, C., & Anderson, D. (2019). Viewing forced climate patterns through an AI lens. 46(22), 13389–13398. https://doi.org/10.1029/2019GL084944

Bertone, E., Burford, M. A., & Hamilton, D. P. (2018). Fluorescence probes for real-time remote cyanobacteria monitoring: A review of challenges and opportunities. *Water Research, 141*, 152–162. https://doi.org/10.1016/j.watres.2018.05.001

Bibri, S. E. (2019). Smart sustainable urbanism: Paradigmatic, scientific, scholarly, epistemic, and discursive shifts in light of big data science and analytics. In: Big Data Science and Analytics for Smart Sustainable Urbanism. Advances in Science, Technology & Innovation. Springer. https://doi.org/10.1007/978-3-030-17312-8_6

Bibri, S. E. (2020). Data-driven environmental solutions for smart sustainable cities: Strategies and pathways for energy efficiency and pollution reduction. *Euro-Mediterranean Journal for Environmental Integration, 5*(66). https://doi.org/10.1007/s41207-020-00211-w

Bibri, S. E. (2021a). Data-driven smart eco-cities and sustainable integrated districts: A best-evidence synthesis approach to an extensive literature review. *European Journal of Futures Research, 9*(1), 16. https://doi.org/10.1186/s40309-021-00181-4

Bibri, S. E. (2021b). The underlying components of data-driven smart sustainable cities of the future: A case study approach to an applied theoretical framework. *European Journal of Futures Research, 9*(13). https://doi.org/10.1186/s40309-021-00182-3

Bibri, S. E., & Krogstie, J. (2020a). Data-driven smart sustainable cities of the future: A novel model of urbanism and its core dimensions, strategies, and solutions. *Journal of Futures Studies, 25*(2), 77–94. https://doi.org/10.6531/JFS.202012_25(2).0009

Bibri, S. E., Alexandre, A., Sharifi, A., & Krogstie, J. (2023a). Environmentally sustainable smart cities and their converging AI, IoT, and big data technologies and solutions: An integrated approach to an extensive literature review. *Energy Informatics, 6*, 9., https://doi.org/10.1186/s42162-023-00259-2

Bibri, S. E., Huang, J., & Krogstie, J. (2024a). Artificial intelligence of things for synergizing smarter eco-city brain, metabolism, and platform: Pioneering data-driven environmental governance. *Sustainable Cities and Society*, 105516. https://doi.org/10.1016/j.scs.2024.105516

Bibri, S. E., Huang, J., Jagatheesaperumal, S. K., & Krogstie, J. (2024b). The synergistic interplay of artificial intelligence and digital twin in environmentally planning sustainable smart cities: A comprehensive systematic review. *Environmental Science and Ecotechnology, 20*, 100433. https://doi.org/10.1016/j.ese.2024.100433

Bibri, S. E., & Huang, J. (2025). Artificial intelligence of sustainable smart city brain and digital twin : A pioneering framework for advancing environmental sustainability. *Environmental Technology and Innovation* (in press).

Bibri, S. E., & Krogstie, J. (2020b). Environmentally data-driven smart sustainable cities: Applied innovative solutions for energy efficiency, pollution reduction, and urban metabolism. *Energy Informatics, 3*(1), 29. https://doi.org/10.1186/s42162-020-00130-8

Bibri, S. E., Krogstie, J., Kaboli, A., & Alahi, A. (2023b). Smarter eco-cities and their leading-edge artificial intelligence of things solutions for environmental sustainability: A comprehensive systemic review. *Environmental Science and Ecotechnlogy, 19*, https://doi.org/10.1016/j.ese.2023.100330

Bowes, B. D., Sadler, J. M., Morsy, M. M., Behl, M., & Goodall, J. L. (2019). Forecasting groundwater table in a flood prone coastal city with long short-term memory and recurrent neural networks. *11*(5), 1098. https://doi.org/10.3390/w11051098

Boza, P., & Evgeniou, T. (2021). Artificial intelligence to support the integration of variable renewable energy sources to the power system. *Applied Energy, 290*, 116754. https://doi.org/10.1016/j.apenergy.2021.116754

Brevini, B. (2020). Black boxes, not green: Mythologizing artificial intelligence and omitting the environment. *Big Data & Society, 7*(2), 2053951720935141. https://doi.org/10.1177/2053951720935141

Brisimi, T. S., Chen, T., Mela, A., Olshevsky, I. C., Paschalidis, & Shi, W. (2018). Federated learning of predictive models from federated electronic health records. *International Journal of Medical Informatics, 112*, 59–67.

Buis, A. (2020). *Study Confirms Climate Models Are Getting Future Warming Projections Right.* https://www.giss.nasa.gov/research/features/202001_accuracy/

Calp, M. H. (2019). A hybrid ANFIS-GA approach for estimation of regional rainfall amount. *Gazi University Journal of Science, 32*, 145–162.

Caprotti, F. (2020). Smart to green: Smart eco-cities in the green economy. In: K. S. Willis & A. Aurigi (Eds.), *The Routledge Companion to Smart Cities* (pp. 200–209). Routledge, London.

Caprotti, F., Cowley, R., Bailey, I., Joss, S., Sengers, F., Raven, R., Spaeth, P., Jolivet, E., Tan-Mullins, M., & Cheshmehzangi, A. (2017). *Smart Eco-City Development in Europe and China.* Policy Directions. University of Exter (SMART-ECO Project, Exter.

Caprotti, F., Cowley, R., Flynn, A., Joss, S., & Yu, L. (2016). *Smart-ECO CITIES in the UK: Trends and CityProfiles 2016.* University of Exeter (SMART-ECO Project, Exeter.

Chang, F.-J., Chang, L.-C., & Chen, J.-F. (2023). Artificial intelligence techniques in hydrology and water resources management. *Water, 15*(10):1846. https://doi.org/10.3390/w15101846

Chen, L., Chen, Z., Zhang, Y., Liu, Y., Osman, A. I., Farghali, M., Hua, J., Al-Fatesh, A., Ihara, I., Rooney, D. W., & Yap, P.-S. (2023). Artificial intelligence-based solutions for climate change: A review. *Environmental Chemistry Letters, 21*, 2525–2557. https://doi.org/10.1007/s10311-023-01556-7

Choubin, B., Borji, M., Mosavi, A., Sajedi-Hosseini, F., Singh, V. P., & Shamshirband, S. (2019). Snow avalanche hazard prediction using machine learning methods. *Journal of Hydrology, 577*, 123929. https://doi.org/10.1016/j.jhydrol.2019.123929

Collins, J. A., & Fauser, B. C. J. M. (2005). Balancing the strengths of systematic and narrative reviews. *Human Reproduction Update, 11*(2), 103–104. https://doi.org/10.1093/humupd/dmh058

Cooper, H. M. (2017). *Research Synthesis and Meta-Analysis: A Step-by-Step Approach.* Sage Publications, London.

Crane-Droesch, A. (2018). Machine learning methods for crop yield prediction and climate change impact assessment in agriculture. *Environmental Research Letters, 13*(11), 114003. https://doi.org/10.1088/1748-9326/aae159

Cugurullo, F. (2020). Urban artificial intelligence: From automation to autonomy in the smart city. *Frontiers in Sustainable Cities, 2*, 38. https://doi.org/10.3389/frsc.2020.00038

Cugurullo, F., Acheampong, R. A., Gueriau, M., & Dusparic, I. (2021). The transition to autonomous cars, the redesign of cities and the future of urban sustainability. *Urban Geography, 42*(6), 833–859.

Cugurullo, F., Caprotti, F., Cook, M., Karvonen, A., McGuirk, P., & Marvin, S. (2024). Artificial *Intelligence and the City: Urbanistic Perspectives on AI*. Taylor & Francis.

Dadashzadeh, M., & Zarezadeh, F. J. (2022). A review of the various advances in smart cities: Application of artificial intelligence and machine learning. In: *4th Intercontinental Geoinformation Days (IGD)*, 20–21 June 2022, Tabriz, Iran.

Das, U. K., Tey, K. S., Seyedmahmoudian, M., Mekhilef, S., Idris, M. Y. I., Van Deventer, W., Horan, B., & Stojcevski, A. (2018). Forecasting of photovoltaic power generation and model optimization: A review. *Renewable and Sustainable Energy Reviews, 81*, 912–928. https://doi.org/10.1016/j.rser.2017.08.017

Dauvergne, P. (2021). The globalization of artificial intelligence: Consequences for the politics of environmentalism. *Globalizations, 18*(2), 285–299. https://doi.org/10.1080/14747731.2020.1785670

Dawoud, S. M., Lin, X., & Okba, M. I. (2018). Hybrid renewable microgrid optimization techniques: A review. *Renewable and Sustainable Energy Reviews, 82*, 2039–2052. https://doi.org/10.1016/j.rser.2017.08.007

De Jong, M., Joss, S., Schraven, D., Zhan, C., & Weijnen, M. (2015). Sustainable-smart-resilient low carbon-eco-knowledge cities; making sense of a multitude of concepts promoting sustainable urbanization. *Journal of Cleaner Production, 109*, 25–38.

Delanoë, P., Tchuente, D., & Colin, G. (2023). Method and evaluations of the effective gain of artificial intelligence models for reducing CO_2 emissions. *Journal of Environmental Management, 331*, 117261. https://doi.org/10.1016/j.jenvman.2023.117261

Demertzis, K., Iliadis, L., & Pimenidis, E. (2021). Geo-AI to aid disaster response by memory-augmented deep reservoir computing. *Integrated Computer-Aided Engineering, 28*(4), 383–398. https://doi.org/10.3233/ICA-210657

Demirci, M., Üneş, F., & Körü, S. (2019). Modeling of groundwater level using artificial intelligence techniques: A case study of Reyhanli region in Turkey. *Applied Ecology and Environmental Research, 17*, 2651–2663. https://doi.org/10.15666/aeer/1702_26512663

Dheeraj, A., Nigam, S., Begam, S., Naha, S., Jayachitra Devi, S., Chaurasia, H. S., Kumar, D., Ritika, Soam, S. K., Srinivasa Rao, N., Alka, A., Sreekanth Kumar, V. V., & Mukhopadhyay, S. C. (2020). Role of artificial intelligence (AI) and internet of things (IoT) in mitigating climate change. In: Ch. Srinivasa, R. T. Srinivas, R. V. S. Rao, N. Srinivasa Rao, S. Senthil Vinayagam, & P. Krishnan (Eds.), *Climate Change and Indian Agriculture: Challenges and Adaptation Strategies* (pp. 325–358). ICAR-National Academy of Agricultural Research Management.

Din, I. U., Guizani, M., Rodrigues, J. J. P. C., Hassan, S., & Korotaev, V. V. (2019). Machine learning in the internet of things: Designed techniques for smart cities. *Future Generation Computer Systems, 100*, 826–843.

Dominiković, I., Ćukušić, M., & Jadrić, M. (2021). The role of artificial intelligence in smart cities: Systematic literature review. In: International Conference on Data and Information (pp. 64–80). Springer. https://doi.org/10.1007/978-3-030-77417-2_5

Efthymiou, I.-P., & Egleton, T. E. (2023). Artificial intelligence for sustainable smart cities. In: *Handbook of Research on Applications of AI, Digital Twin, and Internet of Things for Sustainable Development* (pp. 1–11). IGI Global.

El Himer, S., Ouaissa, M., Ouaissa, M., & Boulouard, Z. (2022). Artificial intelligence of things (AIoT) for renewable energies systems. In: S. El Himer, M. Ouaissa, A. A. A. Emhemed, M. Ouaissa, & Z. Boulouard (Eds.), *Artificial Intelligence of Things for Smart Green Energy Management. Studies in Systems, Decision and Control* (vol. 446). Springer. https://doi.org/10.1007/978-3-031-04851-7_1

Falk, J., Gaffney, O., Bhowmik, A. K., Bergmark, P., Galaz, V., Gaskell, N., Henningsson, S., Höjer, M., Jacobson, L., Jónás, K., Kåberger, T., Klingenfeld, D., Lenhart, J., Loken, B., Lundén, D., Malmodin, J., Malmqvist, T., Olausson, V., Otto, I., Pearce, A., Pihl, E., & Shalit, T. (2020). *Exponential Roadmap 1.5.1.* Future Earth. Sweden..

Fan, C., Zhang, C., Yahja, A., & Mostafavi, A. (2021). Disaster city digital twin: A vision for integrating artificial and human intelligence for disaster management. *International Journal of Information Management, 56*, 102049.

Fang, B., Yu, J., Chen, Z., Osman, A. I., Farghali, M., Ihara, I., Hamza, E. H., Rooney, D. W., & Yap, P.-S. (2023). Artificial intelligence for waste management in smart cities: A review. *Environmental Chemistry Letters*. https://doi.org/10.1007/s10311-023-01604-3

Fatemidokht et al. (2021), Liao et al. (2021), Nikitas et al. (2020); Olayode et al. (2020); Othman (2022), Saleh et al. (2022), Silva et al. (2022), Tyagi and Aswathy (2021)

Fatemidokht, H., Rafsanjani, M. K., Gupta, B. B., & Hsu, C. H. (2021). Efficient and secure routing protocol based on artificial intelligence algorithms with UAV-assisted for vehicular ad hoc networks in intelligent transportation systems. *IEEE Transactions on Intelligent Transportation Systems, 22*, 4757–4769. https://doi.org/10.1109/TITS.2020.3041746

Fathi, S., Srinivasan, R. S., Kibert, C. J., Steiner, R. L., & Demirezen, E. (2020). AI-based campus energy use prediction for assessing the effects of climate change. *Sustainability, 12*, 3223.

Ferreira, L., Duarte, A., Cunha, F., & Fernandes-Filho, E. (2019). Multivariate adaptive regression splines (MARS) applied to daily reference evapotranspiration modeling with limited weather data. *Acta Scientiarum Agronomy, 41*. https://doi.org/10.4025/actasciagron.v41i1.39780

Floridi, L., Cowls, J., King, T. C., & Taddeo, M. (2020). How to design AI for social good: Seven essential factors. *Science and Engineering Ethics, 26*, 1771–1796.

Fujii, M., Fujita, T., Chen, X., Ohnishi, S., & Yamaguchi, N. (2012). Smart recycling of organic solid wastes in an environmentally sustainable society. *Resources, Conservation and Recycling*, 63, 1–8.

García-Magariño, I., Muttukrishnan, R., & Lloret, J. (2019). Human-centric AI for trustworthy IoT systems with explainable multilayer perceptrons. *IEEE Access*, 7, 125562–125574.

Ghonge, M. M., Pradeep, N., Jhanjhi, N., & Kulkarni, P. (Eds.). (2023). Advances in Explainable AI Applications for Smart Cities. IGI Global, Hershey, PA. https://doi.org/10.4018/978-1-6684-6361-1

Ghosh, R., & Sengupta, D. (2023). Smart urban metabolism: A big-data and machine learning perspective. In: R. Bhadouria, S. Tripathi, P. Singh, P. K. Joshi, & R. Singh (Eds.), *Urban Metabolism and Climate Change*. Springer. https://doi.org/10.1007/978-3-031-29422-8_16

González García, C., Núñez Valdéz, E. R., García Díaz, V., Pelayo García-Bustelo, B. C., & Cueva Lovelle, J. M. (2019). A review of artificial intelligence in the internet of things. *International Journal of Interactive Multimedia And Artificial Intelligence*, 5, 1.

Gourisaria, M. K., Jee, G., Harshvardhan, G. M., Konar, D., & Singh, P. K. (2023). Artificially intelligent and sustainable smart cities. In: P. K. Singh, M. Paprzycki, M. Essaaidi, & S. Rahimi (Eds.), *Sustainable Smart Cities. Studies in Computational Intelligence* (vol. 942). Springer. https://doi.org/10.1007/978-3-031-08815-5_14

Granata, F., Gargano, R., & de Marinis, G. (2020). Artificial intelligence based approaches to evaluate actual evapotranspiration in wetlands. *Science of the Total Environment*, 703, 135653. https://doi.org/10.1016/j.scitotenv.2019.135653

Greenfield, A. (2018). *Radical Technologies: The Design of Everyday Life*. Verso Books, London.

Guiqiang, L., Jin, Y., Akram, M. W., Chen, X., & Ji, J. (2018). Application of bio-inspired algorithms in maximum power point tracking for PV systems under partial shading conditions: A review. *Renewable and Sustainable Energy Reviews*, 81, 840–873. https://doi.org/10.1016/j.rser.2017.08.034

Guo, K., Lu, Y., Gao, H., & Cao, R. (2018). Artificial intelligence-based semantic internet of things in a user-centric smart city. *Sensors (Basel)*, 18(5). https://doi.org/10.3390/s18051341

Han, T., Muhammad, K., Hussain, T., Lloret, J., & Baik, S. W. (2020). An efficient deep learning framework for intelligent energy management in IoT networks. *IEEE IoT Journal*, 8, 3170–3179.

Hannan, M. A., Begum, R. A., Al-Shetwi, A. Q., Ker, P. J., Al Mamun, M. A., Hussain, A., Basri, H., & Mahlia, T. M. I. (2020). Waste collection route optimisation model for linking cost saving and emission reduction to achieve sustainable development goals. *Sustainable Cities and Society*, 62, 102393. https://doi.org/10.1016/j.scs.2020.102393

Hashem, I. A. T., Chang, V., Anuar, N. B., Adewole, K., Yaqoob, I., Gani, A., Ahmed, E., & Chiroma, H. (2016). The role of big data in smart city. *International Journal of Information Management*, 36(5), 748–758.

Hawkins, J., & Nurul Habib, K. (2019). Integrated models of land use and transportation for the autonomous vehicle revolution. *Transport Reviews, 39*(1), 66–83. https://doi.org/10.1080/01441647.2018.1449033

Herath, H., & Mittal, M. (2022). Adoption of artificial intelligence in smart cities: A comprehensive review. *International Journal of Information Management Data Insights, 2*(1), 100076. https://doi.org/10.1016/j.jjimei.2022.100076

Hoang, T. V. (2024). Impact of integrated artificial intelligence and internet of things technologies on smart city transformation. *Journal of Technical Education Science, 19*(Special Issue 1), 64–73. https://doi.org/10.54644/jte.2024.1532

Hoffmann, A. L. (2019). Where fairness fails: Data, algorithms, and the limits of antidiscrimination discourse. *Information, Communication & Society, 22*(7), 900–915. https://doi.org/10.1080/1369118X.2019.1573912

Huntingford, C., Jeffers, E. S., Bonsall, M. B., Christensen, H. M., Lees, T., & Yang, H. (2019). Machine learning and artificial intelligence to aid climate change research and preparedness. *Environmental Research Letters, 14*(12), 124007.

Iris-Panagiota, E., & Egleton . T. E. (2023). Artificial intelligence for sustainable smart cities. In: B. Kishore Mishra (Eds.), *Handbook of Research on Applications of AI, Digital Twin, and Internet of Things for Sustainable Development* (pp. 1–11). IGI Global. https://doi.org/10.4018/978-1-6684-6821-0.ch001

Isabelle, D. A., & Westerlund, M. (2022). A review and categorization of artificial intelligence-based opportunities in wildlife. *Ocean and Land Conservation. Sustainability, 14*, 1979. https://doi.org/10.3390/su14041979

Ise, T., & Oba, Y. (2019). Forecasting climatic trends using neural networks: An experimental study using global historical data. *Original Research, 6*, 32. https://doi.org/10.3389/frobt.2019.00032

Ivanova, S., Ivanova, E., &Medarov, M. (2023). Guidelines for the application of artificial intelligence in the study of the influence of climate change on transport infrastructure. In: Conference: VIII International Scientific Conference "Industry 4.0", Summer Session, 28 June–1 July 2023, Varna (vol. 1, pp. 29–33).

Jagatheesaperumal, S. K., Bibri, S. E., Ganesan, S., & Jeyaraman, P. (2023). Artificial Intelligence for road quality assessment in smart cities: A machine learning approach to acoustic data analysis. *Computers in Urban Science, 3*, 28. https://doi.org/10.1007/s43762-023-00104-y

Jagatheesaperumal, S. K., Bibri, S. E., Huang, J Ganesan, S., & Jeyaraman, P. (2024). Artificial intelligence of things for smart cities: Advanced solutions for enhancing transportation safety. *Computers, Environment and Urban Systems, 4*, 10. https://doi.org/10.1007/s43762-024-00120-6

Jahani, A., & Rayegani, B. (2020). Forest landscape visual quality evaluation using artificial intelligence techniques as a decision support system. *Stochastic Environmental Research and Risk Assessment, 34*(10), 1473–1486. https://doi.org/10.1007/s00477-020-01832-x

Jain, H., Dhupper, R., Shrivastava, A., Kumar, D., & Kumari, M. (2023). AI-enabled strategies for climate change adaptation: Protecting communities, infrastructure, and businesses from the impacts of climate change. *Computational Urban Science, 3*, Article 25. https://doi.org/10.1007/s43762-023-00041-6

Javed, A. R., Ahmed, W., Pandya, S., Maddikunta, P. K. R., Alazab, M., & Gadekallu, T. R. (2023). A survey of explainable artificial intelligence for smart cities. *Electronics, 12*(4):1020. https://doi.org/10.3390/electronics12041020

Jeon, D. J., Ki, S. J., Cha, Y., Park, Y., & Kim, J. H. (2018). New methodology of evaluation of best management practices performances for an agricultural watershed according to the climate change scenarios: A hybrid use of deterministic and decision support models. *Ecological Engineering, 119*, 73–83. https://doi.org/10.1016/j.ecoleng.2018.05.006

Jha, S. K., Bilalovic, J., Jha, A., Patel, N., & Zhang, H. (2017). Renewable energy: Present research and future scope of Artificial Intelligence. *Renewable and Sustainable Energy Reviews, 77*, 297–317. https://doi.org/10.1016/j.rser.2017.04.018

Ji, L., Wang, Z., Chen, M., Fan, S., Wang, Y., & Shen, Z. (2019). How much can AI techniques improve surface air temperature forecast? A report from AI challenger 2018 global weather forecast contest. *Journal of Meteorological Research, 33*(5), 989–992. https://doi.org/10.1007/s13351-019-9601-0

Jiang, W., & Zhang, L. (2019). Geospatial data to images: A deep-learning framework for traffic forecasting. *Tsinghua Science and Technology, 24*(1), 52–64. https://doi.org/10.26599/TST.2018.9010033

Jin, W., Atkinson, T. A., Doughty, C., Neupane, G., Spycher, N., McLing, T. L., Dobson, P. F., Smith, R., & Podgorney, R. (2022). Machine-learning-assisted high-temperature reservoir thermal energy storage optimization. *Renewable Energy, 197*, 384–397. https://doi.org/10.1016/j.renene.2022.07.118

Jones, E. C., & Leibowicz, B. D. (2019). Contributions of shared autonomous vehicles to climate change mitigation. Transportation Research Part D: *Transport and Environment, 72*, 279-298. https://doi.org/10.1016/j.trd.2019.05.005.

Joseph, A., Chandra, J., & Siddharthan, S. (2021). Genome analysis for precision agriculture using artificial intelligence: a survey. *Data Science and Security, 132*, 221–226. https://doi.org/10.1007/978-981-15-5309-7_23

Kabir, M. H., Hasan, K. F., Hasan, M. K., & Ansari, K. (2021). Explainable artificial intelligence for smart city application: A secure and trusted platform. In M. Ahmed, S. R. Islam, A. Anwar, N. Moustafa, & A. S. K. Pathan (Eds.), *Explainable Artificial Intelligence for Cyber Security* (Vol. 1025, pp. 213–237). Springer. https://doi.org/10.1007/978-3-030-96630-0_11.

Kalmykov, L. V., & Kalmykov, V. L. (2015). A solution to the biodiversity paradox by logical deterministic cellular automata. *Acta Biotheoretica, 63*(2), 203–221. https://doi.org/10.1007/s10441-015-9257-9

Kamal Paul, S., & Bhaumik, P. (2022). Disaster management through integrative AI. In: 23rd International Conference on Distributed Computing and Networking (ICDCN '22) (pp. 290–293). https://doi.org/10.1145/3491003.3493235.

Kamrowska-Załuska, D. (2021). Impact of AI-based tools and urban big data analytics on the design and planning of cities. *Land, 10*, 1209. https://doi.org/10.3390/land10111209

Kankanamge, N., Yigitcanlar, T., & Goonetilleke, A. (2021). Public perceptions on artificial intelligence driven disaster management: Evidence from Sydney, Melbourne and Brisbane. *Telematics and Informatics, 65*, 101729. https://doi.org/10.1016/j.tele.2021.101729

Kaplan, A., & Haenlein, M. (2019). Siri, Siri, in my hand: Who's the fairest in the land? On the interpretations, illustrations, and implications of artificial intelligence. *Business Horizons, 62*(1), 15–25. https://doi.org/10.1016/j.bushor.2018.08.004

Kassens-Noor, E., & Hintze, A. (2020). Cities of the future? The potential impact of artificial intelligence. *AI, 1*, 192–197. https://doi.org/10.3390/ai1020012

Khalilpourazari, S., & Pasandideh, S. H. R. (2021). Designing emergency flood evacuation plans using robust optimization and artificial intelligence. *Journal of Combinatorial Optimization, 41*(3), 640–677. https://doi.org/10.1007/s10878-021-00699-0

Khan, S., Paul, D., Momtahan, P., & Aloqaily, M. (2018). Artificial intelligence framework for smart city microgrids: State of the art, challenges, and opportunities. In: 2018 Third International Conference on Fog and Mobile Edge Computing (FMEC) (pp. 283–288). IEEE. https://doi.org/10.1109/FMEC.2018.8364080.

Kitchin, R. (2016). The ethics of smart cities and urban science. *Philosophical Transactions of the Royal Society A: Mathematical, Physical and Engineering Sciences, 374*(2083), 1–15. https://doi.org/10.1098/rsta.2016.0115

Koffka, K. (2023). Intelligent Things: Exploring AIoT Technologies and Applications. Kinder Edition.

Kopelias, P., Demiridi, E., Vogiatzis, K., Skabardonis, A., & Zafiropoulou, V. (2020). Connected & autonomous vehicles – Environmental impacts – A review. *Science of The Total Environment, 712*, 135237. https://doi.org/10.1016/j.scitotenv.2019.135237

Koumetio Tekouabou, S. C., Diop, E. B., Azmi, R., & Chenal, J. (2023). Artificial intelligence based methods for smart and sustainable urban planning: A systematic survey. *Archives of Computational Methods in Engineering, 30*, 1421–1438. https://doi.org/10.1007/s11831-022-09844-2

Kramers, A., Wangel, J., & Höjer, M. (2016). Governing the smart sustainable city: The case of the Stockholm Royal Seaport. *ICT for Sustainability, 46*, 99–108.

Lannelongue, L., Grealey, J., & Inouye, M. (2021). Green algorithms: Quantifying the carbon footprint of computation. *Advanced Science, 8*, 2100707. https://doi.org/10.1002/advs.202100707

Larsson, S., & Heintz, F. (2020). Transparency in artificial intelligence. *Internet Policy Review, 9*, 1–12

Leal Filho, W., Wall, T., Mucova, S. A. R., Nagy, G. J., Balogun, A.-L., Luetz, J. M., Ng, A. W., Kovaleva, M., Azam, F. M. S., & Alves, F. (2022). Deploying artificial intelligence for climate change adaptation. *Technological Forecasting and Social Change, 180*, 121662.

Lee, M. F. R., & Chien, T. W. (2020, August). Artificial intelligence and internet of things for robotic disaster response. In: 2020 International Conference on Advanced Robotics and Intelligent Systems (ARIS) (pp. 1–6). IEEE. https://doi.org/10.1109/ARIS50834.2020.9205794

Liao, Z., Taiebat, M., & Xu, M. (2021). Shared autonomous electric vehicle fleets with vehicle-to-grid capability: *Economic viability and environmental co-benefits.* Applied Energy, 302, 117500. https://doi.org/10.1016/j.apenergy.2021.117500

Linardatos, P., Papastefanopoulos, V., & Kotsiantis, S. (2021). Explainable AI: A review of machine learning interpretability methods. *Entropy, 23,* 18.

Liu, H., Chen, C., Lv, X., Wu, X., & Liu, M. (2019). Deterministic wind energy forecasting: A review of intelligent predictors and auxiliary methods. *Energy Conversion and Management, 195,* 328–345.

Liu, Z., Sun, Y., Xing, C., Liu, J., He, Y., Zhou, Y., & Zhang, G. (2022). Artificial intelligence powered large-scale renewable integrations in multi-energy systems for carbon neutrality transition: Challenges and future perspectives. *Energy AI, 10,* 100195. https://doi.org/10.1016/j.egyai.2022.100195

Liyanage, S., Dia, H., Abduljabbar, R., & Bagloee, S. A. (2019). Flexible mobility on-demand: An environmental scan. *Sustainability, 11*(5), 1262.

Lu, H., Li, H., Liu, T., Fan, Y., Yuan, Y., Xie, M., & Qian, X. (2019). Simulating heavy metal concentrations in an aquatic environment using artificial intelligence models and physicochemical indexes. *The Science of the Total Environment, 694,* 133591. https://doi.org/10.1016/j.scitotenv.2019.133591

Lu, H., Li, Y., Chen, M., Kim, H., Serikawa, S. J. M. N., & Applications. (2018). Brain intelligence: Go beyond artificial intelligence. *Mobile Networks Applications, 23*(2), 368–375.

Mahardhika Sakti, P., & Putriani, O. (2023). Deployment and use of Artificial Intelligence (AI) in water resources and water management. IOP Conference Series: Earth and Environmental Science *1195,* 012056.

Mahmood, A., Bennamoun, M., An, S., Sohel, F., Boussaid, F., Hovey, R., Kendrick, G., & Fisher, R. B. (2017). Deep learning for coral classification. In: P. Samui, S. Sekhar, & V. E. Balas (Eds.), *Handbook of Neural Computation* (pp. 383–401). Academic Press.

Manikandan, S., Kaviya, R. S., Hemnath Shreeharan, D., Subbaiya, R., Vickram, S., Karmegam, N., Kim, W., & Govarthanan, M. (2024). Artificial intelligence-driven sustainability: Enhancing carbon capture for sustainable development goals – A review. Sustainable Development. https://doi.org/10.1002/sd.3222

Marasinghe, R., Yigitcanlar, T., Mayere, S., Washington, T., & Limb, M. (2024). Computer vision applications for urban planning: A systematic review of opportunities and constraints. *Sustainable Cities and Society, 100,* 105047. https://doi.org/10.1016/j.scs.2023.105047

Martínez-Santos, P., & Renard, P. (2020). Mapping groundwater potential through an ensemble of big data methods. 58(4), 583–597. https://doi.org/10.1111/gwat.12939

Masoumi, Z., & van Genderen, J. (2023). Artificial intelligence for sustainable development of smart cities and urban land-use management. *Geo-spatial Information Science.* https://doi.org/10.1080/10095020.2023.2184729

Mastorakis, G., Mavromoustakis, C. X., Batalla, J. M., & Pallis, E. (Eds). (2020). *Convergence of Artificial Intelligence and the Internet of Things*. Springer.

Mathur, S., & Modani, U. S. (2016). Smart city- a gateway for artificial intelligence in India. In: 2016 IEEE Students' Conference on Electrical, Electronics and Computer Science (SCEECS), 5–6 March 201 (pp. 1–3). https://doi.org/10.1109/SCEECS.2016.7509291

Mayuri, M., Vasile, P., & Indranath, C. (2023). Explainable AI: Foundations. *Methodologies and Applications*. https://doi.org/10.1007/978-3-031-12807-3

Mehr, A. D., Nourani, V., Kahya, E., Hrnjica, B., Sattar, A. M., & Yaseen, Z. M. (2018). Genetic programming in water resources engineering: A state-of-the-art review. *Journal of Hydrology*, 566, 643–667. https://doi.org/10.1016/j.jhydrol.2018.09.043.

Merizalde, Y. L., Hernández-Callejo, O. & Duque-Perez, V. (2019). Alonso-Gómez maintenance models applied to wind turbines a comprehensive overview. *Energies*, 12(2), 225

Mhlanga, D. (2023). Artificial intelligence and machine learning for energy consumption and production in emerging markets: A review. *Energies*, 16, 745. https://doi.org/10.3390/en16020745

Mi, M., Dordevic, A., & Arsić, A. K. (2017). The optimization of vehicle routing of communal waste in an urban environment using a nearest neighbirs' algorithm and genetic algorithm: Communal waste vehicle routing optimization in urban areas. In: 2017 Ninth International Conference on Advanced Computational Intelligence (ICACI) (pp. 264–271). IEEE.

Mishra, B. K. (Ed.). (2023). *Handbook of Research on Applications of AI, Digital Twin, and Internet of Things for Sustainable Development*. IGI Global. https://doi.org/10.4018/978-1-6684-6821-0

Mishra, P., & Singh, G. (2023). Artificial intelligence for sustainable smart cities. In: *Sustainable Smart Cities*. Springer. https://doi.org/10.1007/978-3-031-33354-5_6

Mishra, R. K., Kumari, C. L., Chachra, S., Krishna, P. S. J., Dubey, A., & Singh, R. B. (Eds.). (2022). Smart cities for sustainable development. In: *Advances in Geographical and Environmental Sciences*. Springer.

Mohamed, E. (2020). The relation of artificial intelligence with internet of things: A survey. *Journal of Cybersecurity and Information Management (JCIM)*, 1(1), 30–34. https://doi.org/10.5281/zenodo.3686810.

Mokhtarzad, M., Eskandari, F., Jamshidi Vanjani, N., & Arabasadi, A. (2017). Drought forecasting by ANN, ANFIS, and SVM and comparison of the models. *Environmental Earth Sciences*, 76(21), 729. https://doi.org/10.1007/s12665-017-7064-0

Mosavi, A., Salimi, M., Faizollahzadeh Ardabili, S., Rabczuk, T., Shamshirband, S., & Varkonyi-Koczy, A. R. (2019). State of the art of machine learning models in energy systems: A systematic review. *Energies*, 12(7), 1301. https://www.mdpi.com/1996-1073/12/7/1301

Mounadel, A., Ech-Cheikh, H., Lissane, E., Saâd, R., Ahmed, S. M., & Abdellaoui, B. (2023). Application of artificial intelligence techniques in municipal solid waste management: A systematic literature review. *Environmental Technology Reviews*, *12*(1), 316–336, https://doi.org/10.1080/21622515.2023.2205027

Muhammad, K., Lloret, J., & Baik, S. W. (2019). Intelligent and energy-efficient data prioritization in green smart cities: Current challenges and future directions. *IEEE Communications Magazine*, *57*(2), 60–65. https://doi.org/10.1109/MCOM.2018.1800371

Mukhopadhyay, S. C., Tyagi, S. K. S., Suryadevara, N. K., Piuri, V., Scotti, F., & Zeadally, S. (2021). Artificial intelligence-based sensors for next generation IoT applications: A review. *IEEE Sensors Journal*, *21*, 24920–24932. https://doi.org/10.1109/JSEN.2021.3055618.

Najafzadeh, M., & Ghaemi, A. (2019). Prediction of the five-day biochemical oxygen demand and chemical oxygen demand in natural streams using machine learning methods. *Environmental Monitoring and Assessment*, *191*(6), 380. https://doi.org/10.1007/s10661-019-7446-8

Nasir, I., & Aziz Al-Talib, G. A. (2023). Waste classification using artificial intelligence techniques: *Literature Review. Technium: Romanian Journal of Applied Sciences and Technology*, *5*, 49–59. https://doi.org/10.47577/technium.v5i.8345

Navarathna, P. J., & Malagi, V. P. (2018). Artificial Intelligence in Smart City Analysis International Conference on Smart Systems Inventive Technology (ICSSIT), pp. 44–47, https://doi.org/10.1109/ICSSIT.2018.8748476

Neo, E. X., Hasikin, K., Lai, K. W., Mokhtar, M. I., Azizan, M. M., Hizaddin, H. F., Razak, S. A., & Yanto. (2023). Artificial intelligence-assisted air quality monitoring for smart city management. PeerJ Computer Science, 9, e1306. https://doi.org/10.7717/peerj-cs.1306

Nikitas, A., Michalakopoulou, K., Njoya, E. T., & Karampatzakis, D. (2020). Artificial Intelligence, transport and the smart city: definitions and dimensions of a new mobility era. *Sustainability*, *12*(7), 2789. https://doi.org/10.3390/su12072789

Nishant, R., Kennedy, M., & Corbett, J. (2020). Artificial intelligence for sustainability: Challenges, opportunities, and a research agenda. *International Journal of Information Management*, *53*, 102104.

Nti, E. K., Cobbina, S. J., Attafuah, E. E., Opoku, E., & Gyan, M. A. (2022). Environmental sustainability technologies in biodiversity, energy, transportation and water management using artificial intelligence: A systematic review, *Sustainable Futures*, *4* 100068.

Nunes, J. A. C., Cruz, I., Nunes, A., & Pinheiro, H. T. (2020). Speeding up coral reef conservation with AI-aided automated image analysis. *Nature Machine Intelligence*, *2*(6), 292–292.

Ochoa, V., & Urbina-Cardona, N. (2017). Tools for spatially modeling ecosystem services: Publication trends, conceptual reflections and future challenges. *Ecosystem Services*, *26*, 155–169.

Olabi, A. G., Abdelghafar, A. A., Maghrabie, H. M., Sayed, E. T., Rezk, H., Radi, M. A., Obaideen, K., & Abdelkareem, M. A. (2023). Application of artificial intelligence for prediction, optimization, and control of thermal energy storage systems. *Thermal Science and Engineering Progress*, 39, 101730. https://doi.org/10.1016/j.tsep.2023.101730

Olabi, A. G., Abdelghafar, A. A., Maghrabie, H. M., Sayed, E. T., Rezk, H., Radi, M. A., Obaideen, K., & Abdelkareem, M. A. (2023). Application of artificial intelligence for prediction, optimization, and control of thermal energy storage systems. *Thermal Science and Engineering Progress*, 39, 101730. https://doi.org/10.1016/j.tsep.2023.101730

Olayode, O. I., Tartibu, L. K., & Okwu, M. O. (2020). Application of artificial intelligence in traffic control system of non-autonomous vehicles at signalized road intersection. *Procedia CIRP*, 91, 194–200. https://doi.org/10.1016/j.procir.2020.02.167

Osman, A. I., Chen, L., Yang, M., Msigwa, G., Farghali, M., Fawzy, S., Rooney, D. W., & Yap, P.-S. (2022). Cost, environmental impact, and resilience of renewable energy under a changing climate: A review. *Environmental Chemistry Letters*, 21, 741–764. https://doi.org/10.1007/s10311-022-01532-8

Othman, K. (2022). Exploring the implications of autonomous vehicles: A comprehensive review. *Innovative Infrastructure Solutions*, 7, 76. https://doi.org/10.1007/s41062-022-00763-6

Oyebode, O., & Stretch, D. (2019). Neural network modeling of hydrological systems: A review of implementation techniques. *Natural Resource Modeling*, 31(4), e12189. https://doi.org/10.1111/nrm.12189.

Page, M. J., McKenzie, J. E., Bossuyt, P. M., Boutron, I., Hoffmann, T. C., Mulrow, C. D., Shamseer, L., Tetzlaff, J. M., Akl, E. A., Brennan, S. E., Chou, R., Glanville, J., Grimshaw, J. M., Hróbjartsson, A., Lalu, M. M., Li, T., Loder, E. W., Mayo-Wilson, E., McDonald, S., McGuinness, L. A., Stewart, L. A., Thomas, J., Tricco, A. C., Welch, V. A., Whiting, P., & Moher, D. (2021). The prisma 2020 statement: An updated guideline for reporting systematic reviews. *International Journal of Surgery*, 88, 105906. https://doi.org/10.1016/j.ijsu.2021.105906.

Paracha, A., Arshad, J., Ben Farah, M., & Ismail, K. (2024). Machine learning security and privacy: A review of threats and countermeasures. *EURASIP Journal on Information Security*, 2024(10). https://doi.org/10.1186/s13635-024-00158-3

Parihar, V., Malik, A., Bhawna, Bhushan, B., & Chaganti, R. (2023). From smart devices to smarter systems: The evolution of artificial intelligence of things (AIoT) with characteristics, architecture, use cases, and challenges. In: *AI models for Blockchain-based Intelligent Networks in IoT Systems: Concepts, Methodologies, Tools, and Applications* (pp. 1–28). Springer.

Park, J.-h., Salim, M., Jo, J., Sicato, J., Rathore, S., & Park, J. (2019). CIoT-Net: A scalable cognitive IoT based smart city network architecture. *Human-centric Computing and Information Sciences*, 9, 29.

Peponi, A., Morgado, P., & Kumble, P. (2022). Life cycle thinking and machine learning for urban metabolism assessment and prediction. *Sustainable Cities and Society*, 80, 103754. https://doi.org/10.1016/j.scs.2022.103754

Perera, C., Qin, Y., Estrella, J. C., Reiff-Marganiec, S., & Vasilakos, A. V. (2017). Fog computing for sustainable smart cities: A survey. *ACM Comput Surv (CSUR), 50*(3), 1–43.

Petit, M. (2018). Toward a critique of algorithmic reason. A state-of-the-art review of artificial intelligence, its influence on politics and its regulation. *Quaderns del CAC, 44*(Vol. XXI, July 2018). https://ssrn.com/abstract=3279470.

Piasecki, A., Jurasz, J., & Adamowski, J. F. (2018). Forecasting surface water-level fluctuations of a small glacial lake in Poland using a wavelet-based artificial intelligence method. *Acta Geophysica, 66*(5), 1093–1107.

Popescu, S. M., Mansoor, S., Wani, O. A., Kumar, S. S., Sharma, V., Sharma, A., Arya, V. M., Kirkham, M. B., Hou, D., Bolan, N., & Chung, Y. S. (2024). Artificial intelligence and IoT driven technologies for environmental pollution monitoring and management. *Frontiers in Environmental Science, 12*, 1336088. https://doi.org/10.3389/fenvs.2024.1336088

Poul, A. K., Shourian, M., & Ebrahimi, H. (2019). A comparative study of MLR, KNN, ANN, and ANFIS models with wavelet transform in monthly stream flow prediction. *Water Resources Management, 33*, 2907–2923. https://doi.org/10.1007/s11269-019-02273-0

Priya, A. K., Devarajan, B., Alagumalai, A., & Song, H. (2023). Artificial intelligence enabled carbon capture: A review. *Science of the Total Environment, 886*, 163913. https://doi.org/10.1016/j.scitotenv.2023.163913

Puri, V., Srivastava, G., Rishiwal, V., & Singh, R. K. (2019). A hybrid artificial intelligence and Internet of Things model for generation of renewable resource of energy. *IEEE Access, 7*, 111181–111191. https://doi.org/10.1109/ACCESS.2019.2934228 P

Puri, V., Van Le, C., Kumar, R., & Jagdev, S. S. (2020). Fruitful synergy model of artificial intelligence and internet of things for smart transportation system. *International Journal of Hyperconnectivity and the Internet of Things (IJHIoT), 4*(1), 43–57. https://doi.org/10.4018/IJHIoT.2020010104

Qiao, Q., Tao, F., Wu, H., Yu, X., & Zhang, M. (2020). Optimization of a capacitated vehicle routing problem for sustainable municipal solid waste collection management using the PSO-TS algorithm. *International Journal of Environmental Research and Public Health, 17*(6), 2163.

Qin, M., Hu, W., Qi, X., & Chang, T. (2024). Do the benefits outweigh the disadvantages? Exploring the role of artificial intelligence in renewable energy. *Energy Economics, 131*, 107403. https://doi.org/10.1016/j.eneco.2024.107403

Raj, E. F. I., Appadurai, M., & Athiappan, K. (2021). Precision farming in modern agriculture. In *Smart Agriculture Automation Using Advanced Technologies* (pp. 294–329). Springer. https://doi.org/10.1007/978-981-16-6124-2_4

Raji, M., Dashti, A., Amani, P., & Mohammadi, A. H. (2019). Efficient estimation of CO2 solubility in aqueous salt solutions. *Journal of Molecular Liquids, 283*, 804–815.

Raman, R., Pattnaik, D., Lathabai, H. H., Kumar, C., Govindan, K., & Nedungadi, P. (2024). Green and sustainable AI research: An integrated thematic and topic modeling analysis. *Journal of Big Data, 11*, 55. https://doi.org/10.1186/s40537-024-00920-x

Rane, N., Choudhary, S., & Rane, J. (2024). Artificial Intelligence and machine learning in renewable and sustainable energy strategies: A critical review and future perspectives. SSRN. https://ssrn.com/abstract=4838761, https://doi.org/10.2139/ssrn.4838761

Raseman, W. J., Kasprzyk, J. R., Rosario-Ortiz, F. L., Stewart, J. R., & Livneh, B. (2017). Emerging investigators series: A critical review of decision support systems for water treatment: Making the case for incorporating climate change and climate extremes. *Environmental Science: Water Research & Technology, 3*(1), 18–36.

Rathore, M. M., Ahmad, A., Paul, A., & Rho, S. (2016). Urban planning and building smart cities based on the Internet of Things using Big Data analytics. *Computer Networks, 101*, 63–80.

Rathore, M. M., Paul, A., Hong, W. H., Seo, H., Awan, I., & Saeed, S. (2018). Exploiting IoT and big data analytics: Defining smart digital city using real-time urban data. *Sustainable Cities and Society, 40*, 600–610.

Rashid, A., & Kumari, S. (2023). Performance evaluation of ANN and ANFIS models for estimating velocity and pressure in water distribution networks. *Water Supply, 23*(9), 3925–3949. https://doi.org/10.2166/ws.2023.224

Raza, M., Awais, M., Ali, K., Aslam, N., Paranthaman, V. V., Imran, M., & Ali, F. (2020). Establishing effective communications in disaster affected areas and artificial intelligence based detection using social media platform. *Future Generation Computer Systems, 112*, 1057–1069.

Reddy, K. S. P., Roopa, Y. M., L. N., K. R., & Nandan, N. S. (2020). IoT based smart agriculture using machine learning. In *2020 Second International Conference on Inventive Research in Computing Applications (ICIRCA)* (pp. 130–134). IEEE. https://doi.org/10.1109/ICIRCA48905.2020.9183373

Rodríguez-Soto, C., Velazquez, A., Monroy-Vilchis, O., Lemes, P., & Loyola, R. (2017). Joint ecological, geographical and cultural approach to identify territories of opportunity for large vertebrates conservation in Mexico. *Biodiversity and Conservation, 26*(8), 1899–1918.

Rojek, I., & Studzinski, J. (2019). Detection and localization of water leaks in water nets supported by an ICT system with artificial intelligence methods as a way forward for smart cities. *Sustainability, 11*(2), 518. https://doi.org/10.3390/su11020518

Rolnick, D., Donti, P. L., Kaack, L. H., Kochanski, K., Lacoste, A., Sankaran, K., Ross, A. S., Milojevic-Dupont, N., Jaques, N., & Waldman-Brown, A. (2022). Tackling climate change with machine learning. *ACM Computing Surveys (CSUR), 55*(2), 1–96.

Sahoo, B. B., Jha, R., Singh, A., & Kumar, D. (2019). Application of support vector regression for modeling low flow time series. *KSCE Journal of Civil Engineering, 23*(2), 923–934

Salcedo-Sanz, S., Cuadra, L., & Vermeij, M. J. A. (2016). A review of Computational Intelligence techniques in coral reef-related applications. *Ecological Informatics, 32*, 107–123. https://doi.org/10.1016/j.ecoinf.2016.01.008

Saleem, S., & Mehrotra, M. (2022). Emergent use of artificial intelligence and social media for disaster management. In: Proceedings of International Conference on Data Science and Applications: ICDSA 2021 (vol. 2, pp. 195–210). Springer. https://doi.org/10.1007/978-981-16-5348-3_15

Saleh, M., Milovanoff, A., Posen, I. D., MacLean, H. L., & Hatzopoulou, M. (2022). Energy and greenhouse gas implications of shared automated electric vehicles. Transportation Research Part D: *Transport and Environment, 105*, 103233. https://doi.org/10.1016/j.trd.2022.103233

Salim, H. K., Padfield, R., Hansen, S. B., Mohamad, S. E., Yuzir, A., Syayuti, K., Tham, M. H., & Papargyropoulou, E. (2018). Global trends in environmental management system and ISO14001 research. *Journal of Cleaner Production, 170*, 645–653. https://doi.org/10.1016/j.jclepro.2017.09.017

Salluri, D. K., Bade, K., & Madala, G. (2020). Object detection using convolutional neural networks for natural disaster recovery. *International Journal of Safety and Security Engineering, 10*(2), 285–291. https://doi.org/10.18280/ijsse.100217

Samadi, S. (2022). The convergence of AI, IoT, and big data for advancing flood analytics research. *Frontiers in Water, 4*, 786040. https://doi.org/10.3389/frwa.2022.786040

Samaei, S. R., & Hassanabad, M. G. (2024). The crucial interplay of seas, marine industries, and artificial intelligence in sustainable development. In: *Eighth International Conference on Technology Development in Oil, Gas, Refining and Petrochemicals* (vol. 8). https://www.researchgate.net/publication.

Sanchez, T. W. (2023). Planning on the verge of AI, or AI on the verge of planning. *Urban Science, 7*, 70. https://doi.org/10.3390/urbansci7030070

Santangeli, A., Chen, Y., Kluen, E., Chirumamilla, R., Tiainen, J., & Loehr, J. (2020). Integrating drone-borne thermal imaging with artificial intelligence to locate bird nests on agricultural land. *Scientific Reports, 10*(1), 10993. https://doi.org/10.1038/s41598-020-67898-3

Schultz, M. (2022). Artificial intelligence for air quality. *Project Repository Journal, 12*, 70–73. https://www.europeandissemination.eu/article/artificial-intelligence-for-air-quality/18384

Schürholz, D., Kubler, S., & Zaslavsky, A. (2020). Artificial intelligence-enabled context-aware air quality prediction for smart cities. *Journal of Cleaner Production, 271*, 121941. https://doi.org/10.1016/j.jclepro.2020.121941

Schwartz, R., Dodge, J., Smith, N. A., & Etzioni, O. (2020). Green AI. *Communications of the ACM, 63*(12), 54–63. https://doi.org/10.1145/3381831

Sellak, H., Ouhbi, B., Frikh, B., & Palomares, I. (2017). Toward next-generation energy planning decision-making: An expert-based framework for intelligent decision support. *Renewable and Sustainable Energy Reviews, 80*, 1544–1577. https://doi.org/10.1016/j.rser.2017.07.013

Sempere-Payá, V., & Santonja-Climent, S. (2012). Integrated sensor and management system for urban waste water networks and prevention of critical situations. *Computers, Environment and Urban Systems, 36*, 65–80.

Seng, K. P., Ang, L. M., & Ngharamike, E. (2022). Artificial intelligence Internet of Things: A new paradigm of distributed sensor networks. *International Journal of Distributed Sensor Networks*, 18(3). https://doi.org/10.1177/15501477211062835

Seuring, S., & Gold, S. (2012). Conducting content-analysis based literature reviews in supply chain management. *Supply Chain Management*, 17(5), 544–555. https://doi.org/10.1108/13598541211258609

Şeyma, M. K. M. (2023). Utilizing artificial intelligence (AI) for the identification and management of marine protected areas (MPAs): A review. *Journal of Geoscience and Environment Protection*, 11, 118–132. https://doi.org/10.4236/gep.2023.119008

Shaamala, A., Yigitcanlar, T., Nili, A., & Nyandega, D. (2024). Algorithmic green infrastructure optimization: Review of artificial intelligence driven approaches for tackling climate change. *Sustainable Cities and Society*, 101, 105182. https://doi.org/10.1016/j.scs.2024.105182

Shahrokni, H., lazarevic, D., & Brandt, N. (2015). Smart urban metabolism: Toward a real-time understanding of the energy and material flows of a city and its citizens. *Journal of Urban Technology*, 22(1), 65–86.

Sharps, K., Masante, D., Thomas, A., Jackson, B., Redhead, J., May, L., Prosser, H., Cosby, B., Emmett, B., & Jones, L. (2017). Comparing strengths and weaknesses of three ecosystem services modelling tools in a diverse UK river catchment. *Science of the Total Environment*, 584-585, 118–130. https://doi.org/10.1016/j.scitotenv.2016.12.160

Shen, C. (2018). A transdisciplinary review of deep learning research and its relevance for water resources scientists. *Water Resources Research*, 54(11), 8558–8593. https://doi.org/10.1029/2018WR022643

Shen, R., Huang, A., Li, B., & Guo, J. (2019). Construction of a drought monitoring model using deep learning based on multi-source remote sensing data. *International Journal of Applied Earth Observation and Geoinformation*, 79, 48–57. https://doi.org/10.1016/j.jag.2019.03.006

Shi, F., Ning, H., Huangfu, W., & Wei, D. (2020). Recent progress on the convergence of the Internet of Things and artificial intelligence. *IEEE Network*, 34(5), 8–15. https://doi.org/10.1109/MNET.011.2000070

Shin, W., Han, J., & Rhee, W. (2021). AI-assistance for predictive maintenance of renewable energy systems. *Energy*, 221, 119775. https://doi.org/10.1016/j.energy.2021.119775

Silva, Ó., Cordera, R., González-González, E., & Nogués, S. (2022). Environmental impacts of autonomous vehicles: A review of the scientific literature. *Science of The Total Environment*, 830, 154615. https://doi.org/10.1016/j.scitotenv.2022.154615

Silvestro, D., Goria, S., Sterner, T. et al. (2022). Improving biodiversity protection through artificial intelligence. *Nature Sustainability*, 5, 415–424. https://doi.org/10.1038/s41893-022-00851-6

Sinaeepourfard, A. J., Garcia, X. M.-B., Marín-Tordera, E., Cirera, J., Grau, G., & Casaus, F. (2016). Estimating smart city sensors data generation current and future data in the city of Barcelona. In: *Proceedings of Conference: The 15th IFIP Annual Mediterranean Ad Hoc Networking Workshop.*

Skiba, M., Mrówczyńska, M., & Bazan-Krzywoszańska, A. (2017). Modeling the economic dependence between town development policy and increasing energy effectiveness with neural networks. Case study: The town of Zielona Góra. *Applied Energy, 188,* 356–366. https://doi.org/10.1016/j.apenergy.2016.12.006

Sleem, A., & Elhenawy, I. (2023). Survey of Artificial Intelligence of Things for Smart Buildings: A closer outlook. *Journal of Intelligent Systems and Internet of Things, 8*(2), 63–71. https://doi.org/10.54216/JISIoT.080206

Solaymani, S. (2019). CO2 emissions patterns in 7 top carbon emitter economies: the case of transport sector. *Energy, 168,* 989–1001. https://doi.org/10.1016/j.energy.2018.11.145

Son, T. H., Weedon, Z., Yigitcanlar, T., Sanchez, T., Corchado, J. M., & Mehmood, R. (2023). Algorithmic urban planning for smart and sustainable development: Systematic review of the literature. *Sustainable Cities and Society, 94,* 104562.

Sonetti, G., Naboni, E., & Brown, M. (2018). Exploring the potentials of ICT tools for human-centric regenerative design. *Sustainability, 10*(4), 1217. https://doi.org/10.3390/su10041217

Sowmya, V., & Ragiphani, S. (2022). Air quality monitoring system based on artificial intelligence. In: P. Kumar Jain, Y. Nath Singh, R. P. Gollapalli, & S. P. Singh (Eds.), *Advances in Signal Processing and Communication Engineering* (p. 26). Springer. https://doi.org/10.1007/978-981-19-5550-1_26

Späth, P., & Rohracher, H. (2011). The "eco-cities" Freiburg and Graz: The social dynamics of pioneering urban energy and climate governance. In: H. Bulkeley, V. Castan-Broto, M. Hodson, & S. Marvin (Eds.), *Cities and Low Carbon Transitions. Routledge Studies in Human Geography* (pp. 88–106). Routledge.

Späth, P., Hawxwell, T., John, R., Li, S., Löffler, E., Riener, V., & Utkarsh, S. (2017). *Smart-eco Cities in Germany: Trends and City Profiles.* University of Exeter Press.

Stein, A. L. (2020). Artificial intelligence and climate change. *Yale Journal on Regulation, 37,* 890.

Su, S., Yan, X., Agbossou, K., Chahine, R., & Zong, Y. (2022). Artificial intelligence for hydrogen-based hybrid renewable energy systems: A review with case study. *Journal of Physics: Conference Series, 2208,* 012013.

Sun, W., Bocchini, P., & Davison, B. D. (2020). Applications of artificial intelligence for disaster management. *Natural Hazards, 103*(3), 2631–2689. https://doi.org/10.1007/s11069-020-04124-3

Swarna, K. P., & Bhaumik, P. (2022). Disaster management through integrative AI. In: *Proceedings of the 23rd International Conference on Distributed Computing and Networking* (pp. 290–293). https://doi.org/10.1145/3491003.3493235

Szpilko, D., Naharro, F. J., Lăzăroiu, G., Nica, E., & Gallegos, A. D. L. T. (2023). Artificial Intelligence in the smart city: A literature review. *Engineering Management in Production and Services*, *15*(4), 53–75. https://doi.org/10.2478/emj-2023-0028.

Talan, G., & Sharma, G. D. (2019). Doing well by doing good: A systematic review and research agenda for sustainable investment. *Sustainability*, *11*(2), 353. https://doi.org/10.3390/su11020353.

Tan, L., Guo, J., Mohanarajah, S., & Zhou, K. (2021). Can we detect trends in natural disaster management with artificial intelligence? A review of modeling practices. *Natural Hazards*, *107*(3), 2389–2417.

Tarek, S. (2023). Smart eco-cities conceptual framework to achieve UN-SDGs: A case study application in Egypt. *Civil Engineering and Architecture*, *11*(3), 1383–1406. https://doi.org/10.13189/cea.2023.110322.

Thakker, D., Mishra, B. K., Abdullatif, A., Mazumdar, S., & Simpson, S. (2020). Explainable artificial intelligence for developing smart cities solutions. *Smart Cities*, *3*, 1353–1382.

Thamik, H., Cabrera, J. D. F., & Wu, J. (2024). The digital paradigm: Unraveling the impact of artificial intelligence and internet of things on achieving sustainable development goals. In: S. Misra, K. Siakas, & G. Lampropoulos (Eds.), *Artificial Intelligence of Things for Achieving Sustainable Development Goals*. Lecture Notes on Data Engineering and Communications Technologies (vol. 192). Springer. https://doi.org/10.1007/978-3-031-53433-1_2

Toivonen, T., Heikinheimo, V., Fink, C., Hausmann, A., Hiippala, T., Järv, O., ... & Di Minin, E. (2019). Social media data for conservation science: A methodological overview. *Biological Conservation*, *233*, 298–315. https://doi.org/10.1016/j.biocon.2019.01.023.

Tomazzoli, C., Scannapieco, S., & Cristani, M. (2020). Internet of Things and artificial intelligence enable energy efficiency. *Journal of Ambient Intelligence and Humanized Computing*, *14*, 4933–4954. https://doi.org/10.1007/s12652-020-02151-3

Tyagi, A. K., & Aswathy, S. U. (2021). Autonomous intelligent vehicles (AIV): Research statements, open issues, challenges, and road for future. *International Journal of Intelligent Networks*, *2*, 83–102. https://doi.org/10.1016/j.ijin.2021.07.002

Tyralis, H., Papacharalampous, G., & Langousis, A. (2019). A brief review of random forests for water scientists and practitioners and their recent history in water resources. *11*(5), 910. https://www.mdpi.com/2073-4441/11/5/910

Ullah, A., Anwar, S. M., Li, J., Nadeem, L., Mahmood, T., Rehman, A., & Saba, T. (2023). Smart cities: The role of Internet of Things and machine learning in realizing a data-centric smart environment. *Complex Intelligent Systems*, *10*, 1607–1637. https://doi.org/10.1007/s40747-023-01175-4.

Ullah, A., Salehnia, N., Kolsoumi, S., Ahmad, A., & Khaliq, T. (2018). Prediction of effective climate change indicators using statistical downscaling approach and impact assessment on pearl millet (*Pennisetum glaucum* L.) yield through genetic algorithm in Punjab, Pakistan. *Ecological Indicators*, *90*, 569–576. https://doi.org/10.1016/j.ecolind.2018.03.053

Ullah, Z., Al-Turjman, F., Mostarda, L., & Gagliardi, R. (2020). Applications of artificial intelligence and machine learning in smart cities. *Computer Communications, 154*, pp.313–323.

Valizadeh, N., Mirzaei, M., Allawi, M. F., Afan, H. A., Mohd, N. S., Hussain, A., & El-Shafie, A. (2017). Artificial intelligence and geo-statistical models for stream-flow forecasting in ungauged stations: State of the art. *Natural Hazards, 86*(3), 1377–1392. https://doi.org/10.1007/s11069-017-2740-7

Vázquez-Canteli, J., Ulyanin, S., Kämpf, J., & Nagy, Z. (2018). Fusing TensorFlow with building energy simulation for intelligent energy management in smart cities. *Sustainable Cities and Society, 45*, 243–257. https://doi.org/10.1016/j.scs.2018.11.021

Velev, D., & Zlateva, P. (2023). Challenges of artificial intelligence application for disaster risk Management. *The International Archives of the Photogrammetry, Remote Sensing and Spatial Information Sciences, 48*, 387–394. https://doi.org/10.5194/isprs-archives-XLVIII-M-1-2023-387-2023

Villon, S., Mouillot, D., Chaumont, M., Darling, E. S., Subsol, G., Claverie, T., & Villéger, S. (2018). A deep learning method for accurate and fast identification of coral reef fishes in underwater images. *Ecological Informatics, 48*, 238–244. https://doi.org/10.1016/j.ecoinf.2018.09.007

Vinuesa, R., Azizpour, H., Leite, I., Balaam, M., Dignum, V., Domisch, S., ... & Fuso Nerini, F. (2020). The role of artificial intelligence in achieving the Sustainable Development Goals. *Nature Communications, 11*(1), 233. https://doi.org/10.1038/s41467-019-14108-y

Vu, H. L., Bolingbroke, D., Ng, K. T. W., & Fallah, B. (2019). Assessment of waste characteristics and their impact on GIS vehicle collection route optimization using ANN waste forecasts. *Waste Management, 88*, 118–130.

Walsh, T., Evatt, A., & de Witt, C. S. (2020). Artificial intelligence & climate change. *Supplementary Impact Report, Oxford, 1*, 1–15

Wang, P., Yao, J., Wang, G., Hao, F., Shrestha, S., Xue, B., Xie, G., & Peng, Y. (2019). Exploring the application of artificial intelligence technology for identification of water pollution characteristics and tracing the source of water quality pollutants. *Science of the Total Environment, 693*, 133440. https://doi.org/10.1016/j.scitotenv.2019.07.246

Wang, S., & Qu, X. (2019). Blockchain applications in shipping, transportation, logistics, and supply chain. In: X. Qu, L. Zhen, R. Howlett, & L. Jain (Eds.), *Smart Transportation Systems 2019* (Vol. 149, pp. 225–231). Springer, Singapore. https://doi.org/10.1007/978-981-13-8683-1_23.

Wang, Y Q., Yao, J. T. Kwok, & L. M. Ni (2020). Generalizing from a few examples: A survey on few-shot learning. *ACM Computing Surveys, 53*(3), 1–34, .

Wang, Y., Zhang, B., Ma, J., & Jin, Q. (2023). Artificial intelligence of things (AIoT) data acquisition based on graph neural networks: A systematic review. *Concurrency and Computation: Practice and Experience, e7827*. doi:https://doi.org/10.1002/cpe.7827.

Wang, Z., & Liao, H.-T. (2020). Towards the eco-design of Artificial Intelligence and Big Data applications: A bibliometric analysis of related research. *IOP Conference Series: Materials Science and Engineering, 806*(1), 012039.

Wang, Z., & Srinivasan, R. S. (2017). A review of artificial intelligence-based building energy use prediction: Contrasting the capabilities of single and ensemble prediction models. *Renewable and Sustainable Energy Reviews*, 75, 796–808. https://doi.org/10.1016/j.rser.2016.10.079

Ward, D., Melbourne-Thomas, J., Pecl, G. T., Evans, K., Green, M., Mccormack, P. C., Novaglio, C., Trebilco, R., Bax, N., Brasier, M. J., Cavan, E. L., Edgar, G., Hunt, H. L., Jansen, J., Jones, R., Lea, M.-A., Makomere, R., Mull, C., Semmens, J. M., Shaw, J., Tinch, D., van Steveninck, T. J., & Layton, C. (2022). Safeguarding marine life: Conservation of biodiversity and ecosystems. *Reviews in Fish Biology and Fisheries*, 32(1), 65–100. https://doi.org/10.1007/s11160-022-09700-3

Wazid, M., Das, A. K., Chamola, V., & Park, Y. (2022). Uniting cyber security and machine learning: Advantages, challenges and future research. *ICT Express*, 8(3), 313–321. https://doi.org/10.1016/j.icte.2022.04.007

Wei, M. C. F., Maldaner, L. F., Ottoni, P. M. N., & Molin, J. P. (2020). Carrot yield mapping: A precision agriculture approach based on machine learning. *AI*, 1(2), 229–241. https://doi.org/10.3390/ai1020015

Willcock, S., Martínez-López, J., Hooftman, D. A. P., Bagstad, K. J., Balbi, S., Marzo, A., ... & Athanasiadis, I. N. (2018). Machine learning for ecosystem services. *Ecosystem Services*, 33, 165–174. https://doi.org/10.1016/j.ecoser.2018.04.004

Woo, T. H. (2019). Global warming analysis for greenhouse gases impacts comparable to carbon-free nuclear energy using neuro-fuzzy algorithm. 17(2), 219–233. https://doi.org/10.1504/ijgw.2019.097862

Yang, M., Chen, L., Wang, J., Msigwa, G., Osman, A. I., Fawzy, S., Rooney, D. W., & Yap, P.-S. (2023). Circular economy strategies for combating climate change and other environmental issues. *Environmental Chemistry Letters*, 21, 55–80. https://doi.org/10.1007/s10311-022-01499-6

Yang, T.-H., & Liu, W.-C. (2020). A general overview of the risk-reduction strategies for floods and droughts. *Sustainability*, 12(7), 2687.

Yeung, P. S., Fung, J. C.-H., Ren, C., Xu, Y., Huang, K., Leng, J., & Wong, M. M.-F. (2020). Investigating future urbanization's impact on local climate under different climate change scenarios in MEGA-urban regions: A case study of the Pearl River Delta, China. *Atmosphere*, 11(7), 771. https://doi.org/10.3390/atmos11070771

Yigitcanlar, T., Desouza, K. C., Butler, L., & Roozkhosh, F. (2020). Contributions and risks of artificial intelligence (AI) in building smarter cities: Insights from a systematic review of the literature. *Energies*, 13(6), 1473.

Yin, Z., Feng, Q., Yang, L., Deo, R. C., Wen, X., Si, J., & Xiao, S. (2017). Future projection with an extreme-learning machine and support vector regression of reference evapotranspiration in a mountainous inland watershed in North-West China. 9(11), 880. https://doi.org/10.3390/w9110880

Youssef, A., El-Telbany, M., & Zekry, A. (2017). The role of artificial intelligence in photo-voltaic systems design and control: A review. *Renewable and Sustainable Energy Reviews*, 78, 72–79. https://doi.org/10.1016/j.rser.2017.04.046

Yu, H., Yang, Z., & Sinnott, R. (2018). Decentralized big data auditing for smart city environments leveraging blockchain technology. *IEEE Access, 7*, 6288–6296. https://doi.org/10.1109/ACCESS.2018.2888940

Zaidi, A., Ajibade, S.-S. M., Musa, M., & Bekun, F. V. (2023). New insights into the research landscape on the application of artificial intelligence in sustainable smart cities: A bibliometric mapping and network analysis approach. *International Journal of Energy Economics and Policy, 13*(4), 287–299. https://doi.org/10.32479/ijeep.14683

Zhang, A.-B., Hao, M.-D., Yang, C.-Q., & Shi, Z.-Y. (2017). Barcoding R: An integrated r package for species identification using DNA barcodes. *8*(5), 627–634. https://doi.org/10.1111/2041-210X.12682

Zhang, J., & Letaief, K. B. (2019). Mobile edge intelligence and computing for the internet of vehicles. *Proceedings of IEEE, 108*, 246–261.

Zhang, J., & Tao, D. (2021). Empowering things with intelligence: A survey of the progress, challenges, and opportunities in artificial intelligence of things. *IEEE Internet of Things Journal, 8*(10), 7789–7817. https://doi.org/10.1109/JIOT.2020.3039359.

Zhang, Y., Ruohan, Z., Shang, L., Zeng, H., Yue, Z., & Wang, D. (2023, August). On optimizing model generality in ai-based disaster damage assessment: A subjective logic-driven crowd-ai hybrid learning approach. In: *International Joint Conferences on Artificial Intelligence Organization*. IEEE. https://doi.org/10.24963/ijcai.2023/701

Zhang, Z., Zheng, Y., Qian, L., Luo, D., Dou, H., Wen, G., Yu, A., & Chen, Z. (2022). Emerging trends in sustainable CO_2-management materials. *Advanced Materials, 34*, 2201547. https://doi.org/10.1002/adma.202201547

Zhao, C., Dong, K., Wang, K., & Nepal, R. (2024). How does artificial intelligence promote renewable energy development? The role of climate finance. *Energy Economics,* 133, 107493. https://doi.org/10.1016/j.eneco.2024.107493

Zhou, T., Han, G., Xu, X., Han, C., Huang, Y., & Qin, J. (2019). A learning-based multimodel integrated framework for dynamic traffic flow forecasting. *Neural Processing Letters, 49*(1), 407–430. https://doi.org/10.1007/s11063-018-9804-x

Zhu, M., Wang, J., Yang, X., Zhang, Y., Zhang, L., Ren, H., Wu, B., & Ye, L. (2022). A Review of the Application of Machine Learning in Water Quality Evaluation. *Eco-Environment & Health, 1*, 107-116. https://doi.org/10.1016/j.eehl.2022.06.001

Zhuang, F., Qi, Z., Duan, K., Xi, D., Zhu, Y., Zhu, H., Xiong, H., He, Q., & Xiong, Y. (2020). A comprehensive survey on transfer learning. *Proceedings of IEEE, 109*(1), 43–ll76. https://doi.org/10.1109/JPROC.2020.3004555.

CHAPTER 4

Artificial Intelligence of Things for Advancing Smart Sustainable Cities

A Synthesized Case Study Analysis of City Brain, Digital Twin, Metabolism, and Platform

Abstract

Smart sustainable cities are increasingly embracing cutting-edge technologies and collaborative models to enhance environmental outcomes and interventions through the evolving planning and governance systems driven by Artificial Intelligence of Things (AIoT). Despite this notable progress, the current scholarly landscape lacks comprehensive studies on the synergistic integration of these systems for the advancement of smart sustainable cities. To address this critical gap, this chapter investigates the transformative potential and intricate interplay of City Brain, Urban Digital Twin (UDT), Smart Urban Metabolism (SUM), and platform urbanism—as prominent urban planning and governance systems—within the framework of "Artificial Intelligence of

Smart Sustainable City Things." This investigation is guided by four specific objectives. First, it analyzes and delineates the roles played by City Brain, UDT, SUM, and platform urbanism in advancing data-driven environmental urban planning and governance. Second, it explores how the distinct functions of these systems can be seamlessly integrated. Third, it examines the contribution of AIoT functionalities and capabilities to the integrated operation of these systems. Finally, it identifies the key anticipated challenges and obstacles in the integration of these systems' functions and proposes strategies to address them. This chapter adopts a research design integrating case study analysis and literature review, allowing for a thorough and nuanced exploration of real-world implementations and their foundational underpinnings to enhance the depth and breadth of the study. The findings extend beyond theoretical enrichment, providing invaluable insights that empower urban planners, policymakers, and practitioners to adopt integrated and holistic solutions. By exploring the potentials, synergies, and opportunities demonstrated by the real-world implementation of City Brain, UDT, SUM, and platform urbanism, city stakeholders acquire the essential knowledge for making informed decisions, including implementing effective strategies and designing policies that prioritize the collective intelligence of emerging data-driven planning and governance systems, thereby advancing sustainable urban development.

4.1 INTRODUCTION

In recent years, smart sustainable cities have surged to prominence as a paradigm shift in urbanism aimed at advancing the objectives of sustainable urban development and navigating the complexities inherent in urban planning and governance. These cities strategically employ and leverage innovative technologies and sustainable strategies to harmonize the environmental, social, and economic dimensions of sustainability (Ahvenniemi et al. 2017; Bibri and Krogstie 2020a). Representing a transformative trajectory in urban development, they place a distinct emphasis on bolstering environmental sustainability strategies (Bibri and Krogstie 2020b; Formisano et al. 2022; Corsi et al. 2022; Thornbush and Golubchikov 2019; Yigitcanlar and Cugurullo 2020)—i.e., the responsible and balanced utilization of natural resources and ecosystems to meet present needs without compromising the ability of future generations to meet their own needs. This paradigm shift seeks to achieve a seamless integration of data-driven technologies with environmental technologies to enhance intelligence and efficiency while maintaining a focus on the long-term objectives of environmentally sustainable urban development (Bibri 2020, Bibri et al. 2023a). The primary focus of this study lies in the imperative task of advancing these objectives through innovative approaches to urban planning and governance, harnessing the key capabilities of Artificial Intelligence (AI) and Artificial Intelligence of Things (AIoT) for the resilience and enduring vitality of urban landscapes.

Moreover, the landscape of urban planning and governance systems has experienced significant metamorphosis, largely propelled by the convergence of AI, the Internet of Things (IoT), and Big Data, collectively known as AIoT. This convergence has fundamentally altered the way cities are planned and governed, ushering in a new era of innovation and efficiency in sustainable urban development (e.g., Ajuriaguerra et al. 2022; Bibri et al. 2024a, b; Nishant et al. 2020; Koumetio et al. 2023; Marasinghe et al. 2024; Samuel et al. 2023; Sanchez et al. 2023; Son et al. 2024; Wang et al. 2022), especially sustainable smart cities. It signifies a groundbreaking approach, highlighting the synergistic potential of these advanced technologies in reshaping urban strategies and policies to particularly respond to and mitigate environmental challenges. Indeed, it has demonstrated a significant impact on various fronts. These pertain to optimizing or enhancing resource management, energy efficiency, waste management, water management, transportation management, biodiversity protection, CO_2 emissions reduction, and climate change mitigation (e.g., Alahi et al. 2023; Bibri et al. 2023a, b; Chen et al. 2023; El Himer et al. 2022; Jagatheesaperumal et al. 2024; Jain et al. 2023; Leal Filho et al. 2022; Samadi 2022; Rane et al. 2024; Seng et al. 2022; Zhang and Tao 2021). AIoT stands on the brink of a major turning point, with substantial transformations and innovations ready to unfold, aimed at fostering more sustainable, efficient, resilient, and environmentally conscious urban environments.

While the application of AI and AIoT may vary across cities and domains, their widespread adoption is giving rise to what has been identified as large-scale AI and AIoT systems, gaining increasing momentum and prevalence worldwide. These data-driven systems are evolving into urban planning and governance paradigms, presenting promising solutions for the development of smarter and more sustainable cities. Among these notable paradigms are City Brain (Bibri et al. 2024a; Kierans et al. 2023; Liu et al. 2022; Xu 2024; Wang et al. 2022), Urban Digital Twin (UDT)

(Beckett's 2022; Bibri 2024b; Li et al. 2021), Smart Urban Metabolism (SUM) (Bibri 2024a; Ghosh and Sengupta 2023; Peponi et al. 2022), and platform urbanism (ITU 2022; Haveri and Anttiroiko 2023; Katmada et al. 2023). Notably, these paradigms present synergistic opportunities to advance city planning and redefine city governance, aiming to influence environmental outcomes and actions through real-time monitoring and resource management, data-driven decision-making, and collaborative and participatory approaches to governance. For example, acting as a mirror reflecting the environmental dynamics of the city, UDT serves as a powerful tool not only for urban planning but also adaptive governance strategies (Dar et al. 2023; Deng et al. 2021; Weil et al. 2023).

Despite notable progress in emerging urban planning and governance systems, a critical imperative persists in understanding and harnessing their synergistic integration for the advancement of smart sustainable cities. Building on this premise, the linchpin resides in the transformative potential of AI and AIoT. The intricate interplay between these AIoT-powered systems forms a complex tapestry that requires nuanced insights into their interconnected functions and roles in advancing environmental sustainability. This endeavor is essential to weave a cohesive and intelligent urban ecosystem that transcends the silos of individual planning and governance systems, aligning with the overarching goal of, and the holistic approach to, environmental sustainability. At the heart of this effort is the visionary concept of "Artificial Intelligence of Smart Sustainable City Things" (Bibri and Huang 2024; Jagatheesaperumal et al. 2024). This approach entails the convergence of AI models and techniques and the collective network of physical objects associated with the brain and digital twin of smart sustainable cities, resulting in an intelligently interwoven urban ecosystem. This visionary concept signifies a paradigm shift, leveraging the power of AIoT to not only enhance the individual functions of the planning and governance systems of smart sustainable cities but also to interconnect them dynamically, fostering a holistic approach to urban development.

This chapter aims to investigate the innovative potential and intricate interplay of City Brain, UDT, SUM, and platform urbanism—prominent urban planning and governance systems within the framework of Artificial Intelligence of Smart Sustainable City Things. This is motivated by the clear synergy in their operations, aiming to propel the planning and governance of smart sustainable cities to new heights. To guide this investigation, the following four questions are formulated:

RQ1: How do City Brain, UDT, SUM, and platform urbanism contribute to the evolution of urban planning and governance within the framework of Artificial Intelligence of Smart Sustainable City Things?

RQ2: How can the distinct functions of these planning and governance systems be seamlessly integrated?

RQ3: In what ways do the functionalities and capabilities of AI and AIoT contribute to this integration?

RQ4: What challenges and obstacles are anticipated in exploring the synergistic integration of the functions of these systems enabled by AI and AIoT, and how can they be addressed and overcome?

By addressing these questions, this multifaceted exploration enriches the body of knowledge base guiding future urban planning and governance initiatives, with a particular focus on environmental sustainability. Specifically, the study's noteworthy contributions include:

- Offering an in-depth understanding of the diverse roles played by the brain, digital twin, metabolism, and platform of smart sustainable cities in advancing their environmental planning and governance on the basis of AI and AIoT technologies.

- Providing nuanced insights into the seamless integration of the distinct functions of these data-driven planning and governance systems, leveraging the computational and analytical capabilities of AI and AIoT.

- Unveiling how the functionalities and capabilities of AI and AIoT contribute to the cohesive integration of these systems, fostering technological synergy for sustainable urban development.

- Identifying anticipated challenges and obstacles in exploring not only the synergy in the technical operation of these systems but also the integration of their functions and proposing future directions to address them. This contributes to propelling the planning and governance of smart sustainable cities to new levels of effectiveness in navigating environmental challenges.

In essence, the study contributes actionable and novel insights for advancing smart sustainable urban development, providing a foundation for informed decision-making and policy formulation in the pursuit of environmental sustainability. These contributions collectively expand the rapidly evolving research landscape, empowering diverse city stakeholders with invaluable information to grasp the existing intricacies and future trajectories of urban planning and governance.

The remainder of this chapter unfolds as follows: Section 4.2 describes and justifies the applied research design and methodology. Section 4.3 introduces and discusses the conceptual framework of Artificial Intelligence of Smart Sustainable City Things. Section 4.4 presents the results of the case study analysis and aggregative synthesis. Section 4.5 discusses key challenges, obstacles, knowledge gaps, and future research directions. Section 4.6 concludes the study with a summary of findings and implications.

4.2 RESEARCH METHODOLOGY

This study adopted a qualitative research design to explore the integration of the brain, digital twin, metabolism, and platform of smart sustainable cities through the unifying power of AI and AIoT technologies. This methodological approach is deemed most appropriate to capture the intricacies, contexts, and multifaceted dynamics inherent in the symbiotic relationship enabled by these technologies and between these emerging planning and governance systems. The study employed a case study methodology to investigate a varied and balanced spectrum of real-world cities in terms of their evolving environmental planning and governance practices. Carefully selected, the case studies provided context-specific insights into the application of City Brain, UDT, SUM, and platform urbanism, mainly in European cities. The empirical evidence and practical insights derived from the case studies were supported by a comprehensive literature review to ensure that this research is grounded in both real-world implementations and theoretical knowledge. As shown in Figure 4.1, the study methodology consists of eight stages.

To lay a robust foundation for this study, a comprehensive review of the extant literature was conducted, focusing on City Brain, UDT, SUM, and platform urbanism. This review encompassed a diverse range of peer-reviewed articles, conference proceedings, reports, policy documents, and other documentation providing insights into the development, implementation, and impact of these planning and governance systems. The goal was to gather valuable insights, establish a nuanced understanding of the subject matter, gain a deeper understanding of theoretical frameworks and contextual landscapes, and identify potential challenges and gaps. This foundational step is crucial for informing the case study

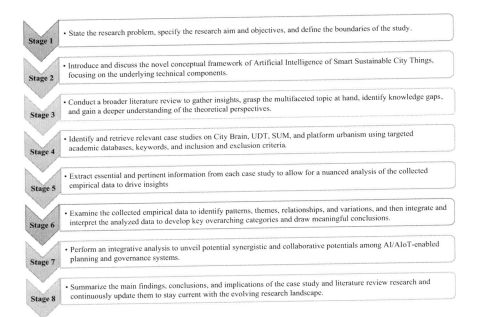

FIGURE 4.1 A flow diagram outlining the eight methodological stages of the study.

research in terms of both analysis and synthesis, ensuring that the study builds upon existing knowledge while contributing novel perspectives to the evolving discourse on the integration of AI and AIoT technologies in urban planning and governance. For what it entailed, the literature review set the stage for the subsequent data search and extraction processes.

The insights gained from the broader literature review served as a basis to narrow down the focus as to identifying and retrieving the relevant case studies as the primary sources for understanding the real-world implementation of City Brain, UDT, SUM, and platform urbanism in the realm of smart sustainable cities. The search process encompassed a structured approach to identify these case studies. It involved utilizing academic databases, including Web of Science (WoS), Scopus, and ScienceDirect, as they are considered to be more reliable and trustworthy sources of academic literature. A set of chosen keywords related to each city planning and governance system, namely "City Brain," "Urban Digital Twin," "Smart Urban Metabolism," and "Platform Urbanism," combined with "Smart Cities," "Smart Sustainable Cities," "Artificial Intelligence," and "Artificial

Intelligence of Things," ensured specificity in the search and retrieval of pertinent literature.

The inclusion criteria filtered studies based on their relevance, reliability, language, publication date, and publication type (article, conference paper, or book chapter), providing definitive primary information. Exclusion criteria were applied to remove studies unrelated to the focal topics and irrelevant to the research aim and questions. Based on these predefined inclusion/exclusion criteria, the selection process involved initial screening based on titles and abstracts, followed by a detailed full-text review for eligibility. Accordingly, the study compiled a set of relevant case studies offering diverse perspectives and ensuring the generation of nuanced insights contributing to the overarching aim of the study.

The data extraction process aimed to gather pertinent information from each identified case study. Details such as city planning or governance system, linkages between them, linkages between AI and AIoT and planning and governance, and knowledge gaps were recorded. Information on how AI and AIoT were integrated into the functionalities of the identified city planning and governance systems was captured. The extraction also covered insights into how the integration influenced urban planning and governance, challenges faced, opportunities presented, and any findings related to environmental sustainability. Overall, the obtained insights involved identifying key themes, patterns, and variations in the development and implementation of City Brain, UDT, SUM, and platform urbanism in the context of smart sustainable cities. It ensured a comprehensive analysis, allowing for meaningful comparisons across different cases with respect to their commonalities and differences in the context of urban planning and governance.

The process of data analysis involved a systematic examination of the collected empirical data, aiming to identify patterns, themes, variations, and insightful nuances. The qualitative nature of the research captured the nuances of the functions of City Brain, UDT, SUM, and platform urbanism in relation to planning and governance. The data synthesis phase encompassed the integration, interpretation, and development of overarching themes from the analyzed data. This entailed categorizing information based on commonalities and differences, formulating conceptual and descriptive categories that encapsulated the essence of the data, and constructing a narrative that interwove key findings and insights.

Following the thematic analysis and synthesis, an integrative analysis was conducted to deepen the understanding of the unified central topic.

This involved amalgamating, consolidating, and organizing information derived from both the literature review and case study research. The objective was to produce integrative insights, thereby contributing to a more cohesive and insightful interpretation of the subject matter. This pertains to the synergistic and collaborative nature of the emerging planning and governance systems within the framework of Artificial Intelligence of Smart Sustainable City Things, with a particular focus on environmental sustainability. The integrative analysis added depth to understanding the interconnected functions of the brain, digital twin, metabolism, and platform for smart sustainable cities, highlighting the various contributions of the computational and analytical capabilities of the technical components of AIoT to environmentally sustainable urban development goals.

4.3 ARTIFICIAL INTELLIGENCE OF SMART SUSTAINABLE CITY THINGS: A NOVEL CONCEPTUAL FRAMEWORK FOR ENVIRONMENTAL URBAN PLANNING AND GOVERNANCE

Artificial Intelligence of Smart Sustainable City Things represents a conceptual framework where AI is integrated into smart sustainable city planning and governance processes to drive their intelligence and sustainability. In this context, the term "things" encompasses the brain, digital twin, metabolism, and platform of smart sustainable cities as Cyber-Physical Systems of Systems (CPSoS). The integration of AI aims to enhance the functionality, efficiency, and sustainability of these systems, with a particular emphasis on their interconnectedness facilitated by IoT. Specifically, the conceptual framework embodies an innovative approach to environmental urban planning and governance in terms of the essential elements it encompasses.

Concerning Cyber-Physical Systems (CPS), they are engineered to interact with the physical world through sensors and actuators, facilitating a seamless blend of digital and physical processes. The primary objective of CPS is to enable real-time monitoring, coordination, and control of complex systems in various domains of smart cities (Juma and Shaalan 2020; Khan et al. 2021; Rajawat et al. 2022; Singh et al. 2023). The efficacy of CPS relies on several critical technical components, each contributing uniquely to the overall functionality and performance of these systems. These components include sensing, communication network, computation and control component, actuation system, software integration, and security. The seamless integration of these technical components is fundamental to the performance, reliability, and security of CPS, enabling them to transform various domains by providing real-time monitoring and control.

The integration of AIoT represents a significant evolution in the field of CPS. This convergence aims to harness the power of AI to enhance the capabilities of interconnected physical and computational components, enabling more intelligent, autonomous, and efficient systems. In the context of smart sustainable cities, to better understand the impact of traditional CPS, a comparative analysis of their technical components, commonalities, and differences is essential (see Chapter 5 for a detailed account). This approach sheds light on how AIoT extends and enhances the foundational structures of CPS, providing a comprehensive view of the progression from conventional systems to more advanced integrations.

AIoT and CPS are crucial components in the development and operation of advanced urban planning and governance systems such as City Brain, UDT, and SUM—all categorized under CPSoS. By integrating both AIoT and CPS, these systems can leverage the strengths of AI, IoT, and CPS to enhance urban planning and governance comprehensively. City Brain leverages AIoT to integrate and analyze extensive real-time data from multiple urban sensors, facilitating dynamic urban management and informing planning. Meanwhile, UDT utilizes CPS to construct a digital twin of urban settings, which models real-world conditions, processes, and responses to potential changes before they are physically enacted. The virtual modeling in UDT provides a safe and cost-effective environment for testing urban development strategies, reducing the risk of real-world implementation errors. This functionality is enhanced by AIoT through its predictive analytics and ML/DL models, offering more profound insights and precise predictions. Such predictive capabilities enable urban planners to anticipate problems and optimize city functions before issues arise. Likewise, SUM uses both CPS and AIoT to monitor, analyze, and optimize the flow of energy, materials, and resources within city infrastructures, transforming urban management through efficiency and sustainability. The integration of these technologies in SUM not only streamlines resource management but also significantly enhances urban governance processes. Incorporating AIoT within these three CPSoS frameworks enables cities to evolve beyond mere smartness into self-aware ecosystems capable of substantial autonomous decision-making and adaptation, representing a new phase in urban technological evolution. This transformation toward self-aware cities is a key indicator of the future direction of urban living, where technology and data drive decision-making processes, creating more resilient and adaptable urban environments.

In smart sustainable cities, urban planning emerges as a multifaceted and systematic process crucial for designing, regulating, and managing land use and built environments, as well as their interconnected infrastructure (Bibri 2021). This process ensures orderly development, enhances public services, and promotes sustainable growth. Urban planning encompasses a wide array of aspects to create sustainable, livable, and well-functioning cities. It addresses the challenges associated with urbanization, such as resource depletion, ecological degradation, and social equity. By integrating technological advancements and data-driven insights, urban planning in smart sustainable cities aims to improve efficiency, reduce environmental impact, and enhance the quality of life for all residents (see Chapter 2 for a detailed account).

Urban governance denotes the frameworks, policies, processes, and mechanisms through which urban areas are managed and regulated. It encompasses the roles and interactions of various stakeholders—government agencies, institutions, private entities, community organizations, and citizens—in organizing, overseeing, and funding urban environments, as well as in decision-making and policy implementation. It "involves a continuous process of negotiation and contestation over the allocation of social and material resources and political power" (Avis 2016), "where conflicting or diverse interests may be accommodated, and cooperative action can be taken" (UN-Habitat 2023). Its key function is to make and implement policy (Pierre and Peters 2005) and execute public authority (Rydin 2012) or multi-layered political authority (Bridge and Perreault 2009). One of the key roles of urban policy—as a set of plans, laws, rules, regulations, and actions—lies in aligning and mobilizing different urban actors. Urban policy—plans, laws, regulations, and actions—urban policy is adopted by local governments, mediated by civic institutions, and performed by city agents, and this requires a huge coordination effort between a range of actors and networks.

The relationship between urban planning and governance is deeply interwoven and mutually dependent. Urban planning serves as the cornerstone by providing plans and strategies for urban development, while urban governance is tasked with executing and overseeing the implementation of these plans and strategies, ensuring adherence to established policies. Essentially, urban planning functions as a policy-driven approach to governance (Schmitt and Wiechmann 2018), an activity of governance

(Stead 2021), or even as an institutional technology (Rivolin 2012). Urban governance, on the other hand, represents a policy process directed at formulating sectoral coordination and integration and guiding stakeholder inclusion in planning practices (Schmitt and Danielzyk 2018). As regards environmental planning, it involves the assessment, management, and optimization of natural resources and the environment to foster sustainable development, reduce environmental impact, and enhance community well-being. It aims to create cities that are environmentally responsible, resource-efficient, and resilient to climate change. Concurrently, environmental governance entails coordinating policies, institutions, and actions through collaborative efforts among multiple stakeholders to influence environmental actions and outcomes (de Wit 2020; Lemos and Agrawal 2006; UNEP 2018; Vatn 2016).

Bibri (2019) examines the transformative shifts occurring in smart sustainable urbanism, propelled by the integration of big data science and analytics. Through a comprehensive exploration, the author elucidates the paradigmatic, scientific, scholarly, epistemic, and discursive changes shaping the landscape of smart sustainable urbanism, providing valuable insights for researchers, policymakers, and practitioners alike. The discursive changes in the field provide a nuanced perspective on the evolving discourse surrounding smart sustainable urbanism. This insightful analysis equips these stakeholders with the knowledge needed to navigate the complexities of urban planning and governance. Overall, this work underscores the profound implications of big data science and analytics for emerging technologies. It serves as a vital resource for understanding how the integration of big data science and analytics implicitly paves the way for the advancement of AI and AIoT solutions in shaping the future of sustainable urban development.

It is equally essential to establish clear definitions for AI and AIoT. AI refers to the development of computer systems capable of performing tasks that typically require human intelligence, or of simulating human-like cognitive processes, such as learning from experience, understanding natural language, recognizing patterns, solving problems, and making decisions. AIoT merges AI's computational and analytical capabilities with the ubiquitous network of interconnected IoT sensors and devices, enabling the collection, analysis, and interpretation of real-time data from various urban systems. The wealth of these data allows for extracting actionable insights in the form of applied intelligence. AIoT equips IoT systems with intelligence, context-awareness, and decision-making capabilities (e.g., Seng et al. 2022; Shi et al. 2020; Zhang and Toa 2021). Through the integration of AI, especially Machine Learning (ML) and Deep Learning (DL), into IoT devices and systems, AIoT empowers real-time data analytics, predictive modeling, and adaptive responses, thereby optimizing the overall performance and efficiency of IoT applications. Parihar et al. (2023) discuss the evolution of AIoT in terms of its features, architectures, applications, and challenges.

Regarding the conceptual framework underpinning the functionality of planning and governance processes in the ecosystem of smart sustainable cities, the studies reviewed showcase the diversity and complexity of AIoT architectures, each tailored to meet specific contextual requirements and objectives. Zhang and Tao (2021) propose a tri-tier architecture focusing on cloud/fog/edge computing layers, highlighting the distributed nature of edge things and fog nodes in contrast to the centralized cloud. Bibri et al. (2023b) outlines five interconnected computational processes, namely sensing, perception, learning, visualizing, and acting emphasizing the role of ML and DL in enhancing adaptability and performance. Samadi (2022) discusses the integration of IoT data analytics systems, Application Programming Interfaces (APIs), and AI/ML systems to enable real-time data monitoring and decision-making, while Seng et al. (2022) provide a comprehensive overview of AIoT advancements and challenges. Important to note is that early research on AIoT has largely dealt with technical aspects, focusing on edge computing (Fragkos et al. 2020; Yang et al. 2020; Yang et al. 2021b). Edge computing is central to the functioning of AIoT because it moves data processing from the network edge to as close as possible to IoT devices and systems. This strengthens AIoT applications thanks to edge intelligence, which is key for deploying advanced AIoT-powered planning and governance systems in smart sustainable cities (Bibri et al. 2024a; Bibri and Huang 2024). Overall, these studies collectively contribute to a deeper understanding of AIoT architectures and their applications, highlighting the importance of adaptability, integration, and continuous feedback loops within the data science cycle. These insights are valuable for informing the design and implementation of AIoT systems across diverse domains and applications, driving innovation and advancement in the field of smart sustainable cities.

It is worth noting that this study is one of the initial endeavors, if not the first, to address the visionary concept of Artificial Intelligence of Smart Sustainable City Things in the context of environmental planning and governance. The main focus here is on the technical components of this

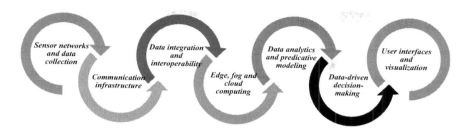

FIGURE 4.2 Key technical components of a novel framework of Artificial Intelligence of Smart Sustainable City Things.

framework as distilled based on the synthesis of the aforementioned studies in the context of City Brain, UDT, SUM, and platform urbanism as AIoT-driven CPS (Figure 4.2). Furthermore, this framework builds upon the foundations laid by two earlier studies on AI for the Internet of City Things by Bibri and Jagatheesaperumal (2023) and "Artificial Intelligence of Smart City Things for Transportation Safety" by Jagatheesaperumal et al. (2024). By integrating and expanding upon the concepts and findings of these earlier works, the framework presented here aims to provide a comprehensive understanding of AIoT architectures and their implications for smart sustainable city development.

1. Sensor networks and data collection: In the context of AI-driven smart sustainable city things, robust sensor networks play a pivotal role in their connection. The brain, digital twin, metabolism, and platform of smart sustainable cities leverage a wide range of interconnected sensors and devices to collect real-time data on various aspects of urban life. These sensors and devices facilitate comprehensive data collection, covering various environmental parameters associated with the planning and governance of smart sustainable cities.

2. Communication infrastructure: The success of Artificial Intelligence of Smart Sustainable City Things relies on a robust communication infrastructure, including low-latency networks and protocols, to ensure the seamless and reliable connectivity and transmission of data between the interconnected sensors, devices, systems, and infrastructure of these things and centralized systems. Communication technologies like 5G enhance the responsiveness of these things, ensuring timely communication. This enhances the synergic interaction between these things, fostering a planning and governance ecosystem that is more cohesive.

3. Data integration and interoperability: The effective functioning of smart sustainable city things relies on seamless data integration and interoperability. AI ensures that data from diverse sources, including City Brain, UDT, SUM, and platform urbanism, can be coalesced and integrated cohesively. This integration enhances the overall understanding of urban dynamics and conditions, and fostering interoperability among different urban systems for more informed decision-making pertaining to planning and governance.

4. Edge, fog, and cloud Computing: The convergence of edge, fog, and cloud computing is integral to the implementation of AIoT in smart sustainable cities and its incorporation into their brain, digital twin, metabolism, and platform. Cloud computing leverages expansive computing resources to store, manage, and analyze the massive amounts of data generated by the interconnected IoT sensors and devices pertaining to these four entities. Edge computing brings computation and storage closer to the source, reducing latency and bandwidth usage, improving real-time responsiveness, and enabling faster responses to events by executing complex analytics locally. Fog computing extends edge computing by creating a hierarchical architecture that includes multiple edge devices and gateways associated with City Brian, UDT, SUM, and platform urbanism. Similarly, fog nodes are strategically placed in proximity to data sources to perform intermediate processing, data filtering, and preliminary analytics before transmitting relevant data to the cloud. All together, they form a powerful computing model supporting the computational demands of smart sustainable city things.

5. Data analytics and predicative modeling: AI-driven analytics processes the vast datasets generated by the brain, digital twin, metabolism, and platform of smart sustainable cities. ML and DL models and algorithms identify patterns, trends, and anomalies, providing valuable insights. Predictive modeling entails developing predictive models that use historical data and real-time inputs to forecast environmental trends, potential risks, and areas of concern for proactive decision-making.

6. Data-driven decision-making: AI augments decision-making processes by interpreting insights from data analysis and predictive modeling and analytics to inform and guide actions for decision-making.

Smart sustainable city administrators, planners, and policymakers can leverage AI-generated knowledge to make well-informed or evidence-based decisions related to infrastructure development, resource allocation, and environmental sustainability. Data-driven decision-making is a key outcome of incorporating AI into smart sustainable city things concerning environmental planning and governance.

7. User interfaces and visualization: Smart sustainable city things are designed with user-friendly interfaces and visualizations that offer varied forms of real-time displays of data and insights for different planning and governance purposes. These interfaces serve as accessible platforms for city officials, administrators, planners, and citizens to actively monitor and engage with various aspects of urban life. By enhancing transparency and participation, these user interfaces contribute to the effective functioning of the interconnected and intelligent urban ecosystem.

The notion of Artificial Intelligence (AI) of Smart Sustainable City Things encapsulates the dynamic evolution of the Internet, transitioning beyond its conventional role as a network connecting computers and things to a more comprehensive concept known as the Internet of Everything and the Internet of Cities. This paradigm shift represents a transformative vision that extends the reach of connectivity and intelligence to encompass all aspects of urban life. Within this framework, AI serves as a catalyst for innovation and efficiency, driving progress toward sustainable urban development. Central to this concept is the idea of the city as a living organism, where AI functions as its cognitive brain, processing vast amounts of data to optimize resource allocation, enhance environmental performance, improve service delivery, and ameliorate the quality of life for cities. By harnessing the power of AI-driven cyber-physical systems of City Brain, UDT, SUM, and platform urbanism, cities can unlock new possibilities for sustainability, resilience, and inclusivity, shaping a future where urban living is not smarter and more sustainable.

4.4 RESULTS

This section presents the results derived from addressing three key research questions that elucidate the innovative potentials and anticipated challenges of integrating City Brain, UDT, SUM, and platform urbanism within the framework of Artificial Intelligence of Smart Sustainable City Things. The empirical analysis emphasizes the distinctive contributions of these CPS to urban planning and governance in the context of smart sustainable cities. The theoretical analysis focuses on the integrative aspects of these systems, facilitated by the functionalities of AIoT in terms of its underlying architectural layers. It is important to highlight the relevance of placing the outcomes of the case study analysis within the conceptual framework of Artificial Intelligence of Smart Sustainable City Things. The objective of the integrative analysis is to highlight potential synergies and opportunities arising from the collaborative integration of these things, thereby going beyond the individual evaluation of the case studies. Recognizing the symbiotic relationships among City Brain, UDT, SUM, and platform urbanism is paramount for understanding how their combined efforts can drive advancements in urban development practices.

4.4.1 Case Study Analysis

This subsection unveils the outcomes of our comprehensive investigation into the functions of smart sustainable city brain, digital twin, metabolism, and platform. The diverse insights distilled from a spectrum of case studies shed light on key emergent patterns and profound implications. The former involve recognizing differences and commonalities and uncovering fertile insights that may not be immediately evident but become apparent through the analysis of various case studies. These patterns, in turn, contribute to a deeper understanding of the potential impacts resulting from the integration of City Brain, UDT, SUM, and platform urbanism within smart sustainable cities. The latter indicate transformative effects on the different aspects of environmental urban planning and governance. They encompass significant changes in the way smart sustainable cities function, make decisions, and address environmental challenges.

4.4.1.1 City Brain

City Brain, an advanced urban management system, integrates AIoT to transform city operations and services.

As an AIoT-driven platform, it is designed to manage urban environments through real-time data integration from various city systems, intelligent decision-making, and advanced automation. City Brain not only optimizes existing urban functions but also provides valuable data-driven insights for urban planning and governance. Through real-time data processing and analysis, it offers a comprehensive understanding of urban dynamics and challenges, aiding in informed decision-making for city planning and governance. Its capacity to synthesize diverse data sources enhances efficiency across various urban domains, marking a significant step toward smarter, more sustainable urban environments.

Urban operational management is the day-to-day administration and coordination of urban infrastructure and services to ensure efficient functioning and a high quality of life for citizens. This includes managing transportation systems, waste collection, energy distribution, public safety, and other essential services. The primary goal is to optimize the performance of these systems to meet the needs of citizens effectively and sustainably. In the context of AIoT, urban operational management, planning, and governance are becoming increasingly interconnected and data-driven. AI and AIoT technologies provide advanced tools for real-time data collection, analysis, and decision-making, thereby enhancing efficiency, predictive analytics, citizen engagement, and stakeholder collaboration. The synergistic relationship among urban operational management, planning, and governance is evident as data-driven insights from operational management inform urban planning, which in turn shapes governance policies. Effective urban governance relies on the seamless integration of planning and management to address complex urban challenges and promote sustainability. Leveraging AI and IoT technologies, City Brain processes vast amounts of data in real-time to optimize urban operations, inform planning decisions, and enhance governance.

4.4.1.1.1 Operational Urban Management

In the domain of operational city management, the advent of City Brain marks a paradigm shift, leveraging sophisticated AIoT capabilities to revolutionize the day-to-day functioning of smart cities. Specifically designed to optimize efficiency and responsiveness, City Brain contributes to steering smart cities toward a path of environmental sustainability in terms of enhancing essential services, infrastructure, and resource utilization. In this light, by employing the technical layers of AIoT architecture (e.g., Bibri et al. 2024a; Parihar et al. 2023; Seng et al. 2022), City Brain facilitates real-time data collection, processing, and analysis, supporting informed decision-making and efficient urban systems management. City Brain harnesses AI and IoT technologies for cognition, optimization, prediction, decision-making, and intervention in various urban domains (Zhang et al. 2019). Depending on the specific domain where it can be applied, City Brain's construction and implementation involve a varied array of technical components (Liu et al. 2018, 2021; Zhang et al. 2019) whose integration is based on the needs, objectives, and challenges of each city (e.g., Gao et al. 2021; Kierans et al. 2023; Liu et al. 2022; Wang et al. 2022; Xu 2022; Xu et al. 2024). This adaptability allows City Brain to address a wide range of urban issues, from traffic management and energy efficiency to waste management and environmental monitoring, ultimately leading to more sustainable and efficient urban environments.

There are diverse engaging case studies of City Brain, showcasing its transformative impact on transport management, traffic optimization, public safety, environmental monitoring, and urban planning. As cities around the world seek innovative solutions to enhance efficiency, sustainability, and the overall quality of life for their citizens, Kierans et al. (2023) present a comparative analysis of two projects in Basel City and Shenzhen City that involve the transition of their public transportation systems from fossil-fueled vehicles to electric vehicles. Central to the contribution of the case study is the concept of a "Smart City Brain for Sustainable Development," a visionary platform that is proposed as a means to achieve the SDGs within cities. The platform acts as a sharable and collaborative system, facilitating the collection and aggregation of data from individual projects. By fostering data sharing, the platform enables the creation of easily understandable visualizations for Key Performance Indicators (KPIs) and other essential sustainability indicators. The case study also includes product material data, an expanded approach that supports circularity within cities, promoting the effective and sustainable use of resources and materials. In this regard, the "Smart City Brain for Circular Economy" is characterized by (Kierans et al. 2023):

- Circular economy integration: A centralized platform utilizing data-driven technologies to manage resources and waste in a closed-loop system.
- Real-time resource tracking: Utilizing sensors and devices to monitor the flow of resources, energy, and waste in real-time for efficient and sustainable management.
- AI-driven resource optimization: Leveraging AI and ML to optimize resource allocation, minimize waste, and predict future needs.
- Stakeholder collaboration: Facilitating collaboration and coordination to ensure the adoption of circular economy practices city-wide.

Hangzhou, the birth of City Brain, serves as a compelling case study of how this platform has advanced traffic management and reduced emissions (e.g., Caprotti and Liu 2022; Wang et al. 2022). City Brain analyzes real-time traffic patterns, identifies congestion hotspots, and dynamically adjusts traffic signal timings. This proactive approach has resulted in a significant reduction in travel times, improved overall traffic flow, and minimized fuel consumption, making Hangzhou a model for other cities looking to tackle urban traffic congestion. The case study carried out by Wang et al. (2022) establishes an intelligent modeling and visualization system driven by Intelligent Transportation Systems (ITS) traffic data to map real-time, high-resolution on-road vehicle emissions and traffic states in Hangzhou's Xiaoshan District. ITS is a system that uses data-driven technologies to improve transportation safety, efficiency, and sustainability. City Brain, which utilizes a High-Performance Computing platform and an agile Web Geographic Information System, identifies emission hotspots and visualizes the impact of traffic control policies, offering valuable insights for optimizing air quality management (Wang et al. 2022). Kuala Lumpur has strategically incorporated City Brain into various domains such as transportation management and environmental protection (Zhang et al. 2019). This integration aims to optimize resource allocation, minimize environmental impacts, and enable swift responses to accidents, disasters, and potential security threats.

4.4.1.1.2 Urban Planning and Governance

In the ever-evolving landscape of urban planning and governance, the integration of cutting-edge technologies has become imperative for creating sustainable, efficient, and resilient cities. Within this context, the emergence of City Brain represents a transformative force, offering unprecedented opportunities for enhancing decision-making processes and optimizing urban functions. As an advanced large-scale AIoT system, City Brain operates as the analytical nucleus of smart cities, assimilating vast datasets, processing real-time data, and providing actionable insights. Here, the focus is on the pivotal role of City Brain in shaping the contours of contemporary urban planning and governance, exploring its multifaceted contributions and the implications it holds for smart sustainable cities of the future.

City Brain stands at the forefront of optimizing urban management operations and services through advanced technologies, data analytics, and real-time monitoring, thereby enhancing planning and governance processes for more efficient and sustainable urban development. Through data-driven insights into key environmental factors and parameters, such as energy consumption, transport efficiency, traffic patterns, waste generation, and pollution control, City Brain equips urban planners and city officials with the tools to implement measures that effectively mitigate environmental impacts (Bibri et al. 2024a). City officials can leverage these insights to identify interventions, formulate effective strategies, responsive policies, and make more informed decisions. Originally designed for transportation management and traffic forecasting, City Brain has evolved into a large-scale AIoT system with broader applications, extending its reach to encompass urban planning and governance (Cugurello 2020).

In addition, according to various real-world implementations (e.g., Caprotti and Liu 2022; Kierans et al. 2023; Liu et al. 2022; Xie 2020), City Brain includes elements of urban planning and governance by involving both public and private stakeholders in enhancing, coordinating, and managing a wide range of urban systems, infrastructures, and resources. For example, Kierans et al. (2023) highlight the role played by City Brain in enhancing stakeholder collaboration and facilitating coordination among government agencies, businesses, and citizens to ensure the adoption of environmental sustainability practices. Furthermore, in analyzing the case of Haidian District, Liu et al. (2022) employ the City Brain concept to analyze the avenues for enhancing and modernizing urban governance, emphasizing the importance of its conceptual advancement, multi-city approach, and process digitalization. Its implementation in Haidian District has shifted the governance mode from passive to active, fostering knowledge in urban operation, evaluation, and development. City Brain has implemented over 50 innovative application scenarios and integrated data operations from more than 40 information systems. It actively drives innovation across transportation, urban management, ecological environmental protection, and smart energy. Technology-driven innovations are central to planning and governance mechanisms.

The primary objective of City Brain is to address urban planning and governance challenges (Zhang et al. 2019; Xu et al. 2024). This focus is instrumental in advancing the urban landscape toward digital, intelligent, and modern development (Liu et al. 2022). Amid the rising call for cutting-edge smart eco-city solutions, City Brain stands as a transformative force, driving innovation, optimizing efficiency, and minimizing environmental footprints while enhancing the overall quality of life for citizens. However, it is crucial that its design and deployment prioritize privacy, security, fairness, accountability, and transparency considerations.

Overall, City Brain represents a significant advancement in urban operational management. By integrating and analyzing real-time data, it supports more efficient, sustainable, and inclusive urban planning and governance. This platform ultimately contributes to the development of smart sustainable cities by ensuring that urban environments are managed effectively, resources are allocated efficiently, and decision-making processes are inclusive and transparent.

4.4.1.2 Urban Digital Twin

UDT is a sophisticated technological concept denoting a comprehensive virtual 3D representation of a city's physical and functional aspects, mirroring its structures, systems, and dynamics in real-time. In this context, UDT acts as an intelligent, data-driven counterpart to the physical city, employing advanced IoT sensors, data analytics, and AI and ML/DL models and algorithms. AIoT enhances UDT by embedding AI computational and analytical functionalities into IoT devices, facilitating real-time data acquisition and analysis while equipping IoT systems of intelligence and decision-making capabilities. This integration empowers UDT to simulate, monitor, and optimize urban processes and flows with a high level of precision, fostering a dynamic and responsive framework for informed decision-making. The seamless integration and collaboration among AI, AIoT, and UDT plays a pivotal role in creating intelligent and environmentally sustainable urban ecosystems. Bibri et al. (2024b) conduct a systematic review to uncover the intricate relationship among AI, AIoT, UDT, data-driven planning, and environmental sustainability in the dynamic context of smart sustainable cities. The authors reveal the nuanced dynamics and untapped synergies among these technologies, models, and domains in these urban environments. Their integration is revolutionizing data-driven planning strategies and practices, thereby supporting the achievement of environmental sustainability goals. The study identifies and synthesizes the theoretical and practical foundations that support the convergence of AI, AIoT, UDT, data-driven planning, and environmental sustainability into a comprehensive framework. It also investigates how the integration of AI and AIoT transforms data-driven planning to enhance the environmental performance of smart sustainable cities. Additionally, the study explores how AI and AIoT can augment UDT capabilities to improve data-driven environmental planning processes. The integration of AI, AIoT, and UDT technologies significantly enhances data-driven environmental planning processes, providing innovative solutions to complex environmental challenges and promoting sustainable urban development.

4.4.1.2.1 Case Studies on UDT

The tangible impacts of UDT on sustainable urban development have been underscored by a number of recent case studies conducted on real-world cities, often focusing on IoT and big data analytics.

The DT of Zurich City in Switzerland, founded on 3D spatial data, enhances planning by offering a digital representation of the city (Schrotter and Hürzeler 2020). It supports applications, collaborative platforms, and simulations that inform decision-making pertaining to urban issues. It expands the current spatial data infrastructure by incorporating 3D spatial data and continuously enhancement of their inventory. It has essential components for urban planning, namely:

- The 3D spatial data form the foundation for linking additional spatial data and connecting digital space to the real world.
- Elements of reality are appropriately linked to the digital world in real time.
- The data provides optimal conditions for the presentation and shaping of public space.
- Facilitating various analyses such as visibility, noise propagation, solar potential, and flood simulations.
- Enabling planned development through metadata and lifecycle management.
- Real-time updates are feasible, enhancing responsiveness to changes.
- Direct connections to sensors and other components in real space for certain applications.
- Changing processes in digitalized real space, consistently capturing and storing the third dimension.

The digital twin will serve to test various development scenarios for future planning, such as changes in density and their impacts on urban climate, traffic, and mobility. The outcomes of these tests will create new opportunities for decision-making in urban planning and facilitate discussions with pertinent stakeholders. Furthermore, it will introduce novel avenues for citizens to learn about city projects and issues, enabling their active involvement.

The case study in Ålesund City in Norway explores the application of a graphical digital twin to a district and demonstrates its effectiveness for urban mobility planning (Major et al. 2022). It aims to enhance related practices by addressing challenges associated with coordinating diverse stakeholders with varying expertise and reliance on ad-hoc modeling by costly expert domains. This practice hinders the incorporation of new knowledge into planning over time and prevents stakeholders from easily changing simulations and visualizing the resultant impacts. To overcome these challenges, the graphical digital twin is used to automate model configuration and data integration, ensuring flexibility and scalability for large-scale planning based on a multi-city approach. It also facilitates automatic updates of models when input data changes. Furthermore, it enables stakeholder interaction through user interfaces, allowing the exploration of insightful what-if scenarios for well-informed decisions.

The DT of Herrenberg City in Germany serves as an intricate data model facilitating collaborative processes (Dembski et al. 2020). The prototype encompasses, in addition to urban mobility simulation, a 3D model of the built environment, a street network model utilizing space syntax theory, wind flow simulation, and empirical data based on volunteered geographic information (VGI). Implemented on a virtual reality visualization platform and presented in public participatory processes, the digital twin demonstrates a significant potential for aiding participatory and collaborative decision-making in planning. Tested with real-life scenarios and potential solutions, it offers insights into addressing urban challenges and fostering consensus in public decision-making. This tool proves valuable for urban planners, designers, and the public, enhancing collaboration, communication, and decision support in the context of smart sustainable cities. It showcases a democratized approach to urban data, preserving data sovereignty for cities and citizens, while visualizing complex "invisible" urban data and simulations to support citizen participation and expert collaboration in sustainable urban development.

Lausanne City in Switzerland serves as a dynamic test bed and living laboratory for the ongoing Blue City Project, which aims to develop a responsive AI-powered UDT functioning as an open platform (Huang et al. 2024). This platform is designed to empower citizens and policymakers, fostering collective, evidence-based decision-making to enhance sustainability, resilience, quality of life, well-being, and ecological value (ENAC 2024). At its core, this initiative focuses on building an integrated, open-source platform to scrutinize emerging patterns and enable informed, proactive decisions in urban planning. The Blue City involves a transdisciplinary consortium unified in its mission to map the multi-layered, interconnected network of flows within the city, spanning energy, materials, mobility, goods, biodiversity, and waste. This collaborative approach, blending AI, architecture, urban planning, engineering, and environmental science, underpins an extensive analysis and modeling of flows and infrastructures, offering a holistic understanding of their impact on the urban environment (ETH 2024). The integrated platform will enable: (1) the detection of correlations and evidence for causalities across flow categories over time and space; (2) seeing emerging patterns not discernible with traditional analytical tools; (3) testing intervention hypotheses and scenario prediction; and (4) leveraging engineering and AI tools for planning a more resilient city (ETH 2024). Ultimately, the Blue City Project embodies a pioneering effort toward a smarter, more sustainable future for Lausanne City.

The Helsinki Digital Twin project explores innovative digital modeling solutions, specifically employing the cityGML standard for constructing virtual 3D models (Hämäläinen 2021). This experimentation with cityGML has empowered urban designers to assess a new 3D modeling standard within a realistic urban design setting, gaining insights into its effectiveness and practical applications. Additionally, Helsinki's urban developers have utilized dynamic 3D city models to simulate energy-related data, educating property owners and citizens on energy efficiency and carbon-neutral behavior. The successful results of the cityGML experimentation in Helsinki have laid the groundwork for broader digital twin applications, enabling the creation of virtual replicas for specific city districts or use cases. Helsinki has experienced enhanced stakeholder communication, transparency, and openness in city development and design processes through the integration of digital twin platforms (Figure 4.3). Moreover, dynamic 3D city models have played a crucial role in improving governance and the practical implementation of various urban planning projects in Helsinki.

Ricciardi and Callegari (2022) address the urgency of responding to climate change impacts in urban areas and the need to achieve carbon neutrality and resilience targets as outlined by the Paris Agreement. The authors highlight the vast amount of data available at the city scale and

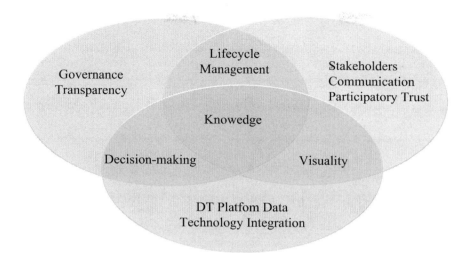

FIGURE 4.3 The dynamic digital twin of the Helsinki Smart City and its implications. Adapted from Hämäläinen (2021).

its potential to support climate action planning. Among the technologies emerging in the Industry 4.0 realm, the concept of UDT has gained traction in urban planning and design. They analyze the development of UDT in Europe over recent years, specifically focusing on their consideration of climate change issues. They examine the current state of the art, explore applications developed for mitigating climate change, and assess the level of experimentation. The main contribution of the study lies in providing insights and guidelines for building UDT as tools to support the creation of climate-neutral and resilient cities.

Cureton and Hartley (2023) explore the transition from City Information Models (CIMs) to UDTs through a case study of Lancaster. CIMs, which integrate urban GIS/BIM in GeoBIM, serve as precursors for UDTs and broader smart city initiatives. The study identifies four socio-technical research challenges: technical training for adoption, acquisition strategies for aerial data, data licensing types, and stakeholder engagement. Using back-casting futuring methods and co-designed workshops, the authors establish a framework for CIMs and explore the Lancaster City Information Model (LCIM) prototype. The LCIM engages architects, planners, and stakeholders with 3D urban models and urban analytics, addressing digital inequality in planning and data fragmentation. The LCIM results in the UK's most extensive 3D open urban dataset, including 1" buildings and infrastructure and 7.5 cm aerial imagery. The study's outcomes highlight the need for precision in terminology and systems between CIMs and UDTs. Additionally, it addresses applied challenges in digital transformation and future socio-technical relationships, providing insights for advancing the transition from CIMs to UDTs.

The focus of the study carried out by Gholami et al. (2024) is on developing a microclimate UDT for Imola, providing complementary roles in collecting and processing micrometeorological data, automating microclimate modeling, and representing climatic interactions virtually. As for the results, the study proposes a methodology for real-time simulation of the cooling effects of trees and other green systems on pedestrian-level thermal comfort. It also evaluates the applicability of the microclimate DT through its implementation in Imola, demonstrating its potential for urban planning and policymaking. The study's contributions lie in advancing smart green planning by providing a precise methodology for urban microclimate simulation and supporting policy decisions with empirical data. Additionally, it enhances the understanding of the potential and intentions of developing urban microclimate DT, offering insights into their conceptual framework and applicability in different urban settings.

Lu et al. (2021) focus on implementing DT systems within smart city frameworks, specifically targeting smart transportation and energy grids. The authors detail the design and real-world applications of DTs, demonstrating their potential to address operational challenges and improve system efficiency during events like the 2021 winter Texas power crisis. Their findings suggest that digital twins can provide critical insights for emergency response and infrastructure resilience. The study offers valuable insights into the deployment of DT systems for enhancing the robustness of smart city infrastructures. The practical applications shown during the Texas power crisis highlight the importance of digital twins in ensuring continuity and efficiency in critical services during emergencies.

The examination of various UDT projects and initiatives underscores the transformative impact of advanced city modeling and simulation on urban planning and governance, with a distinct focus on enhancing environmental sustainability. These projects and initiatives leverage advanced technologies such as reinforcement learning, IoT, 3D modeling, and advanced analytical tools to create sophisticated virtual representations of cities. UDT enables advanced analysis, prediction, and optimization of various aspects of urban life. The tangible outcomes of UDT implementation have been evidenced in improved stakeholder communication, transparency, and openness in city development and design processes. Additionally, UDT has facilitated participatory decision-making, educated citizens on environmental issues, and provided a platform for collaborative urban development initiatives. These diverse applications collectively contribute to the overarching goal of creating resilient, sustainable, and well-informed cities. The success stories and lessons learned from these UDT endeavors serve as valuable insights for future urban planning and governance endeavors, highlighting the potential for technology-driven solutions to address complex urban challenges.

This selective set of projects and initiatives, while not primarily incorporating AI and AIoT, lay a solid foundation for understanding how integrating these advanced technologies could further advance UDT capabilities. AI and AIoT are crucial components that can greatly enhance existing UDT systems. AI algorithms can enable these systems to analyze vast amounts of data collected from IoT sensors and other sources, providing deeper insights into urban dynamics and facilitating more accurate predictions and optimizations. Furthermore, the integration of AI and AIoT into UDT can improve stakeholder communication and decision-making processes. AI-driven analytics can generate actionable insights from UDT data, allowing city planners and policymakers to make informed decisions based on real-time information. Moreover, AI-powered UDTs can provide interactive platforms for citizens to explore and understand urban data, fostering greater awareness of sustainability issues and encouraging behavior change. AIoT systems can facilitate citizen feedback and participation in urban planning and governance processes, thereby promoting inclusivity and transparency in decision-making. Lastly, AI and AIoT can enable UDT to support collaborative urban development initiatives more effectively. By analyzing data from diverse sources and stakeholders, UDT can identify common goals and priorities by facilitating coordination and cooperation among different actors in the urban ecosystem. Overall, while existing UDT projects and initiatives have demonstrated significant benefits for urban planning and governance, integrating AI and AIoT can further enhance UDT capabilities.

4.4.1.2.2 Case Studies on AI-Driven UDT

Case studies focusing on AI-driven UDT highlight its innovative potential in enhancing urban management, planning, and infrastructure resilience. By integrating AI with UDT, these studies demonstrate significant improvements in predictive analytics, resource optimization, and real-time decision-making capabilities. This approach supports sustainable urban development practices.

Salunke (2023) presents a pioneering approach called Reinforcement Learning Empowered Digital Twins (RLEDs), which merges DT technology with reinforcement learning algorithms to establish dynamic virtual models of urban systems. This framework aims to revolutionize smart city development by optimizing various aspects of urban dynamics, with a particular emphasis on traffic flow in the context of Chhattisgarh. By using this case study, the author aims to demonstrate the effectiveness and applicability of RLEDs in addressing real-world urban challenges, particularly in the domains of transportation management, energy management, and urban planning. This approach allows for a detailed examination of the potential benefits and limitations of RLEDs in a specific geographical and socioeconomic context. By focusing on traffic flow optimization, the research addresses the critical issue of congestion and inefficiency in urban areas. Through the integration of reinforcement learning algorithms, RLEDs are capable of learning and adapting to complex urban environments, ultimately enhancing the efficiency and sustainability of city operations.

Kamal et al. (2024) propose a DT-based deep reinforcement learning approach for adaptive traffic signal control. Their study demonstrates that digital twins can optimize traffic signals to reduce CO_2 emissions and fuel consumption in urban areas, using simulations from a neighborhood in Amman, Jordan. The results show significant improvements in traffic flow efficiency and air quality. Additionally, their approach highlights the potential for scaling the solution to other cities facing similar traffic management challenges. The practical insights from their study indicate that implementing DT-based traffic signal systems can lead to substantial environmental benefits and improved urban mobility. The approach not only optimizes signal timings but also provides a scalable framework that other urban areas can adopt to address their unique traffic issues.

The study by Austin et al. (2020) aims to enable next-generation smart city systems through the widespread adoption of sensing and communication technologies embedded within urban environments. The study proposes a smart city digital twin architecture that integrates semantic knowledge representation and ML to enhance data collection, processing, event identification, and automated decision-making. The results, demonstrated through a case study on energy usage analysis in buildings in the Chicago Metropolitan Area, showcase the complementary roles of semantic and ML components in improving smart city operations. This contribution highlights the potential of digital twin architectures to advance the efficiency and functionality of smart city systems.

Sabri et al. (2023) discuss designing UDTs for smart water infrastructure and flood management. The authors use case studies from Orange County and Victoria to illustrate the importance of accurate GIS data and GeoAI in managing water resources and mitigating flood impacts in smart cities. Their research shows that UDTs can significantly enhance predictive capabilities and disaster preparedness. The practical insights underscore the necessity of integrating advanced data analytics and real-time GIS data in managing urban water resources. The implementation of UDT has proven effective in improving flood prediction and management, thereby supporting the creation of more resilient urban environments against water-related disasters.

Overall, the diverse applications—from traffic signal optimization to flood management—highlight the versatility and effectiveness of UDT in addressing a wide range of urban challenges. As cities continue to grow and evolve, the insights gained from these case studies provide a foundation for the widespread adoption and implementation of AI-driven UDT, paving the way for smarter, more efficient, and more resilient urban environments.

4.4.1.3 Smart Urban Metabolism
Urban metabolism refers to the dynamic flow of energy, materials, and resources through a city. Kennedy et al. (2007, p. 44) define it as "the total sum of the technical and socioeconomic processes that occur in cities, resulting in growth, production of energy, and elimination of waste." According to Currie and Musango (2017), urban metabolism encompasses a diverse array of socio-technical-ecological processes and operations, which generate a multitude of people, energy, by-products, waste, emissions, raw materials, data, and information flows circulating within the city and its surrounding environment. The complexity of urban metabolism, coupled with the substantial data needed for its comprehensive analysis, has led to the emergence of SUM. This advanced model collects and integrates colossal amounts of high-spatial and temporal resolution data from various urban systems, allowing for real-time processing, analysis, and visualization (Bibri et al. 2024a). SUM involves the application of data-driven, technology-enhanced approaches—based on or integrating AI, IoT, and Big Data—to monitor, analyze, and optimize resource flows,

energy consumption, waste production, and overall sustainability within a city (e.g., Bibri and Krogstie 2020b; D'Amico et al. 2021; Peponi et al. 2022; Ghosh and Sengupta 2023; Shahrokni et al. 2015a, b), i.e., how resources are consumed, processed, and disposed of. The implementation of its principles ensures that smart sustainable cities become resilient and environmentally friendly.

The analytical framework of smart metabolism circularity in the chosen case studies takes into account a range of flows specific to each city or district. While it comprehensively addresses various sustainability dimensions, the analysis and comparison here place a particular emphasis on environmental sustainability. This exploration seeks to uncover distinctive insights into how cities navigate the complexities of SUM, aiming to enhance not only their overall sustainability but also their environmental efficiency and resilience. D'Amico et al. (2022) focus on five circular cities, namely Porto, The Hague, Prato, Oslo, and Kaunas, which constitute the Urban Agenda Partnership on Circular Economy (UAPCE). Table 4.1 provides an overview of the case study analysis and comparison of digitalized

TABLE 4.1 Analysis and Comparison of Five Self-Defined Circular Cities

City	Urban Metabolism Flows	Key Strategies and Initiatives	Digital Technologies	Monitoring and Evaluation
Porto	Mobility, Construction, Food	ScaleUp Porto's Manifesto, Web Sharing Platforms	Real-time tracking sensors, GPS Monitoring, Web Sharing Platforms	Optimization of vehicle circulation, Reduction in emissions
The Hague	Food, Energy, social business	Circular Netherlands in 2050 program	Real-time tracking sensors, GPS Monitoring, Circular Economy Initiatives	Entrepreneurship, Strengthening social inclusion
Prato	Mobility, Construction, Waste, Water	Prato Smart City, Prato Circular City project	Automatic Vehicle Monitoring, GIS Technology, Smartphone Applications	Improved water and energy flows, tailored public services
Oslo	Economy, Waste	Initiatives for Recycling and Reusing	Web Sharing Platforms, Circular Economy Standards	Improved sharing and recycling of goods and services
Kaunas	Governance, Transport, Mobility	Strategic Development Plan	Digital Vigilance Cameras, Real-time Vigilance Devices	Enhanced vehicular flow knowledge

metabolism circularity in these cities. It examines their approaches and outcomes in the context of sustainable urban development.

The SUM model was first implemented in Stockholm City with the vision to enable a better understanding of the correlations and causalities that govern urban development. Shahrokni et al. (2015a) introduce the concept of SUM as a real-time model for understanding energy and material flows within a city and its households. The authors specifically detail the application of SUM in the Stockholm Royal Seaport (SRS) district through a research and development project. The prototype, applied in the initial construction phase, focuses on assessing GHG emissions resulting from the consumption of resources (energy and water consumption) and the generation of waste.

The focus of the study conducted by Shahrokni et al. (2015a) is to explore the practical implementation of SUM in the Smart City SRS project. The SUM framework operates at high temporal and spatial resolutions, generating four KPIs in real-time: kilowatt-hours per square meter, carbon dioxide equivalents per capita, kilowatt-hours of primary energy per capita, and the percentage share of renewables. The real-time and in-depth feedback on the resulting metabolic flows related to the household, building, and district scales serve to inform policymakers, decision-makers, urban managers, and urban planners to identify and implement intervention measures based on system inefficiencies. The authors identify accessing and integrating siloed data as a significant barrier, emphasizing the need for perceived value to overcome this challenge. They also discuss long-term opportunities, suggesting that SUM could provide new insights into urban causality and offer valuable feedback to citizens and city officials about the consequences of their choices.

The case studies by Beck et al. (2023) and Wiedmann et al. (2021) offer distinct, yet interconnected, perspectives on advancing urban sustainability. Beck et al. (2023) investigate how to improve Suzhou City's urban socioeconomic metabolism by reducing the environmental burden and increasing circularity through technological interventions in urban infrastructure, specifically focusing on the nitrogen metabolism. The authors evaluate 15 scenarios with promising technologies in water and waste management. They introduce Multi-sectoral Systems Analysis model to examine various sectors in Suzhou's urban infrastructure and economy. Three Metabolic Performance Metrics are used to assess the environmental impact of the technological interventions. In a complementary initiative, Wiedmann et al. (2021) delve into the "Metabolism of Cities Data

Hub," an online platform dedicated to urban metabolic data collection, processing, visualization, and data-driven insights. Investigating challenges related to mining urban metabolism data, the authors concentrate on Lausanne and Geneva, aiming to improve environmental monitoring and formulate strategies for mitigating environmental impacts in cities. Important to note is that SUM in different case studies provides strategic stakeholders with useful instruments to handle and evaluate the circularity of not only natural resources but also social and economic ones to improve their efficiency and allocation.

In terms of real-world implementations, SUM is currently in its nascent stages of integrating AI and AIoT technologies, suggesting that there is considerable potential for further advancements in leveraging these tools for enhancing urban sustainability and resilience. The ongoing exploration of AI and AIoT within the framework of SUM signifies an exciting frontier where innovative solutions are poised to revolutionize urban management and resource optimization. In this context, Peponi et al. (2022) address the complexity of urban systems, characterized by nonlinear, dynamic, and interconnected processes, necessitating improved management for enhanced sustainability. They propose a novel, evidence-based methodology to understand and manage these complexities, aiming to bolster the resilience of urban processes under the concept of smart and regenerative urban metabolism. By integrating Life Cycle Thinking and Machine Learning, they assess the metabolic processes of Lisbon's urban core using multidimensional indicators, including urban ecosystem service dynamics. They developed and trained a multilayer perceptron (MLP) network to identify key metabolic drivers and predict changes by 2025. The model's performance was validated through prediction error standard deviations and training graphs. Results indicate significant drivers of urban metabolic changes include employment and unemployment rates (17%), energy systems (10%), and various factors such as waste management, demography, cultural assets, and air pollution (7%), among others. This research framework serves as a knowledge-based tool to support policies for sustainable and resilient urban development.

To sum up, the case studies on SUM reveal its transformative potential in enhancing environmental efficiency and resilience in urban areas. The findings underscore both the opportunities and challenges of its practical implementation, emphasizing its role in informing urban policymakers, decision-makers, and planners about system inefficiencies. Accordingly, SUM emerges as a transformative tool for environmental urban planning and governance, offering real-time insights into the dynamic flow and circularity of resources in urban environments. In doing so, it equips city stakeholders with a comprehensive understanding of environmental impact across various domains through monitoring and analyzing a multitude of urban metabolic flows. These invaluable insights allow for identifying targeted interventions to enhance resource efficiency, minimize waste, reduce emissions, and promote sustainable practices. SUM's capacity to generate KPIs at different spatial scales enables evidence-based decision-making tailored to a varied set of urban forms and diverse stakeholders. Moreover, the continuous monitoring and evaluation enabled by SUM empower cities to adapt and optimize their environmental strategies over time, thereby fostering the development of resilient and ecologically sustainable urban environments. As a tool that integrates advanced technologies, data management, and analytics models, as well as environmental science, SUM plays a pivotal role in shaping the future of urban planning and governance toward greater environmental stewardship and well-being.

While the examined SUM projects and initiatives show promising prospects, there is a clear opportunity to enhance their capabilities by integrating AI and AIoT. These technologies can significantly improve SUM systems by enabling more sophisticated data analysis and decision-making processes. AI algorithms play a crucial role in enhancing its computational and analytical capabilities. They allow SUM systems to analyze large volumes of data collected from IoT sensors and other sources, providing deeper insights into urban metabolic flows and enabling more accurate predictions and optimizations. With AI, SUM systems can better understand complex urban dynamics, leading to more effective resource management and sustainability strategies. In addition to AI, the integration of AIoT into SUM systems can greatly benefit urban planning and governance (Bibri et al. 2024a). AIoT-powered SUM can engage citizens by providing interactive platforms for exploring urban data and fostering awareness of sustainability issues, encouraging behavior change, and promoting inclusivity and transparency in decision-making processes. Moreover, AI and AIoT can empower SUM to support collaborative urban development initiatives more effectively. SUM can facilitate coordination and cooperation among different actors in the urban ecosystem. All in all, integrating AI and AIoT into SUM systems holds significant potential for advancing urban planning and governance. These technologies can enhance data analysis, stakeholder engagement, and decision-making processes, which can ultimately lead to more resilient, sustainable, and well-informed cities.

4.4.1.4 Platform Urbanism

Platform urbanism refers to a rapidly evolving urban development landscape characterized by the pervasive influence of digital platforms in shaping various aspects of urban life, governance, and sustainability. This prominence has arisen due to the practice of platformization (Poell, Nieborg, and Van Dijck 2019; Seibt 2024) and its underlying AIoT processes, namely "digital instrumentation, digital hyper-connectivity, datafication, and algorithmization" (Bibri et al. 2024a). As a set of digitally enabled socio-political intermediations and socio-technical assemblages, platform urbanism is becoming the central organizing interface for smart cities (e.g., Caprotti et al. 2022; Repette et al. 2021), smart sustainable cities (e.g., ITU 2022; Noori et al. 2020), and smarter eco-cities (Bibri et al. 2024a).

At the core of the interactions and exchanges between various city stakeholders lies the notion of "digital mediation" (Barns 2019) or "algorithmic mediation" (Bibri 2023). This is facilitated by AI and AIoT-driven systems that leverage advanced real-time data processing, analytics, and decision-making capabilities for effectively addressing environmental urban challenges and promoting environmentally sustainable development goals (Bibri et al. 2024a). AI and AIoT play key roles in shaping the digital mediation of cities and hence data-driven governance structures and processes. The integration of AI and AIoT into governance platforms enables stakeholders to collaborate, innovate, and co-create solutions for urban challenges.

The integration of AI and AIoT in digital platforms provides urban stakeholders with actionable insights and supports collaborative decision-making to create interconnected ecosystems that enhance the efficiency and responsiveness of urban systems. In this context, platforms are digital infrastructures that facilitate the exchange of information, services, and resources among diverse urban stakeholders, including citizens, communities, businesses, and governmental entities. At its core, platform urbanism relies on decentralized networks of various stakeholders to achieve better outcomes, maximize value, and share resources and services. This has been demonstrated in real-world cities by the establishment of interconnected ecosystems that facilitate data-driven, participatory, adaptive, and collaborative processes and practices.

In light of the above, Morell and Espelt (2018) investigate the extent to which digital collaborative models in various spheres, beyond their initial domains, maintain an open character. Focusing on platform projects in Barcelona, the authors assess the adoption of open collaborative approaches in technological, knowledge, and governance aspects. Findings from empirical analysis reveal that open modalities in collaborative digital platforms are not predominant, with approximately one-third of the sample of 100 cases in Barcelona demonstrating openness in the analyzed dimensions. Moreover, the study identifies varying levels of diffusion in different areas, emphasizing a correlation between open practices in technological, data, and knowledge policies and the adoption of an open and democratic collaborative economy model. The results underscore the significance of open technology and knowledge in adopting fostering an open and democratic collaborative model.

Noori et al. (2020) examine and compare smart city projects in Amsterdam and Barcelona in terms of their pathways for implementation. The findings of this cross-case analysis show Amsterdam's focus on business-driven approaches with innovation and Barcelona's focus on social inclusion. Concerning innovation, several comparative studies demonstrate that these two cities vary in their institutional arrangements (Noori et al. 2020; Putra et al. 2018). Barcelona's innovation policy is mainly based on the real open innovation approach (Gascó et al. 2016), cyclic, and cross-cutting innovation model (Ferrer 2017). As regards social inclusion, the city council of Barcelona established "Decidim Barcelona," a participatory democracy platform aimed at empowering citizens, and another platform for sharing information generated by individuals and organizations (Bibri and Krogstie 2021).

The Amsterdam Smart City (ASC) is a collaborative platform that brings together various private and public actors to connect ideas and share challenges. The goal is to develop and implement smart solutions for sustainability by means of accelerating collaborative learning and doing (Noori et al. 2020). The ASC platform exemplifies a cohesive network connecting the municipal administration, academics, start-up investors, entrepreneurs, citizens, and others through a common information flow. It facilitates several projects, including smart energy grids, electric vehicle charging infrastructure, traffic and mobility management, and intelligent waste management. By integrating technology, data, and citizen participation, it aims to create a sustainable and climate-resilient urban environment. Data-driven technology serves as the "Data Hub Manager" of the "urban sustenance lab," enabling smart data collection, storage, and synthesis for future planning (Mora and Bolici 2017). Amsterdam's engagement strategy involves the "Smart Citizen" program, where residents actively participate as data representatives, contributing to sustainability knowledge. This collaborative approach highlights the collective contribution of people, technology, and ecology in advancing the smart city of Amsterdam. Furthermore, the ASC platform is governed by many collaborating partners (Mora et al. 2017), and the role of the city administration goes beyond being an initiator to include facilitation and finance (Van Winden et al. 2017).

Amsterdam and Barcelona are part of the European Union's scientific DECODE project pertaining to testing the technological approach to return data sovereignty to citizens. Adding to the DECODE project, there are "Data Commons Barcelona" and "City Data Analytics Office" as strategic initiatives for data protection and regulation (Calzada 2018). The former serves as an open-source policy toolkit to enable "cities to develop digital policies that put citizens at the center and make governments more open, transparent, and collaborative" (Noori et al. 2020). Serving as an "Internet of Cities" platform for sharing solutions across diverse cities and collaborative learning from each other, the "City Protocol" is a collaborative governance project between Amsterdam and Barcelona. Both City Data in Amsterdam and CityOS in Barcelona are openly accessible through the Internet, where the public can freely use data (Bibri and Krogstie 2020b; Noori et al. 2020). These case studies demonstrate how the integration of platform urbanism and smart urbanism can contribute to environmental sustainability. By leveraging digital platforms, data analytics, and citizen engagement, these initiatives enable more efficient resource management, promote renewable energy adoption, and foster sustainable urban development.

Furthermore, the study conducted by Jiang, Geertman, and Witte (2022), which analyzes the Helsinki Smart City, reveals diverse governance modes and advanced ICT functionalities, emphasizing the significance of context in shaping socio-technical governance for specific smart city challenges. In Helsinki's smart city development, governance as a platform plays a crucial role, using an integrative innovation platform, Forum Virium Helsinki, to co-produce the city's smart initiatives with universities, companies, and citizens. Civic engagement and collaboration are key features, fostering co-innovation and co-creation through online and offline platforms, such as living labs, to address urban challenges collectively. The democratic culture and bottom-up decision-making process enable wide collaboration between governments, businesses, citizens, and research institutions, contributing to sustainable and citizen-centered smart living. Also, Haveri and Anttiroiko (2023) analyze urban platform governance, exploring the role of platforms in local public governance. Based on theoretical analyses and empirical insight from three urban platforms in Finland's largest cities, the authors highlight platforms as emerging hybrids, blending features of networks, markets, and hierarchies. These unique characteristics lead to their classification as a distinct fourth mode of governance. Platform logic extends the concept of network governance, emphasizing broader connections, multiple logics orchestration, and ecosystem thinking. The findings contribute to understanding this novel approach to public governance.

To sum up, the role of platform urbanism in governance is transformative, offering new avenues for citizen engagement, data-driven decision-making, and collaborative problem-solving. While the emphasis is on the multifaceted role of platform urbanism in city governance, particularly its implications for fostering environmental sustainability in the face of contemporary urban challenges, both platform urbanism and smart urbanism contribute to advancing environmental urban governance. Smart urbanism provides the foundation with its focus on technology-driven solutions, whereas platform urbanism creates interconnected and collaborative ecosystems that promote data-driven, participatory, and adaptive environmental governance practices. By embracing these two approaches, cities can move closer to achieving their environmental sustainability goals and fostering resilient, livable, and environmentally responsible urban environments. Through platform urbanism, smart sustainable cities can tap into the collective intelligence and coordinative potential of their stakeholders, fostering innovation, optimizing resources, and enabling sustainable practices. All in all, platform urbanism fosters collaboration, coordination, innovation, and co-creation among multiple stakeholders thanks to scalable and efficient digital ecosystems, enabling the development and sharing of solutions to address urban challenges.

While the examined platform urbanism projects and initiatives present significant potential in enhancing urban governance, there is a clear opportunity to further advance its functionalities through AI- and AIoT-powered digital platforms. AI and AIoT technologies can significantly improve the effectiveness of platform urbanism in addressing environmental challenges and fostering sustainable development through innovative approaches to data-driven governance. Moreover, AI- and

AIoT-powered digital platforms can facilitate citizen engagement by providing interactive experiences, fostering greater awareness of sustainability issues, and encouraging behavior change.

4.4.2 An Integrative Analysis

In the dynamic landscape and increasing complexity of urban planning and governance, there is a growing need to adopt and leverage technology-driven and collaborative solutions. Within the framework of Artificial Intelligence of Smart Sustainable City Things, a thorough integrative analysis becomes of high relevance and importance to emphasize the collective roles of the brain, digital twin, metabolism, and platform of emerging smart sustainable cities in enhancing their planning and governance systems performance. Drawing on both the outcomes of the case studies and the evolving body of literature, this integrative analysis unravels the interconnections and synergies among City Brain, UDT, SUM, and platform urbanism concerning their contribution to urban planning and governance. It explores how this amalgamation—enabled by the unifying power of AIoT—can transform the urban landscape to become more environmentally conscious and technologically advanced.

4.4.2.1 Harnessing Synergies among AI/AIoT, City Brain, UDT, and SUM for Data-Driven Management and Planning

The synergies among City Brain, UDT, and SUM for city management and planning, enabled by AIoT, offer enhanced capabilities through advanced data analytics and real-time decision-making. City Brain and UDT can synergistically benefit from each other in terms of urban management and planning. Besides its primary role in planning, UDT has found extensive applications in urban management (Ferré-Bigorra et al. 2022) in the context of smart sustainable cities in the field of environmental sustainability (Weil et al. 2023), including water management (Pedersen et al. 2021), transportation and traffic management (Wu et al. 2022), wind energy management, farms and livestock management (Fuller et al. 2020), and resource management (Almusaed and Yitmen 2023). Some of these studies and most of recent studies (e.g., Agostinelli et al. 2021; Almusaed and Yitmen 2023; Beckett's 2022; Bibri 2024b; Li et al. 2021; Wang et al. 2022b; Ziakkas et al. 2024) underscore the significant role of AI or AIoT in enhancing the capabilities of UDT in optimizing urban systems.

Given the above, City Brain can harness the real-time data and insights provided by UDT to enhance urban operations. As noted by Zhang et al.

(2019), the objective of City Brain is to extract valuable insights from the city's vast and diverse data using AI and IoT technologies, alongside rapidly expanding computing capabilities, from cognition to optimization, decision-making, prediction, and intervention. This process also underscores the significance of predictive modeling, which can be facilitated by UDT, particularly in the realm of environmental sustainability (Bibri et al. 2024b; Bibri and Huang 2024; Ye et al. 2023; Weil et al. 2023). In turn, the real-time data gathered by City Brain can be fed into UDT, thereby improving its precision and applicability in testing real-world scenarios and potential solutions for urban planning. UDT provides a dynamic platform for simulating these scenarios and solutions, allowing City Brain to optimize responses and strategies in a more informed and context-aware manner. This integration creates a powerful ecosystem where the precision of UDT complements the agility of City Brain, leading to more effective urban management, planning, and decision-making.

Concerning the link between AI and UDT, the synergistic integration of AI with UDT has sparked a transformative wave in the realm of smart sustainable cities, bringing about significant advancements in how these urban environments can be planned and designed. In this context, Zvarikova et al. (2022) propose UDT algorithms leveraging 3D spatiotemporal simulations, ML, and DL for accurate urban modeling and simulation. Austin et al. (2020) combine semantic knowledge representation and ML in their UDT architecture, showcasing the practical impact of AI in smart city contexts. Beckett (2022) underscores the significant potential of integrating AI and UDT in terms of enhancing urban design and planning strategies through the integration of 3D modeling, visualization tools, and spatial cognition algorithms. A comprehensive systematic review conducted by Bibri et al. (2024b) explore the foundations of AI, AIoT, UDT, urban planning, and environmental sustainability in the context of smart sustainable cities. The authors investigate how AI and AIoT reshape urban planning for enhancing environmental sustainability and how they augment UDT predictive capabilities in this context, as well as propose an innovative conceptual framework for the seamless integration of these technologies to advance the environmental performance of smart sustainable cities. These studies collectively highlight the transformative potential of AI in advancing UDT, thereby shaping sustainable, resilient, efficient, and environmentally conscious urban landscapes and profoundly impacting the trajectory of sustainable urban development.

Furthermore, the integration of City Brain with SUM establishes a symbiotic relationship, enhancing the efficiency and sustainability of urban environments. This synergy entails leveraging City Brain's data-driven insights and combining them with SUM's capabilities in monitoring and optimizing resource flows. City Brain's analytical outcomes contribute valuable information to SUM's assessments of energy and resource dynamics, facilitating more informed strategies for environmentally sustainable urban development. Reciprocally, SUM's insights into resource utilization and environmental impact feed back into City Brain, influencing decisions that prioritize resource efficiency and reduce ecological footprints. Utilizing real-time data and AI and ML techniques (Ghosh and Sengupta 2023; Peponi et al. 2022) or AIoT (Bibri et al. 2024a), SUM enhances the efficiency and environmental performance of urban systems and infrastructure. This reciprocal collaboration establishes a dynamic and comprehensive framework for optimizing both operational efficiency and resource optimization in urban settings.

The synergy between UDT and SUM holds significant potential for advancing environmental sustainability in urban planning. While both UDT and SUM play roles in urban planning (Dar et al. 2023; Ferre-Bigorra et al. 2022; Oliveira and Vaz 2021; Schrotter and Hürzeler 2020), they function as distinct entities, each with unique capabilities in decision support, data integration, visualization, real-time monitoring, and environmental sustainability and resilience. UDT, acting as a digital replica of the physical city by capturing intricate details of the built environment, gains enhanced capabilities when integrated with SUM. This collaboration enables a more nuanced understanding of how the city's structures and functions impact resource consumption and waste generation.

SUM leverages the dynamic visualization platform offered by UDT to conduct real-time simulations and analyses, systematically evaluating the environmental implications associated with diverse urban scenarios. This strategic approach empowers informed decision-making within the realm of sustainable urban development. UDT further bolsters SUM's capabilities by providing essential data for continuous monitoring and analysis of resource flows within urban settings. This collaboration significantly contributes to resource efficiency and facilitates data-driven decision-making. In essence, UDT plays a significant role in supporting SUM, providing the necessary data to optimize resource flows and elevate the overall sustainability of urban environments. The synergistic relationship between UDT and SUM encompasses data integration, resource optimization, environmental impact assessment, scenario testing, decision support, and long-term planning. This reciprocal collaboration ensures a holistic understanding of the intricate resource dynamics within urban landscapes. Furthermore, SUM reciprocates by offering valuable insights into the complex resource flows inherent in urban environments. These insights, encompassing aspects such as energy consumption, waste generation, and material flows, serve as a substantial data source for UDT. Through the integration of SUM's detailed understanding of these resource dynamics, UDT refines its accuracy and comprehensiveness, presenting a more nuanced and realistic portrayal of the urban landscape. This reciprocal flow of information between SUM and UDT establishes a symbiotic relationship, forging a powerful toolset that significantly enhances the efficacy of urban planning and decision-making processes. Ultimately, this collaborative synergy leads to more robust, informed, and sustainable urban development strategies.

4.4.2.2 The Role of City Brain, UDT, and SUM in Environmental Governance

The collaborative integration of City Brain, UDT, and SUM plays a key role in advancing environmental governance through the provision of real-time data, analytics, and insights for collective decision-making. This collective contribution facilitates citizen engagement, stakeholder collaboration, and the development and implementation of effective policies. In essence, these advanced solutions empower cities and authorities to tackle environmental challenges, fostering evidence-based decisions in environmental governance. City Brain, as a sophisticated platform for data-driven decision-making, is instrumental in shaping more effective and responsive policies. Through its insights into crucial factors such as energy consumption, transport efficiency, traffic congestion, waste generation, and pollution control, City Brain equips policymakers with the tools to implement measures that effectively mitigate environmental impacts. The AIoT-driven capabilities of City Brain further support the formulation of regulations aimed at promoting sustainable practices, enabling city officials to make informed decisions in alignment with environmental sustainability goals.

Furthermore, UDT holds the potential to reshape governance structures and processes beyond planning (Deng et al. 2021; Lei et al. 2023; Weil et al. 2023). SUM contributes to environmental governance by providing real-time insights into urban resource flows, allowing for informed decision-making and targeted interventions to enhance environmental sustainability. It enables cities to identify inefficiencies, minimize waste, reduce emissions, and promote sustainable practices by monitoring and analyzing urban metabolic flows. This data-driven approach empowers policymakers and urban planners to develop evidence-based strategies for environmental management and governance, leading to more resilient and ecologically sustainable urban environments (Bibri et al. 2024a). Conversely, governance, being a key dimension of SUM, plays a crucial role in enhancing the efficiency of urban metabolism circularity, aligning with the overarching goals of environmental sustainability (D'Amico et al. 2022). The intersection of City Brain, UDT, and SUM and their collective impact on environmental governance underscores their significance in fostering sustainable urban development and environmental stewardship.

4.4.2.3 Governance Platforms

City Brain, UDT, and SUM represent notable examples of the evolving landscape of platform urbanism. This trend is underscored by the increasing adoption of governance platforms by public institutions, facilitated by diverse actors operating across various geographical scales and addressing a spectrum of issues, including sustainable development and environmental sustainability (Ansell and Miura 2019). The exploration of platform governance as a nascent dimension of local public governance by Haveri and Anttiroiko (2023) highlights the convergence of networks and hierarchies into a distinctive fourth mode of governance, characterized by its intrinsic features. This hybrid model exhibits traits that defy complete attribution to conventional governance frameworks in terms of structures and processes (Ansell and Gash 2018).

Governance platforms represent a novel organizing framework that encourages distributed participation (Ansell and Miura 2019). These platforms enable cross-cutting innovation (Hodson et al. 2021; Putra et al. 2018) and support scalable and adaptive governance in the realm of public innovation and collaborative governance (Ansell and Gash 2017; Ansell and Miura 2019). For example, participation platforms can significantly enhance citizen involvement in policy deliberation (Blasio and Selva 2019) and urban policy and decision-making. Moreover, multi-stakeholder platforms foster productive interactions among various expert groups (ITU 2022). Overall, the rise of governance platforms signifies a profound transformation in public governance paradigms, providing innovative solutions for tackling complex urban challenges and advancing inclusive decision-making processes.

The successful mitigation of environmental sustainability challenges in urban areas relies heavily on the synergy between effective governance and well-designed urban planning, forming a cohesive and comprehensive approach (Asadzadeh et al. 2023; Kramers et al. 2016). Platform urbanism in urban planning primarily focuses on the design, development, and organization of both physical and digital infrastructure within a city, exemplified by initiatives like City Brain, UDT, and SUM. In contrast, platform urbanism in urban governance involves leveraging digital platforms for citizen engagement, initiating open data initiatives, coordinating government agencies, and integrating policies (Barns 2019; Bibri et al. 2024a; Brynskov et al. 2018; Haveri and Anttiroiko 2023; Hodson et al. 2021; Katmada et al. 2023; Repette et al. 2021). Together, platform planning and governance contribute synergistically to the creation of smarter, more efficient, and responsive urban environments.

4.4.2.4 The Technical Logic behind the Integration of City Brain, UDT, SUM, and Platform Urbanism

Integrating City Brain, UDT, SUM, and platform urbanism within the framework of Artificial Intelligence of Smart Sustainable City Things necessitates a systematic approach to harnessing their collective potential for environmental urban planning and governance. Each component plays a distinct role in enhancing urban operational functioning, planning, and governance. The technical logic of their integration involves:

- Adopting standardized data formats and protocols across City Brain, UDT, SUM, and platform urbanism to ensure seamless information exchange.
- Developing interconnected APIs for each component, facilitating integration and interoperability, and enabling real-time data sharing and communication between planning and governance systems.
- Leverage AI and ML/DL algorithms within City Brain, UDT, SUM, and platform urbanism to continuously enhance the accuracy of predictions, optimizations, and practices. Feedback from each component informs and refines these algorithms.
- Establishing a robust feedback mechanism where insights from each component influence the others, which is crucial in the development of smart sustainable cities. For example, data-driven insights

from SUM can inform planning strategies and governance policies aimed at enhancing environmental sustainability through UDT and platform urbanism. UDT simulations guide optimization efforts in urban planning and infrastructure development. Additionally, platform urbanism aligns with real-time insights, facilitating the implementation of sustainability innovations and practices across urban domains. Moreover, integrating City Brain into the feedback mechanism can enhance urban planning and governance. City Brain's ability to provide actionable insights enables policymakers to make informed decisions regarding infrastructure development and resource allocation enhanced by UDT. This integration enhances the overall effectiveness of urban planning and governance, leading to more responsive urban development. Overall, establishing a feedback mechanism that integrates insights from City Brain, SUM, UDT, and platform urbanism is essential for promoting AIoT-driven planning and governance systems for addressing the complex challenges of environmental sustainability.

Within the framework as a cohesive whole, each of these components interacts synergistically, contributing to the overarching goal of advancing smart sustainable cities. City Brain serves as the foundational element, initialized with fundamental urban data. UDT overlays onto City Brain, enhancing its capabilities for simulation and monitoring by creating a dynamic virtual replica of the city. The data generated by SUM are integrated into City Brain and UDT, provide valuable insights into resource flows and environmental impacts, operational optimization, and scenario prediction, further enriching the comprehensive understanding of urban dynamics. Platform urbanism is strategically aligned with these three components, ensuring that planning processes and governance structures contribute to fostering more effective decision-making processes. This orchestrated integration establishes a robust foundation for leveraging the collective potential of these components in advancing smart sustainable urban development.

Within the framework of Artificial Intelligence of Smart Sustainable City Things (Figure 4.4), the integration of the key technical components of AIoT architecture (see Section 4.3) plays a critical role in enhancing the functions of City Brain, UDT, SUM, and platform urbanism in terms of their efficiency, responsiveness, and sustainability. This integration establishes a cohesive urban ecosystem, driving the planning and governance

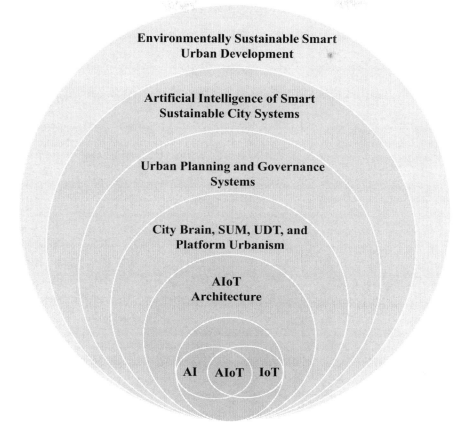

FIGURE 4.4 Artificial Intelligence of Smart Sustainable City Things for data-driven environmental planning and governance.

of smart sustainable cities toward advancing environmental sustainability goals.

Overall, the integration of City Brain, UDT, SUM, and platform urbanism offers a holistic approach to sustainable urban development. By leveraging their complementary capabilities, these systems can effectively bridge the gap between environmental planning and governance in smart sustainable cities through the unifying power of AIoT. This symbiotic relationship allows for the implementation of data-driven, cooperative, coherent, and adaptive strategies and policies to address interconnected environmental challenges. Overcoming obstacles in environmental governance plays a crucial role in addressing challenges in environmental planning (e.g., Asadzadeh et al. 2023; Kramers et al. 2016; Stead 2021).

The interrelation between these challenges highlights the pivotal role of effective governance approaches in facilitating and optimizing planning mechanisms.

4.5 DISCUSSION

Smart sustainable cities are at the forefront of leveraging cutting-edge technologies and collaborative models to enhance environmental outcomes and interventions through the evolving management, planning, and governance systems driven by AIoT. Despite the promising opportunities on the horizon, this ambitious endeavor is not devoid of its challenges and complexities. This chapter investigates the transformative potential and intricate interplay of City Brain, UDT, SUM, and platform urbanism within the framework of "Artificial Intelligence of Smart Sustainable City Things." This section provides a comprehensive discussion, encompassing various dimensions. It focuses on analyzing the roles of these systems in advancing data-driven environmental urban planning and governance, exploring integration possibilities, and identifying key challenges and strategies for overcoming them.

4.5.1 Summary of Findings and Interpretation of Results

The study reveals that the integration of City Brain, UDT, SUM, and platform urbanism as AIoT-driven systems has the potential to enhance the effectiveness of urban management, planning, and governance by providing a comprehensive understanding of urban dynamics and challenges. Comparing their functions and roles highlights their complementarity and demonstrates the importance of collaboration among these systems for achieving sustainable urban development. The study makes a significant contribution to the research landscape by filling a crucial gap in research concerning the synergistic integration of emerging data-driven management, planning, and governance systems. It delves into how these systems can collaborate and complement each other within the framework of AIoT. Instead of viewing these systems in isolation, it analyzes how they can interact and support each other to address complex urban challenges more effectively.

City Brain, for instance, provides real-time data analytics capabilities, allowing cities to monitor and optimize various aspects of urban life. UDT offers a virtual representation of the city, enabling planners to simulate and visualize different scenarios for future development. SUM focuses on understanding resource flows and urban metabolism, providing insights into sustainability and resilience. Platform urbanism, on the other hand, emphasizes governance structures and processes and participatory and collaborative decision-making processes. The case study analysis and literature review highlight that these systems play complementary roles in advancing smart sustainable cities. City Brain's data analytics can be integrated with UDT's simulation capabilities to inform predictive modeling and scenario planning. SUM's insights into resource flows can inform decision-making processes facilitated by platform urbanism, ensuring that governance structures align with sustainability goals.

In sum, the findings underscore the significance of integrating City Brain, UDT, SUM, and platform urbanism within the framework of AIoT to address the complexity of contemporary urban challenges. This integrated approach offers cities a more holistic perspective on urban development, enabling informed decision-making, improving efficiency, and fostering resilience in the face of rapid urbanization and environmental change.

4.5.2 Comparison with Existing Studies

Most studies have addressed City Brain (e.g., Kierans et al. 2023; Liu et al. 2022; Xu 2024; Wang et al. 2022), UDT (Agostinelli et al. 2021; Almusaed and Yitmen 2023; Beckett's 2022; Bibri 2024b; Huang et al. 2024; Li et al. 2021; Wang et al. 2022b; Ziakkas et al. 2024), SUM (Bibri 2024a; Ghosh and Sengupta 2023; Peponi et al. 2022), and platform urbanism (Ansell and Gash 2018; ITU 2022; Haveri and Anttiroiko 2023; Hodson et al. 2021; Katmada et al. 2023; Repette et al. 2021) separately or in isolation. They focus on their individual contributions to specific aspects of urban management, planning, and governance. This study contributes to the landscape of research and practice by addressing the gap concerning the lack of comprehensive studies on the synergistic and collaborative integration of these systems. It highlights the importance of this integration to realize the full potential of data-driven environmental management, planning, and governance in smart sustainable cities. City Brain, UDT, SUM, and platform urbanism each offer unique functionalities that, when seamlessly integrated and effectively harnessed, can provide comprehensive insights and support more effective decision-making processes. This integrated framework offers a more holistic perspective on sustainable urban development, moving beyond fragmented solutions to embrace a comprehensive, interdisciplinary approach. As cities continue to evolve in the face of rapid urbanization and environmental pressures, understanding the synergies among these systems becomes increasingly important for fostering sustainable development and resilience.

Overall, the integration of City Brain, UDT, SUM, and platform urbanism within the framework of AIoT offers promising opportunities for advancing data-driven environmental urban management, planning, and governance. By analyzing their roles, exploring integration possibilities, and addressing key challenges, this chapter provides valuable insights for urban planners, policymakers, and practitioners to adopt holistic solutions that prioritize sustainable urban development.

4.5.3 Challenges and Complexities Surrounding the Integration of City Brain, UDT, SUM, and Platform Urbanism

This subsection presents a varied set of challenges associated with synergizing the brain, digital twin, metabolism, and platform of smart sustainable cities to advance their environmental planning and governance. Table 4.2 is based on the integrative insights distilled from the available scholarly literature on the challenges and the corresponding mitigation strategies pertaining to City Brain (Caprotti and Liu 2022; Kierans et al. 2023; Liu et al. 2022; Wang et al. 2022; Xu 2022), UDT (e.g., Botín-Sanabria et al 2022; Lei et al. 2022; Weil et al. 2023; Wang et al. 2023), SUM (e.g., D'Amico et al. 2020, 2021, 2022; Shahrokni et al. 2015a, b), and platform urbanism (e.g., Ansell and Miura 2019; Barnes 2019, 2020; Caprotti, Chang and Joss 2022; Graham 2020; Hernández et al. 2020; Komninos and Kakderi 2019; Repette et al. 2021; Vadiati 2022) as emerging large-scale AIoT or CPSoS in urban planning and governance.

Additionally, it is essential to acknowledge that the integration of AIoT into smart sustainable city planning and governance systems may face additional obstacles and complexities. Similarly, Table 4.3 is grounded in the integrative insights distilled from the reviewed literature associated with City Brain, UDT, SUM, and platform urbanism.

A thorough understanding of the challenges is essential to pave the way for a successful implementation of the framework of Artificial Intelligence of Smart Sustainable City Things. Addressing technical, computational, ethical, social, and regulatory challenges requires a collaborative effort from stakeholders, including policymakers, urban planners, technologists, businesses, communities, and citizens. The additional obstacles and complexities highlight the multifaceted nature of integrating advanced technologies into urban environments and emphasize the need for holistic and adaptive strategies to overcome potential obstacles. By navigating

TABLE 4.2 Integration Challenges and the Corresponding Mitigation Strategies

Key Categories of Challenges	Corresponding Mitigation Strategies
The amalgamation of City Brain, UDT, SUM, and platform urbanism demands seamless technical interoperability. Challenges arise in integrating diverse systems, ensuring real-time data synchronization, and maintaining robustness in the face of technological failures.	Develop and adhere to standardized data formats and communication protocols to ensure seamless integration among these systems. Implement robust mechanisms for real-time data synchronization, leveraging advanced technologies to minimize latency and enhance responsiveness. Establish continuous monitoring systems to detect and address technological failures promptly.
Handling massive datasets generated by City Brain, UDT, SUM, and platform urbanism poses computational challenges. Efficient processing, analysis, and interpretation of these data are vital for informed decision-making, demanding advanced computational capabilities and algorithms.	Invest in advanced algorithms and computational models to efficiently process and analyze massive datasets generated by these systems. Deploy scalable computational infrastructure to handle the computational demands associated with data processing and analysis.
Ethical considerations pertain to data privacy, surveillance, and responsible AI and AIoT in City Brain, UDT, SUM, and platform urbanism. Balancing the benefits of enhanced planning and governance with ethical concerns requires a nuanced approach to ensure the ethical deployment of these technologies.	Incorporate privacy considerations into the design phase of these systems, ensuring that data privacy is a fundamental aspect of the architecture. Enhance transparency in data usage and obtain explicit consent from citizens for the collection and utilization of their data within the framework of ethical guidelines.
Integrating advanced technologies and models in urban spaces raises social challenges related to inclusivity and accessibility. Ensuring equitable access to technology and fostering community engagement become crucial aspects of implementing City Brain, UDT, SUM, and platform urbanism.	Implement initiatives focused on reducing the digital divide, providing equitable access to technology, and fostering digital literacy to ensure inclusivity. Actively engage with local communities, seeking their input and feedback in the planning and deployment of smart technologies to address concerns and ensure social acceptance.
The regulatory landscape must evolve to accommodate the complexities of integrating City Brain, UDT, SUM, and platform urbanism. Establishing clear frameworks for using AI and AIoT in urban planning and governance becomes paramount to navigate legal and regulatory challenges effectively.	Collaborate with regulatory bodies and government authorities to establish clear frameworks for data governance, privacy laws, and standards for the responsible use of AI in urban governance. Conduct regular audits to ensure compliance with evolving regulatory requirements and make necessary adjustments to align with legal standards.

TABLE 4.3 Obstacles and Complexities and the Corresponding Mitigation Strategies

Obstacles and Complexities	Corresponding Mitigation Strategies
Ensuring seamless interoperability not only on a technical level but also among various stakeholders involved in smart sustainable city initiatives is critical.	Facilitate collaboration and communication among various stakeholders through the establishment of interconnected Application Programming Interfaces (APIs). Unified Standards: Advocate for the development and adoption of unified standards for smart sustainable city technologies to enhance overall interoperability.
Safeguarding the vast amount of sensitive data related to the emerging planning and governance systems of smart sustainable cities is a significant concern.	Implement robust encryption mechanisms and access controls to secure sensitive data from unauthorized access. Employ continuous monitoring systems to detect and respond to potential security threats in real-time.
The deployment of City Brain, UDT, SUM, and platform urbanism involves significant financial investments. Securing funding and allocating resources strategically pose evident challenges, especially for emerging smart sustainable cities with limited financial capabilities.	Foster collaborations with private entities through public-private partnerships to share the financial burden and bring in additional resources. Grant programs and funding: Actively seek participation in grant programs and funding initiatives dedicated to smart sustainable city development. Implement strategies for efficient budget allocation, prioritizing projects based on long-term sustainability and impact.
Paradoxically, the implementation of smart sustainable city technologies and solutions may have unintended environmental consequences.	Emphasize sustainable design practices in the development of smart sustainable city technologies to minimize environmental impact. Implement comprehensive lifecycle management strategies to reduce e-waste and ensure responsible disposal of outdated technologies. Conduct thorough environmental impact assessments before the deployment of new technologies to anticipate and mitigate potential negative consequences.

the challenges, overcoming the obstacles, and unraveling the complexities with a strategic and inclusive approach, it becomes possible to harness the full potential of the emerging planning and governance systems enabled by AI and AIoT technologies, contributing to the realization of environmentally sustainable and technologically advanced cities.

4.5.4 Existing Gaps and Future Research Directions

In the pursuit of unlocking the full potential of AIoT in smart sustainable cities, it is essential to address the identified gaps that have come to light through a nuanced exploration of the interconnected functions of their brain, digital twin, metabolism, and platform. Identifying these gaps serves as a crucial foundation for a more dedicated investigation into the dynamic landscape of smart sustainable cities, shedding light on potential challenges and highlighting key areas for future research endeavors.

The journey toward smarter and more sustainable cities encounters several critical gaps that impede the realization of cohesive and synergistic urban ecosystems. Insufficient integration across city functions leads to fragmented urban systems, hindering comprehensive planning and governance. Future research should investigate strategies and frameworks for harnessing the functions of City Brain, UDT, SUM, and platform urbanism for joined-up and comprehensive urban planning and governance. Moreover, a limited understanding of the synergic potential of AIoT results in missed opportunities for enhancing urban efficiency and sustainability. Conducting in-depth studies to unravel the synergies between AIoT functionalities and planning and governance functions can address this gap. Ethical blind spots in AIoT integration raise concerns about privacy, security, and fairness, necessitating the development of specific ethical and social frameworks. Future research should focus on developing ethical frameworks tailored to AIoT integration in urban planning and governance, addressing issues such as privacy, security, transparency, bias, inequality, and fairness. Furthermore, the absence of interdisciplinary collaboration undermines innovation and effective decision-making in urban planning and governance. Investigating effective collaborative frameworks, cross-disciplinary endeavors, and technology-enabled platforms can foster interdisciplinary engagement in environmental urban planning and governance for smart sustainable cities. Lastly, incomplete integration of data-driven decision-making systems across city domains may lead to suboptimal decision outcomes and missed opportunities for environmentally sustainable urban development. Developing methodologies and frameworks for integrating data-driven decision-making processes, as well as exploring the development of unified platforms or models that holistically consider inputs from all city systems and domains can enhance decision-making accuracy and effectiveness. Through these efforts, cities can pave the way for smarter, more sustainable, and environmentally friendly urban development.

This structured analysis is intended to guide researchers, policymakers, and urban planners in addressing critical areas that require further exploration. The trajectory of future research should be guided by a commitment to fostering beneficial integration, harnessing synergies, fortifying ethical foundations, and enhancing the social outcomes of smart sustainable cities. Through these concerted efforts, the prospect of creating intelligent, inclusive, and ethically sound urban environments may shift from an aspiration to an achievable reality.

4.6 CONCLUSION

The chapter explored the transformative potential and intricate interplay of City Brain, UDT, SUM, and platform urbanism—as prominent urban planning and governance systems—within the framework of Artificial Intelligence of Smart Sustainable City Things. Key findings include role clarification, integration insights, technological synergy, and challenges and obstacles anticipation. Specifically, an in-depth understanding of the diverse roles played by City Brain, UDT, SUM, and platform urbanism was established. Nuanced insights into the seamless integration of the distinct functionalities of these planning and governance systems, leveraging the computational and analytical capabilities of AIoT, were provided based on an integrative analysis. The latter also included new perspectives on how these functionalities and capabilities contribute to the cohesive integration of these systems, fostering technological synergy for sustainable urban development. Anticipated challenges and obstacles in exploring not only the clear synergy in the operation of these systems but also their interconnected functions have been identified. Potential strategies have been proposed to address these challenges and obstacles to propel the planning and governance of smart sustainable cities to new levels of effectiveness and sustainability.

Given the above considerations, the study holds significant implications from diverse perspectives. From a research standpoint, it enriches the existing knowledge base on smart sustainable urban development. It offers a comprehensive understanding of the innovative potential and synergistic opportunities of City Brain, UDT, SUM, and platform urbanism within the novel framework of Artificial Intelligence of Smart Sustainable City Things. Researchers in the field gain valuable insights into the synergies and complexities involved in the amalgamation of these advanced planning and governance systems, providing a foundation for further investigations and in-depth qualitative analyses on cutting-edge urban technologies and collaborative urban models in the context of environmental sustainability. Moreover, the study underscores the importance of fostering interdisciplinary collaboration to address the formidable challenges pertaining to, and harness the full potential of integrated urban planning and governance systems.

The research design, which integrates case study analysis and scholarly literature review, offers a robust approach to comprehensively explore real-world implementations and theoretical foundations. This methodological choice allows for a nuanced understanding of the intricate interplay between City Brain, UDT, SUM, and platform urbanism, contributing to the depth and breadth of the study's insights. The findings emphasize the effectiveness of this interdisciplinary methodology in unraveling the complexities of integrated urban planning and governance systems. Additionally, the study advocates for the adoption of similar methodologies in future research endeavors to advance the understanding and implementation of smart sustainable urban development practices.

For urban practitioners, the study provides practical insights that can be translated into real-world applications. By elucidating the functions of City Brain, UDT, SUM, and platform urbanism and their integration mechanisms, practitioners are equipped with actionable knowledge to enhance the planning and governance of smart sustainable cities. The findings offer practical pathways for the adoption of integrated solutions, enabling city professionals and officials to develop and implement holistic strategies that foster resilience, intelligence, and sustainability in urban landscapes. Moreover, the study underscores the significance of diverse case studies illustrating this emerging trend in various urban contexts, highlighting the adaptability and effectiveness of the proposed strategies.

The implications of this study extend to the realm of urban policymaking. As cities worldwide grapple with the imperative to become smarter and more sustainable, policymakers are confronted with complex decisions. The study's insights offer strategic pathways for decision-makers to formulate policies that align with the principles and goals of environmental sustainability. Policymakers gain a nuanced understanding of how AI and AIoT can be harnessed to optimize the functions of City Brain, UDT, SUM, and platform urbanism, guiding the development of policies that foster technological advancements and their amalgamation to address complex urban challenges.

In the global context, where cities face the necessity to evolve into technologically advanced and environmentally conscious urban environments—as demonstrated by real-world implementations—the ensuing insights are poised to guide different stakeholders across smart and sustainable cities in unraveling the complexities of urban development. Decision-makers globally can draw upon the strategic pathways provided by this research to unlock the opportunities presented by the integration of innovative approaches to urban planning and governance to shape environmental outcomes and actions.

REFERENCES

Agostinelli, S., Cumo, F., Guidi, G., & Tomazzoli, C. (2021). Cyber-physical systems improving building energy management: Digital twin and artificial intelligence. *Energies*, 14(8), 2338.

Ahvenniemi, H., Huovila, A., Pinto-Seppä, I., Airaksinen, M., & Valvontakonsultit Oy, R. (2017). What are the differences between sustainable and smart cities? *Cities*, 60, 234–245. https://doi.org/10.1016/j.cities.2016.09.009

Ajuriaguerra Escudero, M. A., & Abdiu, M. (2022). Artificial intelligence in european urban governance. In: J. Saura, & F. Debasa (Eds.), *Handbook of Research on Artificial Intelligence in Government Practices and Processes* (pp. 88–104). IGI Global. https://doi.org/10.4018/978-1-7998-9609-8.ch006

Alahi, M. E. E., Sukkuea, A., Tina, F. W., Nag, A., Kurdthongmee, W., Suwannarat, K., & Mukhopadhyay, S. C. (2023). Integration of IoT-enabled technologies and artificial intelligence (AI) for smart city scenario: Recent advancements and future trends. *Sensors*, 23(11), 5206.

Almalki, F. A., Alsamhi, S. H., Sahal, R., Hassan, J., Hawbani, A., Rajput, N. S., Saif, A., Morgan, J., & Breslin, J. (2021). Green IoT for eco-friendly and sustainable smart cities: Future directions and opportunities. *Mobile Networks and Applications*. https://doi.org/10.1007/s11036-021-01790-w

Almusaed, A., & Yitmen, I. (2023). Architectural reply for smart building design concepts based on artificial intelligence simulation models and digital twins. *Sustainability*, 15(6), 4955.

Ansell, C., & Gash, A. (2018). Collaborative platforms as a governance strategy. *Journal of Public Administration Research and Theory*, 28(1), 16–32.

Ansell, C., & Miura, S. (2019). Can the power of platforms be harnessed for governance? Public Administration. https://doi.org/10.1111/padm.12636.

Asadzadeh, A., Fekete, A., Khazai, B., Moghadas, M., Zebardast, E., Basirat, M., & Kötter, T. (2023). Capacitating urban governance and planning systems to drive transformative resilience. *Sustainable Cities and Society*, 96, 104637. https://doi.org/10.1016/j.scs.2023.104637

Austin, M., Delgoshaei, P., Coelho, M., & Heidarinejad, M. (2020). Architecting smart city digital twins: Combined semantic model and machine learning approach. *Journal of Management in Engineering*, 36(4), 04020026.

Avis, W. R. (2016). Sustainable Development Goals (Urban governance topic guide). GSDRC, University of Birmingham. GSDRC, Birmingham. http://gsdrc.org/topic-guides/urban-governance/concepts-and-debates/what-is-urban-governance/. Accessed 17 March 2023.

Barns, S. (2019) Negotiating the platform pivot: From participatory digital ecosystems to infrastructures of everyday life. *Geography Compass, 13,* e12464

Barns, S. (2020). Re-engineering the city: Platform ecosystems and the capture of urban big data. *Frontiers in Sustainable Cities, 2.* https://doi.org/10.3389/frsc.2020.00032

Beck, M. B., Chen, C., Walker, R. V., Wen, Z., & Han, X. (2023). Multi-sectoral analysis of smarter urban nitrogen metabolism: A case study of Suzhou, China. *Ecological Modelling, 478,* 110286. https://doi.org/10.1016/j.ecolmodel.2023.110286

Beckett (2022). Smart city digital twins, 3d modeling and visualization tools, and spatial cognition algorithms in artificial intelligence-based urban design and planning. *Geopolitics, History, and International Relations, 14*(1), 123–138.

Bibri, S. E. (2019). Smart sustainable urbanism: Paradigmatic, scientific, scholarly, epistemic, and discursive shifts in light of big data science and analytics. In: *Big Data Science and Analytics for Smart Sustainable Urbanism. Advances in Science, Technology & Innovation.* Springer. https://doi.org/10.1007/978-3-030-17312-8_6

Bibri, S. E. (2020). Data-driven environmental solutions for smart sustainable cities: Strategies and pathways for energy efficiency and pollution reduction. *Euro-Mediterranean Journal for Environmental Integration, 5*(66). https://doi.org/10.1007/s41207-020-00211-w

Bibri, S. E. (2021). Data-driven smart eco-cities and sustainable integrated districts: A best-evidence synthesis approach to an extensive literature review. *European Journal of Futures Research, 9,* 1–43. https://doi.org/10.1186/s40309-021-00181-4.

Bibri, S. E., Alexandre, A., Sharifi, A., & Krogstie, J. (2023a). Environmentally sustainable smart cities and their converging AI, IoT, and big data technologies and solutions: An integrated approach to an extensive literature review. *Energy Informatics, 6,* 9.

Bibri, S. E., & Jagatheesaperumal, S. K. (2023). Harnessing the potential of the metaverse and artificial intelligence for the internet of city things: Cost-effective xreality and synergistic AIOT technologies. *Smart Cities, 6*(5), 2397–2429. https://doi.org/10.3390/smartcities6050109

Bibri, S. E., & Krogstie, J. (2020a). Data-driven smart sustainable cities of the future: A novel model of urbanism and its core dimensions, strategies, and solutions. *Journal of Futures Studies, 25*(2), 77–94. https://doi.org/10.6531/JFS.202012_25(2).0009

Bibri, S. E., & Krogstie, J. (2020b). Environmentally data-driven smart sustainable cities: Applied innovative solutions for energy efficiency, pollution reduction, and urban metabolism. *Energy Informatics, 3*(1), 29. https://doi.org/10.1186/s42162-020-00130-8

Bibri, S. E., & Krogstie, J. (2021). A novel model for data-driven smart sustainable cities of the future: A strategic roadmap to transformational change in the era of big data. *Future Cities and Environment, 7*(1), 3. https://doi.org/10.5334/fce.116

Bibri, S. E., Huang, J., & Krogstie, J. (2024a), Artificial intelligence of things for synergizing smarter eco-city brain, metabolism, and platform: Pioneering data-driven environmental governance. *Sustainable Cities and Society*, 105516. https://doi.org/10.1016/j.scs.2024.105516

Bibri, S. E., Huang, J., Jagatheesaperumal, S. K., & Krogstie, J. (2024b). The synergistic interplay of artificial intelligence and digital twin in environmentally planning sustainable smart cities: A comprehensive systematic review. *Environmental Science and Ecotechnology*, 20, 100433. https://doi.org/10.1016/j.ese.2024.100433

Bibri, S. E., & Huang, J. (2025). Artificial intelligence of sustainable smart city brain and digital twin: A pioneering framework for advancing environmental sustainability. *Environmental Technology and Innovation* (in press).

Bibri, S. E., Krogstie, J., Kaboli, A., & Alahi, A. (2023b). Smarter eco-cities and their leading-edge solutions for environmental sustainability: A comprehensive systemic review. *Environmental Science and Eco-technology*, 19, https://doi.org/10.1016/j.ese.2023.100330

Blasio, E., & Selva, D. (2019). Implementing open government: A qualitative comparative analysis of digital platforms in France, Italy and United Kingdom. *Quality & Quantity: International Journal of Methodology*, 53(2), 871–896. https://doi.org/10.1007/s11135-018-0803-9

Botín-Sanabria, D. M., Mihaita, A.-S., Peimbert-García, R. E., Ramírez-Moreno, M. A., Ramírez-Mendoza, R. A., & Lozoya-Santos, J. D. J. (2022). Digital twin technology challenges and applications: A comprehensive review. *Remote Sensing*, 14(6), 1335. https://doi.org/10.3390/rs14061335

Bridge, G., & Perreault, T. (2009). Environmental governance. In: N. Castree, D. Demeritt, D. Liverman, & B. Rhoads (Eds.), *A Companion to Environmental Geography* (pp. 475–497). Blackwell Publishing.

Brienza, S., Galluzzi, A., Cascavilla, G., Pietroni, E., & Malizia, A. (2018). A tale of ten cities: A content analysis of urban practices across the "wicked problems" of climate change, water management, and air quality. *Sustainability*, 10(9), 3114. https://doi.org/10.3390/su10093114

Brynskov, M., Calvillo, N., & Borup, M. (2018). The urban data platform as city orchestrator: Participatory urbanism in Copenhagen. *European Journal of Cultural Studies*, 21(4), 474–490.

Calzada, I. (2018). (Smart) citizens from data providers to decision-makers? The case study of Barcelona. *Sustainability*, 10, 3252.

Canetta, L., & Lombardi, P. (2012). A methodology for developing smart city projects. In: *2012 IEEE First International Conference on Smart Grid Communications (SmartGridComm)* (pp. 518–523). https://doi.org/10.1109/SmartGridComm.2012.6486026

Caprotti, F., & Liu, D. (2022). Platform urbanism and the Chinese smart city: The co-production and territorialisation of Hangzhou City Brain. *GeoJournal*, 87, 1559–1573. https://doi.org/10.1007/s10708-020-10320-2

Caprotti, F., Chang, I.-C. C., & Joss, S. (2022). Beyond the smart city: A typology of platform urbanism. *Urban Transformations*, 4(4). https://doi.org/10.1186/s42854-022-00033-9

Chen, L., Chen, Z., Zhang, Y., Liu, Y., Osman, A. I., Farghali, M., Hua, J., Al-Fatesh, A., Ihara, I., Rooney, D. W., & Yap, P.-S. (2023). Artificial intelligence-based solutions for climate change: A review. *Environmental Chemistry Letters, 21*, 2525–2557. https://doi.org/10.1007/s10311-023-01556-7

Corsi, A., Pagani, R., Cruz, T. B., Souza, F. F., & Kovaleski, J. L. (2022). Smart sustainable cities: Characterization and impacts for sustainable development goals. *The Journal of Environmental Management, 11*, 1–32.

Cugurullo, F. (2020). Urban artificial intelligence: From automation to autonomy in the smart city. *Frontiers in Sustainable Cities, 2*, 38. https://doi.org/10.3389/frsc.2020.00038

Cureton, P., & Hartley, E. (2023). City information models (CIMs) as precursors for urban digital twins (UDTs): A case study of lancaster. *Frontiers in Built Environment, 9*, 1048510.

Currie, P. K., & Musango, J. K. (2017). African urbanization: Assimilating urban metabolism into sustainability discourse and practice. *Journal of Industrial Ecology, 21*(5), 1262–1276.

D'Amico, G., Arbolino, R., Shi, L., Yigitcanlar, T., & Ioppolo, G. (2021). Digital technologies for urban metabolism efficiency: Lessons from urban agenda partnership on circular economy. *Sustainability, 13*(11), 6043. https://doi.org/10.3390/su13116043

D'Amico, G., Arbolino, R., Shi, L., Yigitcanlar, T., & Ioppolo, G. (2022). Digitalisation driven urban metabolism circularity: A review and analysis of circular city initiatives. *Land Use Policy, 112*, 105819. https://doi.org/10.1016/j.landusepol.2021.105819.

D'Amico, G., Taddeo, R., Shi, L., Yigitcanlar, T., & Ioppolo, G. (2020). Ecological indicators of smart urban metabolism: A review of the literature on international standards. *Ecological Indicators, 118*, 106808.

Dar, M. A., Raina, R., Singh, R., Kaur, S., Singh, H. P., & Batish, D. R. (2023). Role of sustainable urban metabolism in urban planning. In: R. Bhadouria, S. Tripathi, P. Singh, P. K. Joshi, & R. Singh (Eds.), *Urban Metabolism and Climate Change*. Springer. https://doi.org/10.1007/978-3-031-29422-8_8.

de Wit, M. P. (2020). Environmental governance: Complexity and cooperation in the implementation of the SDGs. In: W. Leal Filho, A. Azul, L. Brandli, A. Lange Salvia, & T. Wall (Eds.), *Affordable and Clean Energy: Encyclopedia of the UN Sustainable Development Goals*. Springer. https://doi.org/10.1007/978-3-319-71057-0_25-1.

Dembski, F., Wössner, U., Letzgus, M., Ruddat, M., & Yamu, C. (2020). Urban digital twins for smart cities and citizens: The case study of Herrenberg, Germany. *Sustainability, 12*(6), 2307. https://doi.org/10.3390/su12062307

Deng, T., Zhang, K., & Shen, Z.-J. (2021). A systematic review of a digital twin city: A new pattern of urban governance toward smart cities. *Journal of Management Science and Engineering, 6*, 125–134. https://doi.org/10.1016/j.jmse.2021.03.003.

El Himer, S., Ouaissa, M., Ouaissa, M., & Boulouard, Z. (2022). Artificial intelligence of things (AIoT) for renewable energies systems. In: S. El Himer, M. Ouaissa, A. A. A. Emhemed, M. Ouaissa, & Z. Boulouard (Eds.), *Artificial Intelligence of Things for Smart Green Energy Management. Studies in Systems, Decision and Control* (vol. 446). Springer. https://doi.org/10.1007/978-3-031-04851-7_1

ENAC (2024). *The Blue City Project.* https://www.epfl.ch/schools/enac/blue-city-project/ (Accessed 12 October 2023).

ETH (2024). *The Blue City Project.* https://esd.ifu.ethz.ch/research/research-projects/research-and-theses/bluecity.html (accessed 12 October 2023).

Ferré-Bigorra, J., Casals, M., & Gangolells, M. (2022). The adoption of urban digital twins. *Cities, 131*, 10390.

Ferrer, J. R. (2017). Barcelona's smart city vision: An opportunity for transformation. *Journal of Field Actions, 16*, 70–75.

Formisano, V., Iannucci, E., Fedele, M., & Bonab, A. B. (2022). City in the loop: Assessing the relationship between circular economy and smart sustainable cities. *Sinergie Italian Journal of Management, 40*(2), 147–168.

Fragkos, G., Tsiropoulou, E. E., & Papavassiliou, S. (2020). Artificial intelligence enabled distributed edge computing for Internet of Things applications. In: Proceedings of the 2020 16th international conference on distributed computing in sensor systems (DCOSS) (pp. 450–457). IEEE.

Fuller, A., Fan, Z., Day, C., & Barlow, C. (2020). Digital twin: Enabling technologies. *Challenges and Open Research IEEE Access, 8*, 108952–108971. https://doi.org/10.1109/ACCESS.2020.2998358.

Gao, W., Ma, S., Duan, L., Tian, Y., Xing, P., Wang, Y., Wang, S., Jia, H., & Huang, T. (2021). Digital retina: A way to make the city brain more efficient by visual coding. *IEEE Transactions on Circuits and Systems for Video Technology, 31*(11), 4147–4161.

Gascó, M., Trivellato, B., & Cavenago, D. (2016). How do Southern European cities foster innovation? Lessons from the experience of the smart city approaches of Barcelona and Milan. In: J. Gil-Garcia, T. Pardo, & T. Nam (Eds.), *Smarter as the New Urban Agenda:* Public Administration and Information Technology (vol. 11, pp. 191–206). Springer.

Gholami, M., Torreggiani, D., Barbaresi, A., & Tassinari, P. (2024). Smart green planning for urban environments: The city digital twin of imola. In: F. Belaïd, & A. Arora (Eds.), *Smart Cities (Studies in Energy, Resource and Environmental Economics).* Springer. https://doi.org/10.1007/978-3-031-35664-3_10

Ghosh, R., & Sengupta, D. (2023). Smart urban metabolism: A big-data and machine learning perspective. In: R. Bhadouria, S. Tripathi, P. Singh, P. K. Joshi, & R. Singh (Eds.), *Urban Metabolism and Climate Change.* Springer. https://doi.org/10.1007/978-3-031-29422-8_16.

Gourisaria, M. K., Jee, G., Harshvardhan, G. M., Konar, D., & Singh, P. K. (2023). Artificially intelligent and sustainable smart cities. In: P. K. Singh, M. Paprzycki, M. Essaaidi, & S. Rahimi (Eds.), *Sustainable Smart Cities: Studies in Computational Intelligence* (vol. 942). Springer. https://doi.org/10.1007/978-3-031-08815-5_14

Graham, M. (2020). Regulate, replicate, and resist: The conjunctural geographies of platform urbanism. *Urban Geography, 41*, 453–457. https://doi.org/10.108 0/02723638.2020.1717028

Hämäläinen, M. (2021). Urban development with dynamic digital twins in Helsinki City. *IET Smart Cities, 3*, 201–210. https://doi.org/10.1049/smc2.12015

Haveri, A., & Anttiroiko, A.-V. (2023). Urban platforms as a mode of governance. *International Review of Administrative Sciences, 89*(1), 3–20. https://doi.org/10.1177/00208523211005855.

Hernández, J. L., García, R., Schonowski, J., Atlan, D., Chanson, G., & Ruohomäki, T. (2020). Interoperable open specifications framework for the implementation of standardized urban platforms. *Sensors, 20*(8), 2402. https://doi.org/10.3390/s20082402

Hodson, M., Kasmire, J., McMeekin, A., Stehlin, J. G., & Ward, K. (2021). *Urban Platforms and the Future City. Transformations in Infrastructure, Governance, Knowledge and Everyday Life.* Routledge and Taylor and Francis.

Huang, J. Bibri, S. E., & Keel, P (2025). Generative spatial artificial intelligence for sustainable smart cities: A pioneering large flow foundation model for urban digital twin. *Environmental Science and Ecotechnology* (in press).

Iris-Panagiota, E., & Egleton, T. E. (2023). Artificial intelligence for sustainable smart cities. In: B. K. Mishra (Ed.), *Handbook of Research on Applications of AI, Digital Twin, and Internet of Things for Sustainable Development* (pp. 1–11). IGI Global. https://doi.org/10.4018/978-1-6684-6821-0.ch001

ITU (2022). *The United for Smart Sustainable Cities (U4SSC).* The International Telecommunication Union, https://u4ssc.itu.int (accessed 10 December 2022).

Jagatheesaperumal, S. K., Bibri, S. E., Huang, J Ganesan, S., & Jeyaraman, P. (2024). Artificial intelligence of things for smart cities: Advanced solutions for enhancing transportation safety. *Computers, Environment and Urban Systems, 4*, 10. https://doi.org/10.1007/s43762-024-00120-6

Jain, H., Dhupper, R., Shrivastava, A., Kumar, D., & Kumari, M. (2023). AI-enabled strategies for climate change adaptation: Protecting communities, infrastructure, and businesses from the impacts of climate change. *Computational Urban Science, 3*, Article 25. https://doi.org/10.1007/s43762-023-00041-6

Jiang, H., Geertman, S., & Witte, P. (2022). Smart urban governance: An alternative to technocratic "smartness". *GeoJournal, 87*(7), 1639–1655. https://doi.org/10.1007/s10708-020-10326-w

Juma, M., & Shaalan, K. (2020). Cyberphysical systems in the smart city: Challenges and future trends for strategic research. In: A. E. Hassanien, & A. Darwish (Eds.), *Intelligent Data-Centric Systems: Swarm Intelligence for Resource Management in Internet of Things* (pp. 65–85). Academic Press. https://doi.org/10.1016/B978-0-12-818287-1.00008-5.

Kamal, H., Yánez, W., Hassan, S., & Sobhy, D. (2024). Digital-twin-based deep reinforcement learning approach for adaptive traffic signal control. *IEEE Internet of Things Journal, 11*(12), 21946–21953. https://doi.org/10.1109/JIOT.2024.3377600.

Kamrowska-Załuska, D. (2021). Impact of AI-based tools and urban big data analytics on the design and planning of cities. *Land, 10*, 1209.

Kanishk, C., & Kolbe, T. H. (2016). Integrating dynamic data and sensors with semantic 3D city models in the context of smart cities. *ISPRS Annals of the Photogrammetry, Remote Sensing and Spatial Information Sciences, 4*, 31–38. https://doi.org/10.5194/isprs-annals-IV-2-W1-31-2016

Katmada, A., Katsavounidou, G., & Kakderi, C. (2023). Platform urbanism for sustainability. In: N. A. Streitz, & S. Konomi (Eds.), *Distributed, Ambient and Pervasive Interactions. HCII 2023, Lecture Notes in Computer Science* (vol. 14037). Springer. https://doi.org/10.1007/978-3-031-34609-5_3.

Kennedy, C., Cuddihy, J., & Engel-Yan, J. (2007). The changing metabolism of cities. *Journal of Industrial Ecology, 11*(2), 43–59.

Khan, F., Kumar, R. L., Kadry, S., Nam, Y., & Meqdad, M. N. (2021). Cyber physical systems: A smart city perspective. *International Journal of Electrical and Computer Engineering (IJECE), 11*(4), 3609–3616. https://doi.org/10.11591/ijece.v11i4.pp3609-3616.

Kierans, G., Jüngling, S. & Schütz, D. (2023). Society 5.0 2023. *EPiC Series in Computing, 93*, 82–96

Koffka, K. (2023). *Intelligent Things: Exploring AIoT Technologies and Applications*. Kinder edition.

Komninos, N., & Kakderi, C. (Eds.). (2019). *Smart Cities in the Post-algorithmic Era: Integrating Technologies, Platforms and Governance*. Edward Elgar Publishing

Koumetio, T. S. C., Diop, E. B., Azmi, R., & Chenal, J. (2023). Artificial intelligence based methods for smart and sustainable urban planning: A systematic survey. *Archives of Computational Methods in Engineering, 30*(5), 1421–1438. https://doi.org/10.1007/s11831-022-09844-2

Leal Filho, W., Wall, T., Mucova, S. A. R., Nagy, G. J., Balogun, A.-L., Luetz, J. M., Ng, A. W., Kovaleva, M., Azam, F. M. S., & Alves, F. (2022). Deploying artificial intelligence for climate change adaptation. *Technological Forecasting and Social Change, 180*, 121662.

Lei, B., Janssen, P., Stoter, J., & Biljecki, F. (2023). Challenges of urban digital twins: A systematic review and a Delphi expert survey. *Automation in Construction, 147*, Article 104716. https://doi.org/10.1016/j.autcon.2022.104716.

Lemos, M. C., & Agrawal, A. (2006). Environmental governance. *Annual Review of Environment and Resources, 31*, 297–325. https://doi.org/10.1146/annurev.energy.31.042605.135621

Li, D., Yu, W., & Shao, Z. (2021). Smart city based on digital twins. *Computers, Environment and Urban Systems., 1*(1), 1–11. https://doi.org/10.1007/s43762-021-00005-y

Liu, F., Liu, F. Y., & Shi, Y. (2018). City brain, a new architecture of smart city based on the internet brain. In: *IEEE 22nd International Conference on Computer Supported Cooperative Work in Design* (pp. 9–11). Nanjing. https://doi.org/10.1109/CSCWD.2018.8465164

Liu, F., Ying, L., & Yunqin, Z. (2021). Discussion on the definition and construction principles of city brain. In: *2021 IEEE 2nd International Conference on Big Data, Artificial Intelligence and Internet of Things Engineering (ICBAIE)*, Nanchang, China. https://doi.org/10.1109/ICBAIE52039.2021.9390064

Liu, W., Mei, Y., Ma, Y., Wang, W., Hu, F., & Xu, D. (2022). City brain: A new model of urban governance. In: M. Li, G. Bohács, A. Huang, D. Chang, & X. Shang (Eds.), *IEIS 2021. Lecture Notes in Operations Research*. Springer. https://doi.org/10.1007/978-981-16-8660-3_12

Lu, Q., Jiang, H., Chen, S., Gu, Y., Gao, T., & Zhang, J. (2021). Applications of digital twin system in a smart city system with multi-energy. In: *2021 IEEE 1st International Conference on Digital Twins and Parallel Intelligence (DTPI)* (pp. 58–61). IEEE.

Major, P., da Silva Torres, R., Amundsen, A., Stadsnes, P., & Tennfjord Mikalsen, E. (2022). On the use of graphical digital twins for urban planning of mobility projects: A case study from a new district in Ålesund, Norway. In: *Proceedings of the 36th ECMS International Conference on Modelling and Simulation ECMS 2022*. https://doi.org/10.7148/2022-0236

Marasinghe, R., Yigitcanlar, T., Mayere, S., Washington, T., & Limb, M. (2024). Computer vision applications for urban planning: A systematic review of opportunities and constraints. *Sustainable Cities and Society, 100*, 105047. https://doi.org/10.1016/j.scs.2023.105047

Mishra, P., & Singh, G. (2023). Artificial intelligence for sustainable smart cities. In: *Sustainable Smart Cities*. Springer. https://doi.org/10.1007/978-3-031-33354-5_6

Mohamed, K. S. (2023). Deep Learning for IoT "artificial intelligence of things (AIoT)". In: *Deep Learning-Powered Technologies. Synthesis Lectures on Engineering, Science, and Technology*. Springer. https://doi.org/10.1007/978-3-031-35737-4_3

Mora, L., & Bolici, R. (2017). How to become a smart city: Learning from Amsterdam. In A. Bisello, D. Vettorato, R. Stephens, & P. Elisei (Eds.), *Smart and sustainable planning for cities and regions. SSPCR 2015. Green energy and technology* (pp. 251–266). Springer, Cham. https://doi.org/10.1007/978-3-319-44899-2_15.

Morell, M. F., & Espelt, R. (2018). How much are digital platforms based on open collaboration? An analysis of technological and knowledge practices and their implications for the platform governance of a sample of 100 cases of collaborative digital platforms in Barcelona. In: *OpenSym '18: Proceedings of the 14th International Symposium on Open Collaboration*, Article No.: 26, pp. 1–5.

Nishant, R., Kennedy, M., & Corbett, J. (2020). Artificial intelligence for sustainability: Challenges, opportunities, and a research agenda. *International Journal of Information Management, 53*, 102104.

Noori, N., Hoppe, T., & de Jong, M. (2020). Classifying pathways for smart city development: Comparing design, governance and implementation in Amsterdam, Barcelona, Dubai, and Abu Dhabi. *Sustainability, 12*, 4030.

Parihar, V., Malik, A., Bhawna, Bhushan, B., & Chaganti, R. (2023). From smart devices to smarter systems: The evolution of artificial intelligence of things (AIoT) with characteristics, architecture, use cases and challenges. In: *AI Models for Blockchain-Based Intelligent Networks in Iot Systems: Concepts, Methodologies, Tools, and Applications* (pp. 1–28). Springer.

Pedersen, A. N., Borup, M., Brink-Kjær, A., Christiansen, L. E., & Mikkelsen, P. S. (2021). Living and Prototyping digital twins for urban water systems: Towards multi-purpose value creation using models and sensors. *Water, 13*, 592. https://doi.org/10.3390/w13050592

Peponi, A., Morgado, P., & Kumble, P. (2022). Life cycle thinking and machine learning for urban metabolism assessment and prediction. *Sustainable Cities and Society, 80*, 103754. https://doi.org/10.1016/j.scs.2022.103754

Pierre, J., & Peters, B. G. (2005). *Governing Complex Societies: Trajectories and Scenarios*. Palgrave Macmillan.

Poell, T., Nieborg, D., & Van Dijck, J. (2019). Platformisation. *Internet Policy Review, 8*, 1–13.

Putra, W., Wahidayat, Z. D., & van der Knaap, W. (2018). Urban innovation system and the role of an open web-based platform: The case of Amsterdam smart city. *Journal of Regional and City Planning, 29*(3), 234–249.

Rajawat, A. S., Bedi, P., Goyal, S. B., Shaw, R. N., & Ghosh, A. (2022). Reliability analysis in cyber-physical system using deep learning for smart cities industrial IoT network node. In: V. Piuri, R. N. Shaw, A. Ghosh, & R. Islam (Eds.), *AI and IoT for Smart City Applications (Studies in Computational Intelligence* (Vol. 1002). Springer. https://doi.org/10.1007/978-981-16-7498-3_10.

Rane, N., Choudhary, S., & Rane, J. (2024). Artificial Intelligence and machine learning in renewable and sustainable energy strategies: A critical review and future perspectives. *SSRN*. https://ssrn.com/abstract=4838761, https://doi.org/10.2139/ssrn.4838761

Repette, P., Sabatini-Marques, J., Yigitcanlar, T., Sell, D., & Costa, E. (2021). The evolution of city-as-a-platform: Smart urban development governance with collective knowledge-based platform urbanism. *Land, 10*, 33. https://doi.org/10.3390/land10010033.

Ricciardi, G., & Callegari, G. (2022, June). Digital twins for climate-neutral and resilient cities: State of the art and future development as tools to support urban decision-making. In: *International Conference on Technological Imagination in the Green and Digital Transition* (pp. 617–626). Springer.

Rivolin, U. J. (2012). Planning systems as institutional technologies: A proposed conceptualization and the implications for comparison. *Planning Practice and Research, 27(1)*, 63–85. https://doi.org/10.1080/02697459.2012.661181.

Sabri, S., Alexandridis, K., Koohikamali, M., Zhang, S., & Ozkaya, H. E. (2023, November). Designing a spatially-explicit urban digital twin framework for smart water infrastructure and flood management. In: *2023 IEEE 3rd International Conference on Digital Twins and Parallel Intelligence (DTPI)* (pp. 1–9). IEEE.

Salunke, A. A. (2023). Reinforcement learning empowered digital twins: Pioneering smart cities towards optimal urban dynamics. *EPRA International Journal of Research & Development*. https://doi.org/10.36713/epra13959

Samadi, S. (2022). The convergence of AI, IoT, and big data for advancing flood analytics research. *Frontiers in Water*, 4, 786040. https://doi.org/10.3389/frwa.2022.786040.

Samuel, P., Jayashree, K., Babu, R., & Vijay, K. (2023). Artificial intelligence, machine learning, and iot architecture to support smart governance. In: K. Saini, A. Mummoorthy, R. Chandrika, & N. Gowri Ganesh (Eds.), *AI, IoT, and Blockchain Breakthroughs in E-Governance* (pp. 95–113). IGI Global. https://doi.org/10.4018/978-1-6684-7697-0.ch007

Sanchez, T. W. (2023). Planning on the verge of AI, or AI on the verge of planning. *Urban Science*, 7, 70. https://doi.org/10.3390/urbansci7030070.

Schmitt, P., & Danielzyk, R. (2018). Exploring the planning-governance nexus: Introduction to the special issue. *disP: The Planning Review*, 54(4), 16–20. https://doi.org/10.1080/02513625.2018.1562792.

Schmitt, P., & Wiechmann, T. (2018). Unpacking spatial planning as the governance of place: Extracting potentials for future advancements in planning research. *disP: The Planning Review*, 54(4), 21–33. https://doi.org/10.1080/02513625.2018.1562795.

Schrotter, G., & Hürzeler, C. (2020). The digital twin of the city of zurich for urban planning. *PFG*, 88, 99–112. https://doi.org/10.1007/s41064-020-00092-2

Seibt, D. (2024). Platform organizations and fields: Exploring the influence of field conditions on platformization processes. *Critical Sociology*. https://doi.org/10.1177/08969205231221444

Seng, K. P., Ang, L. M., & Ngharamike, E. (2022). Artificial intelligence Internet of Things: A new paradigm of distributed sensor networks. *International Journal of Distributed Sensor Networks*, 18(3). https://doi.org/10.1177/15501477211062835

Seng, K. P., Ang, L. M., & Ngharamike, E. (2022). Artificial intelligence Internet of Things: A new paradigm of distributed sensor networks. *International Journal of Distributed Sensor Networks*. https://doi.org/10.1177/15501477211062835

Shahrokni, H., Årman, L., Lazarevic, D., Nilsson, A., & Brandt, N. (2015b). Implementing smart urban metabolism in the Stockholm Royal Seaport: Smart city SRS. *Journal of Industrial Ecology*, 19(5), 917–929.

Shahrokni, H., Lazarevic, D., & Brandt, N. (2015a). Smart urban metabolism: Towards a real-time understanding of the energy and material flows of a city and its citizens. *Journal of Urban Technology*, 22(1), 65–86.

Shi, F., Ning, H., Huangfu, W., Zhang, F., Wei, D., Hong, T., & Daneshmand, M. (2020). Recent progress on the convergence of the Internet of Things and artificial intelligence. *IEEE Network*, 34(5), 8–15.

Singh, K. D., Singh, P., Chhabra, R., Kaur, G., Bansal, A., & Tripathi, V. (2023a). Cyber-physical systems for smart city applications: A comparative study. In: *2023 International Conference on Advancement in Computation & Computer Technologies (InCACCT)* (pp. 871–876). Gharuan, India. https://doi.org/10.1109/InCACCT57535.2023.10141719.

Son, T. H., Weedon, Z., Yigitcanlar, T., Sanchez, T., Corchado, J. M., & Mehmood, R. (2023). Algorithmic urban planning for smart and sustainable development: Systematic review of the literature. *Sustainable Cities and Society, 94*, 104562.

Stead, D. (2021). Conceptualizing the policy tools of spatial planning. *Journal of Planning Literature, 36*(3), 297–311. https://doi.org/10.10.1177/0885412221 992283.

Thornbush, M. J., & Golubchikov, O (2019). *Sustainable Urbanism in Digital Transitions-from Low Carbon to Smart Sustainable Cities.* Springer. https://doi.org/10.1007/978-3-030-25947-1

UN- Habitat. (2023). Urban Governance. https://unhabitat.org/topic/urban-governance (accessed 12 March 2023).

UNEP (2018) *The Weight of Cities–Resource Requirements of Future Urbanization.* International Resource Panel Secretariat

Vadiati, N. (2022). Alternatives to smart cities: A call for consideration of grassroots digital urbanism. *Digital Geography and Society, 3*, 100030

van Winden, M., Oskam, W., van den Buuse, I., Schrama, D., va Dijck, W., & Frederiks, E. J. (2016). *Organising Smart City Projects Lessons from Amsterdam.* (vol. 37, p. 118). Hogeschool van Amsterdam.

Vatn, A. (2016). *Environmental Governance: Institutions, Policies, and Actions.* Edward Elgar.

Wang, L., Chen, X., Xia, Y., Jiang, L., Ye, J., Hou, T., Wang, L., Zhang, Y., Li, M., Li, Z. et al. (2022). Operational data-driven intelligent modelling and visualization system for real-world, on-road vehicle emissions: A case study in Hangzhou City, China. *Sustainability, 14*(9), 5434. https://doi.org/10.3390/su14095434

Wang, Y., Su, Z., Guo, S., Dai, M., Luan, T. H., & Liu, Y. (2023). A survey on digital twins: Architecture, enabling technologies, security and privacy, and future prospects. *IEEE Internet of Things Journal, 10*(17), 14965–14987. https://doi.org/10.1109/JIOT.2023.3263909

Weil, C., Bibri, S. E., Longchamp, R., Golay, F., & Alahi, A. (2023). Urban digital twins challenges: A systematic review and perspectives for sustainable smart cities. *Sustainable Cities and Society, 99*, 104862. https://doi.org/10.1016/j.scs.2023.104862.

Wiedmann, N. S., Athanassiadis, A., & Binder, C. R. (2021). Data mining in the context of urban metabolism: A case study of Geneva and Lausanne, Switzerland. *Journal of Physics: Conference Series, 2042*, 012020. https://doi.org/10.1088/1742-6596/2042/1/012020

Wu, J., Wang, X., Dang, Y., & Lv, Z. (2022). Digital twins and artificial intelligence in transportation infrastructure: Classification, application, and future research directions. *Computers and Electrical Engineering, 101*(September), 107983. https://doi.org/10.1016/j.compeleceng.2022.107983.

Xie, J. (2020). Exploration and reference of urban brain construction in xiacheng district of Hangzhou. *Communications World, 31*, 26–27.

Xu, R. (2022). Deeply understand, analyze and draw lessons from Hangzhou's urban brain construction paradigm. *BCP Business & Management, 29*, 441–446. https://doi.org/10.54691/bcpbm.v29i.2310

Xu, Y., Cugurullo, F., Zhang, H., Gaio, A., & Zhang, W. (2024). The emergence of artificial intelligence in anticipatory urban governance: Multi-scalar evidence of China's transition to city brains. *Journal of Urban Technology*. https://doi.org/10.1080/10630732.2023.2292823.

Yang, L., Chen, X., Perlaza, S. M., & Zhang, J. (2020). Special issue on artificial-intelligence-powered edge computing for Internet of Things. *IEEE IoT Journal, 7*(10), 9224–9226. https://doi.org/10.1109/JIOT.2020.3019948 .

Yang, S., Xu, K., Cui, L., Ming, Z., Chen, Z., & Ming, Z. (2021). EBI-PAI: Towards an efficient edge-based IoT platform for artificial intelligence. *IEEE IoT Journal, 8*, 9580–9593. https://doi.org/10.1109/JIOT.2020.3019008.

Yigitcanlar, T., & Cugurullo, F. (2020). The sustainability of artificial intelligence: An urbanistic viewpoint from the lens of smart and sustainable cities. *Sustainability, 12*(20), 8548.

Zhang, J., & Tao, D. (2021). Empowering things with intelligence: A survey of the progress, challenges, and opportunities in artificial intelligence of things. *IEEE Internet of Things Journal, 8*(10), 7789–7817. https://doi.org/10.1109/JIOT.2020.3039359.

Zhang, J., Hua, X. S., Huang, J., Shen, X., Chen, J., Zhou, Q., Fu, Z., & Zhao, Y. (2019). City brain: Practice of large-scale artificial intelligence in the real world. *IET Smart Cities, 1*, 28–37.

Ziakkas, D., St-hilaire, M., & Pechlivanis, K. (2024). The role of digital twins in the certification of the advanced air mobility (AAM) systems. In: T. Ahram, W. Karwowski, D. Russo, & G. Di Bucchianico (Eds.), *Intelligent Human Systems Integration (IHSI 2024): Integrating People and Intelligent Systems* (vol. 119). AHFE International. https://doi.org/10.54941/ahfe1004512

CHAPTER 5

Artificial Intelligence of Sustainable Smart City Brain, Digital Twin, and Metabolism

Pioneering Data-Driven Environmental Management and Planning

Abstract

Recent advances in Artificial Intelligence of Things (AIoT) have catalyzed a major shift from the Internet of Things (IoT) to the Internet of City Things (IoCT) and the Internet of Everything (IoE) in smart cities, revolutionizing environmentally sustainable urban development. In this dynamic landscape, three AIoT-powered systems—Urban Brain (UB), Urban Digital Twin (UDT), and Smart Urban Metabolism (SUM)—have risen to prominence, redefining the trajectory of emerging sustainable smart cities and enhancing their environmental performance. Despite the significant strides in their respective domains, there remains a critical need to harness their synergistic and collaborative integration to propel the evolution of sustainable urban development and navigate its complexities comprehensively. Therefore, this study introduces a pioneering framework—named Artificial Intelligence of Sustainable Smart City Things—that synergistically interconnects UB, UDT, and SUM as forms of Cyber-Physical Systems of Systems (CPSoS). Methodologically, it adopts an integrated approach that combines a thorough literature review with in-depth case studies. Through a detailed analysis and synthesis of a large body of knowledge, supported by empirical insights, this study establishes a solid foundation that elucidates the functionalities and architectures of UB, UDT, SUM, and AIoT and their synergistic interplay in the context of CPSoS. Furthermore, it sparks a discourse on the substantive effects of AIoT on the trajectory of environmentally sustainable urban development, covering opportunities, benefits, and implications, as well as challenges, barriers, limitations, and potential avenues for future research. Its primary contribution lies in the development of a pioneering framework for environmentally sustainable urban development, offering invaluable insights for researchers, practitioners, and policymakers alike. This framework serves as a guiding roadmap for spurring groundbreaking research endeavors, inspiring practical implementations, and informing policymaking decisions in the realm of emerging AIoT-driven, environmentally sustainable urban development

5.1 INTRODUCTION

The rapid pace of urbanization and the increasing depletion of natural resources necessitate a fundamental shift in sustainable urban development practices. While these challenges present opportunities for smart cities, they also create complex dilemmas for policymakers and decision-makers due to the complex nature of cities as socio-ecological-technical systems. Responding to these dynamics, the groundbreaking convergence of Artificial Intelligence (AI) and the Internet of Things (IoT) within AIoT has emerged as a central focus for innovative solutions in smart cities, particularly in the field of environmental sustainability, and beyond (e.g., Alahi et al. 2023; Bibri et al. 2023a, b; Dheeraj et al. 2020; El Himer et al. 2022; Ishengoma et al. 2022; Jagatheesaperumal et al. 2024; Rane et al. 2024; Tomazzoli et al. 2020; Seng et al. 2022; Zhang and Toa 2021). This integration serves as a transformative force in tackling complex urban challenges and shaping urban development strategies. It facilitates real-time data analysis and decision-making processes, thereby enhancing urban efficiency, functionality, sustainability, and the overall quality of life.

Prominent among the emerging AIoT-powered solutions for environmental sustainability are Urban Brain (UB) (e.g., Bibri et al. 2024a; Kierans et al. 2023; Xu 2022; Xu et al. 2024; Liu et al. 2022), Urban Digital Twin (UDT) (e.g., Agostinelli et al. 2021; Austin et al. 2020; Bibri et al. 2024b; Mishra et al. 2023; Shen et al. 2021), and Smart Urban Metabolism (SUM) (e.g., Bibri et al. 2024a; Ghosh and Sengupta 2023; Peponi et al. 2022). These data-driven environmental management and planning systems exemplify synergistic opportunities, driving the advancement of sustainable smart cities through real-time monitoring, resource management, data-driven decision-making, scenario testing, land-use management, infrastructure development, and catering to community needs. The symbiotic relationship between these systems is evident in the reciprocal interaction: data-driven insights from urban management operations inform urban planning processes, while well-designed plans shape effective management strategies. This synergy fosters more resilient, efficient, and responsive environmentally conscious urban environments.

The integration of AIoT functionalities in UB, UDT, and SUM is making these data-driven environmental management and planning systems essential to the development of sustainable smart cities. These urban environments are increasingly utilizing AI and AIoT to tackle complex environmental challenges in urban management and planning (e.g., Bibri et al. 2023a, b; Gourisaria et al. 2022; Efthymiou and Egleton 2023; Fang et al. 2023; Jagatheesaperumal et al. 2023; Mishra 2023; Mishra and Singh 2023; Szpilko et al. 2023; Ullah et all. 2020; Zaidi et al. 2023). AI and AIoT hold the promise of serving as the connective tissue that fosters synergies among emerging data-driven management and planning processes in urban environments, thereby contributing to the achievement of environmental sustainability goals.

As a linchpin technology, AIoT is harnessed across UB, UDT, and SUM, serving as a versatile technological framework with consistent, yet varied, functionalities. It acts as the central intelligence hub in UB (Bibri et al. 2024a; Liu et al. 2018, 2021), a dynamic layer for modeling and simulation in UDT (Beckett 2022; Bibri et al. 2024b; Zvarikova et al. 2022), and an intelligent system optimizing resource and material flows in SUM (Bibri et al. 2024a; Ghosh and Sengupta 2023; Peponi et al. 2022). By harnessing the computational and analytical power of AIoT, these data-driven management and planning systems facilitate proactive decision-making, real-time monitoring, and adaptive urban strategies, fostering the development of more environmentally sustainable, efficient, and resilient cities.

UB, as an advanced urban management system, leverages the power of AIoT to enhance decision-making, optimize resource allocation, and improve services and overall efficiency within a city. It acts as the cognitive center, processing vast datasets in real-time to streamline city operations and inform planning practices (Cugurullo 2020; Kierans et al. 2023; Liu et al. 2022; Xu et al. 2024). UDT is a dynamic virtual replica of a city that mirrors its physical, functional, and spatial aspects in a digital realm. Integrating real-time data from IoT sensors and devices with advanced AI models and algorithms, UDT dynamically monitors, models, and simulates urban processes, infrastructures, systems, and dynamics (e.g., Austin et al. 2020; Beckett 2022; Bibri et al. 2024b; Zayed et al. 2023). SUM represents an AI or AIoT-powered platform aiming to monitor, analyze, and optimize the flow and exchange of energy, materials, resources, and information in cities (Bibri et al. 2024a; Ghosh and Sengupta 2023; Peponi et al. 2022).

The rapid evolution of sustainable smart cities, driven by the integration of IoT, AI, and Cyber-Physical Systems (CPS), marks the dawn of a transformative era in urban management and planning. These technological advancements promise not only enhanced efficiency in city operations but also a more intelligently planned urban environment. The relationship between IoT, AI, and CPS is synergistic, with each component enhancing the functionality and efficiency of the others (Mohamed 2023; Radanliev et al. 2020) in the context of sustainable smart cities (Alahi et al. 2023; Juma and Shaalan 2020; Rajawat et al. 2022; Singh et al. 2023a). CPS, an integration of computation, networking, and physical processes, combine physical components (sensors, actuators) with computational elements to create systems to interact with and manage physical environments in a coordinated and intelligent manner. IoT provides the sensory and communication infrastructure that feeds data into AI systems (Mohamed 2023). AI processes these data to generate actionable insights and decisions (Alowaidi et al. 2023; Bibri and Huang 2024; Sharma and Sharma 2022). Specifically, AI is used to analyze real-time data generated via IoT, enabling intelligent decision-making and adaptive control of physical processes. CPS use these insights to control physical processes, creating a feedback loop that improves efficiency, accuracy, and adaptability. The combination of IoT and AI in CPS allows for advanced automation and intelligent control of complex systems in sustainable smart cities (Bibri and Huang 2024). This integration enhances system performance, optimizes resource use, and improves the reliability and efficiency of operations in sustainable smart cities.

The symbiotic relationship between AIoT and CPS forms a powerful framework that revolutionizes the operation and integration of data-driven management and planning systems in sustainable smart cities. The integration of AIoT and CPS can amplify the transformative capabilities of UB, UDT, and SUM in the management and planning of these urban environments. Singh et al. (2023a) highlight the significant potential of CPS in improving efficiency, environmental management and planning, and decision-making in smart city systems.

Furthermore, UB, UDT, and SUM can be viewed as integral components of the broader framework of Cyber-Physical Systems of Systems (CPSoS) due to the unique characteristics they exhibit in terms of interconnectedness, integration of digital and physical elements, real-time sensing and control, adaptability and resilience, scalability, data-driven decision-making, cross-domain collaboration, user-centric design, and environmental sustainability. CPS integrates infrastructure, embedded devices, intelligent elements, and the physical environment (Masserov and Masserov 2022), and CPSoS expands on them by comprising multiple interconnected computational algorithms and physical processes, each with its own sensing, computation, communication, and networking capabilities. These systems are designed to monitor, control, and interact with various other systems, thereby creating a complex network of cyber-physical interactions. CPSoS often emphasizes characteristics like interoperability, scalability, adaptability, and resilience. They also focus on cross-domain collaboration and integration, which are critical for managing the complexity and achieving the desired outcomes in large-scale applications such as sustainable smart cities (Bibri and Huang 2024). In these urban environments, UB, UDT, and SUM function as integral components of their digital and physical infrastructure, working together through AIoT and CPS components to manage and optimize various aspects of city operation and inform planning and governance. Their integration within the CPSoS framework allows for seamless coordination and interaction, thereby enabling these cities to achieve greater efficiency, resilience, and sustainability. They form a dynamic network of interconnected digital and physical systems that foster adaptive and responsive urban environments capable of addressing complex challenges in real-time.

Despite notable progress in the research, development, and practice of UB, UDT, and SUM as forms of CPSoS, a crucial imperative persists in leveraging their synergistic and collaborative integration to propel the evolution of smart sustainable urban development and navigate and unravel its complexities. Anchored in this foundation, the linchpin lies in the transformative potential of AIoT. The intricate interplay of these AIoT-powered management and planning systems weaves a complex tapestry that requires nuanced insights into their interconnected functions. This endeavor is critical to creating a cohesive and intelligent urban ecosystem that transcends the silos of individual components, fostering a holistic system perspective. There is a growing demand for collaborative knowledge development that embraces a "whole-system view" and facilitates transformative change across multiple scales—an approach that is rarely observed in practical urban contexts (Webb et al. 2018).

The absence of a thorough exploration of how UB, UDT, SUM, AIoT, and CPSoS can be integrated to support data-driven sustainable urban management and planning highlights the need for a focused effort in developing an innovative framework. In this context, the concept of "Artificial Intelligence of Smart Sustainable City Things (AIoSSCT)" signifies a transformative shift that leverages the power of AIoT to not only enhance the individual functions of city operating and organizing systems but also to dynamically interconnect them. This synergy represents a harmonious collaboration, where each city system's distinct strengths complement and amplify the capabilities of the others. Leveraging AIoSSCT can create more integrated and adaptive systems capable of intelligently managing and optimizing resource usage, energy consumption, and infrastructure, as well as monitoring environmental parameters and changes. This holistic approach to smart cities enables more sustainable and efficient management and planning of their systems while also fostering innovation and growth.

Against the preceding background, this study aims to investigate the transformative potential of AIoT in reshaping sustainable urban development practices and introduces a pioneering AIoSSCT framework that synergistically interconnects UB, UDT, and SUM as data-driven management and planning systems. The objectives guiding this study are as follows:

- Explore the nuanced relationship between sustainable smart cities and the transformative impact of AIoT on advancing their management and planning processes.

- Conduct a thorough analysis and synthesis of the existing body of knowledge on the functions and architectures of UB, UDT, and SUM in the context of AIoT.
- Examine the architectural layers of AIoT anchored in the data science cycle and elucidate the interconnections among UB, UDT, SUM, and AIoT within the context of CPSoS.
- Develop an innovative framework for data-driven management and planning, focusing on the synergistic integration and collaborative potential of UB, UDT, and SUM through AIoT.
- Initiate a discussion on the opportunities, benefits, implications, challenges, barriers, limitations, and potential avenues for future research associated with this framework.

The study's primary contribution is the development of an innovative AIoSSCT framework, positioning itself as a pioneering model for sustainable smart urban development. This involves introducing a synergistic and collaborative approach to data-driven urban management and planning, unlocking the transformative potential of AIoT by leveraging the collective intelligence of UB, UDT, and SUM. Beyond offering a deeper understanding of how AIoT can reshape and optimize the management and planning processes of sustainable smart cities, the study contributes to a comprehensive grasp of the architectural and functional dimensions of UB, UDT, and SUM. Valuable insights into the synergy in their operation form a robust foundation for the AIoSSCT framework. Moreover, the study sheds light on the intricate interplay between AIoT and the physical components of urban environments, providing a holistic perspective on technological integration. Finally, it sets the stage for a collective exploration of the multifaceted aspects surrounding the proposed framework, fostering a deeper understanding and paving the way for future advancements. This work offers invaluable insights for researchers, practitioners, and policymakers, guiding future research endeavors, inspiring practical implementations, and informing policymaking decisions in sustainable urban development. This study builds upon and expands on the work presented in Bibri and Huang (2025), titled "Artificial intelligence of sustainable smart city brain and digital twin: A pioneering framework for advancing environmental sustainability." While the article focuses on UB and UDT, this study extends that discussion to include SUM, providing additional analysis and another applied domain in the context of data-driven environmental management and planning.

The remainder of this study is structured as follows: Section 5.2 reviews and establishes connections between the theoretical underpinnings of the proposed framework. Section 5.3 describes and outlines the applied methodology. Section 5.4 presents the results derived from configurative and aggregative synthesis. Section 5.5 engages in a comprehensive discussion on the opportunities, benefits, implications, challenges, limitations, and potential avenues for future research associated with the proposed framework. Prior to this, it provides an interpretation of results and a comparative analysis. Section 5.6 provides a summary of key findings and presents concluding thoughts that encapsulate the key takeaways from the study.

5.2 THEORETICAL BACKGROUND

This section examines the nuanced relationship between sustainable smart cities and the transformative influence of AIoT in advancing data-driven management and planning systems—UB, UDT, and SUM, particularly through the lens of CPS. It navigates through relevant conceptual frameworks, theoretical perspectives, emerging urban computing trends, and sustainability dimensions, illuminating the dynamic interplay between urban technology and urban sustainability. By integrating CPS with AIoT, this analysis highlights how these technological ecosystems drive and refine the mechanisms of urban management and planning, marking a pivotal shift in how cities are designed, monitored, and managed for environmental sustainability.

5.2.1 Sustainable Smart Cities

A smart city is an urban environment that integrates advanced Information and Communication Technologies (ICT) and their applied solutions to optimize the efficiency of city operations, improve the resilience of city systems, enhance the quality of life for its citizens, and promote sustainable practices. This involves the use of data-driven technologies to manage resources, infrastructures, and services in real-time, fostering innovation, connectivity, and responsiveness to the needs and aspirations of communities. These technologies are specifically employed to seamlessly integrate diverse aspects of urban life, striving for the harmonious coexistence of technology, society, and the environment. Indeed, smart city projects often prioritize economic gains over environmental (and social) considerations (Ahvenniemi et al. 2017; Bibri 2019, 2021a; Evans 2019; Toli and Murtagh

2020). This imbalance has recently prompted a heightened emphasis on research and practical endeavors dedicated to achieving either environmental sustainability goals (e.g., Al-Dabbagh 2022; Saravanan and Sakthinathan 2021; Tripathi et al. 2022) or sustainable development goals (e.g., Mishra et al. 2022; Sharifi et al. 2024; Visvizi and del Hoyo 2021). This also pertains to smart sustainable cities (Bibri 2020; Bibri and Krogstie 2020a).

The emergence of sustainable smart cities as a scholarly discipline reflects a significant transformation in urban development, driven by the integration of various enabling technologies. These technologies, including AI, IoT, Big Data, Wireless Sensor Networks (WSN), Edge Computing, DT, and CPS, are shaping cities into more efficient, connected, and livable environments (Bibri et al. 2023a, Bibri and Huang 2024). By leveraging AI, IoT, and other technologies, cities can optimize resource usage, improve infrastructure efficiency, enhance public services, and create better living conditions for residents. IoT devices and WSNs can enable real-time monitoring of environmental conditions, allowing cities to respond more effectively to complex challenges. Edge computing brings processing power closer to the data source, enabling faster response times and reducing latency in smart city applications. DT provides virtual replicas of physical assets, allowing for simulation, monitoring, and optimization of urban systems. CPS integrates physical and digital components to create smarter infrastructure, from transportation systems to energy grids.

The term "sustainable smart cities" encompasses an approach to urban development that emphasizes the integration of technology and the environment and also considers economic prosperity and social well-being (del Mar Martínez-Bravo and Labella-Fernández 2024; Mishra et al. 2022; Sharifi et al. 2024). This holistic approach recognizes that a city's success depends on its ability to balance technological advancements with environmental sustainability, economic growth, and social equity. As this interdisciplinary field continues to evolve, researchers and practitioners are exploring innovative ways to address urban challenges while fostering sustainability and inclusivity. Beyond the conventional focus on technological and environmental efficiency, social sustainability emphasizes creating inclusive and equitable urban environments. This involves fostering community engagement, enhancing citizen well-being, and promoting social cohesion. Socially sustainable smart cities prioritize accessibility, diversity, and participatory decision-making, ensuring that technological advancements contribute to the overall betterment of citizens and communities. Social sustainability, when integrated into the broader framework of sustainable smart cities, ensures that technological innovations contribute positively to the social fabric of urban life, fostering a sense of community and shared responsibility. The concept of sustainable smart cities represents a significant leap beyond traditional smart cities, aspiring to harmonize the environmental, social, and economic dimensions of sustainability. This commitment to sustainability reflects an ongoing endeavor to enhance urban living while minimizing environmental impacts.

While the evolution from smart cities to sustainable smart cities necessitates a holistic approach that considers all three dimensions of sustainability, environmental sustainability, in particular, has remained at the forefront of this transition (see Bibri et al. 2023a for an extensive review). This continued emphasis is key to effectively tackling challenges associated with resource depletion, advocating for optimal resource management, integrating sustainable design principles, and fortifying the overall ecological resilience of urban environments amid dynamic environmental changes and technological advancements (e.g., Almalki et al. 2023; Saravanan and Sakthinathan 2021; Shruti et al. 2022; Tripathi et al. 2022). In this evolving landscape, sustainable smart cities are increasingly embracing and leveraging AI or AIoT to confront intricate challenges, particularly in the realms of environmental sustainability (e.g., Bibri et al. 2023a, 2024b; Gourisaria et al. 2022; Efthymiou and Egleton 2023; Fang et al. 2023; Mishra 2023; Szpilko et al. 2023; Zaidi et al. 2023). This increasing adoption underscores a deliberate effort to harness the transformative potential of AI or AIoT, positioning it as a pivotal force in addressing the multifaceted environmental issues confronting contemporary urban environments. It is also indicative of its expanding significance and rising prominence in shaping the trajectory of urban management and planning systems.

5.2.2 Artificial Intelligence of Things

AIoT is the convergence of AI models and algorithms and the vast network and infrastructure of IoT, presenting a transformative shift with profound implications for urban environments. This amalgamation transforms urban landscapes into sophisticated, interconnected ecosystems, where sensors, devices, systems, and platforms not only collect and exchange data but also possess the autonomy to analyze, learn, perform complex tasks, make decisions, and adapt to changing environments.

By integrating Machine Learning (ML), Deep Learning (DL), Computer Vision (CV), Natural Language Processing (NLP), and other AI models into IoT, AIoT empowers them to process and scrutinize extensive datasets, acquire knowledge, and execute decisions autonomously or with limited human involvement. AI and ML models and techniques continue to be extensively applied in smart cities, including in relation to AIoT (e.g., Alahakoon et al. 2023; Alahi et al. 2023; Bibri et al. 2023b, 2024b; Din et al. 2022; Marasinghe et al. 2024; Seng et al. 2022; Ullah et al. 2020), compared to other AI models. Thamik et al. (2024) discuss the utilization of ML, DL, CV, NLP, and IoT platforms in the applications of AIoT in environmental protection, energy efficiency, renewable energy, water management, and sustainable urban development and communities.

DL is a subset of the broader field of ML, both falling under the umbrella of AI. ML involves various techniques and algorithms enabling computers to learn from and make predictions or decisions based on data without explicit programming. DL focuses on neural networks, especially Deep Neural Networks (DNN), excelling in handling large and complex datasets. It autonomously learns hierarchical representations of data through the integration of multiple layers in neural networks and can automatically discern intricate features from raw data, minimizing the need for manual feature engineering. Multiple studies have explored DL concepts, architectures, applications, challenges, and future directions (e.g., Alhijaj and Khudeyer 2023; Alom et al. 2019; Alzubaidi et al. 2021; Dong et al. 2021; Joshi et al. 2023; Mathew et al. 2021). Furthermore, both ML and DL have been extensively applied in the realm of smart cities, particularly in environmental sustainability and climate change (Almalaq and Zhang 2018; Bibri et al. 2023b; Chen et al. 2023; Dheeraj et al. 2020; El Himer et al. 2022; Huntingford et al. 2019; Kim and Cho 2019; Mosavi et al. 2019; Nishant et al. 2020; Ullah et al. 2020; Rane et al. 2024; Willcock et al. 2018). Studies indicate that ML techniques are prevalent and effective for specific tasks in smart cities and smart eco-cities, including in the context of AIoT (Bibri et al. 2023b).

Nevertheless, the integration of DL, as a subset of ML, into IoT devices and systems shows particular promise, addressing challenges associated with diverse devices, data sources, and data-intensive tasks. While DL proves powerful for specific tasks within the AIoT framework, ML encompasses a broader spectrum of techniques, including those that are not reliant on DNN. ML- and DL-based devices enable edge intelligence in cloud-based IoT, liberating AI algorithms from cloud constraints and ensuring efficiency, scalability, resilience, and robustness in data processing and analysis for low-latency applications (e.g., Seng et al. 2022; Zhang and Tao 2021). This integration empowers "urban systems" across diverse domains such as transport, energy, waste, water, environment, and more to emulate intelligent behavior and make well-informed decisions, thereby reducing the need for human intervention. Overall, the choice between ML and DL often depends on the nature of the problem, the available data, and the computational resources.

In recent years, the realm of urban management and planning has witnessed a significant transformation due to the influential integration of both AI and AIoT (e.g., Bibri et al. 2024b; Bibri and Huang 2024, Kamrowska-Załuska et al. 2021; Koumetio et al. 2023; Nti et al. 2022; Prajna and Okkie 2023; Sanchez 2023; Sanchez et al. 2022; Son et al. 2022; Yigitcanlar et al. 2020).This integration represents an innovative approach, underscoring the synergistic potential of these cutting-edge technologies in reshaping urban operations and practices, with a particular focus on addressing and mitigating environmental challenges. Its profound impact is evident across various dimensions, including optimizing resource management, enhancing energy efficiency, streamlining waste management processes, improving transportation systems, preserving biodiversity, reducing environmental footprints, and mitigating climate change risks (e.g., Bibri et al. 2023b; Chen et al. 2023; El Himer et al. 2022; Fang et al. 2023; Leal Filho et al. 2022; Rane et al. 2024; Seng et al. 2022; Shaamala et al. 2024; Yigitcanlar et al. 2021; Zhang and Tao 2021).

AIoT applications extend to the social realm of sustainability by enhancing public services and promoting well-being through data-driven decision-making and innovative initiatives (e.g., Baker and Xiang 2023; Ishengoma et al. 2022). AIoT finds itself at a crucial juncture, on the verge of significant transformations and innovations, poised to usher in more sustainable, efficient, resilient, and environmentally conscious urban environments. Nevertheless, the adoption of AIoT introduces legitimate concerns related to data privacy, bias, fairness, transparency, accountability, and trust, as well as environmental risks (see Bibri et al. 2023b, 2024b for a detailed discussion). In particular, safeguarding sensitive information and fortifying AIoT systems against potential breaches constitute critical challenges that demand conscientious attention to ensure the sustainable evolution of sustainable smart cities.

In sum the transformative potential of AIoT, positioned as a key driver in steering sustainable smart cities toward enhanced efficiency, resilience, and environmental consciousness, creates the foundation for an in-depth exploration of the functional and architectural aspects of the AIoSSCT framework. This exploration will focus on the framework's essential components—UB, UDT, and SUM as AIoT-powered management and planning systems, and shed light on their roles in advancing sustainable smart cities.

5.2.3 Cyber-Physical Systems and Artificial Intelligence of Things

CPS are engineered to interact with the physical world through sensors and actuators, facilitating a seamless integration of digital and physical processes. The primary objective of CPS is to enable real-time monitoring, coordination, and control of complex systems within various domains of smart cities (Juma and Shaalan 2020; Khan et al. 2021; Rajawat et al. 2022; Singh et al. 2023a). The efficacy of CPS relies on several critical technical components, each contributing uniquely to the overall functionality and performance of urban systems (Alowaidi et al. 2023; Gürkaş Aydın and Kazanç 2023; Radanliev et al. 2020; Sharma and Sharma 2022) in the context of sustainable smart cities (Bibri and Huang 2024).

At the core of CPS is the *sensing component*, which involves a network of sensors that collect data from the physical environment. These sensors can measure various environmental parameters, depending on the application domain. The collected data are then transmitted to the computational unit for processing. The sensors must be accurate, reliable, and capable of operating under different environmental conditions to ensure the integrity of the data.

Following data collection, the *communication network* plays a pivotal role in transmitting the data from sensors to the processing units and vice versa. This network must support high-speed data transfer, low latency, and robust security measures to protect the data from unauthorized access and cyber threats. Communication protocols are commonly used, each offering distinct advantages based on the specific requirements of the CPS application domain.

The *computation and control* component encompasses the processing units, often embedded systems or dedicated processors, which analyze the sensor data using advanced algorithms. These algorithms may involve AI, ML, and real-time data analytics to make informed decisions and control actions. The processing units must have sufficient computational power, memory, and energy efficiency to handle the complex algorithms and ensure timely responses.

Another critical component is the *actuation system*, which converts the computational decisions into physical actions. Actuators such as motors, valves, and relays are used to control physical processes based on the computed instructions. The precision and reliability of these actuators are essential for the accurate execution of control actions, directly affecting the system's performance and safety.

The *integration of software* is also vital in CPS, where middleware and application software manage the coordination between sensors, communication networks, computation units, and actuators. This software layer ensures interoperability, scalability, and ease of deployment across different platforms and environments. It also includes user interfaces that allow operators to monitor and control the CPS in real-time, providing critical insights and diagnostics.

Security and privacy are paramount in CPS, necessitating robust *cybersecurity measures* to protect against potential threats and vulnerabilities. This includes implementing encryption, authentication protocols, intrusion detection systems, and regular security updates. Given the critical nature of many CPS applications, ensuring the system's resilience against cyber-attacks is crucial to maintaining operational integrity and safety.

Overall, CPS represent a sophisticated fusion of digital and physical domains, leveraging advanced sensing, communication, computation, and actuation technologies. The seamless integration of these technical components is fundamental to the performance, reliability, and security of CPS, enabling them to transform various sectors by providing real-time monitoring and control. As CPS continue to evolve, ongoing advancements in these technical components will further enhance their capabilities, paving the way for more intelligent and autonomous systems.

The integration of AI and IoT, known as AIoT, represents a significant evolution in the field of CPS. This convergence aims to harness the power of AI to enhance the capabilities of interconnected physical and computational components, enabling more intelligent, autonomous, and efficient systems. The integration of CPS in smart cities (Juma and Shaalan 2020; Khan et al. 2021; Rajawat et al. 2022; Singh et al. 2023a) requires a better understanding of the impact of AIoT (e.g., Alahi et al. 2023; Bibri 2024b; Hoang et al. 2024; Seng et al. 2022; Zhang and Tao 2021) on traditional CPS (e.g., Mohamed 2023; Radanliev et al. 2020; Sharma and Sharma 2022), including cybersecurity (Alowaidi et al. 2023; Gürkaş Aydın and Kazanç 2023) in the context of emerging sustainable smart cities (Bibri and Huang 2024).

Conducting a comparative analysis of their technical components, commonalities, and differences is essential for understanding these dynamics. This approach sheds light on how AIoT extends and enhances the foundational structures of CPS, providing a comprehensive view of the progression from conventional systems to more advanced integrations. Table 5.1 categorizes each component, explaining how AIoT builds on the foundations of traditional CPS to bring enhanced capabilities, specifically through the use of AI. The commonalities highlight the base functionality shared between the two, while the differences section outlines how AIoT introduces improvements and advancements.

AIoT complements and significantly advances the functionalities of traditional CPS through the integration of AI capabilities. The comparison demonstrates that while the foundational elements remain similar, AIoT introduces enhancements that optimize performance, increase security, and provide greater efficiency. As this technology continues to evolve, the distinction between AIoT and CPS will become even more pronounced, leading to smarter systems capable of more complex and autonomous operations, which are essential for the future of technological innovation and application in sustainable smart cities.

AIoT and CPS are integral to the development and operation of sophisticated urban management and planning systems like UB, UDT, and SUM, all of which are forms of CPSoS. UB utilizes AIoT to integrate and analyze vast amounts of real-time data from various urban sensors, enabling dynamic urban management and informing planning decisions. In contrast, UDT uses CPS to create a virtual replica of urban environments that simulates real-world conditions, processes, and system responses to changes before they are implemented physically. This capability is enhanced by AIoT through its predictive analytics and ML/DL models, which provide deeper insights and more accurate forecasts. Similarly, SUM employs both CPS and AIoT to monitor, analyze, and optimize the flow of energy, materials, and resources across city infrastructures. This approach transforms urban management by promoting efficiency and sustainability, leveraging real-time data to ensure optimal resource allocation and minimize waste. By employing AIoT within these three CPSoS frameworks, cities become more than just smart; they become self-aware ecosystems capable of significant autonomous decision-making and adaptation, reflecting the next wave of urban technological advancement.

TABLE 5.1 Commonalities and Differences between Artificial Intelligence of Things and Cyber-Physical Systems

Technical Component	AIoT	CPS	Commonalities	Differences
Sensing	AI-enabled smart sensors for real-time data analysis	Sensors collect data from the physical environment	Both involve sensors to collect data	AIoT uses AI for smart preprocessing; CPS uses standard sensors
Communication network	AI for efficient data routing and network optimization	High-speed data transfer, low latency, robust security measures	Both require efficient and secure data communication	AIoT optimizes network with AI; CPS relies on established protocols
Computation and control	Advanced AI analytics and ML for decision-making	Processing units analyze sensor data using algorithms	Both involve data processing for decision-making	AIoT uses advanced AI models; CPS uses predefined algorithms
Actuation	Intelligent control algorithms for precise actuation	Actuators convert decisions into physical actions	Both utilize actuators to perform physical actions	AIoT enables more adaptive control; CPS has traditional control systems
Integration of software	AI-driven middleware for seamless interoperability	Middleware and application software manage coordination	Both need software integration for coordination	AIoT enhances middleware with AI; CPS uses conventional middleware
Cybersecurity	AI for real-time threat detection and mitigation	Encryption, authentication protocols, intrusion detection systems	Both need robust security measures	AIoT provides real-time threat mitigation; CPS uses standard security measures

5.3 METHODOLOGY

The methodology adopted in this study was carefully designed to investigate the transformative role of AIoT in synergizing UB, UDT, and SUM within a unified framework, with the specific aim of advancing the management and planning of sustainable smart cities. This exploration followed a systematic approach consisting of six stages (Figure 5.1). By seamlessly integrating a comprehensive scholarly literature review with in-depth case studies, a methodological path was established that is grounded in both theoretical and empirical knowledge. Hence, the AIoSSCT framework represents a well-founded and forward-looking contribution to the field.

In laying the groundwork for the development of this framework, a comprehensive literature review was undertaken, drawing from a diverse array of scholarly literature, including theoretical studies, empirical studies, review studies, and interdisciplinary studies. This multifaceted approach aimed to ensure a well-rounded understanding of the conceptual and theoretical underpinnings and functional and architectural dimensions of the framework, providing critical insights as a foundational step in shaping its conceptualization and pioneering its development. The research design involved a curated selection of case studies to provide empirical evidence. Case studies on UB, UDT, and SUM were conducted to gain in-depth practical insights into their complexities, dynamics, and nuances. Their selection was primarily guided by their relevance to the interconnected fields of urban management and planning.

Figure 5.2 illustrates the Preferred Reporting Items for Systematic Reviews and Meta-analyses (PRISMA) approach for literature search and selection. Academic research indexing databases, namely Scopus, Web of

- **Stage 1**: State the research aim and objectives, delineate the scope of the conceptual framework, and identify the key components to be interconnected as things.
- **Stage 2**: Conduct a comprehensive literature review on the functions and architectures of UB, UDT, and SUM in relation to AIoT, as well as its impact on sustainable smart city management and planning.
- **Stage 3**: Analyze case studies on UB, UDT, and SUM to acquire in-depth practical insights into their complexities, dynamics, and nuances, while ensuring their selection based primarily on their relevance to the interconnected fields of urban management and planning.
- **Stage 4**: Examine the architectural layers of AIoT, grounded in the data-science cycle, and shed light on its link to CPS in regard to UB, UDT, and SUM in the context of the Internet of Everything.
- **Stage 5**: Develop an innovative conceptual framework of AIoSSCT that visually depicts the integration of UB, UDT, and SUM as CPSoS with urban management and planning processes in the realm of sustainable smart cities.
- **Stage 6**: Discuss the opportunities, benefits, implications, challenges, barriers, and limitations associated with the implementation of the conceptual framework and suggest future research directions for areas requiring further investigation.

FIGURE 5.1 A flow diagram outlining the step-by-step process of developing the AIoSSCT framework.

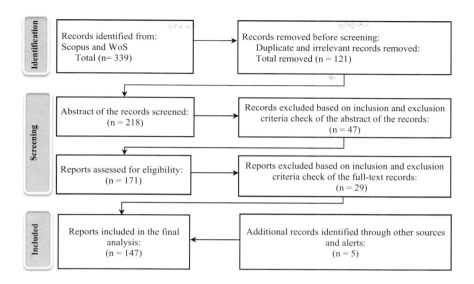

FIGURE 5.2 The PRISMA flowchart for literature search and selection. Source: Adapted from Page et al. (2021).

Science (WoS), and ScienceDirect, were utilized given their broad coverage of the high-quality peer-reviewed studies related to the multifaceted topic of the study. A set of carefully selected keywords and their combination ensured specificity in the search and retrieval of the pertinent scholarly literature. These keywords included "sustainable smart cities," "smart cities AND artificial intelligence OR artificial intelligence of things," "city brain OR urban brain," "city brain AND artificial intelligence," "digital twin AND smart cities," "urban digital twin," "urban digital twin AND artificial intelligence," "smart urban metabolism," "smart urban metabolism AND artificial intelligence," "cyber-physical systems AND artificial intelligence AND internet of things," "cyber-physical systems AND smart cities," and "cyber-physical systems AND urban digital twin." They were used to search against the title, abstract, and keywords of articles to produce initial insights.

The inclusion criteria filtered studies based on their relevance, reliability, language, publication date, and publication type (article, conference paper, or book chapter), providing definitive primary information. Exclusion criteria were applied to remove studies unrelated to the focal topics and irrelevant to the research aim and objectives. Based on these predefined inclusion/exclusion criteria, the selection process involved initial screening based on titles and abstracts, followed by a detailed full-text review for eligibility.

In view of the above, the selection process involved initial screening based on titles and abstracts, followed by a detailed full-text review for eligibility. The search query retrieved a total of 339 records from two databases (Scopus and WoS=339). After removing duplicates and irrelevant documents, 98 records were eliminated. Subsequently, titles and keywords were scrutinized, leading to the exclusion of 23 more records. The remaining 218 records underwent abstract screening against the inclusion and exclusion criteria, resulting in the removal of 47 records. The full-text screening of the remaining 171 records led to the exclusion of an additional 29 records. Ultimately, this process yielded a final selection of 187 publications. Additionally, 5 extra records were included through other sources and alerts, bringing the total number of records included in the final analysis to 147. Throughout the process, critical appraisal was conducted to assess the quality of the selected studies.

The search was limited to the timeframe of the last 7 years to ensure relevance and accuracy, yielding 339 documents in total. Considering the rapid advancements and increasing integration of AI and IoT technologies in urban management and planning, the timespan 2018–2024 is particularly significant for a comprehensive literature review of this study due to several reasons. Firstly, the rapid advancements in both AI and IoT technologies during this period have paved the way for their convergence into AIoT, making it an important era for research and development in this field. Secondly, the concept of sustainable smart cities has gained considerable momentum since 2018, driven by increasing urbanization and the urgent need for innovative solutions. This period has seen a surge in scholarly work and practical implementations related to UB, UDT, SUM, and CPS. Furthermore, the chosen timespan captures the most recent innovations and cutting-edge research, ensuring that the study is grounded in the latest theoretical and practical advancements. By focusing on 2018–2024, the study can provide a thorough analysis and synthesis of contemporary literature, capturing the dynamic evolution of AIoT and its transformative impact on sustainable urban development practices.

Guided by an inductive approach, a content analysis was performed on the included studies to gather the data needed for the analysis and synthesis. The obtained insights enabled the identification of key themes, patterns, and variations, ensuring a comprehensive analysis and synthesis. To develop the AIoSSCT framework, both configurative and aggregative

AI of Sustainable Smart City Brain, Digital Twin, and Metabolism ■ 287

FIGURE 5.3 A flow diagram for data synthesis for developing the AIoSSCT framework.

synthesis techniques were employed as complementary methods through an extensive review of the selected studies (Figure 5.3).

Configurative synthesis entails identifying common themes across synthesized studies and developing a conceptual framework to elucidate interconnections, patterns, and contexts observed in the findings (Gough et al. 2017). Thematic synthesis focuses on understanding underlying arrangements or relationships among research findings, emphasizing interpretation throughout the synthesis process to derive the overall meaning or provide a more nuanced understanding of the topic under review. Conversely, aggregative synthesis, according to Cooper (2017), involves summarizing and combining multiple studies to generate an overall thematic summary of the findings. It provides a comprehensive assessment of research evidence, interpreting the results after the synthesis process within specific conceptual categories, and leveraging evidence to make statements.

5.4 RESULTS: ANALYSIS AND SYNTHESIS

The results section unfolds as a crucial phase in unraveling the intricate landscape of the proposed AIoSSCT framework. Guided by a set of well-defined objectives, the study conducts a thorough analysis and synthesis of existing knowledge, elucidating the functions and architectures of UB, UDT, and SUM in relation to AIoT. The exploration extends to scrutinizing the architectural layers of AIoT, grounded in the data science cycle, while shedding light on its interrelation with CPS. The culmination of these efforts materializes in the formulation of an innovative data-driven management and planning framework.

5.4.1 Functional and Architectural Dimensions

The integration of AIoT into sustainable smart cities has emerged as a primary focus in reshaping their emerging management and planning platforms. This section delves into the functions of UB, UDT, and SUM platforms within the AIoSSCT framework. Exploring the role of AIoT in facilitating these functions and shaping the seamless integration of these platforms, reveals their innovative potential to transform the urban development landscape. Each platform, distinguished by its unique functional features, synergistically contributes to the realization of an efficient, intelligent, and sustainable urban future. From real-time analytical insights to dynamic representations and systematic resource optimization, the AIoT functionalities—intricately interwoven into UB, UDT, and SUM as interconnected things—collectively represent the next frontier in integrating urban management and planning platforms.

5.4.1.1 Urban Brain

In the context of urban management and planning, UB serves as a transformative technological framework, shaping the development of smart cities toward sustainability. Delving into the technical components of UB, its architecture is a critical aspect that slightly varies based on the specific domain of application, such as transportation, traffic, energy, public services, planning, and governance. These components play distinct roles within the UB platform, contributing to the seamless integration of IoT, Big Data, AI, Cloud Computing, Edge Computing, and 5G networks. This combination forms a layered architecture that is instrumental in advancing sustainable urban development practices. The UB architecture encompasses the following layers (Liu et al. 2018, 2021; Zhang et al. 2019):

- The IoT Layer: Acting as the sensory and motor nervous systems, the IoT layer is distributed across the city's operational and organizational frameworks. It encompasses a vast array of connected sensors and devices that continuously gather and transmit data, allowing for real-time monitoring and responsive actions throughout the urban environment.

- The Big Data Layer: This layer plays a key role in collecting, storing, and processing the extensive data generated by the IoT sensors, devices, and systems. It forms the foundational bedrock for applied intelligence in UB by enabling the extraction of meaningful insights and patterns from the vast amounts of raw data. The processed data are then utilized to inform decision-making and strategic planning.

- The Cloud Computing Layer: Serving as the central nervous system, the cloud computing layer orchestrates the control of all other nervous systems. It leverages powerful servers, advanced network operating systems, and sophisticated neural networks, alongside data-driven AI algorithms. This centralized system ensures efficient data management, high computational power, and seamless integration of various urban processes, enabling comprehensive oversight and coordination.

- The Edge Computing Layer: Functioning as the nerve endings, the edge computing layer integrates AI capabilities directly into IoT devices and systems at the periphery of the network. This enhances the sensory and motor functions of UB by allowing for immediate data processing and action at the source. Edge computing reduces latency, improves response times, and enables real-time decision-making, thereby augmenting the overall efficiency and effectiveness of the urban nervous systems.

- The AI Layer: Encompassing a sophisticated suite of AI technologies, the AI layer integrates ML, DL, and CV algorithms to analyze vast amounts of real-time data collected from various urban sensors and IoT devices. By applying this integrated approach to UB's nerve endings, neural networks, and smart terminals, a powerful and cohesive system is created. This fusion enhances the functionalities of all nervous systems, enabling them to operate in a synchronized manner to optimize urban processes and decision-making. The AI layer can identify patterns, predict outcomes, and provide actionable insights. Additionally, it facilitates adaptive learning, allowing UB to continuously improve its recommendations and interventions based on new data and changing urban dynamics.

- The Internet Neural Network Layer: This layer interconnects various components of the urban system, functioning similarly to social networks. It dynamically evolves with the integration and advancement of IoT, Big Data, AI, and Cloud Computing technologies. By facilitating interactions among humans, between humans and things, and among things themselves, this layer enhances the interconnectedness and responsiveness of the urban environment. It supports a continuous flow of information and collaborative interactions, enabling smarter and more adaptive urban processes.

From a broad perspective, as synthesized from various studies (e.g., Bibri et al. 2024a; Kierans et al. 2023; Liu et al. 2022; Wang et al. 2022; Xie 2020; Xu et al. 2024; Zhang et al. 2019), UB typically encompasses a data acquisition layer, incorporating IoT sensors, surveillance cameras, and other city data streams. This is followed by a data integration layer that harmonizes diverse datasets to ensure compatibility and coherence for comprehensive analysis. The subsequent data processing and analytics layer utilizes AI algorithms for real-time data processing, pattern recognition, and decision-making. A visualization and user interface layer provides a user-friendly interface for city officials and stakeholders to interpret data and make informed decisions. Finally, an application layer is implemented, offering specific applications and services based on the insights derived from data analysis, completing the integral components of a comprehensive UB.

This UB architecture aligns with the multilayered architecture of the AIoT-driven urban system, reflecting the synergistic integration of advanced technologies in urban management and planning. UB, a sophisticated and clear illustration of "urban computing and intelligence" (Bibri 2021b; Liu et al. 2017) and a large-scale AIoT system (Cugurullo 2020; Kierans et al. 2023) is characterized by three fundamental capabilities outlined by Zhang et al. (2019). These capabilities include: (1) processing vast and diverse data sources at an ultra-large scale, (2) discerning complex concealed patterns and rules through ML techniques; and (3) formulating overarching optimal strategies that outperform localized, suboptimal decisions made by humans. These are at the core of the AIoT-driven urban system within its data science cycle, encapsulated by various steps, namely sensing, perceiving, learning, visualizing, and acting (e.g., Bibri et al. 2023b) or perceiving, learning, reasoning, and behaving (Zhang and Toa 2021). The slight variations in the number of steps in the AIoT-driven urban system may arise from specific use cases or application domains or the degree of human intervention involved in supporting or automating decision-making processes in these domains.

UB transforms the day-to-day operations of smart cities by optimizing efficiency and responsiveness. It enables real-time data collection, processing, and analysis, fostering informed decision-making and efficient urban systems management. With adaptability to address various urban issues, its applications span traffic management, energy efficiency, waste management, metabolic circularity, environmental monitoring, and more. Various case studies highlight UB's transformative impact on operational city management. Kierans et al. (2023) present projects in Basel and Shenzhen City, emphasizing the transition to electric vehicles for sustainable development. The "Smart City Brain for Circular Economy" platform integrates circular economy principles, real-time resource tracking, AI-driven resource optimization, and stakeholder collaboration. In Hangzhou, City Brain optimizes traffic management, reducing travel times, improving traffic flow, and minimizing fuel consumption (Wang et al. 2022). Kuala Lumpur strategically incorporates City Brain into domains such as transportation management and environmental protection, optimizing resource allocation and enabling swift responses to accidents and disasters (Zhang et al. 2019).

In the domain of urban planning and governance, UB plays a significant role in shaping contemporary urban development. It operates as the analytical nucleus of smart cities, assimilating vast datasets, processing real-time data, and providing actionable insights for informed decision-making in environmental planning and governance (Bibri et al. 2024a, Bibri and Huang 2024). The UB platform extends beyond its original purpose for transportation management, evolving into a large-scale AIoT system with broader applications in urban planning and governance (Cugurullo 2020). Real-world implementations of UB involve both public and private stakeholders, enhancing coordination and management of urban systems, infrastructures, and resources (Caprotti and Liu 2022; Kierans et al. 2023). UB transforms governance modes from passive to active, driving innovation across transportation, urban management, ecological environmental protection, and smart energy (Liu et al. 2022). Overall, UB is also designed to solve urban planning and governance challenges, propelling the urban landscape toward digital, intelligent, and modern development. However, careful consideration of privacy, security, fairness, accountability, and transparency is crucial in the design and deployment of UB.

5.4.1.2 Urban Digital Twin
UDT has introduced transformative advancements in city modeling and simulation, presenting a state-of-the-art approach to sustainable urban development. UDT aims to serve as a robust tool for informed decision-making, enhancing urban management and planning processes to advance sustainable development goals. These advancements enable more accurate and dynamic simulations of urban environments, facilitating better resource allocation, infrastructure development, environmental management, and resilience (e.g., Caprari et al. 2022; Gkontzis et al. 2024; Ferré-Bigorra et al. 2022; Weil et al. 2023). The creation of UDT involves leveraging various advanced technologies, especially IoT, AI, and Big Data. The technical functionalities of UDT are demonstrated by its real-time

data processing and analysis capabilities, which have been significantly enhanced by the recent integration of AIoT into its operational framework (Bibri et al. 2024b; Mishra 2023). This integration involves the application of ML and DL algorithms, which enable predictive modeling, scenario planning, and the generation of valuable insights. AIoT-driven UDT not only fosters dynamic decision-making in sustainable smart cities but also enables them to effectively address challenges associated with population growth, infrastructure demands, and environmental changes.

Recent research has integrated both AI with UDT (e.g., Agostinelli et al. 2021; Austin et al. 2020; Bibri et al. 2024b; Mishra et al. 2023; Zayed et al. 2023) and IoT with DT, which has led to the emergence of the Internet of Digital Twins (IoDT) (Wang et al. 2023). This integration is anticipated to give rise to the concept of AIoT-enhanced UDT (AIoUDT) as these technologies mature, following the convergence of AI and IoT within the AIoT framework (Bibri et al. 2024b). AIoUDT, or AIoT-enhanced UDT, represents the integration of AI with the Internet of Urban Digital Twin (IoUDT). This concept involves combining advanced AI algorithms with various interconnected digital replicas of urban environments enabled by IoUDT and enhanced by IoT devices that offer real-time data to augment the functionalities and capabilities of AIoT applications. This integration plays a key role in enhancing environmental management and planning in sustainable smart cities. It supports sustainable development by enabling real-time monitoring, predictive modeling, resource optimization, scenario simulation, and data-driven decision-making.

Zvarikova et al. (2022) propose UDT algorithms leveraging 3D spatiotemporal simulations, ML, and DL for accurate urban modeling. The spatiotemporal intelligence of UDT enhances precision, providing a nuanced understanding of urban complexities. Beckett's (2022) study on the integration of smart city DT, 3D modeling, visualization tools, and spatial cognition algorithms within the context of AI-based urban design and planning offers a nuanced understanding of the evolving landscape of urban development. The author accentuates the crucial role of AI in UDT, demonstrating how its integration enhances urban design and planning strategies. As evidence of 3D virtual simulation technology, data-driven urban analytics, and real-time decision support systems continues to grow, it has become of crucial importance to explore the integration of IoT-based DT with AI-based urban design and planning along with real-time urban data. This will propel urban planning processes to new heights of sophistication. The progress in predictive accuracy, resource optimization, and informed planning underscores the crucial role of AIoT-driven UDT in steering toward environmentally sustainable and smart urban futures through advanced data-driven planning and design processes (Bibri et al. 2024b).

In the evolving landscape of urban development, the integration of UDT and AI has emerged as a transformative force. This powerful amalgamation is redefining how environmental scenario planning, resource management, mobility, and land use are being approached in urban settings. Through the detailed representations offered by UDT and the analytical capabilities of AI, achieving precise forecasting of urban development impacts, including factors such as energy consumption, air quality, noise levels, and mobility, has become achievable (Ye et al. 2023). Moreover, the application of AI-driven UDT extends beyond scenario planning to optimizing resource management, efficiently monitoring and allocating resources such as energy, water, and waste (Almusaed and Yitmen 2023). Shen et al. (2021) highlight the utilization of ML and AI in UDT to expedite environmental sustainability within Positive Energy Districts (PED). Their focus lies in integrating diverse systems and infrastructures to optimize interactions among buildings, mobility, energy, and advanced technologies. The authors explore the working principles, platforms, and applications of this integration to enhance the environmental performance of urban areas.

Furthermore, the integration of AI and UDT has proven effective in addressing urban mobility challenges by predicting traffic patterns, optimizing public transportation routes, and proposing intelligent traffic management strategies (Wang et al. 2023). Additionally, AI-driven UDT supports informed decision-making in land use planning, simulating scenarios to evaluate potential environmental impacts and guide choices aligning with sustainability objectives (Park and Yang 2020). A comprehensive systematic review conducted by Bibri et al. (2024b) explores the foundations of AI, AIoT, UDT, urban planning, and environmental sustainability in the context of sustainable smart cities. The authors investigate how AI/AIoT reshapes urban planning for enhancing environmental sustainability and augments UDT predictive capabilities in this context, as well as propose an innovative conceptual framework for the seamless integration of these technologies to advance the environmental performance of sustainable smart cities.

Fuller et al. (2020) explore DT technologies in smart cities. The UDT applications highlighted in their study are focused on local infrastructure, automotive technology, traffic management, monitoring wind energy consumption, optimizing smart farms and livestock management, and facilitating the development and maintenance of smart cities. Wu et al. (2022) conduct an in-depth analysis of the intelligent development of transportation infrastructure in the context of smart cities, with a particular focus on the extensive application of AI. The authors emphasize the advantages of DT and AI in the classification and management of transportation infrastructure, providing insights into functional design, intelligent development, and effective integration with new media. The collective efforts showcased in these studies underscore the potential of AI-driven UDT in shaping resilient, environmentally conscious urban landscapes and provide a glimpse into the profound impact on the trajectory of sustainable urban development.

Presenting detailed studies on the integration of UDT and AI focusing on environmental sustainability can provide in-depth insights into how these advanced technologies can be leveraged to enhance urban management and planning, particularly in addressing critical environmental challenges. By exploring the synergies between UDT and AI, these studies can uncover innovative solutions for monitoring, analyzing, and mitigating issues like pollution, climate change, and resource depletion. This focus is essential for developing smarter, more sustainable cities that are resilient to future environmental pressures and capable of supporting long-term urban well-being.

With the above in mind, Agostinelli et al. (2020) discuss the potential of integrating a DT model with AI systems, focusing on its application in optimizing energy management and efficiency in residential districts. The authors present a case study of Rinascimento III, a residential area in Rome, which comprises 16 eight-floor buildings with a total of 216 apartment units. The buildings have an impressive 70% self-renewable energy production, classifying them as Near Zero Energy Buildings (NZEBs). The implemented DT model is designed as a comprehensive three-dimensional data system capable of intelligent optimization and automation of building systems. It allows for real-time monitoring and analysis of energy consumption, production, and distribution in the residential district. By integrating AI systems, it can predict energy demand patterns, optimize energy usage, and automate energy-related processes to enhance overall efficiency. The study highlights the importance of NZEBs in achieving sustainable energy goals and reducing environmental impact. It emphasizes the significance of on-site renewable energy production, which contributes to the overall energy self-sufficiency of the building complex. The integration of DT with AI offers a promising approach to further improve the performance of NZEBs by enabling proactive and data-driven energy management strategies. Overall, the study demonstrates the potential of AI-driven DT models in optimizing energy systems and promoting sustainability in residential areas. It underscores the importance of leveraging advanced technologies to address the challenges of urban energy management and achieve energy efficiency targets.

Kamal et al. (2024) introduce an innovative method for adaptive traffic signal control, integrating DT with deep reinforcement learning. The primary objective of their study is to combat the environmental challenges posed by urban vehicle CO_2 emissions and fuel consumption resulting from fossil fuel dependency. The study tackles the problem of reducing air pollution and energy consumption by optimizing traffic signal adjustments using a DT-based approach. DT allows for the creation of virtual replicas of physical systems, enabling real-time monitoring and control. The authors employ the Multi-Agent Deep Deterministic Policy Gradient (MADDPG) algorithm to enable agents to learn and adapt traffic signal control strategies. By utilizing synthetic and real-world traffic datasets, the proposed approach is evaluated through quantitative simulations. Results demonstrate a significant reduction in CO_2 emissions and fuel consumption, demonstrating the effectiveness of the system in mitigating environmental impact. Notably, even with a basic reward function focusing on reducing the number of stopped vehicles, the approach proves to be successful in achieving its objectives. The contributions of this study lie in the intersection of DT technology, deep reinforcement learning, and traffic management. By combining these domains, the research offers a novel solution to address the pressing issue of urban air pollution and fossil fuel consumption. Moreover, the findings provide valuable insights into the potential of advanced computational methods to optimize traffic control systems for environmental sustainability. This research expands the understanding of traffic signal optimization and offers practical implications for urban planning and environmental policymaking.

Salunke (2023) presents a pioneering approach called Reinforcement Learning Empowered Digital Twins (RLEDs), which merges DT technology with reinforcement learning algorithms to establish dynamic virtual models of urban systems. This framework aims to revolutionize smart city development by optimizing various aspects of urban dynamics, with a particular emphasis on traffic flow. Based on case study analysis, the author aims to demonstrate the effectiveness and applicability of RLEDs in addressing real-world urban challenges, particularly in the domains of transportation management, energy management, and urban planning. This approach allows for a detailed examination of the potential benefits and limitations of RLEDs in a specific geographical and socioeconomic context. By focusing on traffic flow optimization, the research addresses the critical issue of congestion and inefficiency in urban areas. Through the integration of reinforcement learning algorithms, RLEDs are capable of learning and adapting to complex urban environments, ultimately enhancing the efficiency and sustainability of city operations.

The study by Austin et al. (2020) aims to facilitate next-generation smart city systems by widely adopting sensing and communication technologies embedded within urban environments. It proposes a smart city DT architecture that combines semantic knowledge representation with machine learning to improve data collection, processing, event identification, and automated decision-making. Demonstrated through a case study on energy usage analysis in buildings in the Chicago Metropolitan Area, the results highlight the complementary roles of semantic and machine learning components in enhancing smart city operations. This contribution underscores the potential of DT architectures to significantly boost the efficiency and functionality of smart city systems.

Lausanne City in Switzerland serves as a dynamic testing ground for the ongoing Blue City Project, aiming to create a responsive AI-powered UDT (Huang et al. 2024). This UDT functions as an integrated, open-source platform that comprehensively maps the interconnected network of city flows, encompassing people, energy, materials, mobility, goods, biodiversity, and waste. The platform's design empowers citizens to engage in collective, evidence-based decision-making, enhancing sustainability, resilience, quality of life, and ecological value. The Blue City Project establishes a foundation for examining emerging patterns and facilitates more informed and proactive urban planning decisions.

Beyond AI-driven UDT, additional case studies provide clear and tangible evidence of UDT's effectiveness in optimizing various urban systems and strengthening urban resilience in real-world applications. These studies highlight the practical benefits and real-world applications of UDT in enhancing urban infrastructure, improving resource management, and increasing the adaptability of cities to various challenges. The Helsinki Smart City DT project explores digital modeling using the cityGML standard to enhance stakeholder communication, transparency, and openness in city development (Hämäläinen 2021). Dynamic 3D city models in Helsinki simulate energy-related data, educating citizens on energy efficiency. The success of cityGML experimentation lays the groundwork for broader DT applications, contributing to improved governance and urban planning projects.

The case study in Ålesund City, Norway, examines the application of a graphical UDT to a district, demonstrating its efficacy for urban mobility planning and addressing challenges related to stakeholder coordination and ad-hoc modeling (Major et al. 2022). This UDT automates model configuration, integrates data, ensures flexibility and scalability, and facilitates stakeholder interaction for well-informed decision-making. In Herrenberg City, Germany, UDT serves as a complex data model supporting collaborative processes, encompassing urban mobility simulation, a 3D model of the built environment, street network modeling, wind flow simulation, and empirical data (Dembski et al. 2020). Presented in a virtual reality platform, it aids participatory decision-making, offering insights into urban challenges and fostering consensus in public decision-making. The tool enhances collaboration, communication, and decision support in the context of smart, sustainable cities.

The integration of empirical evidence from case studies underscores the innovative role of UDT in addressing contemporary urban challenges, making it an invaluable tool for fostering smart, sustainable, and resilient cities. It showcases the significant impact of UDT on urban management, planning, and governance, leveraging AI, IoT, and 3D modeling for precise analysis, prediction, and optimization. The outcomes include improved stakeholder communication, transparency, openness, participatory decision-making, and collaborative urban development.

From a technical standpoint, considering the preceding analysis and discussion, the architecture of UDT can exhibit variations depending on its applied urban domains or their coordination. These variations are tailored to address the specific requirements and challenges of different urban systems, ensuring effective integration and functionality within the diverse contexts of urban management and planning. Different urban contexts and specific use cases may require tailored UDT architectures to effectively address their unique challenges and objectives. The variations in the UDT architecture can arise from differences in the types of data sources, the complexity of urban systems, the scale of deployment, and the specific goals of the application domain. Customization of the UDT architecture allows it to accommodate diverse urban scenarios, ensuring that the digital representation aligns with the complexities of the targeted city. In light of the various theoretical and empirical studies discussed earlier, the architecture of UDT consists of several components that enable its functionality. These components include:

- Data acquisition and integration layer: This layer is responsible for collecting and integrating data from various sources, such as IoT devices and sensors, GIS mapping, satellite imagery, and other urban data repositories. It ensures a comprehensive and up-to-date representation of the urban environment.

- Modeling and simulation engine: As the core of UDT, this engine uses advanced modeling and simulation techniques to replicate the real-world urban environment. It enables dynamic simulations, scenario planning, and predictive modeling.

- AI and analytics module: Incorporating AI and analytics, this module processes real-time data, identifies patterns, and generates insights. ML/DL algorithms are employed for predictive analysis, anomaly detection, and optimization.

- Visualization and user interface: This component provides a user-friendly interface for stakeholders to interact with UDT. It often includes real-time visualization tools, allowing users to explore urban dynamics and scenarios.

- Communication protocols: To facilitate data exchange between various components, communication protocols are employed. This ensures seamless connectivity and interoperability within UDT architecture.

- Security infrastructure: Given the sensitivity of urban data, a robust security infrastructure is essential, which safeguards data integrity, privacy, and protects against cyber threats.

- Semantic knowledge representation: In some cases, semantic knowledge representation is integrated to enhance the understanding of urban features and relationships, aiding in more intelligent decision-making.

- Feedback mechanism: A feedback loop allows UDT to continuously learn and adapt. Insights generated from real-world data may be used to refine models, improving the accuracy of predictions over time.

- Scalability and flexibility: The UDT architecture is designed to be scalable, accommodating the expansion of the urban environment in terms of both size and complexity. It should also be flexible to incorporate emerging technologies and adapt to the evolving urban challenges.

These components work in concert to create a dynamic and intelligent UDT, supporting informed decision-making in urban management and planning processes. Worth noting is that the synergies between the architectures of UDT and UB are evident in their shared functions of enhancing management and planning processes. Both systems leverage real-time data processing, advanced analytics, and visualization techniques to support dynamic decision-making. The integration of AIoT with UDT and UB enhances predictive accuracy, optimizes resource management, collectively contributing to the realization of environmentally sustainable and intelligent urban futures.

5.4.1.3 Smart Urban Metabolism

The evolution from traditional urban metabolism to the contemporary concept of SUM represents a significant shift in the understanding and management of urban systems. Initially rooted in ecological sciences, urban metabolism primarily focuses on the static analysis of energy and material flows within cities, reflecting the intricate web of technical and socioeconomic processes and operations generating various outputs (Kennedy et al. 2007), including energy, by-products, waste, emissions, raw materials, data, and information. It encompasses the intricate networks of transportation, utilities, activities, and services that sustain the city's functionality, emphasizing efficiency, sustainability, and adaptability (Bibri and Krogstie 2020b; Shahrokni et al. 2015). Its adoption in resource management, planning, and design involves data collection, analytics, and stakeholder collaboration (Baccini and Brunner 2012; Kennedy and Hoornweg 2012; Kennedy et al. 2011). The complex nature of urban metabolism, coupled with extensive data requirements for a thorough analysis of social, ecological, and technical systems, led to the emergence of SUM. This evolution is, in turn, driven by the imperative for cities to become smarter, more efficient, and more sustainable in the face of complex urban challenges.

SUM, as a sophisticated model, can gather and integrate vast quantities of high-spatial and temporal resolution data from diverse urban systems

(Shahrokni et al. 2015), facilitating real-time processing, analysis, and visualization of urban dynamics. This transition introduces a more dynamic and technology-driven approach, incorporating IoT, Big Data (Bibri and Krogstie 2020b; D'Amico et al. 2020; Shahrokni et al. 2015), AI (Ghosh and Sengupta 2023; Peponi et al. 2022), and AIoT (Bibri et al. 2024a) to monitor and optimize resource flows. SUM goes beyond the static examination of resource flows, embracing data-driven insights for responsive and adaptive urban planning. It serves as a comprehensive framework for understanding the complex interactions between various urban entities and facilitates informed decision-making, engaging stakeholders, fostering knowledge sharing, and implementing flexible multi-level governance.

The SUM architecture is designed to capture and process real-time data, utilizing AI and ML algorithms to derive actionable insights that enhance decision-making in urban contexts (Ghost Ghosh and Sengupta 2023; Peponi et al. 2022). The infusion of AIoT in SUM further enhances its capabilities, allowing for predictive modeling, scenario planning, and optimization of resource utilization in urban environments. The incorporation of AI into SUM is the path ahead, offering substantial benefits in understanding real-time city functions (Bibri et al. 2024a). AI-driven SUM is seen as instrumental for urban city planners and governors, empowering them to make well-informed decisions and implement policies with intelligence and efficiency (Ghosh and Sengupta 2023; Peponi et al. 2022).

The effectiveness and relevance of the SUM architecture depend on various key factors in addressing the complexities of urban environments. In light of the examined theoretical and empirical studies, key considerations shaping the design of the SUM architecture encompass: urban context, city or district size and complexity, data requirements, stakeholder requirements, technological infrastructure, budgetary constraints, regulatory and compliance considerations, available skillsets, sustainability goals, citizen involvement, interoperability with existing systems, future scalability and flexibility, and global best practices. While specific SUM architectures may vary, common interconnected components are found across different architectures to effectively capture, process, and analyze urban data for informed decision-making and sustainable urban development. Therefore, drawing on the synthesis of various studies (e.g., Bibri et al. 2024a; Bibri and Krogstie 2020b; D'Amico et al. 2020, 2021, 2022; Ghost Ghosh and Sengupta 2023; Peponi et al. 2022; Shahrokni et al. 2015; Wiedmann and Athanassiadis 2021), the layers comprising the SUM architecture may vary in specifics, depending on distinct urban contexts, objectives, spatial scales, and technological capacities, among others.

In the data acquisition layer, IoT sensors are strategically placed throughout the city to capture real-time data on various urban processes, resource utilization, and environmental impacts. This includes monitoring transportation networks, traffic patterns, air quality, energy consumption, waste generation, and other factors that significantly influence urban life. This layer involves defining local system boundaries and identifying real-time data sources for representing Key Performance Indicators (KPIs) specific to the systems. The IoT layer ensures the seamless integration of sensor data, maintaining a continuous stream of valuable information. Building on this foundation, the connectivity infrastructure layer establishes a robust and scalable communication framework, facilitating seamless data transmission between sensors, IoT devices, and central data processing systems. This ensures the efficiency of the entire data acquisition process. Moving to the data integration layer, the emphasis is on data harmonization to ensure compatibility and coherence among diverse datasets for comprehensive analysis. Centralized or distributed databases are employed for efficient and secure data storage, while data quality assurance mechanisms validate and uphold the accuracy and reliability of collected data.

The subsequent data processing and analytics layer leverages AI algorithms, especially ML/DL, to process real-time data, uncovering patterns, making predictions, and deriving meaningful insights into resource flows, environmental conditions, and urban dynamics. This layer entails the development of a metabolic flow calculation engine for generating specific KPIs in real-time based on system boundaries and the corresponding Application Programming Interfaces (APIs) for managing access to these indicators for feedback. The visualization and user interface layer provide user-friendly interfaces, including dashboards, smartphones, computer terminals, and other visualization platforms tailored for city officials, planners, citizens, and stakeholders based on their specific goals and interests. This layer ensures that complex data are presented in an accessible manner. In the decision support layer, actionable insights are integrated into decision support systems, empowering urban planners, policymakers, and stakeholders to make informed decisions and identify intervention measures regarding resource allocation, infrastructure development, and environmental management. The application layer introduces specific applications and services as tailored solutions derived from data analysis.

Scenario planning tools aid urban planners in simulating and evaluating different development scenarios, while optimization enhances resource utilization and efficiency in urban processes.

The stakeholder involvement and collaboration layer employs platforms for engaging city authorities, citizens, organizations, and institutions in the design, implementation, and monitoring of SUM initiatives. Knowledge-sharing mechanisms foster collaborative learning and information exchange, while multi-level governance structures support iterative co-design and adaptive urban management. Open data platforms and communication channels support transparency and engagement in urban planning. This layer involves evaluating report findings in collaboration with stakeholders and following up the outcomes of interventions in real-time. At the policy and governance layer, policies and governance structures are pivotal in implementing sustainable urban practices. This involves formulating and enforcing regulations that promote smart and sustainable urban development, necessitating collaboration between government bodies, private sector entities, and the community for successful SUM strategy implementation. Overall, the detailed SUM architecture is designed to holistically address the complex and dynamic nature of urban environments, providing a robust and adaptable framework for cities to enhance their resilience, sustainability, and overall efficiency in urban management and planning.

In the quest for sustainable urban development through SUM, cities are strategically prioritizing resource efficiency and circular economy principles, integrating renewable energy solutions, deploying smart infrastructure and technology, enhancing public transportation and mobility solutions, implementing advanced environmental monitoring and management practices, fostering community engagement and inclusivity, building resilience and adaptability to environmental challenges, and establishing effective policies and governance structures for long-term sustainability (e.g., D'Amico 2020, 2021, 2022; Sharhrokni et al. 2015). These comprehensive efforts underscore the crucial role of SUM in shaping contemporary urban management and planning strategies, where data-driven decision-making, community involvement, and sustainable practices converge to create resilient and adaptive cities.

From an empirical perspective, case studies on SUM provide tangible evidence of its crucial role in optimizing various urban systems and fortifying the overall resilience and sustainability of cities. Shahrokni et al. (2015) introduce the application of SUM in the Stockholm Royal Seaport project, focusing on generating four real-time Key Performance Indicators (KPIs): kilowatt-hours per square meter, CO_2 equivalents per capita, kilowatt-hours of primary energy per capita, and the percentage of renewables. The study not only examines barriers and potential long-term implications but also envisions SUM as providing a novel understanding of urban causality. By emphasizing a feedback loop at three levels (household, building, and district through four interfaces tailored to diverse audiences, the research contributes valuable insights for both citizens and city officials, highlighting the interconnection of SUM with effective urban planning and management strategies.

In their comprehensive study on circular cities, including Porto, Prato, Oslo, and Kaunas, constituting the Urban Agenda Partnership on Circular Economy (UAPCE), D'Amico et al. (2022) conduct a case study analysis that delves into multiple dimensions of smart metabolism circularity. The study scrutinizes the distinct urban metabolism flows of each city or district, emphasizing unique approaches and outcomes in the context of sustainable urban development. Porto focuses on mobility, construction, and food; Prato on mobility, construction, waste, and water; Oslo on circular economy and waste; and Kaunas on governance, transport, and mobility. The monitoring and evaluation of these endeavors include optimization of vehicles, circulation, and emission reduction in Porto; improved water and energy flows, along with tailored public services in Prato; enhanced sharing and recycling of goods and services in Oslo; and heightened knowledge about vehicle flows in Kaunas. While the analytical framework of smart metabolism circularity in these case studies encompasses broader city-specific flows, this study places particular emphasis on environmental aspects. This exploration reveals unique insights into how cities navigate the complexities of SUM to enhance their environmental efficiency and resilience.

Transitioning to another facet of urban sustainability, Beckett (2022) explores strategies in their study, focusing on Suzhou City's urban socioeconomic metabolism. Their objective is to reduce environmental burden and enhance circularity through technological interventions, specifically targeting the nitrogen metabolism. The authors evaluate 15 scenarios incorporating promising technologies in water and waste management, utilizing the Multi-sectoral Systems Analysis (MSA) model to scrutinize various sectors in Suzhou's urban infrastructure and economy. Several Metabolic Performance Metrics (MPMs) are introduced to assess the environmental impact of these interventions. Both studies contribute to the broader discourse on SUM, offering nuanced insights into distinct yet interconnected aspects of optimizing urban systems and enhancing environmental efficiency.

In real-world settings, the integration of AI and AIoT technologies within SUM frameworks is still in its early phases, marking the initial stages of development and implementation. This ongoing process signals a promising trajectory toward more sophisticated and efficient urban management solutions. As SUM continues to evolve, the integration of AI and AIoT technologies holds significant potential to revolutionize how cities are monitored, analyzed, and optimized for enhanced sustainability and resilience. A recent study by Peponi et al. (2022) addresses the complexity of urban systems, characterized by nonlinear, dynamic, and interconnected processes, necessitating improved management for enhanced sustainability. The authors propose a novel, evidence-based methodology to understand and manage these complexities, aiming to bolster the resilience of urban processes under the concept of smart and regenerative urban metabolism. By integrating Life Cycle Thinking and ML, they assess the metabolic processes of Lisbon's urban core using multidimensional indicators, including urban ecosystem service dynamics. The authors developed and trained a multilayer perceptron (MLP) network to identify key metabolic drivers and predict changes by 2025. The model's performance was validated through prediction error standard deviations and training graphs. Results indicate significant drivers of urban metabolic changes include employment and unemployment rates (17%), energy systems (10%), and various factors such as waste management, demography, cultural assets, and air pollution (7%), among others. This research framework serves as a knowledge-based tool to support policies for sustainable and resilient urban development.

5.4.2 Architectural Layers of Artificial Intelligence of Things Grounded in the Data Science Cycle

The preceding section addresses the diverse characteristics and applications of AIoT in the context of sustainable smart urban development and explores its functionalities across UB, UDT, and SUM. In this section, the focus shifts to a detailed examination of the architectural layers of AIoT itself, unraveling the specifics of its design based on the data science cycle.

The domain of AIoT architectures exhibits convergence in their application of the pillars of the data science cycle, emphasizing the iterative and data-driven nature of these structures. While the core principles of collecting, analyzing, and utilizing data for informed decision-making remain consistent, variations emerge in the specific design and implementation of AIoT architectures. These differences arise from a variety of factors, including technological requirements, contextual considerations, and specific objectives within various applications and domains. Additionally, other influential factors contributing to these distinctions may include regulatory frameworks, resource availability, and stakeholder preferences. Consequently, AIoT architectures adapt to the unique demands of their intended use cases, showcasing flexibility and versatility in the convergence of the overarching data science principles.

According to Zhang and Tao (2021), AIoT architecture adopts a tri-tier structure focused on computing layers and is composed of three layers: cloud/fog/edge computing. The edge computing layer serves as the perception layer and smart visual sensing block, supporting control and execution over sensors and actuators. This layer is designed to enhance AIoT systems with perceptive capabilities. The fog computing layer, represented by fog nodes, contributes to learning and reasoning abilities by processing data within networks. The cloud computing layer offers various application services, empowering AIoT systems with extensive data access and computational resources for learning and reasoning. Notably, edge things and fog nodes are distributed, while the cloud remains centralized in the AIoT network topology.

The AIoT system or AIoT-driven architecture discussed by Bibri et al. (2023b) encompasses five interrelated computational processes: sensing, perception, leaning, visualization, and acting. The perception process involves utilizing input data from diverse sensors to deduce different aspects of the world, such as object detection, action recognition, and image classification. The system translates sensory data into meaningful information, aiming to recognize and interpret these data in relation to the real world. The learning process, central to the system, relies on ML, allowing it to learn from experience without explicit programming. ML involves collecting and preparing data for training, building models, and training the system to find patterns or make predictions. Three subcategories of ML—supervised learning, unsupervised/semi-unsupervised learning, and reinforcement learning—contribute to diverse applications. DL plays a vital role in improving the system's adaptability and performance, particularly in handling dynamic and complex environments and accelerating data-driven decision-making, making it more accurate and faster. Data visualization emerges as a critical aspect, conveying complex data through graphical representations, charts, and maps. It facilitates easier comprehension and interpretation of data, allowing decision-makers to gain insights more rapidly. The final stage involves acting to achieve a certain goal, requiring the system to reason, make inferences, and behave. Actuation mechanisms play a crucial role in executing actions to optimize smart systems in areas like power grids, transportation, and waste management. Actuation is essential for AIoT applications, enabling the monitoring, control, and operation of diverse functions in smart cities.

Bibri et al. (2024a) include a sixth step between the processes of learning and visualization in the AIoT system proposed by Bibri et al. (2023b). This step is evaluation, which is crucial to the learning process, involving the assessment of the performance and quality of models developed during the analysis phase. After creating and training ML models, their performance needs to be evaluated to ensure their effectiveness and generalizability to new, unseen data. This entails using metrics (accuracy, precision, recall, etc.), testing, validation, and comparison to ensure the reliability of data-driven models and meaningful insights. Evaluation helps practitioners make informed decisions about the model's utility and potential for deployment in real-world applications. In turn, the AIoT system, through learning, continuously improves its understanding of the data and refines its models based on new information or feedback.

The AIoT system outlined by Samadi (2022) consists of three primary components: IoT data analytics systems, APIs, and AI/ML systems. These components work synergistically to enable real-time data monitoring through IoT devices, offering solutions for decision-making. In this integrated system, AI is embedded into infrastructure components, ensuring interoperability through APIs at device, software, and platform levels. The IoT analytics systems process vast amounts of real-time data from diverse sources, employing various analytics procedures. Application Programming Interfaces (APIs) facilitate data integration within the AIoT system, providing seamless communication between different components. AI systems, including ML and DL approaches, contribute advanced methods for predictive modeling and scenario forecasting. The comprehensive integration of these components empowers the AIoT system to address challenges in real-time data analysis and management, leveraging advanced analytics and intelligent algorithms. While the explicit mention of the data science cycle is absent in this description, its implicit presence is evident in the continuous flow of data, its processing through analytics, and the utilization of AI techniques for decision-making, embodying the fundamental principles of the data science cycle.

The data science cycle approach underscores the importance of leveraging large volumes of heterogeneous data, enabling the extraction of meaningful insights and the development of predictive models. It emphasizes the continuous feedback loop, reflecting the dynamic nature of AIoT applications, where data continuously contribute to the intelligence and effectiveness of the system.

Furthermore, Seng et al. (2022) present a comprehensive overview of the recent advancements in AIoT, delving into various computational frameworks. The authors not only identify opportunities but also address challenges associated with the effective deployment of AIoT across diverse applications. Their discussion encompasses different architectures, techniques, and hardware platforms, emphasizing the three fundamental layers of AIoT: (1) sensors, devices, and energy approaches; (2) communication and networking; and (3) applications. Additionally, they shed light on the integration of AIoT with edge, fog, and cloud computing, along with the incorporation of 5G/6G networks.

It is crucial to note that the early research within AIoT has predominantly centered around technical aspects, with a particular focus on edge computing (Fragkos et al. 2020; Yang et al. 2020, 2021). The significance of edge computing lies in its ability to relocate data processing closer to IoT devices and systems, enhancing AIoT applications through edge intelligence. This aspect is vital for deploying next-generation AIoT applications, especially in the management and planning of smart, sustainable cities. The synergy between AI and IoT within AIoT architectures establishes a dynamic ecosystem where AI enhances the capabilities of IoT, and IoT enriches AI's data-driven knowledge. This collaboration drives groundbreaking advancements in data-driven solutions for sustainable smart cities.

5.4.3 The Dynamic Interplay of Artificial Intelligence of Things and Cyber-Physical Systems of Systems

5.4.3.1 Artificial Intelligence of Every Things and Cyber-Physical Systems of Systems

In the context of sustainable smart cities, Artificial Intelligence of Everything (AIoE) epitomizes the comprehensive integration of AI technologies across all facets of urban life, spanning various sectors of urban management and planning. The primary objective of AIoE is to harness AI's capabilities to augment the intelligence and decision-making of IoT systems, thereby automating, optimizing, and enhancing city operations and services. This approach places a strong emphasis on strategic sustainability, aiming to create smarter, more efficient urban environments that prioritize long-term ecological balance and resource efficiency. Key elements of AIoE in sustainable smart cities encompass data analytics driven by diverse city sensors and IoT devices for real-time decision-making and predictive analytics, automation of routine tasks to reduce human errors and boost efficiency, and seamless connectivity between different urban systems to ensure effective communication and coordination across domains. AIoE drives sustainable smart cities toward enhanced efficiency, environmental sustainability, resilience, adaptability, and an improved quality of life. The interaction between AIoE and sustainable smart cities creates a more interconnected, intelligent, and sustainable urban ecosystem (Bibri and Huang 2024). This study emphasizes UB, UDT, and SUM, which represent large-scale components of AIoE as forms of CPSoS in sustainable smart cities. The synergy between AIoE and these cities and systems is crucial in shaping the future of urban living, making cities more efficient, resilient, and adaptable. This holistic approach addresses contemporary urban challenges and sets the stage for future innovations.

The emergence of AIoE marks a significant advancement in IoE, propelled by the dynamic evolution from the conventional IoT to include broader spheres like the Internet of City Things. The framework of AIoSSCT is central to this shift, strategically focusing on how AI can interconnect and elevate the functionalities of urban systems across urban management, planning, and environmental sustainability. The focus here is on the AI of the data-driven management and planning systems of sustainable smart cities, a more comprehensive network of interconnected large-scale AIoT-powered systems. Masserov and Masserov (2022) envision a future of a highly interconnected "Internet of Everything on a Smart Cyber-Physical Earth," with IoT and CPS driving a new smart revolution. The authors highlight the challenge of processing vast amounts of data with current computing power and suggest that AI and data science offer solutions. They argue that IoT combined with AI could bring significant advancements, not only in cost-saving but also in enhancing human life. In a similar vein, Singh et al. (2023b) explore the role of intelligent systems and IoE in developing smart cities. The authors explain the complexities of CPS and smart city infrastructure, underscoring how AI and ML are crucial in enhancing the efficiency, connectivity, and responsiveness of these urban environments. They cover key areas such as the physical layer design of smart city infrastructure, smart sensors and actuators, and the broader applications and challenges of IoE in smart cities. These transformations manifest in sustainable smart cities through their brain, digital twin, and metabolism as AIoT-powered data-driven management and planning systems, which encapsulate the principles of CPS and represent CPSoS.

CPS leverages IoT for deployment by integrating sensor data, connectivity, and real-time monitoring, with AI algorithms analyzing the collected data to enable intelligent decision-making and automation (Mohamed 2023). They are embedded in the principles of embedded systems, control mechanisms, real-time processing, and wireless sensor networks (Mohamed 2023). The integration of AIoT amplifies IoT deployment by infusing AI algorithms into the network of interconnected IoT devices as part of CPS and hence CPSoS, fostering advanced data analysis, real-time insights, intelligent decision-making, and autonomous control in real-time urban environments (Bibri and Huang 2024). Both CPS and AIoT strive for real-time adaptation and response to system changes, with AIoT offering more sophisticated adaptive responses based on analyzed data. The symbiotic relationship between CPS and AIoT forms a powerful framework that transforms traditional urban management and planning.

AIoT plays a foundational role in the operation and functioning of UB, UDT, and SUM as forms of CPSoS, aimed at enhancing data-driven management and planning in sustainable smart cities. These CPSoS leverage sensors, computational tools, actuators, and communication networks to enable the exchange of information and coordination between physical processes and computational systems. UB, UDT, and SUM exemplify CPSoS specifically designed for urban environments due to their unique characteristics in terms of system size, complexity, and requirements. UB serves as a central intelligence hub that orchestrates the functioning of various urban systems, such as transportation, traffic, energy, and waste. It connects with the physical infrastructure of the city through sensors and actuators, creating a feedback loop between the digital and physical worlds. UDT relates to CPSoS through its role in creating a 3D representation of physical urban environments, allowing for real-time monitoring, analysis, and optimization of various urban systems (Agostinelli et al. 2020; Bibri and Huang 2024; Mylonas et al. 2021; Nica et al. 2023; Somma et al. 2023). SUM integrates computational modeling with the physical dynamics of urban metabolism and represents CPSoS through its focus on optimizing resource flows in urban environments, such as energy, water, and waste. This study underscores the synergistic and collaborative nature of CPSoS, taking a step forward in unraveling the intricacies of UB, UDT, and SUM in light of their distinctive features and functions to augment the efficacy of data-driven management and planning in sustainable smart cities. The interplay of AIoT, CPS, UB, UDT, and SUM is crucial, as these components are intricately interconnected and mutually reinforce one another in shaping these urban environments.

5.4.3.2 Foundational Elements of the Synergistic Interplay of Artificial Intelligence of Things and Cyber-Physical Systems of Systems for Data-Driven Environmental Management and Planning

The integration of AI, AIoT, UB, UDT, and SUM forms a complex framework of CPSoS in data-driven operational management and planning. These components, as critical aspects of CPSoS, facilitate effective urban transformation in the context of sustainable smart cities, by addressing several foundational elements:

- Interconnected systems: UB, UDT, and SUM are advanced CPSoS elements that facilitate seamless data integration and communication among urban systems. UB acts as the central nerve center, processing real-time data for immediate decision-making. UDT provides a digital mirror of physical assets, enabling simulation and analysis. SUM analyzes the flow of energy and materials through urban ecosystems to improve resource efficiency and reduce waste. This interconnected network is vital for enabling informed, data-driven decisions that are essential for urban management and planning.

- Data-driven decision-making: AIoT are fundamental in enhancing the capacity of CPSoS by improving data acquisition, analysis, and interpretation. AI algorithms process complex data sets to uncover patterns and insights, while AIoT extends this capability to IoT devices across the cityscape, providing a granular level of data on various aspects of urban life. This enriched data landscape supports robust decision-making processes that are crucial for effective and proactive urban management and planning.

- Dynamic simulation and modeling: UDT and UB utilize AI and AIoT to perform dynamic simulations and modeling of urban scenarios, which are crucial for planning and response strategies. This capability allows city planners to visualize the impacts of potential decisions and adapt to them before implementing them in the real world. By simulating different scenarios, these tools provide accurate predictive insights that help mitigate risks and forecast future conditions, guiding urban development.

Environmental sustainability as a central principle: Incorporating AI and AIoT into CPSoS frameworks under the guiding principle of environmental sustainability ensures that all technological deployments promote resource efficiency, pollution reduction, and ecological balance. This principle is operationalized through the integration of sustainable practices into all layers of urban management and planning, ensuring that technological advances contribute positively to the city's long-term goals of environmental sustainability.

Adaptive urban planning and management: The responsive nature of CPSoS, powered by AI and AIoT, enables urban systems to dynamically adjust to new data and changing conditions. This flexibility is crucial for adapting urban management and planning strategies quickly in response to unforeseen events or new environmental data, ensuring that the city can respond effectively to both gradual changes and sudden emergencies.

Inclusive and participatory management and planning: CPSoS fosters an inclusive approach to urban development by leveraging data-driven technologies and digital platforms to engage a broad spectrum of stakeholders. This participatory approach ensures that the planning process is democratized, allowing input from diverse community members, which leads to more equitable and tailored urban solutions. Digital platforms facilitate this by providing forums for feedback and collaborative decision-making.

Technological synergies: The synergistic interaction among the components of the framework of CPSoS—UB, UDT, AI, and AIoT—enhances the overall system's efficiency and effectiveness. Each component amplifies the capabilities of the others. For instance, AI enhances the predictive accuracy of UDT models, and AIoT extends the sensory reach of UB, creating a more integrated, responsive, and intelligent urban management system. These synergies drive improvements in environmental management and planning, making cities more resilient, efficient, and sustainable.

These dimensions collectively establish a robust foundation for understanding how the integration of CPSoS in the form of UB, UDT, and

SUM can profoundly contribute to advancing data-driven environmental management and planning in sustainable smart cities. They offer valuable insights into the transformative impact of this integration on environmentally sustainable urban development.

5.4.4 An Innovative Data-Driven Management and Planning Framework

As an emerging technological framework, AIoT plays an important role in shaping the landscape of urban management and planning within the realm of sustainable smart cities. This section explores how the architectural layers of AIoT interconnect within the overarching framework of AIoSSCT, emphasizing their synergistic integration with UB, UDT, and SUM. This exploration provides an infrastructure underlying the operational functioning and planning of sustainable smart cities.

5.4.4.1 Essential Technical Components of the Proposed Framework

AIoSSCT constitutes a framework wherein AIoT becomes intricately woven into the fabric of urban management and planning systems, propelling the intelligence of sustainable smart cities. Within this framework, the term "things" encompasses UB, UDT, and SUM as an integrated set of CPSoS. The integration of AIoT is designed to elevate the functionality, efficiency, and performance of these systems, with a strong emphasis on their interconnectedness, coordination, and operational synergy. This innovative approach takes a holistic view of leveraging the prowess of AI and the connectivity of IoT to optimize urban management and planning processes. Drawing insights from the previously analyzed and synthesized literature on AIoT, CPS, UB, UDT, and SUM architectures, the technical components of AIoSSCT are distilled and contextualized, as illustrated in Figure 5.4.

> Data collection: UB utilizes various IoT sensors and devices to collect real-time data on urban life aspects, such as traffic patterns, transport networks, energy consumption, air quality, and public services. UDT gathers data through IoT sensors and devices to create a detailed virtual replica of the city, capturing its physical and operational attributes and mapping and monitoring changes in real-time. SUM: Utilize IoT sensors and devices to monitor resource flows, energy consumption, waste generation, and environmental parameters.

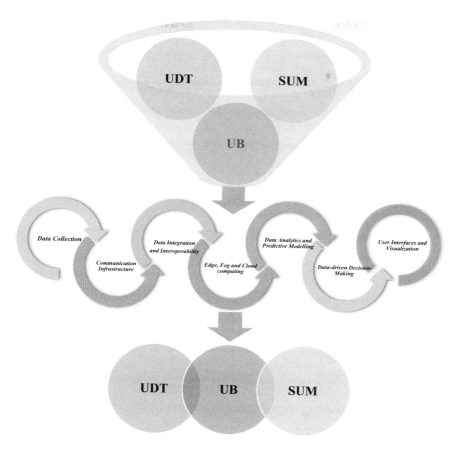

FIGURE 5.4 Essential technical components of AIoSSCT framework.

Communication infrastructure: UB leverages high-speed, low-latency communication networks to ensure real-time data transmission from sensors. UDT establishes communication protocols for seamless interaction between its digital model and the physical urban environment. SUM enables efficient communication between its components to synchronize data and feedback loops. In addition, fog computing is included to enhance the efficiency of communication by optimizing data processing at the edge.

Edge computing: UB Edge processes data locally at the edge for quick response times and reduces the load on the central platform. UDT Edge implements edge computing for on-site processing of data, especially critical for scenarios requiring real-time simulation and interaction. SUM Edge employs edge computing to analyze resource flows locally and optimize processes without relying solely on central servers. Additionally, fog computing is leveraged to enhance the efficiency of edge processing, allowing for distributed computation and analysis at the edge of the network for UB, UDT, and SUM.

Cloud infrastructure: UB cloud services utilize cloud infrastructure for storage, scalability, centralized processing, and efficient execution of complex AI algorithms. UDT cloud services leverage cloud resources for storing and managing large-scale urban models and simulations. SUM cloud services implement cloud-based services for processing and analyzing extensive datasets related to urban metabolism. Additionally, fog computing is integrated within the cloud infrastructure for UB, UDT, and SUM to enhance the distribution of computational resources and optimize data processing in a more decentralized manner.

Data integration and interoperability: UB implements data integration strategies that accommodate the specific data needs and formats, ensuring seamless interoperability with UDT and SUM. UDT develops interoperability protocols that allow it to efficiently exchange data with UB and SUM, fostering integration for synchronized urban modeling and analysis. SUM ensures it can coalesce data from UB and UDT, establishing interoperability to enhance the overall understanding of urban dynamics for optimized management and planning.

Data analytics and predictive modeling: UB employs AI-driven analytics, including ML and DL, to generate insights relevant to UB, enhancing decision-making and predictive modeling. UDT implements predictive modeling, specifically utilizing ML and DL techniques, for real-time simulation and analysis of changes in the physical attributes of urban entities. SUM employs data analytics, including ML and DL, utilizing AI for predictive modeling of resource flows, energy consumption, and waste generation, contributing to sustainable urban planning.

Data-driven decision-making: UB facilitates data-driven decisions by presenting AI-generated insights to city officials and administrators for effective urban management. UDT supports decision-making based on real-time AI-driven UDT insights, enabling urban planners to understand the consequences of different choices in the physical environment. SUM aids decision-makers by providing AI-generated insights derived from real-time data on resource flows and urban processes, contributing to sustainable decision-making.

Application (user interfaces and visualizations): UB develops user-friendly and intuitive applications powered by AI, providing visualizations and insights relevant and tailored to city officials and citizens. These applications enhance transparency, engagement, and decision-making within the urban management framework. UDT creates applications that visualize real-time changes in urban entities, providing urban planners and decision-makers with tools to monitor and interact with the 3D representation of the city's physical and functional aspects. These applications leverage AI-driven simulations and analyses specific to these attributes. SUM design interfaces and applications that present dynamic visualizations of resource flows, environmental parameters, and other relevant data. These applications engage various stakeholders in understanding and managing urban metabolic flows for sustainable urban planning.

APIs integration hub: APIs are essential to facilitate seamless communication and integration between UB, UDT, and SUM. APIs act as intermediaries that allow different systems to exchange information and functionalities. They can enable UB to share relevant insights and data with UDT and SUM, fostering collaboration and ensuring that each component can leverage the strengths of the others. They are crucial for UDT to communicate with UB and SUM, enabling the exchange of real-time data and simulations. This ensures that UDT remains synchronized with both the central decision-making hub (UB) and SUM. APIs facilitate the integration of SUM with UB and UDT, allowing SUM to receive and provide data that contributes to the overall understanding of urban dynamics. This integration enhances the effectiveness of SUM in contributing to optimized management and planning. The integration of all layers within the AIoSSCT framework synergistically enhances the functionality of UB, UDT, and SUM, creating a mutually reinforcing relationship among these components.

5.4.4.2 Synergizing the Framework Components through Collaboration and Enhancement

The AIoSSCT architecture, as elucidated above, underscores the seamless integration of UB, UDT, and SUM as CPSoS—empowered by AIoT. This strategic alignment proves crucial, unlocking and maximizing the collective potential of these components to streamline and enhance the management and planning systems of sustainable smart cities. The interplay of the underlying interconnected layers and systems constitutes a comprehensive framework, as depicted in Figure 5.5. Through their collaboration, these components contribute synergistically to informed decision-making, resource optimization, operational efficiency, and effective and holistic planning. UB serves as the cornerstone, initialized with fundamental urban data. UDT seamlessly integrates with UB, augmenting its capabilities for sophisticated simulation and real-time monitoring through the creation of a dynamic virtual replica of the city. The data streams processed by SUM are subsequently incorporated into UB, a process characterized

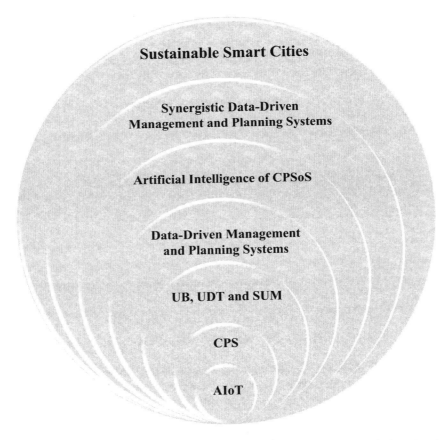

FIGURE 5.5 A novel framework for data-driven management and planning in sustainable smart cities.

by the direct interplay between SUM and UDT. This integration provides valuable insights into intricate resource flows and environmental impacts, which also contribute to the operational functioning optimized by UB and the strategic planning enhanced by UDT. This collaborative approach significantly enriches the holistic understanding of urban dynamics, establishing a robust foundation for informed decision-making in the realm of sustainable smart urban development.

When integrated, UB, UDT, and SUM form a powerful CPSoS for urban management and planning. In this CPSoS framework, each system plays a distinct yet complementary role in optimizing the environmental performance and sustainability of urban environments. UB provides real-time operational data and AI-driven insights for immediate decision-making, while UDT offers detailed virtual models for long-term strategic planning. SUM manages the flow of resources—such as energy, water, and materials—through the city, ensuring a sustainable urban metabolism. Together, these components create an interconnected urban ecosystem capable of managing real-time operations and predicting future needs through continuous data exchange and simulations.

UB is the real-time decision-making hub within this CPSoS, acting as a centralized system that collects and processes data from IoT sensors embedded throughout the urban infrastructure. By analyzing these data using AI algorithms, UB delivers immediate operational insights that can adjust systems like traffic, pollution control, and energy consumption in real time. For instance, UB can dynamically adjust traffic signals to optimize flow or shift energy loads based on current demand. This ability to regulate city operations in real time is critical to a city's functioning. The physical components, such as traffic lights, power grids, and pollution monitors, are continuously adjusted based on data processed in the "cyber" space, allowing UB to optimize urban systems instantly and responsively. These real-time adjustments are key to ensuring that the city operates efficiently and sustainably on a day-to-day basis.

While UB focuses on the real-time management of the city, UDT functions as a high-fidelity virtual model that represents the city in the digital domain. UDT serves as a predictive tool, enabling urban planners and decision-makers to simulate the effects of various strategies, policies, and infrastructure changes before they are implemented in the real world. This is crucial for long-term urban planning, where the impacts of changes—such as new transportation systems, environmental regulations, or urban expansion—must be understood in advance.

UDT is constantly updated with real-time data from UB, ensuring that its simulations and predictions are always based on the most accurate and current information. Conversely, UDT's simulations and scenario-based forecasts can guide UB's real-time decision-making by identifying strategies that will yield long-term sustainability, resilience, and resource efficiency. This bi-directional feedback loop between UB and UDT allows for a closed-loop system where short-term operational actions are informed by long-term strategic goals, and vice versa.

SUM adds another layer to the CPSoS by managing the flow of critical resources through the city—energy, water, materials, and waste. Like its biological counterpart, it entails the processes by which a city consumes, transforms, and expels resources. It continuously monitors these resource flows, using data from UB to adjust consumption patterns in real time while providing valuable insights to UDT for long-term planning. For instance, SUM tracks energy consumption across different domains of the city. By analyzing these data, it identifies opportunities for improving energy efficiency, reducing wastage, and transitioning to renewable energy sources. In a similar manner, SUM helps manage water distribution, waste collection, and recycling processes. This constant monitoring and optimization of resource flows are crucial for promoting a circular economy (see Chapter 8 for a detailed account and discussion), where waste is minimized, and resources are reused, leading to more sustainable urban systems.

In the context of CPSoS, SUM plays a critical role by aligning the real-time operational management of resources (handled by UB) with long-term sustainability strategies (simulated by UDT). It ensures that resources are used efficiently in the present and managed with future needs in mind.

The integration of UB, UDT, and SUM is further enhanced by AIoT, which acts as a bridge between the physical infrastructure and its digital counterparts. AIoT connects the vast array of IoT sensors spread across the urban environment to UB, UDT, and SUM, allowing for seamless data collection, analysis, and real-time application.

AIoT ensures that UB receives constant streams of data from physical systems. These data are then processed by AI algorithms to produce actionable insights, which UB uses to adjust urban operations. Simultaneously, AIoT feeds these real-time data into UDT to update its virtual models, ensuring that future simulations and predictions are grounded in current conditions.

The role of AIoT within this CPSoS framework is essential for maintaining a real-time feedback loop between the physical and digital layers of the city. It enables UB to make informed decisions based on both immediate data and long-term forecasts from UDT while optimizing resource flows managed by SUM.

UB, UDT, and SUM are interconnected, forming a system of systems where each component contributes to the overall functioning of the city. This integration allows for more sophisticated urban management and planning, where the real-time monitoring of physical systems (by UB) is continually informed by predictive simulations (from UDT), while resource flows (managed by SUM) are optimized for both immediate efficiency and long-term sustainability.

This CPSoS framework allows sustainable smart cities to operate as adaptive, resilient ecosystems capable of responding to both short-term challenges and long-term pressures. The integration of real-time operational management, strategic planning, and resource flow optimization creates a cohesive system that is responsive, sustainable, and capable of continuous improvement.

In implementing the integration of UB, UDT, and SUM, a standardized approach is paramount to ensuring consistent data formats and protocols across these components. This standardization facilitates seamless information exchange, forming a foundation for effective collaboration. The integration process incorporates advanced AI and ML/DL algorithms within each component, continuously enhancing decision accuracy, predictions, and optimizations. A crucial aspect of this integration is the establishment of a feedback loop. This loop plays a fundamental role, as insights gleaned from each component are systematically fed back into the overall system. This dynamic process refines the learning and performance of the entire framework, creating a symbiotic relationship where the collective intelligence of each component influences and augments the capabilities of the others. These mechanisms improve decision-making, optimize resource use, and contribute to cohesive urban management and planning. The feedback loop's iterative nature fosters adaptive and responsive systems in this regard, aligning with the evolving landscape of sustainable smart cities.

UB, UDT, and SUM, as CPSoS, form a comprehensive framework for responsive and sustainable urban management and planning. This CPSoS allows sustainable smart cities to balance real-time operational needs with long-term strategic goals, creating urban environments that are more efficient, resilient, and sustainable. Through AIoT, these systems communicate and coordinate seamlessly, ensuring that the city's physical and digital layers operate in harmony to advance environmentally sustainable development goals.

5.5 DISCUSSION

This study explored the transformative potential of AIoT-powered systems in reshaping sustainable urban development practices and developed an innovative framework for advancing data-driven management and planning through the synergistic and collaborative integration of UB, UDT, and SUM. AIoT involves deploying interconnected IoT devices and systems with AI capabilities across these data-driven systems, each exhibiting functional variations yet offering complementary strengths. It holds significant promise for creating more resilient, efficient, and environmentally conscious smart cities by leveraging real-time data, predictive analytics, and intelligent automation and decision-making. However, there are challenges, barriers, and limitations that need to be overcome to realize the full potential of the proposed framework.

5.5.1 Summary of Findings and Interpretation of Results

This chapter has illuminated the transformative potential of integrating UB, UDT, and SUM within the framework of AIoSSCT. By methodically combining a thorough literature review with in-depth case studies, it has provided a comprehensive understanding of how these AIoT-powered systems function individually and synergistically. The findings underscore that while UB, UDT, and SUM each significantly contribute to urban sustainability, their combined integration within the AIoT-driven CPSoS framework creates a more robust and dynamic system capable of addressing the multifaceted challenges of sustainable urban development.

The literature review revealed that UB excels in data analytics and predictive modeling, UDT provides a virtual replica for real-time urban planning, and SUM focuses on optimizing resource flows and environmental impacts. Case studies demonstrated that cities employing these systems individually achieved marked improvements in specific areas, particularly in environmental sustainability. However, the integration of these systems facilitated a more holistic approach, enabling sustainable smart cities to leverage comprehensive data insights for proactive and adaptive urban management and planning.

5.5.2 Comparative Analysis

When comparing the standalone applications of UB (Bibri et al. 2024a; Kierans et al. 2023; Xu 2022; Xu et al. 2024; Liu et al. 2022), UDT (Agostinelli et al. 2021; Almusaed and Yitmen 2023; Bibri et al. 2024b; Mishra et al. 2023; Shen et al. 2021), and SUM (Bibri et al. 2024a, b; Peponi et al. 2022; Ghosh and Sengupta 2023) to their integrated implementation within the proposed AIoT framework, several key advantages emerge. The standalone systems, while advanced and effective in their respective domains, often operate in silos, which limits their ability to fully capitalize on their synergistic potential, collaborative integration, as well as the interconnected nature of urban environments. For instance, UB's predictive analytics are significantly enhanced when fed real-time data from UDT, and SUM's efficiency models become more precise with insights from both UB and UDT. Similarly, UDT's virtual simulations and scenario planning are enriched by integrating the vast datasets from UB, allowing for more accurate and dynamic modeling of urban processes. This enables more effective anticipation of urban challenges, proactive planning, and informed infrastructure management.

The integrated AIoT framework leverages these synergistic interactions, amplifying the capabilities of each system to foster a more cohesive and responsive urban management approach. This integration allows for real-time monitoring, resource management, and data-driven decision-making, thus facilitating scenario planning, land-use management, infrastructure development, and catering to community needs in a more effective manner.

The novelty of this study lies in its comprehensive analysis of the synergistic potential unlocked by integrating UB, UDT, and SUM within an AIoT framework. This integration not only enhances the individual functionalities of these systems but also creates new opportunities for advanced urban management and planning that were previously unattainable. For example, the comparative analysis of case studies from various cities demonstrated that those adopting the integrated AIoT framework exhibit superior performance in sustainability metrics, such as reduced carbon emissions, improved energy distribution, effective mobility, and enhanced public services, compared to cities using these data-driven systems in isolation.

The integrated approach maximizes resource efficiency and enhances urban resilience, demonstrating the transformative potential of AIoT-driven management and planning systems. Moreover, the collective intelligence generated by the integration of UB, UDT, and SUM can drive innovation in urban planning, leading to sustainable development practices that are more adaptable to future challenges.

Overall, the study establishes that the synergistic and collaborative integration and application of UB, UDT, and SUM within the AIoT framework is crucial for advancing sustainable urban development. It offers a compelling vision for future cities, where data-driven decision-making, real-time virtual simulations, and optimized resource management converge to create smarter, more sustainable urban environments. The findings underscore the importance of moving beyond isolated applications to embrace an integrated, holistic approach to urban management and planning that leverages the full potential of interconnected data systems.

5.5.3 Benefits and Opportunities

The AIoSSCT framework unfolds a multitude of benefits and opportunities spanning operational, environmental, and social dimensions. Operationally, the collaborative integration of UB, UDT, and SUM elevates the efficiency of urban management, fostering informed decision-making, optimizing resource allocation, and enhancing overall governance through collaborative and adaptive data-driven approaches. The predictive capabilities embedded in UDT, combined with the analytical prowess of UB, create pathways for proactive planning and effective mitigation strategies for potential environmental risks. These encompassing challenges, uncertainties, or adverse events that may impact urban development, can be identified and addressed through the collaborative and anticipatory approach enabled by the integration of UB, UDT, and SUM. Environmentally, the incorporation of SUM ensures sustainable resource utilization, harnessing insights from UB and UDT to actively contribute to the reduction of greenhouse gases by advancing energy management systems, predicting renewable energy demands, and enhancing the efficiency of energy and waste management systems. The collaborative integration of UB, UDT, and SUM establishes a holistic approach to environmental sustainability. The real-time data and analytics provided by UB and UDT empower decision-makers to optimize resource flows, monitor environmental parameters, and implement measures that align with environmental goals. This collaborative synergy supports a resilient urban ecosystem by fostering a balance between urban development and environmental preservation.

On a social scale, the framework promotes citizen engagement through real-time data accessibility, promoting a sense of community empowerment and active participation in urban decision-making processes. UDT, serving as a 3D representation of the city's dynamics, enhances public awareness of environmental conditions and infrastructure developments (Weil et al. 2022). This awareness, coupled with community-based participatory approaches, can empower residents to actively contribute their perspectives, needs, and concerns, fostering a sense of ownership in the decision-making processes related to urban planning and development (e.g., Leclercq and Rijshouwer 2022; Pansera et al. 2023; Raes et al. 2021). Additionally, SUM contributes to citizen engagement by providing insights into resource flows and environmental impacts (D'Amico et al. 2021; Shahrokni 2015), further encouraging informed decision-making and collaboration between residents and local authorities. Moreover, UB plays a key role in orchestrating these components by serving as the central hub for integrating and processing information from UDT, SUM, and other urban systems. Through its advanced analytics and decision-making capabilities, UB enhances overall citizen engagement and stakeholder collaboration by providing meaningful insights and facilitating informed participation in urban governance processes (Caprotti and Liu 2022; Kierans et al. 2023; Liu et al. 2022). It acts as a key enabler for creating a responsive and intelligent urban environment that aligns with the needs and preferences of the community.

From another perspective, the AIoSSCT framework serves as a transformative catalyst, expanding the conventional scope of AIoT from a focus on individual districts or city systems to the integration of diverse city systems. This integration spans not only urban management and planning but also extends to governance, impacting larger entities like regions. While conventional AIoT concentrates on the interconnection of devices within a single application domain, AIoSSCT broadens this perspective by incorporating the collective intelligence of UB, UDT, and SUM. This extension allows for a more comprehensive and scalable approach, deploying this collective intelligence not only at the device or system level but also at the district, city, and regional levels. The framework promotes a holistic view of AIoT, emphasizing collaborative decision-making on a macroscale for intelligent and sustainable urban development at citywide and regional levels.

5.5.4 Implications for Research, Practice, and Policymaking

In the domain of research, the AIoSSCT framework presents a compelling call for interdisciplinary collaboration among scholars, transcending traditional boundaries between urban management, urban planning, computer science, data science, environmental science, and social sciences, among others. Research endeavors can explore the intersectionality of these domains, fostering a thorough understanding of the framework's implications on sustainable urban development. Scholars can engage in enhancing and fine-tuning computational algorithms, addressing environmental sustainability concerns, and exploring ethical considerations in the deployment of integrated AIoT-driven solutions. Additionally, they can focus on refining the integration dynamics, developing standardized protocols for data governance, and assessing the long-term socio-environmental impacts. The framework's potential to expand existing AIoT frameworks across various spatial scales—districts, cities, and regions—opens avenues for research on scalable and adaptive urban technologies.

In the realm of practical application, the AIoSSCT framework assumes a transformative role in optimizing the deployment, operation, and functioning of urban systems. Capitalizing on the intelligence embedded in UB, UDT, and SUM, the framework unlocks novel opportunities for urban practitioners. It effectively tackles inherent constraints that often impede integrated urban development approaches, including fragmented decision-making, duplicated efforts and resources, data silos, stifled innovation, increased operational costs, decreased scalability, and slower responsiveness. In practical terms, the framework streamlines urban operations, augments resource efficiency, and expedites responsiveness, leading to substantial benefits for cities. Urban planners can also leverage the framework to bolster infrastructure planning, improve environmental management, and enhance citizen services.

At the policymaking level, the AIoSSCT framework advocates for the promotion and adoption of integrated solutions to optimize efficiency, enhance sustainability, and improve inclusivity in alignment with Sustainable Development Goals (SDGs). Policymakers can play a significant role in establishing frameworks that foster collaboration and data sharing among diverse urban sectors. Policies should prioritize the development of standardized protocols, ethical guidelines, and frameworks that support the responsible deployment of AIoT technologies in urban contexts. By aligning policy objectives with the principles embedded in the AIoSSCT framework, policymakers can significantly contribute to the realization of more sustainable, resilient, and inclusive cities.

5.5.5 Challenges, Barriers, and Limitations

While the AIoSSCT framework holds immense promise, it is confronted with a range of complex challenges, spanning computational complexities, environmental concerns, ethical considerations, social implications, regulatory hurdles, and financial constraints. These challenges are distilled from the analyzed and synthesized studies, revealing the multifaceted nature of these considerations not only in AIoT (e.g., Alahi et al. 2023; Bibri et al. 2023a, b; Koffka 2023; Parihar et al. 2023); but also in the implementation of UB (e.g., Caprotti and Liu 2022; Cugurellu 2020; Kierans et al. 2023), UDT (e.g., Bibri et al. 2024b; Charitonidou 2022; Lei et al. 2022; Raes et al. 2021; Weil et al. 2023), and SUM (Bibri et al. 2024a; D'Amico et al. 2020, 2022; Ghosh and Sengupta 2023; Shahrokni 2015).

Concerning computational complexities, a notable challenge within the AIoSSCT framework lies in the heterogeneity of data sources and platforms, introducing complexities in data integration and interoperability. The diverse array of data formats, structures, and sources, ranging from IoT devices to governmental databases, demands the deployment of sophisticated computational algorithms to ensure seamless collaboration among UB, UDT, and SUM. Additionally, the imperative for real-time data processing and analysis imposes a computational strain, necessitating advanced algorithms capable of handling the influx of data streams while ensuring accuracy and responsiveness. Moreover, challenges related to data governance and standardization must be acknowledged. The diversity in data sources, formats, and structures may hinder the establishment of uniform data governance practices, potentially leading to inconsistencies and interoperability issues. Furthermore, the framework's reliance on real-time data assumes a continuous and reliable data stream, introducing vulnerabilities related to data interruptions, inaccuracies, or cyber threats. These challenges may compromise the robustness of the framework, impacting its effectiveness in dynamic urban environments.

Environmental sustainability risks must be carefully navigated within the AIoSSCT framework. The intensive computational processes of AIoT associated with current urban management and planning applications, coupled with the deployment of UB, UDT, and SUM technologies and infrastructures, inadvertently contribute to e-waste, intensive energy and water consumption, and increased GHG emissions. Balancing the benefits of data-driven decision-making with the environmental costs of maintaining computational infrastructure becomes a crucial consideration. Mitigating these risks involves the development of energy-efficient algorithms, sustainable data storage solutions, and strategies for responsible end-of-life disposal of electronic components to minimize the environmental footprint of the AIoSSCT framework.

The integration of UB, UDT, and SUM introduces a spectrum of ethical, social, and regulatory challenges that demand careful consideration. Ethically, privacy and security concerns within the AIoSSCT framework necessitate robust safeguards to protect sensitive data. The vast amounts of data generated, ranging from personal information to critical infrastructure details, require encryption, secure data transmission protocols, and stringent access controls. Establishing frameworks for data anonymization and implementing privacy-by-design principles become imperative to address ethical concerns and comply with data protection regulations. Moreover, the potential for cyber threats and attacks poses a significant risk, necessitating continuous monitoring, threat detection mechanisms, and proactive cybersecurity measures to safeguard the integrity and confidentiality of urban data (e.g., Paracha et al. 2024; Rosenberg et al. 2021; Goldblum et al. 2022).

The social implications of the AIoSSCT framework extend beyond technological considerations, addressing crucial aspects such as transparency, accountability, biased decision-making, fairness, social inequality, and trust. These issues are intricately linked to AI explainability, responsible AI practices, and effective AI governance. Ensuring the equitable distribution of benefits across diverse communities is a paramount concern, emphasizing inclusivity in the framework's implementation. From a regulatory standpoint, challenges arise in developing robust frameworks that govern the ethical use of AI in urban contexts. This includes addressing potential biases inherent in algorithms and establishing clear guidelines for transparent decision-making processes. The regulatory framework should be designed to mitigate biases that might arise from historical data, ensuring that the AIoSSCT system operates fairly and without perpetuating existing social inequalities. Striking a delicate balance between fostering innovation and upholding ethical responsibility is imperative for building and maintaining trust among citizens and stakeholders. The transparency of decision-making processes, the accountability of AI systems, and their explainability contribute to fostering trust and confidence in the framework. Responsible AI practices, which involve continuous monitoring and assessment of the AIoSSCT's impact on society, are crucial to minimize negative consequences and maximize positive outcomes. Implementing effective AI governance mechanisms ensures that the framework aligns with ethical standards and societal values, promoting a socially responsible and sustainable urban development trajectory.

The adoption of the innovative AIoSSCT framework is contingent on substantial financial investments, posing challenges related to accessibility and equity across urban landscapes. The initial implementation costs, including infrastructure development, sensor deployment, and AI system integration, may create disparities in the availability of advanced technologies among different cities and regions. Addressing these challenges requires strategic financial planning, public-private partnerships, and policies that prioritize equitable access to the benefits of the framework. Additionally, ensuring that the financial burden does not disproportionately impact economically disadvantaged communities is crucial for fostering inclusivity and minimizing urban socioeconomic divides.

It is important to acknowledge certain limitations of the AIoSSCT framework. One of these limitations lies in its generalizability and adaptability to diverse urban contexts. The effectiveness of the framework may vary based on the physical, infrastructural, socioeconomic, and socio-cultural disparities among cities. Also, the reliance on advanced technologies assumes a certain level of technological maturity in urban areas, potentially excluding less developed districts, cities, and regions from reaping its benefits for sustainable urban development.

5.5.6 Suggestions for Future Research Directions

Future research endeavors should aim to enhance the adaptability of the AIoSSCT framework to diverse urban landscapes and varying resource availabilities. A crucial aspect involves refining the framework through the development of customizable modules or frameworks. These customizable solutions can be tailored to specific socioeconomic and socio-cultural contexts, accommodating cities and regions with different levels of technological maturity. This tailored approach ensures alignment with the unique characteristics and needs of each urban setting.

Exploring the socio-cultural implications of widespread technology adoption, particularly in terms of citizen engagement and acceptance, remains a pivotal research avenue. Longitudinal studies are essential to assess the evolving adoption and impact of the AIoSSCT framework on urban resilience, sustainability, and overall livability. These studies should extend their focus to examine how the framework shapes the evolution of urban governance structures, influencing decision-making processes, policy design, and civic participation over time.

In-depth investigations into societal aspects of technology adoption will be instrumental for understanding citizen perceptions, concerns, and acceptance of the AIoSSCT framework. This research domain contributes valuable insights for designing user-friendly interfaces, addressing privacy concerns, and promoting inclusive engagement. Additionally, innovative funding models and policy frameworks must be explored to support the widespread adoption of the framework in diverse urban environments. These efforts are crucial for ensuring equitable access to and benefits from this transformative technology.

Furthermore, future research should tackle data governance challenges within the AIoSSCT framework. This involves investigating standardized protocols, metadata frameworks, and governance models that facilitate seamless data interoperability. The reliability and consistency of integrated urban data associated with UB, UDT, and SUM are essential considerations. Additionally, researchers can explore the development of adaptive technologies that mitigate the impact of intermittent data streams, enhancing the framework's resilience in the face of disruptions.

5.6 CONCLUSION

In an era marked by rapid urbanization and increasing environmental degradation, developing more effective and efficient data-driven planning and management systems is crucial. The integration of AIoT with UB, UDT, and SUM as CPSoS offers new opportunities in this regard. However, there is currently a lack of research on how such integration could occur. This chapter aimed to bridge this critical knowledge gap by investigating the transformative potential of AIoT in redefining sustainable urban development practices. An innovative data-driven management and planning framework—named AIoSSCT—was introduced, synergistically integrating UB, UDT, and SUM within the framework of CPSoS. The study objectives were meticulously addressed and successfully achieved, providing significant insights and contributions.

The study unveiled the nuanced relationship between sustainable smart cities and the transformative impact of AIoT, laying the groundwork for a framework that synergizes their data-driven management and planning systems. The analysis and synthesis of the existing body of knowledge regarding the functionalities and architectures of UB, UDT, and SUM in relation to AIoT provided a solid foundation for the development of the proposed framework. Examining the architectural layers of AIoT based on the data science cycle and its connection with CPS has illuminated the interplay between AI and IoT technologies and the physical aspects of urban environments, contributing to the evolution of AIoT and data-driven management and planning systems. Sparking a discourse on the substantive impacts of AIoT on the future trajectory of sustainable urban development and the multifaceted aspects related to the proposed framework, the discussion provided a comprehensive perspective on the potentials and complexities involved in integrating UB, UDT, and SUM through AIoT. This entailed various aspects, ranging from opportunities, benefits, and implications to challenges, barriers, and limitations, in addition to suggestions for research directions.

The framework has the potential to evolve into a responsible, sustainable, and inclusive approach to urban development, harnessing computational technology advancements and innovative platform solutions. It serves to bridge the gap between UB, UDT, and SUM and hence align their functions, providing a holistic approach to urban management and planning in the context of environmental sustainability. Specifically, integrating UB, UDT, and SUM with AIoT and CPS can leverage the strengths of each of these data-driven systems, creating synergies that enhance the effectiveness and efficiency of urban management and planning practices. These systems are all centered around data-driven decision-making. By integrating them, sustainable smart cities can harness the power of data to gain deeper insights into urban systems, identify trends, and make informed decisions for sustainable development. Together, these systems provide a comprehensive understanding of urban dynamics, including technological, environmental, and social factors. Moreover, this innovative integration can enable cities to optimize resource allocation, infrastructure management, and service delivery, leading to more resilient and sustainable urban systems capable of adapting to changing conditions and challenges. Additionally, the framework fosters innovation by encouraging collaboration and knowledge exchange among stakeholders. It also enhances operational efficiency by streamlining processes and reducing redundancy. Therefore, concentrated effort in developing an innovative data-driven management and planning framework is essential in the rapidly evolving landscape of sustainable smart cities.

The innovative AIoSSCT framework, tailored for data-driven management and planning, presents a promising model for sustainable urban development. By orchestrating the collaborative and synergistic integration of UB, UDT, and SUM, the framework harnesses their collective intelligence through AIoT, marking a significant advancement and deepening understanding in guiding future research endeavors in the rapidly evolving landscape of sustainable smart cities. In essence, the study not only enriches the academic discourse on AIoT-driven sustainable urban development but also establishes the groundwork for practical applications that can shape the future of cities. As the exploration unfolds, collaboration among scholars, practitioners, and policymakers becomes crucial in unlocking the full potential of the framework and ushering in a new era of urban landscapes.

REFERENCES

Agostinelli, S., Cumo, F., Guidi, G., & Tomazzoli, C. (2021). Cyber-physical systems improving building energy management: Digital twin and artificial intelligence. *Energies*, 14(8), 2338.

Ahad, M. A., Paiva, S., Tripathi, G., & Feroz, N. (2020). Enabling technologies and sustainable smart cities. *Sustainable Cities and Society*, 61, 102301.

Ahvenniemi, H., Huovila, A., Pinto-Seppä, I., Airaksinen, M., & Valvontakonsultit Oy, R. (2017). What are the differences between sustainable and smart cities? *Cities*, 60, 234–245. https://doi.org/10.1016/j.cities.2016.09.009

Alahakoon, D., Nawaratne, R., Xu, Y., De Silva, D., Sivarajah, U., & Gupta, B. (2023). Self-building artificial intelligence and machine learning to empower big data analytics in smart cities. *Information Systems Frontiers*, 25(1), 221–240. https://doi.org/10.1007/s10796-020-10056-x.

Alahi, M. E. E., Sukkuea, A., Tina, F. W., Nag, A., Kurdthongmee, W., Suwannarat, K., & Mukhopadhyay, S. C. (2023). Integration of IoT-enabled technologies and artificial intelligence (AI) for smart city scenario: Recent advancements and future trends. *Sensors*, 23(11), 5206.

Al-Dabbagh, R. H. (2022). Dubai, the sustainable, smart city. *Renewable Energy and Environmental Sustainability*, 7(3), 12.

Alhijaj, J. A., & Khudeyer, R. S. (2023). Techniques and applications for deep learning: A review. *Journal of Al-Qadisiyah for Computer Science and Mathematics*, 15(2), 114–126. https://doi.org/10.29304/jqcm.2023.15.2.1236

Almalaq, A., & Zhang, J. J. (2018). Evolutionary deep learning-based energy consumption prediction for buildings. *IEEE Access*, 7, 1520–1531.

Almalki, F. A., Alsamhi, S. H., Sahal, R., Hassan, J., Hawbani, A., Rajput, N. S., Saif, A., Morgan, J., & Breslin, J. (2021). Green IoT for eco-friendly and sustainable smart cities: Future directions and opportunities. *Mobile Networks and Applications*. https://doi.org/10.1007/s11036-021-01790-w.

Almusaed, A., & Yitmen, I. (2023). Architectural reply for smart building design concepts based on artificial intelligence simulation models and digital twins. *Sustainability*, 15(6), 4955.

Alom, M. Z., Taha, T. M., Yakopcic, C., Westberg, S., Sidike, P., Nasrin, M. S., Hasan, M., Van Essen, B. C., Awwal, A. A., & Asari, V. K. (2019). A state-of-the-art survey on deep learning theory and architectures. *Electronics, 8*(3), 292.

Alowaidi, M., Sharma, S. K., AlEnizi, A., & Bhardwaj, S. (2023). Integrating artificial intelligence in cyber security for cyber-physical systems. *Electronic Research Archive, 31*(4), 1876–1896. https://doi.org/10.3934/era.2023097

Alzubaidi, L., Zhang, J., Humaidi, A. J., Al-Dujaili, A., Duan, Y., Al-Shamma, O., Santamaría, J., Fadhel, M. A., Al-Amidie, M., & Farhan, L. (2021). Review of deep learning: Concepts, CNN architectures, challenges, applications, future directions. *Journal of Big Data, 8*(1). https://doi.org/10.1186/s40537-021-00444-8

Austin, M., Delgoshaei, P., Coelho, M., & Heidarinejad, M. (2020). Architecting smart city digital twins: Combined semantic model and machine learning approach. *Journal of Management in Engineering, 36*(4), 04020026.

Baccini, P., & Brunner, P. H. (2012). *Metabolism of the Anthroposphere: Analysis, Evaluation, Design* (2nd ed.). MIT Press.

Baker, S., & Xiang, W. (2023). Artificial intelligence of things for smarter healthcare: A survey of advancements, challenges, and opportunities. *IEEE Communications Surveys & Tutorials.* https://doi.org/10.1109/COMST.2023.3256323

Beckett, S. (2022). Smart city digital twins, 3d modeling and visualization tools, and spatial cognition algorithms in artificial intelligence-based urban design and planning. *Geopolitics, History, and International Relations, 14*(1), 123–138. https://www.jstor.org/stable/48679657

Bibri, S. E. (2019). Smart sustainable urbanism: Paradigmatic, scientific, scholarly, epistemic, and discursive shifts in light of big data science and analytics. In: *Big Data Science and Analytics for Smart Sustainable Urbanism. Advances in Science, Technology & Innovation.* Springer. https://doi.org/10.1007/978-3-030-17312-8_6

Bibri, S. E. (2020). Data-driven environmental solutions for smart sustainable cities: Strategies and pathways for energy efficiency and pollution reduction. *Euro-Mediterranean Journal for Environmental Integration, 5*(66). https://doi.org/10.1007/s41207-020-00211-w

Bibri, S. E. (2021a). The underlying components of data-driven smart sustainable cities of the future: A case study approach to an applied theoretical framework. *European Journal of Futures Research, 9*(13). https://doi.org/10.1186/s40309-021-00182-3

Bibri, S. E. (2021b). Data-driven smart sustainable cities of the future: Urban computing and intelligence for strategic, short-term, and joined-up planning. *Computational Urban Science, 1*, 1–29.

Bibri, S. E., & Krogstie, J. (2020a). Data-driven smart sustainable cities of the future: A novel model of urbanism and its core dimensions, strategies, and solutions. *Journal of Futures Studies, 25*(2), 77–94. https://doi.org/10.6531/JFS.202012_25(2).0009

Bibri, S. E., & Krogstie, J. (2020). Environmentally data-driven smart sustainable cities: Applied innovative solutions for energy efficiency, pollution reduction, and urban metabolism. *Energy Informatics*, 3, 29. https://doi.org/10.1186/s42162-020-00130-8

Bibri, S. E., Alexandre, A., Sharifi, A., & Krogstie, J. (2023a). Environmentally sustainable smart cities and their converging AI, IoT, and big data technologies and solutions: An integrated approach to an extensive literature review. *Energy Informatics*, 6, 9.

Bibri, S. E., Huang, J., & Krogstie, J. (2024a). Artificial intelligence of things for synergizing smarter eco-city brain, metabolism, and platform: Pioneering data-driven environmental governance, *Sustainable Cities and Society*, 105516. https://doi.org/10.1016/j.scs.2024.105516

Bibri, S. E., Huang, J., Jagatheesaperumal, S. K., & Krogstie, J. (2024b). The synergistic interplay of artificial intelligence and digital twin in environmentally planning sustainable smart cities: A comprehensive systematic review. *Environmental Science and Ecotechnology*, 20, 100433. https://doi.org/10.1016/j.ese.2024.100433

Bibri, S. E., & Huang, J. (2025). Artificial intelligence of sustainable smart city brain and digital twin: A pioneering framework for advancing environmental sustainability. *Environmental Technology and Innovation* (in press).

Bibri, S., Krogstie, J., Kaboli, A., & Alahi, A. (2023b). Smarter eco-cities and their leading-edge artificial intelligence of things solutions for environmental sustainability: A comprehensive systemic review. *Environmental Science and Ecotechnlogy*, 19, https://doi.org/10.1016/j.ese.2023.100330

Botín-Sanabria, D. M., Mihaita, A.-S., Peimbert-García, R. E., Ramírez-Moreno, M. A., Ramírez-Mendoza, R. A., & Lozoya-Santos, J. D. J. (2022). Digital twin technology challenges and applications: A comprehensive review. *Remote Sensing*, 14(6), 1335. https://doi.org/10.3390/rs14061335

Caprari, G., Castelli, G., Montuori, M., Camardelli, M., & Malvezzi, R. (2022). Digital twin for urban planning in the green deal era: A state of the art and future perspectives. *Sustainability*, 14(10), 6263.

Caprotti, F., & Liu, D. (2022). Platform urbanism and the Chinese smart city: The co-production and territorialisation of Hangzhou City Brain. *GeoJournal*, 87, 1559–1573. https://doi.org/10.1007/s10708-020-10320-2

Chang, Z., Liu, S., Xiong, X., Cai, Z., & Tu, G. (2021). A survey of recent advances in edge-computing-powered artificial intelligence of things. *IEEE Internet Things Journal*, 8(18), 13849–13875. https://doi.org/10.1109/JIOT.2021.3088875

Charitonidou, M. (2022). Urban scale digital twins in data-driven society: Challenging digital universalism in urban planning decision-making. *International Journal of Architectural Computing*, 20(2), 238–253.

Chen, L., Chen, Z., Zhang, Y., Liu, Y., Osman, A. I., Farghali, M., Hua, J., Al-Fatesh, A., Ihara, I., Rooney, D. W., & Yap, P.-S. (2023). Artificial intelligence-based solutions for climate change: A review. *Environmental Chemistry Letters*, 21, 2525–2557. https://doi.org/10.1007/s10311-023-01556-7

Cugurullo, F. (2020). Urban artificial intelligence: From automation to autonomy in the smart city. *Frontiers in Sustainable Cities, 2*, 38. https://doi.org/10.3389/frsc.2020.00038

D'Amico, G., Arbolino, R., Shi, L., Yigitcanlar, T., & Ioppolo, G. (2021). Digital technologies for urban metabolism efficiency: Lessons from urban agenda partnership on circular economy. *Sustainability, 13*(11), 6043. https://doi.org/10.3390/su13116043

D'Amico, G., Arbolino, R., Shi, L., Yigitcanlar, T., & Ioppolo, G. (2022). Digitalisation driven urban metabolism circularity: A review and analysis of circular city initiatives. *Land Use Policy, 112*, 105819. https://doi.org/10.1016/j.landusepol.2021.105819

D'Amico, G., Taddeo, R., Shi, L., Yigitcanlar, T., & Ioppolo, G. (2020). Ecological indicators of smart urban metabolism: A review of the literature on international standards. *Ecological Indicators, 118*, 106808.

del Mar Martínez-Bravo, M., & Labella-Fernández, A. (2024). Sustainable smart cities: A step beyond. In: F. Theofanidis, O. Abidi, A. Erturk, S. Colbran, & E. Coşkun (Eds.), *Digital Transformation and Sustainable Development in Cities and Organizations* (pp. 125–140). IGI Global https://doi.org/10.4018/979-8-3693-3567-3.ch006

Dembski, F., Wössner, U., Letzgus, M., Ruddat, M., & Yamu, C. (2020). Urban digital twins for smart cities and citizens: The case study of herrenberg, Germany. *Sustainability, 12*(6), 2307.

Dheeraj, A., Nigam, S., Begam, S., Naha, S., Jayachitra Devi, S., Chaurasia, H. S., Kumar, D., Ritika, Soam, S. K., Srinivasa Rao, N., Alka, A., Sreekanth Kumar, V. V., & Mukhopadhyay, S. C. (2020). Role of artificial intelligence (AI) and internet of things (IoT) in mitigating climate change. In: Ch. Srinivasa, R. T. Srinivas, R. V. S. Rao, N. Srinivasa Rao, S. Vinayagam, & P. Krishnan (Eds.), *Climate Change and Indian Agriculture: Challenges and Adaptation Strategies* (pp. 325–358). ICAR-National Academy of Agricultural Research Management.

Din, I. U., Guizani, M., Rodrigues, J. J. P. C., Hassan, S., & Korotaev, V. V. (2019). Machine learning in the internet of things: Designed techniques for smart cities. *Future Generation Computer Systems, 100*, 826–843.

Dong, S., Wang, P., & Abbas, K. (2021). A survey on deep learning and its applications. *Computer Science Review, 40*. https://doi.org/10.1016/j.cosrev.2021.100379

Efthymiou, I.-P., & Egleton, T. E. (2023). Artificial intelligence for sustainable smart cities. In: B. K. Mishra (Ed.), *Handbook of Research on Applications of AI, Digital Twin, and Internet of Things for Sustainable Development* (pp. 1–11). IGI Global.

El Himer, S., Ouaissa, M., & Boulouard, Z. (2022). Artificial intelligence of things (AIoT) for renewable energies systems. In: S. El Himer, M. Ouaissa, A. A. A. Emhemed, M. Ouaissa, & Z. Boulouard (Eds.), *Artificial Intelligence of Things for Smart Green Energy Management (Studies in Systems, Decision and Control* (vol. 446, pp. 1–19). Springer. https://doi.org/10.1007/978-3-031-04851-7_1

Evans, J., Karvonen, A., Luque-Ayala, A., Martin, C., McCormick, K., Raven, R., & Palgan, Y. V. (2019). Smart and sustainable cities? Pipedreams, practicalities and possibilities. *Local Environment, 24*(7), 557–564. https://doi.org/10.1080/13549839.2019.1624701

Fang, B., Yu, J., Chen, Z., Osman, A. I., Farghali, M., Ihara, I., Hamza, E. H., Rooney, D. W., Yap, P.-S. (2023). Artificial intelligence for waste management in smart cities: A review. *Environmental Chemistry Letters, 21*(6), 1959–1989. https://doi.org/10.1007/s10311-023-01604-3

Ferré-Bigorra, J., Casals, M., & Gangolells, M. (2022). The adoption of urban digital twins. *Cities, 131*, 10390.

Fragkos, G., Tsiropoulou, E. E., & Papavassiliou, S. (2020). Artificial intelligence enabled distributed edge computing for Internet of Things applications. In: Proceedings of the 2020 16th International Conference on Distributed Computing in Sensor Systems (DCOSS) (pp. 450–457). IEEE.

Fuller, A., Fan, Z., Day, C., & Barlow, C. (2020). Digital twin: Enabling technologies: Challenges and open research. *IEEE Access, 8*, 108952–108971. https://doi.org/10.1109/ACCESS.2020.2998358

Ghosh, R., & Sengupta, D. (2023). Smart urban metabolism: A big-data and machine learning perspective. In: R. Bhadouria, S. Tripathi, P. Singh, P. K. Joshi, & R. Singh (Eds.), *Urban Metabolism and Climate Change*. Springer. https://doi.org/10.1007/978-3-031-29422-8_16

Gkontzis, A. F., Kotsiantis, S., Feretzakis, G., & Verykios, V. S. (2024). Enhancing urban resilience: Smart city data analyses, forecasts, and digital twin techniques at the neighborhood level. *Future Internet, 16*(2), 47. https://doi.org/10.3390/fi16020047

Goldblum, M., Tsipras, D., Xie, C., Chen, X., Schwarzschild, A., Song, D., Madry, A., Li, B., & Goldstein, T. (2022). Dataset security for machine learning: Data poisoning, backdoor attacks, and defenses. *IEEE Transactions on Pattern Analysis and Machine Intelligence, 45*(2), 1563–1580.

Gough, D., Oliver, S., & Thomas, J. (2017). *An Introduction to Systematic Reviews*. SAGE

Gourisaria, M. K., Jee, G., Harshvardhan, G., Konar, D., & Singh, P. K. (2022). Artificially intelligent and sustainable smart cities. In: P. K. Singh, M. Paprzycki, M. Essaaidi, & S. Rahimi (Eds.), *Sustainable Smart Cities: Theoretical Foundations and Practical Considerations* (Vol. 942, pp. 277–294). Springer, Cham. https://doi.org/10.1007/978-3-031-08815-5_14.

Gürkaş Aydın, Z., & Kazanç, M. (2023). Using artificial intelligence in the security of cyber physical systems. *Alphanumeric Journal, 11*(2), 193-206. https://doi.org/10.17093/alphanumeric.1404181

Hämäläinen, M. (2021). Urban development with dynamic digital twins in Helsinki City. *IET Smart Cities, 3*, 201–210. https://doi.org/10.1049/smc2.12015

Hoang, T. V. (2024). Impact of integrated artificial intelligence and internet of things technologies on smart city transformation. *Journal of Technical Education Science, 19*(Special Issue 1), 64–73. https://doi.org/10.54644/jte.2024.1532

Huang, J. Bibri, S. E., & Keel, P. (2025), Generative spatial artificial intelligence for sustainable smart cities: A pioneering large flow foundation model for urban digital twin. *Environmental Science and Ecotechnology* (in press).

Huntingford, C., Jeffers, E. S., Bonsall, M. B., Christensen, H. M., Lees, T., & Yang, H. (2019). Machine learning and artificial intelligence to aid climate change research and preparedness. *Environmental Research Letters, 14*(12), 124007.

Ishengoma, F. R., Shao, D., Alexopoulos, C., Saxena, S., & Nikiforova, A. (2022). Integration of artificial intelligence of things (AIoT) in the public sector: Drivers, barriers and future research agenda. *Digital Policy, Regulation and Governance, 24*(5), 449–462.

Jagatheesaperumal, S. K., Bibri, S. E., Ganesan, S., & Jeyaraman, P. (2023). Artificial Intelligence for road quality assessment in smart cities: A machine learning approach to acoustic data analysis. *Computers in Urban Science, 3*, 28. https://doi.org/10.1007/s43762-023-00104-y

Jagatheesaperumal, S. K., Bibri, S. E., Huang, J Ganesan, S., & Jeyaraman, P. (2024). Artificial intelligence of things for smart cities: Advanced solutions for enhancing transportation safety. *Computational Urban Science, 4*(10). https://doi.org/10.1007/s43762-024-00120-6

Joshi, K., Kumar, V., Anandaram, H., Kumar, R., Gupta, A., & Krishna, K. H. (2023). A review approach on deep learning algorithms in computer vision. In: N. Mittal, A. K. Pandit, M. Abouhawwash, & S, Mahajan (Eds.), *Intelligent Systems and Applications in Computer Vision* (1st ed., p. 15). CRC Press.

Juma, M., & Shaalan, K. (2020). Cyberphysical systems in the smart city: Challenges and future trends for strategic research. In: A. E. Hassanien & A. Darwish (Eds.), *Intelligent Data-Centric Systems: Swarm Intelligence for Resource Management in Internet of Things* (pp. 65–85). Academic Press. https://doi.org/10.1016/B978-0-12-818287-1.00008-5

Kamal, H., Yánez, W., Hassan, S., & Sobhy, D. (2024). Digital-twin-based deep reinforcement learning approach for adaptive traffic signal control. *IEEE Internet of Things Journal, 99*, 1–1. https://doi.org/10.1109/JIOT.2024.3377600

Kamrowska-Załuska, D. (2021). Impact of AI-based tools and urban big data analytics on the design and planning of cities. *Land, 10*, 1209.

Kennedy, C., & Hoornweg, D. (2012). Mainstreaming urban metabolism. *Journal of Industrial Ecology, 16*, 780–782. https://doi.org/10.1111/j.1530-9290.2012.00548.x.

Kennedy, C., Cuddihy, J., & Engel-Yan, J. (2007). The changing metabolism of cities. *Journal of Industrial Ecology, 11*(2), 43–59.

Kennedy, C., Pincetl, S., & Bunje, P. (2011). The study of urban metabolism and its applications to urban planning and design. *Environmental Pollution, 159*(8–9), 1965–1973.

Khan, F., Kumar, R. L., Kadry, S., Nam, Y., & Meqdad, M. N. (2021). Cyber physical systems: A smart city perspective. *International Journal of Electrical and Computer Engineering (IJECE), 11*(4), 3609–3616. https://doi.org/10.11591/ijece.v11i4.pp3609-3616.

Kierans, G., Jüngling, S., & Schütz, D. (2023). Society 5.0 2023. *EPiC Series in Computing, 93*, 82–96.

Kim, J.-Y., & Cho, S.-B. (2019). Electric energy consumption prediction by deep learning with state explainable autoencoder. *Energies, 12*(4), 739.

Koffka, K. (2023). *Intelligent Things: Exploring AIoT Technologies and Applications.* Kinder Edition.

Komninos, N., & Kakderi, C. (Eds.). (2019). *Smart Cities in the Post-algorithmic Era: Integrating Technologies, Platforms and Governance.* Edward Elgar

Koumetio, T. S. C., Diop, E. B., Azmi, R. , & Chenal, J. (2023). Artificial intelligence based methods for smart and sustainable urban planning: A systematic survey. *Archives of Computational Methods in Engineering, 30*(5), 1421–1438. https://doi.org/10.1007/s11831-022-09844-2

Leal Filho, W., Wall, T., Mucova, S. A. R., Nagy, G. J., Balogun, A.-L., Luetz, J. M., Ng, A. W., Kovaleva, M., Azam, F. M. S., & Alves, F. (2022). Deploying artificial intelligence for climate change adaptation. *Technological Forecasting and Social Change, 180*, 121662.

Leclercq, E. M., & Rijshouwer, E. A. (2022). Enabling citizens' right to the smart city through the co-creation of digital platforms. *Urban Transformations, 4*(1), 2.

Lei, B., Janssen, P., Stoter, J., & Biljecki, F. (2023). Challenges of urban digital twins: A systematic review and a Delphi expert survey. *Automation in Construction, 147*, 104716. https://doi.org/10.1016/j.autcon.2022.104716

Liu, F., Liu, F. Y., & Shi, Y. (2018). City brain, a new architecture of smart city based on the internet brain. In: IEEE 22nd International Conference on Computer Supported Cooperative Work in Design, Nanjing (pp. 9–11). https://doi.org/10.1109/CSCWD.2018.8465164

Liu, F., Ying, L., & Yunqin, Z. (2021). Discussion on the definition and construction principles of city brain. In: 2021 IEEE 2nd International Conference on Big Data, Artificial Intelligence and Internet of Things Engineering (ICBAIE), Nanchang, China. https://doi.org/10.1109/ICBAIE52039.2021.9390064

Liu, W., Cui, P., Nurminen, J. K., & Wang, J. (2017). Special issue on intelligent urban computing with big data. *Machine Vision and Applications, 28*, 675–677. https://doi.org/10.1007/s00138-017-0877-8

Liu, W., Mei, Y., Ma, Y., Wang, W., Hu, F., & Xu, D. (2022). City brain: A new model of urban governance. In: M. Li, G. Bohács, A. Huang, D. Chang, & X. Shang (Eds.), *IEIS 2021*. Lecture Notes in Operations Research. Springer. https://doi.org/10.1007/978-981-16-8660-3_12

Lv, Z., Li, Y., Feng, H., & Lv, H. (2021). Deep learning for security in digital twins of cooperative intelligent transportation systems. *IEEE Transactions on Intelligent Transportation Systems, 23*(9), 16666–16675.

Major, P., da Silva Torres, R., Amundsen, A., Stadsnes, P., & Tennfjord Mikalsen, E. (2022). On the use of graphical digital twins for urban planning of mobility projects: A case study from a new district in Ålesund, Norway. In: Proceedings of the 36th ECMS International Conference on Modelling and Simulation ECMS 2022. https://doi.org/10.7148/2022-0236

Marasinghe, R., Yigitcanlar, T., Mayere, S., Washington, T., & Limb, M. (2024). Computer vision applications for urban planning: A systematic review of opportunities and constraints. *Sustainable Cities and Society, 100*, 105047. https://doi.org/10.1016/j.scs.2023.105047

Masserov, D. A., & Masserov, D. D. (2022). Applying artificial intelligence to the internet of things. *Russian Journal of Resources, Conservation and Recycling, 9*(2). https://resources.today/PDF/05ITOR222.pdf (in Russian), https://doi.org/10.15862/05ITOR222

Mathew, A., Amudha, P., & Sivakumari, S. (2021). Deep learning techniques: An overview. In: *Advances in Intelligent Systems and Computing* (pp. 599–608). https://doi.org/10.1007/978-981-15-3383-9_54

Mishra, B. K. (Ed.). (2023). *Handbook of Research on Applications of AI, Digital Twin, and Internet of Things for Sustainable Development*. IGI Global. https://doi.org/10.4018/978-1-6684-6821-0

Mishra, P., & Singh, G. (2023). Artificial Intelligence for Sustainable Smart Cities. In: *Sustainable Smart Cities*. Springer. https://doi.org/10.1007/978-3-031-33354-5_6

Mishra, R. K., Kumari, C. L., Chachra, S., Krishna, P. S. J., Dubey, A., & Singh, R. B. (Eds.). (2022). *Smart Cities for Sustainable Development: Advances in Geographical and Environmental Sciences*. Springer.

Mohamed, K. S. (2023). Deep learning for IoT "artificial intelligence of things (AIoT)". In: *Deep Learning-Powered Technologies. Synthesis Lectures on Engineering, Science, and Technology*. Springer. https://doi.org/10.1007/978-3-031-35737-4_3

Mosavi, A., Salimi, M., Faizollahzadeh Ardabili, S., Rabczuk, T., Shamshirband, S., & Varkonyi-Koczy, A. R. (2019). State of the art of machine learning models in energy systems: Systematic review. *Energies, 12*(7), 1301. https://www.mdpi.com/1996-1073/12/7/1301

Mylonas, G., Kalogeras, A., Kalogeras, G., Anagnostopoulos, C., Alexakos, C., & Muñoz, L. (2021). Digital twins from smart manufacturing to smart cities: A survey. *IEEE ACCESS, 9*, 143222–143249.

Nica, E., Popescu, G. H., Poliak, M., Kliestik, T., & Sabie, O.-M. (2023). Digital twin simulation tools, spatial cognition algorithms, and multi-sensor fusion technology in sustainable urban governance networks. *Mathematics, 11*(9), 1981. https://doi.org/10.3390/math11091981

Nishant, R., Kennedy, M., & Corbett, J. (2020). Artificial intelligence for sustainability: Challenges, opportunities, and a research agenda. *International Journal of Information Management, 53*, 102104.

Nti, E. K., Cobbina, S. J., Attafuah, E. E., Opoku, E., & Gyan, M. A. (2022). Environmental sustainability technologies in biodiversity, energy, transportation and water management using artificial intelligence: A systematic review. *Sustainable Futures, 4*, 100068. https://doi.org/10.1016/j.sftr.2022.100068

Page, M. J., McKenzie, J. E., Bossuyt, P. M., Boutron, I., Hoffmann, T. C., Mulrow, C. D., Shamseer, L., Tetzlaff, J. M., Akl, E. A., Brennan, S. E., Chou, R., Glanville, J., Grimshaw, J. M., Hróbjartsson, A., Lalu, M. M., Li, T., Loder, E. W., Mayo-Wilson, E., McDonald, S., McGuinness, L. A., Stewart, L. A., Thomas, J., Tricco, A. C., Welch, V. A., Whiting, P., & Moher, D. (2021). The prisma 2020 statement: An updated guideline for reporting systematic reviews. *International Journal of Surgery, 88*, 105906. https://doi.org/10.1016/j.ijsu.2021.105906.

Pansera, M., Marsh, A., Owen, R., Flores López, J. A., & De Alba Ulloa, J. L. (2023). Exploring citizen participation in smart city development in Mexico City: An institutional logics approach. *Organization Studies*, 44(10), 1679– 1701.

Paracha, A., Arshad, J., Ben Farah, M., & Ismail, K. (2024). Machine learning security and privacy: A review of threats and countermeasures. *EURASIP Journal on Information Security*, 2024(10). https://doi.org/10.1186/s13635-024-00158-3

Parihar, V., Malik, A., Bhawna, Bhushan, B., & Chaganti, R. (2023). From smart devices to smarter systems: The evolution of artificial intelligence of things (AIoT) with characteristics, architecture, use cases and challenges. In: B. Bhushan, A. K. Sangaiah, & T. N. Nguyen (Eds.), *AI Models for Blockchain-Based Intelligent Networks in IoT Systems (Engineering Cyber-Physical Systems and Critical Infrastructures* (vol. 6). Springer. https://doi.org/10.1007/978-3-031-31952-5_1

Peponi, A., Morgado, P., & Kumble, P. (2022). Life cycle thinking and machine learning for urban metabolism assessment and prediction. *Sustainable Cities and Society*, 80, 103754. https://doi.org/10.1016/j.scs.2022.103754

Prajna, M. S., & Okkie, P. (2023). Deployment and use of Artificial Intelligence (AI) in water resources and water management. *IOP Conference Series: Earth and Environmental Science*, 1195, 012056. https://doi.org/10.1088/1755-1315/1195/1/012056

Radanliev, P., Roure, D. C. D., Van Kleek, M., Santos, O., & Ani, U. D. (2020). Artificial intelligence in cyber physical systems. *SSRN*. https://ssrn.com/abstract=3692592, https://doi.org/10.2139/ssrn.3692592

Raes, L., Michiels, P., Adolphi, T., Tampere, C., Dalianis, A., McAleer, S., & Kogut, P. (2021). Duet: A framework for building interoperable and trusted digital twins of smart cities. *IEEE Internet Computing*, 26(3), 43–50.

Rajawat, A. S., Bedi, P., Goyal, S. B., Shaw, R. N., & Ghosh, A. (2022). Reliability analysis in cyber-physical system using deep learning for smart cities industrial IoT network node. In: V. Piuri, R. N. Shaw, A. Ghosh, & R. Islam (Eds.), *AI and IoT for Smart City Applications (Studies in Computational Intelligence* (vol. 1002). Springer. https://doi.org/10.1007/978-981-16-7498-3_10

Rane, N., Choudhary, S., & Rane, J. (2024). Artificial Intelligence and machine learning in renewable and sustainable energy strategies: A critical review and future perspectives. *SSRN*. https://ssrn.com/abstract=4838761, https://doi.org/10.2139/ssrn.4838761

Rosenberg, I., Shabtai, A., Elovici, Y., & Rokach, L. (2021). Adversarial machine learning attacks and defense methods in the cyber security domain. *ACM Computing Surveys*, 54(5), 1–36.

Salunke, A. A. (2023). Reinforcement learning empowered digital twins: Pioneering smart cities towards optimal urban dynamics. *EPRA International Journal of Research & Development (IJRD)*. https://doi.org/10.36713/epra13959

Samadi, S. (2022). The convergence of AI, IoT, and big data for advancing flood analytics research. *Frontiers in Water*, 4, 786040. https://doi.org/10.3389/frwa.2022.786040

Sanchez, T. W. (2023). Planning on the verge of AI, or AI on the verge of planning. *Urban Science*, 7, 70. https://doi.org/10.3390/urbansci7030070

Sanchez, T. W., Shumway, H., Gordner, T., & Lim, T. (2022). The prospects of artificial intelligence in urban planning. *International Journal of Urban Science*, 27(2), 179–194.

Saravanan, K., & Sakthinathan, G. (2021). *Handbook of Green Engineering Technologies for Sustainable Smart Cities*. CRC Press.

Schrotter, G., & Hürzeler, C. (2020). The digital twin of the city of Zurich for urban planning. *PFG–Journal of Photogrammetry, Remote Sensing and Geoinformation Science*, 88(1), 99–112.

Seng, K. P., Ang, L. M., & Ngharamike, E. (2022). Artificial intelligence Internet of Things: A new paradigm of distributed sensor networks. *International Journal of Distributed Sensor Networks*. https://doi.org/10.1177/15501477211062835

Shaamala, A., Yigitcanlar, T., Nili, A., & Nyandega, D. (2024). Algorithmic green infrastructure optimisation: Review of artificial intelligence driven approaches for tackling climate change. *Sustainable Cities and Society*, 101, 105182. https://doi.org/10.1016/j.scs.2024.105182

Shahat, E., Hyun, C. T., & Yeom, C. (2021). City digital twin potentials: A review and research agenda. *Sustainability*, 13(6), 3386. https://doi.org/10.3390/su13063386

Shahrokni, H., Årman, L., Lazarevic, D., Nilsson, A., & Brandt, N. (2015). Implementing smart urban metabolism in the Stockholm Royal Seaport: Smart city SRS. *Journal of Industrial Ecology*, 19(5), 917–929.

Sharifi, A., Allam, Z., Bibri, S. E., & Khavarian-Garmsir, A. R. (2024). Smart cities and sustainable development goals (SDGs): A systematic literature review of co-benefits and trade-offs. *Cities*, 146, 104659. https://doi.org/10.1016/j.cities.2023.104659

Sharma, R., & Sharma, N. (2022). Applications of artificial intelligence in cyber-physical systems. *Cyber Physical Systems*, 1–14. https://doi.org/10.1201/9781003202752-1

Shen, J., Saini, P. K., & Zhang, X. (2021). Machine learning and artificial intelligence for digital twin to accelerate sustainability in positive energy districts. In: X. Zhang (Ed.), *Data-driven Analytics for Sustainable Buildings and Cities: From Theory to Application* (pp. 411–422).

Shi, F., Ning, H., Huangfu, W., Zhang, F., Wei, D., Hong, T., & Daneshmand, M. (2020). Recent progress on the convergence of the internet of things and artificial intelligence. *IEEE Network*, 34(5), 8–15.

Shruti, S., Singh, P. K., Ohri, A., & Singh, R. S. (2022). Development of environmental decision support system for sustainable smart cities in India. *Environmental Progress & Sustainable Energy*, 41(5), e13817.

Singh, K. D., Singh, P., Chhabra, R., Kaur, G., Bansal, A., & Tripathi, V. (2023a). Cyber-physical systems for smart city applications: A comparative study. In: 2023 International Conference on Advancement in Computation & Computer Technologies (InCACCT) (pp. 871–876). Gharuan, India. https://doi.org/10.1109/InCACCT57535.2023.10141719

Singh, T., Solanki, A., Sharma, S. K., & Hachimi, H. (2023b). Smart sensors and actuators for Internet of Everything based smart cities: Application, challenges, opportunities, and future trends. *Intelligent Systems for IoE Based Smart Cities, 1*, 61. https://doi.org/10.2174/9789815124965123010006

Somma, A., De Benedictis, A., Zappatore, M., Martella, C., Martella, A., & Longo, A. (2023, December). Digital twin space: The integration of digital twins and data spaces. In: 2023 IEEE International Conference on Big Data (BigData) (pp. 4017–4025). IEEE.

Son, T. H., Weedon, Z., Yigitcanlar, T., Sanchez, T., Corchado, J. M., & Mehmood, R. (2023). Algorithmic urban planning for smart and sustainable development: Systematic review of the literature. *Sustainable Cities and Society, 94*, 104562.

Song, M., Tan, K. H., Wang, J., & Shen, Z. (2022). Modeling and evaluating economic and ecological operation efficiency of smart city pilots. *Cities, 124*, 103575.

Szpilko, D., Naharro, F. J., Lăzăroiu, G., Nica, E., & Gallegos, A. D. L. T. (2023). Artificial intelligence in the smart city: A literature review. *Engineering Management in Production and Services, 15*(4), 53–75. https://doi.org/10.2478/emj-2023-0028

Thamik, H., Cabrera, J. D. F., & Wu, J. (2024). The digital paradigm: Unraveling the impact of artificial intelligence and internet of things on achieving sustainable development goals. In: S. Misra, K. Siakas, & G. Lampropoulos (Eds.), *Artificial Intelligence of Things for Achieving Sustainable Development Goals*. Lecture Notes on Data Engineering and Communications Technologies (vol. 192). Springer. https://doi.org/10.1007/978-3-031-53433-1_2

Toli, A. M., & Murtagh, N. (2020). The concept of sustainability in smart city definitions. *Frontiers in Built Environment, 6*, 77. https://doi.org/10.3389/fbuil.2020.00077

Tomazzoli, C., Scannapieco, S., & Cristani, M. (2020). Internet of Things and artificial intelligence enable energy efficiency. *Journal of Ambient Intelligence and Humanized Computing, 14*, 4933–4954. https://doi.org/10.1007/s12652-020-02151-3

Tripathi, S. L., Ganguli, S., Kumar, A., & Magradze, T. (2022). *Intelligent Green Technologies for Sustainable Smart Cities*. John Wiley & Sons.

Ullah, Z., Al-Turjman, F., Mostarda, L., & Gagliardi, R. (2020). Applications of artificial intelligence and machine learning in smart cities. *Computer Communications, 154*, pp.313–323.

Visvizi, A., & del Hoyo, R. P. (2021). *Smart Cities and the Un Sdgs*. Elsevier.

Wang, L., Chen, X., Xia, Y., Jiang, L., Ye, J., Hou, T., Wang, L., Zhang, Y., Li, M., Li, Z. et al. (2022). Operational data-driven intelligent modelling and visualization system for real-world, on-road vehicle emissions: A case study in Hangzhou city, China. *Sustainability, 14*(9), 5434. https://doi.org/10.3390/su14095434.

Wang, Y., Su, Z., Guo, S., Dai, M., Luan, T. H., & Liu, Y. (2023). A survey on digital twins: Architecture, enabling technologies, security and privacy, and future prospects. *IEEE Internet of Things Journal*, *10*(17), 14965–14987. https://doi.org/10.1109/JIOT.2023.3263909

Webb, R., Bai, X., Smith, M. S., Costanza, R., Griggs, D., Moglia, M., Neuman, M., Newman, P., Newton, P., Norman, B., & Ryan, C. (2018). Sustainable urban systems: Co-design and framing for transformation. *Ambio*, *47*, 57–77.

Weil, C., Bibri, S. E., Longchamp, R., Golay, F., & Alahi, A. (2023). Urban digital twin challenges: A systematic review and perspectives for sustainable smart cities. *Sustainable Cities and Society*, *99*, 104862.

Wiedmann, N. S., Athanassiadis, A., & Binder, C. R. (2021). Data mining in the context of urban metabolism: A case study of Geneva and Lausanne, Switzerland. *Journal of Physics: Conference Series*, *2042*, 012020. https://doi.org/10.1088/1742-6596/2042/1/012020

Willcock, S., Martínez-López, J., Hooftman, D. A. P., Bagstad, K. J., Balbi, S., Marzo, A., ... & Athanasiadis, I. N. (2018). Machine learning for ecosystem services. *Ecosystem Services*, *33*(Part B), 165–174. https://doi.org/10.1016/j.ecoser.2018.08.001.

Wu, J., Wang, X., Dang, Y., & Lv, Z. (2022). Digital twins and artificial intelligence in transportation infrastructure: Classification, application, and future research directions. *Computers and Electrical Engineering*, *101*, 107983. https://doi.org/10.1016/j.compeleceng.2022.107983

Xia, H., Liu, Z., Efremochkina, M., Liu, X., & Lin, C. (2022). Study on city digital twin technologies for sustainable smart city design: A review and bibliometric analysis of geographic information system and building information modeling integration. *Sustainable Cities and Society*, *84*, 104009. https://doi.org/10.1016/j.scs.2022.104009

Xie, J. (2020). Exploration and reference of urban brain construction in Xiacheng District of Hangzhou. *Communications World*, *31*, 26–27.

Xu, R. (2022). Deeply understand, analyze and draw lessons from Hangzhou's urban brain construction paradigm. *BCP Business & Management*, *29*, 441–446. https://doi.org/10.54691/bcpbm.v29i.2310

Xu, Y., Cugurullo, F., Zhang, H., Gaio, A., & Zhang, W. (2024). The emergence of artificial intelligence in anticipatory urban governance: Multi-scalar evidence of China's transition to city brains. *Journal of Urban Technology*, 1–25. https://doi.org/10.1080/10630732.2023.2292823

Yang, L., Chen, X., Perlaza, S. M., Junshan Zhang (2020). Special issue on artificial-intelligence-powered edge computing for Internet of Things. *IEEE IoT Journal*, *7*(10), 9224–9226.

Yang, S., Xu, K., Cui, L., Ming, Z., Chen, Z., & Ming, Z. (2021). EBI-PAI: Toward an efficient edge-based IoT platform for artificial intelligence. *IEEE Internet of Things Journal*, *8*(12), 9580–9593. https://doi.org/10.1109/JIOT.2020.3019008.

Ye, X., Du, J., Han, Y., Newman, G., Retchless, D., Zou, L., Ham, Y., & Cai, Z. (2023). Developing human-centered urban digital twins for community infrastructure resilience: A research agenda. *Journal of Planning Literature*, *38*(2), 187–199. https://doi.org/10.1177/08854122221118765.

Yigitcanlar, T., Desouza, K. C., Butler, L., & Roozkhosh, F. (2020). Contributions and risks of artificial intelligence (AI) in building smarter cities: Insights from a systematic review of the literature. *Energies*. https://doi.org/10.3390/en13061473

Yigitcanlar, T., Mehmood, R., & Corchado, J. M. (2021). Green artificial intelligence: Toward an efficient. *Sustainable and Equitable Technology for Smart Cities and Futures*, *13*(16), 8952. https://www.mdpi.com/2071-1050/13/16/8952

Zaidi, A., Ajibade, S.-S. M., Musa, M., & Bekun, F. V. (2023). New insights into the research landscape on the application of artificial intelligence in sustainable smart cities: A bibliometric mapping and network analysis approach. *International Journal of Energy Economics and Policy*, (4), 287.

Zayed, S. M., Attiya, G. M., El-Sayed, A., & Hemdan, E. E.-D. (2023). A review study on digital twins with artificial intelligence and internet of things: Concepts, opportunities, challenges, tools and future scope. *Multimedia Tools and Applications, 82*, 47081–47107. https://doi.org/10.1007/s11042-023-15611-7.

Zhang, J., & Tao, D. (2021). Empowering things with intelligence: A survey of the progress, challenges, and opportunities in artificial intelligence of things. *IEEE Internet of Things Journal*, *8*(10), 7789–7817. https://doi.org/10.1109/JIOT.2020.3039359

Zhang, J., Hua, X. S., Huang, J., Shen, X., Chen, J., Zhou, Q., Fu, Z., & Zhao, Y. (2019). City brain: Practice of large-scale artificial intelligence in the real world. *IET Smart Cities*, *1*, 28–37.

Zvarikova, K., Horak, J., & Downs, S. (2022). Digital twin algorithms, smart city technologies, and 3d spatio-temporal simulations in virtual urban environments. *Geopolitics, History and International Relations*, *14*(1), 139–154.

CHAPTER 6

Artificial Intelligence of Things for Harmonizing Smarter Eco-City Brain, Metabolism, and Platform

An Innovative Framework for Data-Driven Environmental Governance

Abstract

Emerging smarter eco-cities, inherently intertwined with environmental governance, function as experimental sites for testing novel technological solutions and implementing environmental reforms aimed at addressing complex challenges. However, despite significant progress in understanding the distinct roles of emerging data-driven governance systems—namely City Brain, Smart Urban Metabolism (SUM), and platform urbanism—enabled by

DOI: 10.1201/9781003536420-6

Artificial Intelligence of Things (AIoT), a critical gap persists in systematically exploring the untapped potential stemming from their synergistic and collaborative integration in the context of environmental governance. To fill this gap, this chapter aims to explore the linchpin potential of AIoT in seamlessly integrating these data-driven governance systems to advance environmental governance in smarter eco-cities. Specifically, it introduces a pioneering framework that effectively leverages the synergies among these AIoT-powered governance systems to enhance environmental sustainability practices in smarter eco-cities. In developing the framework, the study employs configurative and aggregative synthesis approaches through an extensive literature review and in-depth case study analysis of publications spanning from 2018 to 2023. The study identifies key factors driving the co-evolution of AI and IoT into AIoT and specifies technical components constituting the architecture of AIoT in smarter eco-cities. A comparative analysis reveals commonalities and differences among City Brain, SUM, and platform urbanism within the frameworks of AIoT and environmental governance. These data-driven systems collectively contribute to environmental governance in smarter eco-cities by leveraging real-time data analytics, predictive modeling, and stakeholder engagement. The proposed framework underscores the importance of data-driven decision-making, optimization of resource management, reduction of environmental impact, collaboration among stakeholders, engagement of citizens, and formulation of evidence-based policies. The findings unveil that the synergistic and collaborative integration of City Brain, SUM, and platform urbanism through AIoT presents promising opportunities and prospects for advancing environmental governance in smarter eco-cities. The framework not only charts a strategic trajectory for stimulating research endeavors but also holds significant potential for practical application and informed policymaking in the realm of environmental urban governance. However, ongoing critical discussions and refinements remain imperative to address the identified challenges, ensuring the framework's robustness, ethical soundness, and applicability across diverse urban contexts.

6.1 INTRODUCTION

In the face of escalating ecological degradation and rapid urbanization, there is an urgent imperative to explore innovative solutions for environmental governance in urban environments. This imperative entails embracing advanced technologies and fostering collaborative models in urban governance to effectively tackle environmental challenges. These challenges encompass resource depletion, energy consumption, transport inefficiency, traffic congestion, infrastructure flaws, waste generation, pollution, biodiversity loss, and climate change. Traditional approaches to environmental governance often fall short in effectively managing these complex issues (de Wit 2020; Green and Hadden 2021; Underdal 2010; Van Assche et al. 2020). In response, the groundbreaking convergence of Artificial Intelligence (AI) and the Internet of Things (IoT)—under the umbrella of AIoT—has emerged as a new frontier for innovative solutions in modern cities. Among the prominent AIoT-powered solutions of smart cities are City Brain (Kierans et al. 2023; Liu et al. 2022; Xu 2024), Smart Urban Metabolism (SUM) (e.g., Bibri et al. 2024a; Ghosh and Sengupta 2023; Peponi et al. 2022), and platform urbanism (Caprotti et al. 2022; Haveri and Anttiroiko 2023; Repette et al. 2021). These AIoT-driven governance systems offer synergistic opportunities to advance environmental governance in emerging smarter eco-cities. They collectively facilitate real-time monitoring, resource management, data-driven decision-making, and collaborative and participatory approaches to address environmental challenges and promote sustainability.

Platformization facilitates the implementation of City Brain and SUM, as manifestations of platform urbanism, by providing the digital infrastructure and data integration needed for real-time monitoring, analysis, and optimization of urban systems. It is the process and practice by which various technological, economic, social, and cultural activities are increasingly mediated by digital platforms (Poell et al. 2019). These platforms act as intermediaries that facilitate interactions, transactions, and data exchange between various stakeholders. Platform urbanism extends platformization to the urban context. It describes how digital platforms influence and shape urban life and development. This includes the way cities are planned, governed, and experienced through the integration of digital technologies in transportation, mobility, energy, public services, and other urban systems. Platformization is the broader phenomenon of digital platforms mediating activities, while platform urbanism and hence

City Brain and SUM focus specifically on their impact on urban space and living, including environmental governance and sustainability.

Environmental governance involves coordinating policies, institutions, and actions while actively engaging multiple stakeholders, especially political actors, to shape environmental actions and outcomes (de Wit 2020; Vatn 2016). This multifaceted approach encompasses diverse strategies and initiatives aimed at optimizing resource management, promoting ecological resilience, mitigating environmental impacts, and enhancing quality of life. At the core of smart eco-cities, which prioritize environmental sustainability, is environmental governance (Bibri 2021a; Kramers et al. 2016). These cities distinguish themselves by integrating data-driven IoT and environmental technologies (Bibri 2020; Bibri and Krogstie 2020a; Caprotti et al. 2017; Späth et al. 2017; Tarek 2023). However, the evolution toward smarter eco-cities signifies a leap forward, utilizing AI and AIoT technologies to optimize sustainable systems, integrate them with smart systems, and synergize their functions to advance environmental sustainability goals (Bibri et al. 2023a). This progression aims to create intelligent ecosystems that not only prioritize environmental considerations but also leverage cutting-edge technologies to achieve optimal outcomes.

In the realm of smarter eco-cities, effective environmental governance by local governments hinges on harnessing AI and AIoT technologies to optimize sustainability initiatives and engage stakeholders in shaping environmentally conscious urban policies and actions. The burgeoning adoption of AI in local governments in smart cities (Mishra and Singh 2023; Yigitcanlar et al. 2021, 2022) is transforming various domains and addressing complex environmental challenges in smarter eco-city management and planning (Bibri et al. 2023a). In recent years, there has been a noticeable increase in the deployment of AI in smart cities and city governance (Cugurullo et al. 2024). This trend underscores a concerted effort to leverage AI technologies in local government settings to bolster efficiency, effectiveness, and innovation across diverse domains (Brand 2022; Yigitcanlar et al. 2023a, 2024a, b). The impact of smart city movement on local governments, coupled with the increasing adoption of AI and AIoT, has introduced new applications capable of executing tasks with remarkable precision and potentially paving the way for significant societal shifts (Ishengoma et al. 2022; van Noordt and Misuraca 2022).

However, AI adoption in local governments, aligned with the smart city agenda, still primarily focuses on optimizing efficiency and improving urban services (Yigitcanlar et al. 2024a, b). The main priority continues to be on resource allocation and the overall quality of life for citizens, with environmental sustainability remaining largely underemphasized in the context of AI adoption. Government agencies worldwide are particularly leveraging AI technologies to streamline operations, enhance citizen services, and make data-driven decisions (Gracias et al. 2023; Yigitcanlar et al. 2024a, b). Nevertheless, the development of various AI technologies continues to evolve to address specific challenges (Benbya et al. 2020; Jan et al. 2023). This underscores the dynamic nature of AI's role in shaping the future of societal and environmental governance alike.

The relatively limited focus on environmental sustainability, compounded by its complex and multidimensional nature, within the realm of AI and AIoT adoption by local governments necessitates the exploration of the multifaceted and integrated capabilities of City Brain, SUM, and platform urbanism. These innovative data-driven governance systems can collectively address these concerns in local government initiatives, particularly in relation to emerging smarter eco-cities. As AI and AIoT technologies continue to advance and proliferate across various domains (Bibri et al. 2023a, b; Bibri 2024b; Bibri and Huang 2024; Hoang 2024; Ishengoma et al. 2022; Jagatheesaperumal et al. 2023, 2024; Kuguoglu et al. 2021; Habbal et al. 2024), they are becoming increasingly critical for local governments navigating complex technological landscapes to fulfill their objectives (Madan and Ashok 2022; Habbal et al. 2024) and address their societal needs. This growing reliance paves the way for shaping the future of environmental governance in smarter eco-cities. It goes beyond individual AI- and AIoT-driven governance and planning systems, marking the advent of an era characterized by the growing interest in and adoption of integrated solutions.

The premise underlying the collaborative integration of City Brain, Smart Urban Metabolism (SUM), and platform urbanism to advance environmental governance in smarter eco-cities is grounded in the untapped potential inherent in the synergistic capabilities of AIoT. Numerous "fast, smart, green, and safe AIoT applications are expected to deeply reshape our world" (Zhang and Tao 2021). AI and AIoT represent a paradigm shift in urban computing, marking the advent of an era marked by innovation, efficiency, and sustainability in urban governance and planning (e.g., Ben and Peng 2020; Bibri et al. 2024b; Bibri and Huang 2024; Liu et al. 2022; Koumetio et al. 2023; Marasinghe et al. 2024; Sanchez 2023; Samuel et al. 2023; Son et al. 2022; Xu et al. 2024).

AI and AIoT underpin and drive the functioning of City Brain, SUM, and platform urbanism, enabling their dynamic convergence to establish a holistic approach to urban governance in the pursuit of environmental sustainability. The symbiotic relationship among these AI- or AIoT-driven governance systems establishes a dynamic and interconnected framework, propelling the advancement of environmental governance. This integrated approach underscores a forward-thinking vision, contributing to the evolution of smarter and more sustainable urban environments. It aligns well with the environmental reforms and technological innovations being introduced in smart eco-cities as sites of experimentation (e.g., Caprotti et al. 2017; Späth et al. 2017; Tan Mullins et al. 2017). This alignment stems from the symbiotic nature and shared objective of City Brain, SUM, and platform urbanism in terms of harnessing data-driven approaches to enhance urban efficiency, sustainability, resilience, and stakeholder engagement. Their integration is propelled by the imperative to protect ecosystems, conserve natural resources, reduce pollution, and combat climate change, serving as key strategies for environmental governance.

The central focus of this study lies in the limited attention given to integrated approaches to environmental governance in the context of AI and AIoT adoption by local governments, despite the maturation of emerging data-driven governance systems. This highlights the imperative to develop holistic approaches for advancing environmental governance in emerging smarter eco-cities. Indeed, there is a growing demand for collaborative knowledge development that embraces a "whole-system view" and facilitates transformative change across multiple scales—an approach that is rarely observed in practical urban contexts (Webb et al. 2018). This arises from the complexities of urban environments and the critical need for more effective solutions to attain the Sustainable Development Goals (SDGs). These constitute a form of governance by goals (de Wit 2020).

While advancements have been made in understanding the individual roles and functions of City Brain, SUM, and platform urbanism as emerging data-driven governance systems, coupled with the increasing adoption of AI and AIoT across various domains of local governments, a critical gap persists in the current landscape of research. This gap pertains to the unexplored potential of their synergistic and collaborative integration for advancing environmental urban governance. To the best of our knowledge, no study has comprehensively explored how these emerging AI- or AIoT-powered solutions can be effectively utilized and synergistically harnessed to drive environmental governance forward in the dynamic landscape of smarter eco-cities.

To address the identified gap, this chapter aims to explore the linchpin potential of AIoT in seamlessly integrating City Brain, SUM, and platform urbanism to advance environmental governance in smarter eco-cities. Specifically, it introduces a pioneering framework that effectively leverages the synergies among these emerging AIoT-driven governance systems to enhance environmental sustainability practices in smarter eco-cities. The study is guided by the following four research questions:

RQ1: What are the underlying factors driving the co-evolution of AI and IoT into AIoT, and what specific technical components constitute the architecture of AIoT in the context of smarter eco-cities?

RQ2: What are the commonalities and differences among City Brain, SUM, and platform urbanism within the frameworks of AIoT and environmental governance in smarter eco-cities?

RQ3: How do City Brain, SUM, and platform urbanism collectively contribute to data-driven environmental governance in the dynamic context of smarter eco-cities?

RQ4: How can an innovative framework be devised to effectively utilize AIoT for synergizing the functions of City Brain, SUM, and platform urbanism, thereby advancing environmental governance in emerging smarter eco-cities?

Additionally, the study provides a detailed discussion, covering an interpretation of results, a comparative analysis, implications, limitations, and suggestions for future research directions.

The motivation for this study arises from the pressing need to confront the increasing complexity of environmental challenges resulting from rapid urbanization and ecological degradation. This necessitates the adoption of innovative approaches to environmental governance. Especially, there is a growing imperative to leverage advanced technologies in local government initiatives to tackle environmental governance challenges in smarter eco-cities. These urban environments, given their symbiotic relationship with governance, inherently serve as experimental grounds for trialing innovative technological solutions and implementing institutional reforms aimed at effectively addressing these challenges.

By examining the transformative potential of AIoT and its integration with emerging data-driven governance paradigms, the study contributes to enriching environmental governance practices in smarter eco-cities. Through an exploration of the synergistic and collaborative integration of City Brain, SUM, and platform urbanism, it contributes to spurring the development of innovative integrated solutions for advancing environmental governance in smarter eco-cities. Accordingly, this chapter offers invaluable insights that can guide policymakers, urban planners, and stakeholders in their endeavors to build more sustainable, resilient, and technologically advanced urban environments.

The remainder of this chapter is structured as follows: Section 6.2 describes and establishes connections between the essential components of the proposed framework. Section 6.3 provides a survey of related work. Section 6.4 describes and outlines the applied methodology. Section 6.5 presents the results based on configurative and aggregative synthesis approaches. Section 6.6 provides a detailed discussion, covering an interpretation of results, a comparative analysis, implications, limitations, and suggestions for future research. This chapter ends, in Section 6.7, by providing a summary of findings and some final thoughts.

6.2 ESSENTIAL COMPONENTS OF THE PROPOSED INTEGRATED FRAMEWORK: INTERLINKAGES AND CONCEPTUAL DEFINITIONS

This section introduces and interlinks the essential components of the integrated framework proposed in this study. Figure 6.1 depicts the synergistic and collaborative integration of these components. At the core of this dynamic framework resides AIoT, a catalytic force that underpins the functions of City Brain, City Metabolism, and City Platform. These elements collectively drive the governance of smarter eco-cities toward greater environmental sustainability.

These components shape the development trajectory of smarter eco-cities. As illustrated in Figure 6.1 and will be exemplified and analytically underpinned, AIoT serves as the linchpin technology, unifying the identified models for urban computing and development. Its enabling influence is evident in seamlessly integrating the functionalities of City Brain, SUM, and platform urbanism to enhance environmental governance strategies and practices. While the specifics and nuances of its computational and analytical functionalities may vary across these models, its overarching goal remains consistent: leverage real-time data processing

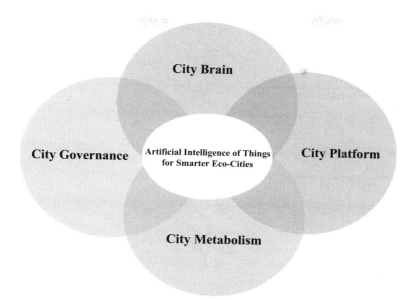

FIGURE 6.1 Essential components of the proposed integrated framework and their interlinkages.

and analysis to generate data-driven insights that inform decision-making in environmental governance. Its integrative potential forms the cornerstone for supporting City Brain's decision-making processes, optimizing SUM's resource management, and enabling the adaptability of platform urbanism. Essentially, AIoT acts as the connective tissue, orchestrating a harmonious blend of technological advancements and institutional transformations within the dynamic context of smarter eco-cities. This holistic convergence underscores the framework's capacity to propel urban development toward heightened intelligence, sustainability, and resilience.

AI and IoT have traditionally functioned as separate technological domains, with AI focusing on cognitive capabilities and IoT centered on connecting physical objects to the internet for data exchange and automation. AI refers to the simulation of human intelligence processes by machines, particularly computer systems. These processes include learning from experience, reasoning, problem-solving, language understanding, and autonomous decision-making. IoT is a network of interconnected physical objects or "things" embedded with sensors, actuators, and communication capabilities. These devices can sense and collect data from their environment, communicate with other devices or systems over the internet or local networks, and actuate or trigger actions based on predefined conditions or commands. IoT enables the seamless exchange of data between the physical and digital worlds, leading to improved efficiency, automation, and insights across various domains of smart, sustainable cities.

AIoT is the integration of AI models, notably Machine Learning (ML) and Deep Learning (DL), with IoT devices. AIoT enhances the capabilities of IoT by enabling devices and systems to analyze and interpret data more intelligently, make autonomous decisions, and adapt to changing conditions in real-time, particularly through ML and DL techniques (Mohamed 2023; Seng et al. 2022; Zhang and Tao 2021). By leveraging AI algorithms, IoT devices can extract meaningful insights from the vast amounts of data they generate, leading to improved efficiency, predictive maintenance, personalized experiences, and innovative applications across various domains and industries (Parihar et al. 2023). AIoT empowers IoT ecosystems to become smarter, more autonomous, and capable of addressing complex challenges and opportunities in the connected world of smart cities and smarter eco-cities (Bibri et al. 2023a). This synergy marks a pivotal point in the ongoing technological narrative, with profound implications for the future urban landscape.

City Brain, an advanced city management platform, utilizes AIoT to improve and streamline various facets of urban operations and services, while also providing valuable insights for governance and policy. By employing AI and IoT, City Brain facilitates real-time data collection, processing, and analysis, supporting informed decision-making for urban management and governance (Liu et al. 2022; Xu et al. 2024). Although City Brain and platform urbanism share essential functionalities of AIoT, they serve distinct purposes and exhibit unique characteristics (see, e.g., Caprotti and Liu 2022; Haveri and Anttiroiko 2023; Repette et al. 2021; Weissenrieder 2023; Xu et al. 2024). As regards SUM, it represents an AIoT-powered platform that aims to monitor, analyze, and optimize the flows of energy, materials, and resources in cities. It embraces a holistic approach to analyzing cities as complex systems characterized by inputs, processes, and outputs. By viewing cities through this systemic lens, it fosters resource efficiency and minimizes the ecological footprint of urban areas. Overall, AIoT, by integrating AI's capabilities with interconnected IoT devices and sensors across diverse urban systems, enables the extraction of valuable insights derived from AI. In doing so, it enhances decision-making processes and advances environmental governance practices in smarter eco-cities through City Brain, SUM, and platform urbanism.

Furthermore, the interconnected relationship between environmental governance and smarter eco-cities emphasizes the importance of their collaborative efforts in shaping sustainable urban futures through data-driven approaches enabled by AIoT. Together, they work hand in hand and synergize, leveraging each other's strengths and capabilities to create cities that are environmentally conscious, efficient, and resilient. This collaboration signifies a shared commitment to addressing the challenges of urbanization and ecological degradation while promoting sustainable development and improving the quality of life for city dwellers.

Urban governance is the system of values, practices, and institutions governing the planning, management, and regulation of urban areas. Hence, it involves a spectrum of decisions, policies, and actions, engaging diverse stakeholders such as government authorities, institutions, community organizations, and citizens. In this process, "conflicting or diverse interests may be accommodated, and cooperative action can be taken" (UN-Habitat 2023). However, the landscape of urban governance has undergone substantial transformation due to AI and IoT and their impactful convergence under the umbrella of AIoT (e.g., Ajuriaguerra Escudero and Abdiu 2022; Nishant et al. 2020; Samuel et al. 2023). As AI and AIoT continue to evolve, their integration within environmental governance frameworks promises to redefine urban sustainability paradigms and promote more inclusive, resilient, and responsive cities through the integration of City Brain, SUM, and ad platform urbanism.

6.3 A SURVEY OF RELATED WORK

In recent years, local governments worldwide have increasingly embraced AI as a transformative tool to address a myriad of complexities and challenges. This trend stems from a convergence of factors driving the adoption of AI technologies at the local level. Firstly, the pressing need for more efficient and effective public services has compelled local governments to seek innovative solutions. Yigitcanlar et al. (2024a) found that there has been a consistent upward trajectory in the adoption of AI by local governments, with public services as well as information management and transportation management being the leading domains. The authors highlight the growing significance of AI in local governance, with high potential to transform the delivery of public services, reshape decision-making processes, and redefine the interaction between governments and citizens. This assertion is consistent with the findings presented by Ishengoma et al. (2022), especially in their exploration of AIoT in relation to public services.

Secondly, the imperative to enhance citizen engagement, coupled with the need to optimize resource allocation and streamline administrative processes, has further fueled the adoption of AI across various municipal functions. Advancements in AI technology empower local authorities to harness the potential of data-driven decision-making, automation, prediction, and asset management (Sarker 2021; Yigitcanlar et al. 2023b, 2024a). The main AI adoption areas in local governments include planning, analytics, cybersecurity, surveillance, energy, modeling (Yigitcanlar et al. 2024b), policy formulation, infrastructure management (Yigitcanlar et al. 2023a), and administrative efficiency (Vogl et al. 2021).

However, there exists a noticeable disparity in the attention given to AI and AIoT deployment in smart cities and smarter eco-cities. While considerable focus has been directed toward utilizing AI and AIoT for advancing the management and planning practices of these cities (e.g., Bibri et al. 2023b; Bibri et al. 2024b; Koumetio et al. 2023; Marasinghe et al. 2024; Sanchez, 2023; Son et al. 2022), there is a relative neglect of their application in enhancing their governance practices. AI has shown significant potential to revolutionize environmental management and planning in smart cities and smarter eco-cities by tracking pollution sources, optimizing waste management, enhancing energy efficiency, conserving natural resources, aiding climate adaptation, and supporting emergency response. By leveraging AI across these fronts, local governments enact proactive measures, reduce environmental impact, and promote sustainability, safeguarding communities and fostering resilience for future generations (Buyya et al. 2018; Cugurullo et al. 2024). Nonetheless, while efforts are being made to enhance efficiency, functionality, and productivity, there is a notable void in addressing environmental governance-related challenges within the technological frameworks of AI and AIoT.

As local governments in smart cities and smarter eco-cities increasingly embrace AI and AIoT technologies, there is a growing interest in exploring their applications in the realm of environmental governance through emerging data-driven governance systems. Especially noteworthy is the rise of smarter eco-cities, representing a novel paradigm of sustainable urbanism characterized by the fusion of the data-driven technologies and solutions of smart cities with the environmental technologies and strategies of eco-cities (Bibri 2021b; Bibri and Krogstie 2020b; Bibri et al. 2023a). In this dynamic landscape, several studies have explored emerging data-driven governance systems, such as City Brain (Ben and Peng 2020; Liu et al. 2022; Xu 2022; Xu et al. 2024), SUM (e.g., D'Amico et al. 2020

2021; Shahrokni et al. 2015), and platform urbanism (Caprotti and Liu 2022; Haveri and Anttiroiko 2023; Noori et al. 2020), along with AI and AIoT technologies aimed at addressing complex challenges. Specifically, these studies have presented different conceptual frameworks or models for understanding and approaching these challenges in urban contexts, including new ways of thinking about sustainability, governance structures and processes, and planning strategies.

City Brain, SUM, and platform urbanism have emerged as promising approaches to advance governance practices in urban settings. However, existing literature primarily focuses on these individual models rather than integrating them into comprehensive frameworks or integrated solutions for advancing environmental governance practices. For instance, research has investigated the efficacy of City Brain in streamlining city operations and refining urban governance models (Kierans et al., 2023; Liu et al. 2022; Xu et al. 2024). Additionally, other studies have explored the potential of SUM to monitor and optimize resource flows in urban settings, thereby informing governance strategies through the identification and implementation of intervention measures (D'Amico et al. 2021; Shahrokni et al. 2015). Similarly, platform urbanism has been studied for its influence on various aspects of urban life and governance. The evolution of city-as-a-platform (e.g., Repette et al. 2021; Komninos and Kakderi 2019) underscores the intersection of technology, governance, and collective knowledge in smart urban development. Urban platforms exemplify a mode of governance that integrates multi-stakeholder innovation and quadruple helix thinking in the discourse of smart cities (e.g., Haveri and Anttiroiko 2023; Borghys et al. 2020; Ansell and Gash 2018, 2019)

Despite the complementary capabilities of these AI- or AIoT-driven governance systems, there is a noticeable lack of systematic exploration of their synergistic and collaborative potential for addressing the multifaceted challenges of environmental governance. Specifically, no study has comprehensively explored how these emerging AI or AIoT-powered solutions can be effectively utilized and synergistically harnessed to propel environmental governance forward within the framework of smarter eco-cities. Or, how they can collectively address the complex and interconnected environmental challenges faced by the governance of these dynamic urban environments.

The current research landscape underscores the imperative for a comprehensive framework that integrates City Brain, SUM, and platform urbanism to advance environmental governance in smarter eco-cities. By addressing this gap, the current study aims to enrich the rapidly growing body of literature on AI- and AIoT-driven sustainable urban development practices. It serves as a guide for navigating the landscape of emerging AI- or AIoT-driven environmental governance. The development of a comprehensive framework serves as a compass for facilitating the implementation of AIoT-driven environmental governance in smarter eco-cities and lays the foundation for a deeper understanding of the transformative journey ahead. Such a framework provides the necessary structure and guidance for integrating AI and AIoT technologies into environmental urban governance strategies.

6.4 RESEARCH METHODOLOGY

The primary aim of this chapter is to explore the linchpin potential of AIoT in seamlessly integrating City Brain, SUM, and platform urbanism to advance environmental governance in smarter eco-cities. Specifically, it seeks to develop a pioneering framework that effectively leverages the synergies among these emerging AI or AIoT-driven governance systems to enhance environmental sustainability practices in smarter eco-cities. This exploration process adheres to a systematic approach encompassing five stages (Figure 6.2). Significantly, the comprehensive literature review and empirical analysis ensure that this research endeavor is grounded in both theoretical insights and practical foundations.

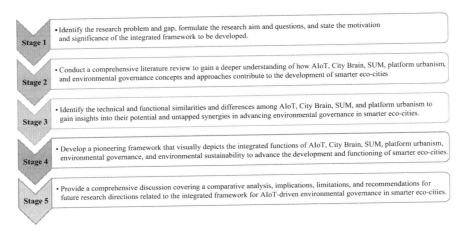

FIGURE 6.2 A flow diagram outlining the step-by-step process of developing the innovative framework for data-driven environmental governance in smarter eco-cities.

To establish a solid groundwork for developing the conceptual framework, a comprehensive review of the existing literature was conducted, focusing on City Brain, SUM, and platform urbanism as well as their interlinkages. In this context, a variety of literature was drawn upon to establish a basis for structuring the research, including theoretical literature, empirical studies, review articles, policy documents, historical literature, and interdisciplinary literature. These types of literature provided insights into the development, implementation, impact, and dynamic interplay of City Brain, SUM, and platform urbanism.

Figure 6.3 illustrates the three-phase flowchart associated with the PRISMA approach. The selection of academic databases Scopus, Web of Science (WoS), and ScienceDirect was deliberate due to their reputation for comprehensive coverage of the 151 high-quality, peer-reviewed studies relevant to the multifaceted topic of this study. To ensure specificity and diversity in the search and retrieval of pertinent scholarly literature, a meticulous selection of keywords related to the core topics and their combinations was employed. These keywords were chosen to accurately reflect the various facets and components of the data-driven governance systems under investigation in relation to environmental governance in the context of smarter eco-cities, thereby optimizing the relevance and precision of the

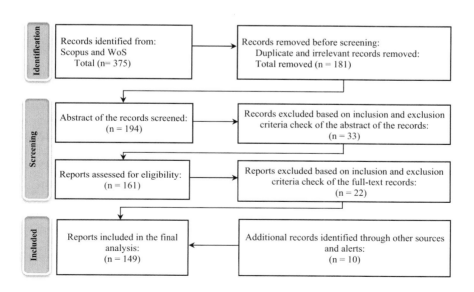

FIGURE 6.3 The PRISMA flowchart for literature search and selection. *Source*: Adapted from Page et al. (2021).

search results. These keywords included "City Brain," "City Brain AND Artificial Intelligence OR Artificial Intelligence of Things," "City Brain AND Platform Urbanism," "Smart Urban Metabolism," "Smart Urban Metabolism AND Artificial Intelligence OR Artificial Intelligence of Things," "Smart Urban Metabolism AND Platform Urbanism," "Platform Urbanism," "Platform Urbanism AND Artificial Intelligence," "Platform Urbanism AND Governance," "Smart Eco-cities AND Governance," "Smart Eco-cities AND Artificial Intelligence OR Artificial Intelligence of Things," "Smart Cities AND Artificial Intelligence OR Artificial Intelligence of Things," and "Environmental Sustainability AND Artificial Intelligence OR Artificial Intelligence of Things." These were used to search against the title, abstract, and keywords of articles to produce initial insights.

The inclusion criteria filtered studies based on their relevance, reliability, language, publication date, and publication type (article, conference paper, or book chapter), providing definitive primary information. Exclusion criteria were applied to remove studies unrelated to the focal topics and their interlinkages and irrelevant to the research aim and questions. Based on these predefined inclusion/exclusion criteria, the selection process involved initial screening based on titles and abstracts, followed by a detailed full-text review to determine eligibility for inclusion in the analysis. The search query retrieved a total of 375 records from three databases. After removing duplicates, 129 records were eliminated. Subsequently, titles and keywords were scrutinized, leading to the exclusion of 52 more records. The remaining 194 records underwent abstract screening against the inclusion and exclusion criteria, resulting in the removal of 33 records. The full-text screening of the remaining 161 records led to the exclusion of an additional 22 records. Ultimately, this process yielded a final selection of 139 publications. Additionally, 10 extra records were included through other sources and alerts, bringing the total number of records included in the final analysis to 149. Through this selection process, the study compiled a diverse set of multiple studies offering varied perspectives. This approach ensured the generation of nuanced and cohesive insights, contributing significantly to the overarching aim and objectives of the study.

The literature search was conducted with a focus on studies published between 2018 and 2023. This timeframe was selected to align with the widespread emergence of the concepts of City Brain, SUM, platform urbanism, and smart(er) eco-cities. The period between 2018 and 2023 was particularly key to marking the foundational phase for integrating these concepts into sustainable urban development strategies to advance environmental goals. It provided a comprehensive view of the evolution of smart eco-cities to smarter eco-cities in the context of AIoT, City Brain, SUM, and platform urbanism. It allowed for an in-depth exploration of the interconnections between these data-driven smart and sustainable city systems and the evolving synergies among smart cities, smart eco-cities, environmental sustainability, and AI and AIoT technologies. In particular, the studies published between 2021 and 2023 made it possible to capture the latest advancements, trends, and insights related to the integration of AI and AIoT technologies and urban development approaches. These include both environmentally sustainable smart cities and smarter eco-cities (see Bibri et al. 2023a, b for a comprehensive systematic review and a detailed bibliometric analysis). Overall, the decision to conduct the literature search between 2018 and 2023 ensures that the research findings are grounded in the most current and relevant scholarly discourse, thereby enriching the analysis and contributing to the advancement of knowledge in the field.

Guided by an inductive approach, a content analysis was conducted on the included studies to collect the necessary data for the analysis and synthesis of the existing literature. Through this process, key themes, patterns, and variations were identified, ensuring a thorough analysis and synthesis. The analysis focused on the fundamental components of the conceptual framework, examining their technical and functional similarities and distinctions, their role in and impact on environmental governance, and the challenges and obstacles associated with their integration.

To develop the conceptual framework, both configurative and aggregative synthesis methods were employed as complementary approaches through an extensive review of the selected studies (Figure 6.4).

Configurative synthesis involved identifying common themes across the synthesized studies and developing a conceptual framework to elucidate the observed interconnections, variations, and contextual nuances (Gough et al. 2017). Thematic synthesis, in this context, is aimed at

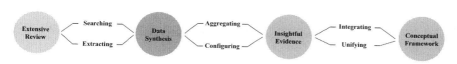

FIGURE 6.4 A flow diagram for data analysis and synthesis for developing the conceptual framework.

understanding the underlying patterns, arrangements, or relationships among research findings, with a focus on interpretation throughout the synthesis process to derive an overarching meaning or provide a nuanced understanding of the subject under review. On the other hand, aggregative synthesis, as described by Cooper (2017), entailed summarizing and consolidating multiple studies to generate an overall thematic summary of the findings. This approach aimed to offer a comprehensive assessment of research evidence, interpret the results after the synthesis process within specific conceptual frameworks, and leverage evidence to make informed statements. Through these complementary synthesis methods, the conceptual framework was developed to provide a robust understanding of the integration of AIoT-driven governance systems in smarter eco-cities in the context of environmental sustainability.

6.5 RESULTS: ANALYSIS AND SYNTHESIS

This section presents the results of the exploration of the linchpin potential of AIoT in seamlessly integrating City Brain, SUM, and platform urbanism to advance environmental governance in smarter eco-cities. Specifically, it introduces a pioneering framework derived from a thorough analysis and synthesis of an extensive literature review and several case studies. It focuses on how AIoT can be effectively leveraged to synergize City Brain, SUM, and platform urbanism for enhancing environmental sustainability practices in smarter eco-cities. Through a detailed analysis and interpretation, this section offers profound insights into the transformative influence of AIoT, City Brain, SUM, and platform urbanism on the trajectory of environmental governance in the complex ecosystem of smarter eco-cities.

6.5.1 Artificial Intelligence of Things

The symbiotic relationship between AI and IoT, encapsulated in the concept of AIoT, emerges as a foundational element shaping the trajectory of smarter eco-cities. AI enhances the capabilities of IoT devices by enabling them to process and interpret data more effectively, extract actionable insights, and adapt to changing environments autonomously. In turn, IoT provides AI systems with a rich source of real-world data for training, validation, and continuous improvement. Together, AI and IoT synergize to create intelligent systems and applications that drive innovation across various domains, including the development of smarter eco-cities.

6.5.1.1 Artificial Intelligence of Things as a Natural Evolution of the Internet of Things

In recent years, the exponential growth of data generated by IoT sensors and devices has posed significant challenges, straining the centralized infrastructure of IoT with increased demands on computing resources and data processing. The convergence of AI and IoT emerges as a key solution to address these challenges, effectively managing the flow and storage of data within the vast IoT network (Zhang and Tao 2021). The goal of AIoT is to make IoT devices and systems transmit data but also enable them to analyze and interpret these data, derive insights, optimize operations, and dynamically adapt to changing environmental conditions in different urban settings (Bibri and Huang 2024). AIoT enhances the capabilities of IoT devices and systems by equipping them with intelligence and decision-making functionalities, overcoming barriers to the widespread adoption of IoT (Seng et al. 2022; Shi et al. 2020). This advancement unlocks the potential for more efficient and rapid data utilization across urban systems (Mastorakis et al. 2020). Parihar et al. (2023) emphasize the necessity of integrating IoT with AI within the AIoT framework, delving into the evolution, architecture, applications, and challenges associated with this integration.

The instrumental role of AI becomes evident in meeting the evolving demands of IoT devices and systems. Concurrently, AI is experiencing a resurgence fueled by the abundant and potent data flow from IoT sensors and devices. This resurgence is facilitated by increased computing storage capacity and the remarkable speed of real-time data processing. Integrating ML and DL into IoT devices holds the potential to bring smart functionalities closer to reality and address challenges posed by the large numbers of heterogeneous devices (Alahakoon et al. 2020; Mahdavinejad et al. 2018; Mohamed 2023). AIoT involves the execution of AI on IoT-enabled devices, empowering them with human-like cognitive and behavioral abilities. The deployment of ML- and DL-based systems introduces edge intelligence in cloud-based IoT, freeing AI algorithms from cloud constraints and addressing the high computing resource demands (Zhang and Tao 2021). As an enabler of Big Data, IoT relies on AI algorithms for interpretation, understanding, and decision-making to achieve optimal outcomes (Baker and Xiang 2023).

AIoT, as a transformative force, fosters mutual benefits for both AI and IoT, ushering in smarter eco-city systems that enhance efficiency, effectiveness, and robustness. Its inherent capabilities position it as a fitting solution for addressing complex environmental challenges. In essence, AIoT has introduced a myriad of opportunities for innovation, scalability, and automation across diverse domains, spanning environmental sustainability, climate change, and urban planning and development (e.g., Alahakoon et al. 2020; Bibri et al. 2023a, b, 2024b; El Himer et al. 2022; Gourisaria et al. 2022; Kuguoglu et al. 2021; Leal Filho et al. 2022; Puri et al. 2020; Samadi 2022; Seng et al. 2022). AIoT not only lays the groundwork for the development of smarter and more sustainable cities but also heralds a new era of urban intelligence. Its underlying synergistic potential emerges as a crucial convergence shaping the broader vision of intelligent and resilient urban areas, charting the crossroads of technology and sustainability.

6.5.1.2 The Architecture of Artificial Intelligence of Things and Its Technical Components

The architecture of AIoT embodies a powerful framework that seamlessly integrates the potential of AI with the ubiquitous connectivity of IoT. At its core, AIoT architecture consists of a sophisticated network of IoT devices and sensors that collect and transmit copious amounts of data from the physical world. This entails an array of coordinated urban systems such as energy, transportation, mobility, waste, water, and metabolism. These data are then processed and analyzed through AI models and algorithms, which enable the IoT devices and systems to learn, reason, and make intelligent decisions autonomously. The synergic interplay between AI and IoT empowers the AIoT architecture to create a dynamic ecosystem where AI augments the computational and analytical capabilities of IoT, and IoT enriches the applied knowledge derived by AI models. It fosters groundbreaking advancements in data-driven solutions for tackling environmental challenges in emerging smarter eco-cities.

Seng et al. (2022) provide an overview of recent advancements in AIoT, exploring various computational frameworks for AIoT. The authors discuss architectures, techniques, and hardware platforms, as well as propose three layers: (1) sensors, devices, and energy approaches; (2) communication and networking; and (3) applications. Similar to Zhang and Tao (2021), they shed light on the integration of AIoT with edge computing, fog computing, cloud computing, 5G/6G networks, among others. Additionally, they discuss the various challenges and issues that need to be addressed and overcome to enable practical deployments of AIoT systems into urban environments. Highlighting the potential of 5G networks in relation to

IoT and AI, Alahi et al. (2023) provide insights into how to create sustainable and productive urban environments by integrating IoT-enabled technologies and AI for smart city scenarios.

Important to note is that early research on AIoT has largely dealt with the technical aspects of AIoT, focusing largely on edge computing (Fragkos et al. 2020; Song et al. 2020; Yang et al. 2020, 2021; Zhang and Tao 2021). This computing model is central to the functioning of AIoT because it moves data processing from the network edge to as close as possible to IoT devices and systems. This strengthens AIoT applications thanks to edge intelligence, which is key to deploying the next-generation AIoT applications for not only the operational management and planning of smarter eco-cities but their governance. As distilled from the synthesized studies, AIoT architecture is composed of the following elements:

1. IoT sensors and devices: Deploying a network of a wide range of interconnected sensors and devices to collect real-time data on environmental parameters and changes.

2. Cloud computing: Leveraging cloud computing resources to store, manage, and analyze the massive amounts of data generated by IoT sensors and devices.

3. Edge computing: Utilizing edge computing to bring computation and storage closer to the source, reducing latency and bandwidth usage, improving real-time responsiveness, and enabling faster responses to environmental events by executing computations locally.

4. Fog computing: Extending edge computing by creating a hierarchical architecture that includes multiple edge devices and gateways. Fog nodes are strategically placed in proximity to data sources to perform intermediate processing, data filtering, and preliminary analytics before transmitting relevant data to the cloud.

5. Data analytics: Utilizing AI algorithms, especially ML/DL, to analyze the enormous amount of environmental data generated by IoT sensors and devices to detect patterns and anomalies and extract meaningful insights.

6. Predictive modeling: Developing predictive models that use historical data and real-time inputs to forecast environmental trends, potential risks, and areas of concern for proactive decision-making.

7. Decision-making: Interpreting insights derived from predictive modeling and analytics to inform and guide proactive actions and decision-making processes in response to environmental changes and challenges.

6.5.2 City Brain as a Large-Scale System of Artificial Intelligence of Things

City Brain stands as a cornerstone in reshaping the governance paradigm of smarter eco-cities by supporting city managers and policymakers in making decisions based on insights gained from AIoT-driven analytics. By harnessing real-time data analytics, predictive modeling, and automated decision-making, City Brain enables local governments and city authorities to optimize resource allocation, enhance service delivery, and mitigate environmental risks. Its integration in smarter eco-cities catalyzes a fundamental shift toward data-driven governance, fostering agility, responsiveness, and resilience in the face of environmental challenges. As such, it structures urban computing and intelligence systems for smarter eco-cities based on ubiquitous sensing, advanced data analytics, and novel visualization tools.

In City Brain, large-scale AIoT can be viewed as the digital nervous system of the IoT-connected sensors, devices, and systems in the city (the body) through multiple networks, with AI being the brain. Akin to humans, the latter represents the control system of the body's functions in terms of sensing, perceiving, decision-making, and behaving, and the nervous system serves as the network that relays messages back and forth from the brain to the different parts of the body. At its essence, as highlighted by Liu et al. (2018), City Brain conceptualizes the city as an organism. It integrates different technologies into the urban organism's system through the urban neuron network. This network resembles the central nervous system, sensory system, and nerve endings, enabling interactions between people and objects while facilitating the organic integration of various urban components. AIoT plays a crucial role in enabling real-time data processing and decision-making in this interconnected urban ecosystem to enhance urban processes and practices. These are associated with the various subsystems of the city, such as transport, mobility, energy, water, waste, environment, and land use. Overall, the IoT infrastructure, multi-modal data analytics, and multi-view AI are combined to monitor, collect, aggregate, analyze, and interpret real-time data to make decisions and predictions pertaining to urban management, planning, and governance.

6.5.2.1 City Brain Pillars Based on the Data Science Cycle

As a sophisticated manifestation of "urban computing and intelligence" (Bibri 2021c; Liu et al. 2017), City Brain encompasses three core capabilities, as delineated by Zhang et al. (2019): (1) processing vast and diverse data sources at an ultra-large scale, surpassing human real-time understanding (termed as global cognition); (2) discerning intricate concealed patterns and rules hitherto unidentified by humans, often achieved through ML techniques; and (3) formulating overarching optimal strategies that outperform localized, suboptimal decisions made by humans, a phenomenon referred to as global coordination.

These metrics form the foundation of an AIoT system grounded in the data science cycle, encapsulated by steps such as sensing, perceiving, learning, visualizing, and acting (Bibri et al. 2023a) or perceiving, learning, reasoning, and behaving (Zhang and Tao 2021). The slight variations in the number of steps characterizing AIoT may stem from the approach taken, which is influenced by specific use cases or the degree of human intervention required, either to support or automate decision-making processes. For instance, the concept of cognition in Zhang et al. (2019), involving the acquisition and understanding of data, is described as perceiving in (Zhang and Tao 2021) and as sensing and perceiving in (Bibri et al. 2023a). As depicted in Figure 6.5, AIoT system consists of six pillars.

The data science cycle is closely linked to the six pillars of AIoT system in the context of City Brain. The connection between each pillar and the data science cycle is outlined as follows:

1. Sensing entails the data acquisition process, where data are collected from various sources in the urban environment. In the context of City Brain, IoT sensors and devices play a crucial role in gathering real-time data on urban systems. The data science cycle begins with sensing, as it relies on diverse and high-quality data inputs.

2. Perceiving involves data preprocessing and understanding to extract meaningful patterns. Once the data are collected from different sources, they undergo preprocessing for cleaning, filtering, and alignment for analysis. In City Brain, this step involves timestamp synchronization, data validation, and data integration to ensure accurate and reliable information for further processing in the data science cycle.

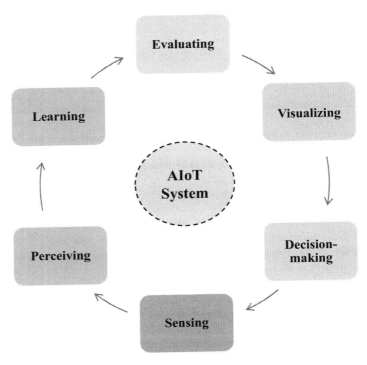

FIGURE 6.5 Six pillars of AIoT system based on the data science cycle. Adapted from Bibri (2023b).

3. Learning is at the heart of the data science cycle, where ML/DL algorithms are applied to extract insights and patterns from the integrated data. In City Brain, learning algorithms analyze the data to identify patterns, make predictions, detect anomalies, and understand urban behavior. Learning helps City Brain make data-driven decisions and optimize urban strategies and urban services.

4. Evaluating, a crucial step in the learning process, involves assessing the performance and quality of models or algorithms developed during the analysis phase. After creating and training ML models using specific algorithms, the performance of these models needs to be evaluated to ensure their effectiveness and generalizability to new, unseen data. This entails using metrics (accuracy, precision, recall, etc.), testing, validation, and comparison to ensure the reliability of data-driven models and meaningful insights. Evaluation helps practitioners make informed decisions about the model's utility and

potential for deployment in real-world applications. In turn, City Brain, through learning, continuously improves its understanding of the data and refines its models based on new information or feedback.

5. Visualizing is essential for communicating the results of decisions made by City Brain in a comprehensible format and a clear and appealing way, allowing stakeholders, including city managers and decision-makers, to easily and effectively interpret the results and make informed decisions. In City Brain, visualizations enable city authorities to gain actionable insights quickly and understand complex urban dynamics by presenting real-time environmental parameters and urban conditions. They bridge the gap between data analysis and decision-making, allowing city stakeholders to monitor, manage, and govern city operations and processes in a real-time manner.

6. Decision-making aligns with the deployment and operational phase of the data science cycle. After data collection, preprocessing, learning, evaluation, and visualization, City Brain reaches a point where it is capable of triggering actions or behaviors in real-time, such as adjusting environmental controls, optimizing processes, or sending alerts to stakeholders, to achieve desired outcomes or responses. The insights gained from data analysis are put into practical use, enabling City Brain to actively contribute to problem-solving, optimization, and decision-making processes in real-world environments.

In sum, the data science cycle is deeply integrated with the six pillars of the AIoT system in City Brain. It forms the backbone of City Brain's capabilities in creating smarter and more sustainable urban environments through improving decision-making processes.

6.5.2.2 Technical Components of City Brain
The construction of City Brain involves a set of components that tend to slightly vary depending on the domain where City Brain is applied. These components include IoT, Big Data, AI, Cloud Computing, Edge Computing, and 5G networks. The City Brain architecture encompasses the following components (Liu et al. 2018, 2021; Xu 2022) in the context of smarter eco-cities:

- IoT for City Brain pertains to the smarter eco-city's sensors and smart driving systems as sensory nervous and motor nervous systems distributed in its operating and organizing systems.

- Big Data for City Brain represents the valuable data generated by the connected IoT, sensors, devices, and systems and transmitted and accumulated in the operation process of the smarter eco-city. These data, which include citizens' everyday life, business operations, government functions, buildings, transport systems, energy systems, water systems, and so on, are the basis for City Brain to generate knowledge in the form of applied intelligence.

- Cloud computing for City Brain is associated with the smarter eco-city's central nervous system, which controls all its nervous systems through the server, network operating system, neural network, and data-driven AI algorithms (ML, DL, etc.). For example, Cloud computing is used to process real-time data gathered from the sensor network to optimize traffic flow by controlling traffic light scheduling or managing the timing of traffic lights.

- Edge Computing, the development of the nerve endings of City Brain, integrates AI into the IoT sensors and systems distributed across the smarter eco-city to make the edge of the sensory nervous and motor nerve systems of City Brain smarter, more robust, and/or autonomous.

- AI for City Brain is the catalyst that enhances the knowledge and wisdom of the smarter eco-city. AI is seamlessly integrated with Big Data using ML and DL algorithms and applied to the City Brain's nerve endings, neural networks, and smart terminals, resulting in a powerful and cohesive system. The fusion of AI with the smarter eco-city enhances the capabilities and functionalities of all nervous systems in a synchronized manner.

- The internet Neural Network for City Brain is a collection of interconnected nodes in a layered structure that connects the different components of the smarter eco-city to City Brain. It is inspired by social networks enabling human-human interactions on the Internet, and whose form is set to change in response to the integration and advancement of IoT, Big Data, AI, and Cloud Computing. This applies also to real-world physical environments as to how everyday objects and smart things interact with each other and with people. Accordingly, social networks go beyond the human-human interaction to include human-human, human-things, and things-things interactions.

6.5.3 Smart Urban Metabolism

Urban Metabolism (UM) refers to the dynamic flow of energy, materials, and resources through a city. Kennedy et al. (2007, p. 44) define it as "the total sum of the technical and socioeconomic processes that occur in cities, resulting in growth, production of energy, and elimination of waste." In the realm of UM, city planners encounter a significant challenge due to inadequate data inventories and inappropriate data management and evaluation solutions, hindering the efficient adaptation of approaches to resource management. In turn, this limitation hampers the capacity of eco-cities to tackle intricate environmental challenges, hindering their efforts to understand and manage their energy and material flows effectively. As open ecosystems, eco-cities are characterized by complex metabolic flows due to the multiple dimensions of urban contexts (Simboli et al. 2019). This multiplicity involves numerous environmental, political, economic, and technological interactions among multiple internal and external actors and factors (D'Amiro et al. 2020, 2021; Kennedy 2011). The current UM framework does not satisfy the needs of eco-cities in response to rapid technological change. Therefore, there is a need for a multidimensional approach to understanding the functioning of eco-cities and optimizing the efficiency of their metabolic processes (Bibri and Krogstie 2020a; Shahrokni et al. 2025). Such an approach is necessary for supporting energy efficiency, renewable energy, transport management, water management, waste management, material usage, land use, air and noise pollution prevention, and environmental management (D'Amiro et al. 2020). It can facilitate the evolutionary co-design of SUM based on stakeholder engagement, knowledge and resource sharing, and flexible multi-level governance (Webb et al. 2018). Consequently, the SUM framework has emerged as a response to the limitations inherent in the UM framework. playing a key role in advancing smart eco-cities (Bibri and Krogstie 2020a; Shahrokni 2015). This evolution marks a pivotal shift as it empowers eco-cities with data-driven insights that transcend conventional boundaries.

6.5.3.1 Essential Characteristics of Smart Urban Metabolism

The SUM framework builds on the conventional UM framework (e.g., Baccini and Brunner 2012; Princetl and Bunje 2009; Kennedy et al. 2007; Kennedy and Hoornweg 2012). It is intended to overcome its inherent limitations concerning methods for energy, water, substance, and material flows analysis. By harnessing advanced technologies and data analytics, smart eco-cities gain a deeper understanding of their energy and material flow, allowing them to optimize resource utilization, enhance overall efficiency, and minimize environmental impact. However, the essential characteristics of the SUM framework lie in that it (Bibri and Krogstie 2020a; Shahrokni et al. 2015):

- Brings together a wide range of data from various sources, eliminating data silos and enabling a holistic view and thus comprehensive understanding of the urban system by considering multiple dimensions and variables.

- Generates real-time updates of the metabolic flows that reflect the dynamic nature of the city, allowing for timely and accurate analysis of energy consumption, material usage, waste generation, and environmental impacts.

- Offers a high spatial resolution, allowing urban metabolic flows to be analyzed and visualized at multiple spatial scales, from the city-wide level down to the district, building, and individual households. This multi-level analysis provides a nuanced understanding of resource flows and their distribution in the urban environment, providing a better basis for decision-making.

- Provides continuous feedback in the form of tailored visualizations, catering to the specific needs and interests of the different stakeholders involved in decision-making processes relating to urban planning and governance. Such feedback ensures that the information is relevant and meaningful for each stakeholder group, empowering them to take appropriate actions and leading to identifying new intervention measures toward fostering sustainable urban development practices.

6.5.3.2 Common Steps of Smart Urban Metabolism

The key features of the SUM framework contribute to a comprehensive understanding of UM to make informed choices, optimize resource utilization, and enhance the sustainability and resilience of cities. This relates to the ability of data-driven technologies to better integrate the knowledge derived from and visualized based on real-time monitoring and analysis of energy and material flows into urban planning and governance processes. As presented in Table 6.1, the sequence of SUM steps can differ depending on the context and application, as observed in the case of the Stockholm Royal Seaport (SRS) District (Shahrokni et al. 2015). However, a general overview of the essential SUM steps can be distilled from various studies

TABLE 6.1 General and Contextual Approaches to SUM

General Approach	Contextual Approach (SRS)
• Gather relevant data on urban processes, resource use, and environmental impacts. • Employ analytical tools and methodologies to process and interpret the collected data. • Develop models that represent the urban system, incorporating data-driven insights. • Use models to simulate urban scenarios and predict potential outcomes. • Identify opportunities for optimizing resource use, energy efficiency, and overall urban performance. • Involve key stakeholders, such as local communities and authorities, in decision-making processes. • Establish a continuous feedback loop to refine models and strategies based on real-time data and evolving urban conditions. • Implement sustainable strategies derived from the analysis to enhance urban resilience and sustainability. • Continuously monitor the implemented strategies and evaluate their effectiveness over time. • Adapt the SUM framework based on new data, changing urban dynamics, and emerging technologies to ensure ongoing effectiveness.	• Define stakeholder goals and interests and the corresponding KPIs and metrics. • Define local system boundaries. • Identify real-time data sources for representing KPIs. • Manage data rights and responsibilities based on consortium. • Develop an integration platform. • Develop a metabolic flow calculation engine for generating KPIs in real-time based on system boundaries. • Develop Application Programming Interfaces (APIs) for managing access to specific KPIs for feedback. • Design and develop visualization interfaces based on stakeholders' goals and interests. • Perform in-depth analytics of the subsystems of urban metabolic flows. • Identify intervention measures based on insights into patterns, correlations, and clusters. • Evaluate report findings in collaboration with stakeholders. • Follow up the outcomes of interventions in real-time.

(e.g., Bibri and Krogstie 2020a; D'Amico et al. 2020; Ghosh and Sengupta 2023; Peponi et al. 2022).

The SUM framework encompasses a series of interconnected steps that collectively contribute to its adaptive and data-driven nature. It enables smart eco-cities to dynamically respond to changing conditions and optimize resource utilization in real-time.

6.5.3.3 Roles and Advantages of Artificial Intelligence of Things-Powered Smart Urban Metabolism

In the context of SUM, the integration of AI and AIoT stands out as a transformative force, playing a pivotal role in enhancing data-driven decision-making, optimizing resource utilization, and fostering sustainability in urban systems. Research has underscored the significance of integrating AI and AIoT into SUM (e.g., Bibri et al. 2024a; Ghosh and Sengupta 2023; Peponi et al. 2022; Yigitcanlar et al. 2020) to equip it with new capabilities for bolstering the environmental performance of smarter eco-cities. The utilization of AI and AIoT technologies in SUM entails organizing and handling diverse and massive datasets, creating intelligent models that autonomously learn and adapt, and extracting valuable insights through predictive analytics (Bibri et al. 2024a; Ghosh and Sengupta 2023; Peponi et al. 2022). The initial step toward constructing a robust ML model within the SUM framework involves, as noted by Ghosh and Sengupta (2023), capturing and storing up-to-date data. These data serve as the foundation for predictive analytics, enabling the ML model to glean insights, identify patterns, and forecast trends crucial for optimizing urban systems and fostering sustainable practices. AI and AIoT technologies have made a substantial contribution to SUM through their capacity to formulate predictive models and optimization algorithms.

The deployment of IoT-connected sensors and devices across the urban environment enables the capture and analysis of data on energy consumption, transportation patterns, water usage, material usage, waste generation, and other crucial environmental parameters. This data-driven approach not only enhances the understanding of UM dynamics but also streamlines the identification of areas for improvement in resource optimization and environmental performance. Empowered by AI and AIoT, SUM facilitates real-time monitoring and analysis of urban metabolic flows, enabling well-informed decision-making concerning resource management, planning, and governance. This data-driven approach has the potential to minimize resource consumption, enhance energy efficiency, optimize waste management, and create more resilient and sustainable cities.

More than a decade ago, Trantopoulos et al. (2011) put forward a roadmap comprising three elements aimed at enhancing the measurement and prediction of UM. These elements encompass the use of digital traces to capture human activities, the integration of these traces into large-scale simulation models, and the incorporation of sensors and IoT to facilitate real-time feedback on human activity. The authors introduced the notion of an "urban nervous system," which connects urban infrastructures, stakeholders, and citizens. They emphasized the importance of establishing feedback systems in this nervous system to enable effective functioning and interaction. Building on the process of implementing the UM framework (Baccini and Brunner 2012; Kennedy and Hoornweg 2012; Princetl and Bunje 2009), coupled with the insights gained from the previous discussion, the implementation of AI- and AIoT-based SUM involves a range of diverse components. These components include sensor networks, IoT devices, AI-driven analytics; connectivity infrastructure, cloud computing; decision support systems, and stakeholder engagement. They work together to create a holistic data-driven approach to managing urban resources, promoting environmental sustainability, and enhancing the overall quality of life in smarter eco-cities. As concluded by Ghosh and Sengupta (2023), the use of AI in SUM is "the way forward that will be principally beneficial in understanding the real-time city functions, as they can help urban city planners or governors to make smart decisions as well as smart implementation and accomplishment of policies."

6.5.4 Platform Urbanism

The interconnectedness of platform urbanism, urban governance, urban sustainability, and urban AIoT forms a nexus that underscores synergistic interplay and collaborative integration. Platform urbanism, empowered by AIoT, not only catalyzes effective urban governance structures and processes but also facilitates collaborative efforts among stakeholders. This collaborative approach is essential for promoting urban sustainability and resilience in urban landscapes.

6.5.4.1 From Smart Urbanism to Platform Urbanism

Platform urbanism and smart urbanism are interconnected concepts that reflect the evolving nature of urban development and governance in the digital revolution. Smart urbanism is understood as the use and application of digital technologies, data-driven approaches, and networked infrastructures in urban environments to improve the efficiency, sustainability, and livability of modern cities. It involves integrating various aspects of urban life into a cohesive and data-driven ecosystem and seeks to leverage technological advancements to address urban challenges, optimize resource allocation, and enhance the overall quality of life. As a progression of smart urbanism, platform urbanism represents a paradigm shift in the way cities are planned, organized, and operated (Caprotti et al. 2022). It takes a more holistic ecosystem approach by creating open and interconnected digital platforms that facilitate collaboration, coordination, and co-creation among various urban stakeholders. It reflects the ways in which the landscape of smart cities develops and prepares for the future (Bibri 2023).

Platform urbanism originates in the accelerated and multifaceted development of smart urbanism as fueled by the marked intensification of the datafication, algorithmization, and platformization practices resulting from the convergence of AI, IoT, and Big Data technologies. Nonetheless, as it modulates the constituent technologies, processes, and practices of smart urbanism, the two models operate and work simultaneously in different spaces rather than necessarily superseding each other (Sadowski 2020). As echoed by Bibri (2023, p. 1353), platform urbanism "represents a form of data-driven smart urbanism, a drastic shift to data-centralized digital systems that are becoming increasingly platformized across smart cities as implemented by civic organizations, public institutions, research institutes, and private companies…, presenting technological solutions to specific problems." The boundaries—whether conceptual, functional, or spatial—between smart urbanism and platform urbanism partially coincide with each other (Caprotti and Liu 2022). Many studies have addressed the relationship between platform urbanism and smart urbanism in terms of commonalities and differences (e.g., Barns 2020; Bibri 2023; Caprotti et al. 2022; Katmada Katsavounidou and Kakderi 2023).

6.5.4.2 Key Differences between Smart Urbanism and Platform Urbanism in Governance

There are key differences between smart urbanism and platform urbanism concerning how cities are governed, which have direct implications for environmental governance in emerging smarter eco-cities. Smart urbanism hinges on city governments being the prime initiators of innovative solutions and on corporate actors being the main providers of these solutions. There is a strong relationship between city authorities and technology corporations in terms of driving smart city activities. Private initiatives and partnerships are enabled by and operated under central governments to serve public governance (Caprotti et al. 2022). Conversely, platform urbanism entails a novel relationship to governance in that it is more directly connected to and interactive with stakeholders as well as more antagonistic to government regulations (Sadowski 2020). It has become central to the dynamic formation of socio-technical constellations of the multiple stakeholders involved in the different spheres of urban life. In view of the above, governance in smart urbanism and platform urbanism represents two distinct approaches to managing and shaping urban environments, each with its own unique and sometimes opposing characteristics. Table 6.2, which presents the key differences between the two approaches, is distilled based on the insights derived from the relevant synthesized studies (e.g., Ansell and Miura 2019; Barns 2019 2020; Brynskov et al. 2018; Caprotti et al. 2022; Haveri and Anttiroiko 2023; Ismagilova et al. 2020; Katmada Katsavounidou and Kakderi 2023; Komninos and Kakderi 2019; Noori et al. 2020; Poell et al. 2019; Repette et al. 2021).

While governance in both smart urbanism and platform urbanism aims to improve urban living, they adopt different approaches to achieve their goals, with varying levels of centralization, participation, and flexibility. Combining elements of both approaches could offer a more comprehensive and balanced approach to environmental governance in smarter

TABLE 6.2 Comparative Dimensions of Governance in Smart Urbanism and Platform Urbanism

Comparative Dimensions	Smart Urbanism	Platform Urbanism
Approach	Smart urbanism focuses on using advanced technologies and data-driven solutions to enhance various aspects of urban life. It often involves top-down planning and implementation by municipal authorities and emphasizes the use of technology to optimize urban systems.	Platform urbanism is centered around the concept of platforms that facilitate collaboration and interaction among various stakeholders, including citizens, communities, businesses, and local authorities. It seeks to create open and inclusive urban environments where individuals and organizations can participate in co-creating solutions for advancing sustainable development.

(Continued)

TABLE 6.2 (*Continued*)   Comparative Dimensions of Governance in Smart Urbanism and Platform Urbanism

Comparative Dimensions	Smart Urbanism	Platform Urbanism
Participatory decision-making	In smart urbanism, decision-making is often driven by data and algorithms, relying on centralized planning and expertise. While citizens may benefit from smart technologies, their active participation in decision-making processes might be limited, thereby the lack of inclusivity, diversity, and accountability.	Platform urbanism encourages a shift toward more participatory and inclusive decision-making processes within the context of urban development. It leverages digital platforms to enable greater engagement and interaction among multiple stakeholders. It empowers citizens to contribute their ideas, suggestions, and feedbacks, fostering a sense of collaboration and involvement in shaping urban initiatives and policies. The diversity of input ensures that urban development plans are more comprehensive, responsive to real needs, and aligned with sustainable and equitable goals. Additionally, platform urbanism can enhance transparency and accountability in governance processes.
Data and privacy	The implementation of smart urban solutions relies heavily on collecting and analyzing vast amounts of data from various sources. This data-driven approach raises concerns about privacy, security, and potential misuse of personal information.	Platform urbanism's emphasis on data sharing, transparency, and user consent plays a crucial role in fostering responsible and ethical use of data in the context of environmental initiatives. Environmental governance relies heavily on data to make informed decisions and implement effective intervention measures. Platforms designed with privacy in mind prioritize data protection and can provide users with more control over their data. This fosters trust and encourages active participation from citizens and stakeholders. With transparent data-sharing mechanisms, individuals can have visibility into the types of data being collected, how it is being processed, and the purposes for which it is used.

(*Continued*)

TABLE 6.2 (*Continued*) Comparative Dimensions of Governance in Smart Urbanism and Platform Urbanism

Comparative Dimensions	Smart Urbanism	Platform Urbanism
Flexibility and adaptability	Smart urban initiatives are typically designed to address specific challenges and may be more rigid in their application. Upgrading or changing these systems can be costly and time-consuming.	The adaptability and agility of platform urbanism offer significant advantages. Designing platforms that can evolve and respond to changing urban needs enable environmental initiatives to adopt a more flexible and iterative approach to urban development. This allows for quicker adjustments to and refinements of strategies, ensuring that responses to environmental challenges remain relevant and effective over time. Such adaptability allows to address emerging issues and embrace innovative solutions.
Scope	The focus of smart urbanism is mainly on optimizing urban systems and services through technology and data-driven solutions.	The broader scope of platform urbanism is advantageous. It goes beyond technological advancements and incorporates social and economic factors, promoting collaboration and inclusivity in environmental initiatives. By considering multiple dimensions, such as citizen engagement, public-private partnerships, and equitable resource allocation, platform urbanism enables more holistic sustainability strategies. This integrative approach fosters community involvement and stakeholder participation, leading to more well-rounded and effective environmental outcomes.

eco-cities. However, technological advances and social transformations continue to shape human interactions, reflected in emerging trends such as the digital revolution, ecosystem thinking, and platformization. Indeed, the pervasive influence of digital platforms becomes apparent through their widespread integration across various societal realms and their transformative impacts on social organizations (van Dijck et al. 2018). Notably, the governance frameworks and institutional dynamics are concurrently being reshaped by the (re-)organization of socio-cultural practices tied to and driven by platformization (Poell et al. 2019). This phenomenon gives rise to the concept of platform governance, encapsulating structures, processes, and mechanisms that oversee interactions and engagements among diverse stakeholders in emerging urban platforms.

6.5.4.3 Platform Urbanism and Governance: Key Characteristics

Platform urbanism and urban governance engage in a symbiotic relationship, with digital platforms serving as key facilitators in enhancing interaction and cooperation among key urban actors. These platforms foster citizen engagement, ensure transparency, and improve decision-making processes, ultimately leading to more effective and inclusive urban governance. This relationship introduces a novel dimension to governance by establishing a more direct connection with stakeholders while challenging traditional government regulations (Sadowski 2020). Fundamentally, platform urbanism relies on decentralized networks to achieve outcomes, create value, and deliver services.

Platform urbanism has emerged as a central element in the dynamic shaping of socio-technical constellations involving various stakeholders across diverse spheres of urban life. This reality is exemplified in real-world cities where interconnected ecosystems have been established to foster data-driven, participatory, adaptive, and collaborative governance practices (Haveri and Anttiroiko 2023; Jiang et al. 2022; Noori et al. 2020). In the realm of platform urbanism, governance embraces and leverages a distinctive approach to managing, planning, and influencing urban environments. Underpinned by the principles of platform urbanism, this approach exhibits unique dimensions when compared to traditional modes of urban governance. As delineated in Table 6.3, these dimensions are derived from insights distilled based on a range of relevant studies on various facets of platform urbanism across diverse urban settings (e.g., Ansell and Gash 2018; Ansell and Miura 2019; Anttiroiko 2016; Brynskov et al. 2018; Caprotti et al. 2022; Katmada Katsavounidou and Kakderi 2023; Komninos and Kakderi 2019; Noori et al. 2020; Soe et al. 2022; Vadiati 2022).

6.5.4.4 Platform Urbanism and Its Underlying AIoT Functions

The relationship between platformization, platform urbanism, and AIoT is characterized by a mutually reinforcing dynamic. Platformization

TABLE 6.3 Key Distinctive Dimensions of Platform Urbanism as a Strategy for Urban Governance

Dimensions	Descriptions
Open connected platforms	Platform urbanism emphasizes the development of open digital platforms that encourage participation and collaboration among stakeholders. These platforms enable seamless data sharing and communication, which foster a more connected and inclusive urban ecosystem.
Ecosystem approach	Platform urbanism views the city as an ecosystem where multiple stakeholders interact, exchange data, and co-create solutions. It recognizes that the success of urban development lies in the collective efforts of various actors working together on shared goals.
Participatory engagement	The platform approach encourages citizen engagement and participation in urban planning and decision-making processes. Citizens can contribute ideas, feedback, and expectations, and collaborate with local authorities to shape the urban environment based on their needs and expectations.
Adaptability and flexibility	As a key feature of platform urbanism, flexibility and adaptability are crucial for responding to dynamic urban challenges. The open nature of digital platforms enables rapid integration of new technologies and services, ensuring the continuous evolution of cities as to meeting changing needs and priorities.
Data-driven governance	Platform urbanism relies heavily on data-driven decision-making and analytics. Digital platforms collect huge amounts of data from diverse sources, enabling policymakers to gain insights into urban patterns and trends and make informed decisions. At the heart of platform urbanism is algorithmic mediation, where domains of urban life as well as stakeholder interactions are managed and regulated by distributed computer systems.

(Poell et al. 2019), underpinned by AIoT processes such as digital instrumentation, digital hyper-connectivity, datafication, and algorithmization (Bibri et al. 2024a; Calvo 2020), amplifies the potential of platform urbanism. This symbiotic interplay influences urban governance structures and processes, fostering a dynamic relationship that drives innovation and efficiency in urban development. At the heart of the interactions and exchanges among stakeholders in urban governance, facilitated by digital platforms that transform urban spaces into dynamic ecosystems, lies the concept of "digital mediation of cities" (Barns 2019). This notion encapsulates the transformative role of digital technologies in shaping the dynamics of urban life and the way cities are governed, facilitating communication, collaboration, and decision-making processes in an increasingly interconnected urban landscape. Digital platforms are "the overlay of unfathomably vast amounts of sophisticated data, providing opportunities for the application of AI to release operators from boring and tough data analysis tasks, e.g., monitoring, regulating, and planning" (Huynh-The et al. 2023).

The integration of AIoT in digital platforms provides urban stakeholders with actionable insights and supports collaborative decision-making to create interconnected ecosystems that enhance the efficiency and responsiveness of urban systems (ITU 2022). The incorporation of AIoT into digital platforms significantly transforms these ecosystems, facilitating seamless data sharing, real-time analysis, predictive insights, adaptive decision-making, and tailored services. This enhances the efficiency, effectiveness, and responsiveness of urban governance processes. Accordingly, AIoT-driven mediation empowers stakeholders to harness the power of data and technology to address complex urban challenges, optimize resource allocation, and improve the quality of life for citizens. The increasing interconnectivity among city stakeholders in urban ecosystems, enabled by AI, brings forth new opportunities to transform cities into more human-centric, intelligent, and sustainable spaces (Repette et al. 2021). In this context, digital platforms function as mediums for diverse stakeholders to engage, share information, exchange services and resources, and participate in decision-making processes. AIoT is a fundamental enabler of platform urbanism in the dynamic and interconnected urban environments of the future. Platform urbanism represents a transformative approach to urban governance, bringing together unique features that advance the development of smarter eco-cities.

6.5.5 Roles of Emerging Data-Driven Governance Systems in Enhancing Environmental Governance Processes and Strategies: A Theoretical and Empirical Analysis

In the dynamic landscape of urban development, the integration of innovative paradigms becomes pivotal for effective environmental governance processes. This section delves into how City Brain, SUM, and platform urbanism contribute to environmental governance while elucidating their combined effects and interconnected roles in shaping a cohesive framework for smarter eco-city development. Through combining a theoretical and empirical analysis of their individual strengths and collaborative potentials, it unravels a profound understanding of how they jointly propel environmental governance in the trajectory of smarter and more resilient and sustainable cities.

6.5.5.1 Environmental Governance in Smarter Eco-Cities

Environmental urban governance, as a multifaceted field, plays a key role in steering the development of smarter eco-cities. Expanding upon the foundational principles set forth by eco-cities and smart cities, the concept of a smart eco-city experiment is emerging as a prominent guiding framework among scholars, planners, and policymakers. The notion of an experimental city (e.g., Evans, Karvonen and Raven 2016; Caprotti and Cowley 2017) aligns with transition theories and governance approaches aimed at fostering low-carbon cities (Bulkeley and Castán Broto 2013). It entails conducting trials and enacting environmental and institutional reforms and instigating transformative shifts in the technological infrastructures that define the urban landscape (Späth et al. 2017; Tan Mullins et al. 2017). Given this context, it is evident that smarter eco-cities, as an advanced model of environmental sustainability transitions, are intrinsically intertwined with governance. This assertion is justified by the fact that effective governance structures and processes are essential for implementing and managing the complex interplay of policies, technologies, and societal changes required to realize the vision of smarter eco-cities. This interdependence, evidenced by a self-reinforcing relationship between environmental governance and eco-cities or smart eco-cities (Bibri 2021a; Deng et al. 2021; Flynn et al. 2016; Kramers et al. 2016; Joss 2011; Caprotti et al. 2017; Späth and Rohracher 2011; Späth et al. 2017), is central to shaping their planning and development strategies.

From a societal perspective, governance denotes "the system of values, policies, and institutions by which a society manages its economic, political, and social affairs through interactions within and among the state, civil society, and private sector" (UNDP 2018). Environmental governance encompasses decision-making, policy implementation, and actions aimed at addressing environmental challenges in societies. This multifaceted concept involves a complex interplay of structures, processes, and institutions that enable political actors to exert influence over environmental activities and outcomes (de Wit 2020; Lemos and Agrawal 2006; Meuleman and Niestroy 2015). Structures are the organizational arrangements, frameworks, and hierarchies that define how decisions are made, authority is distributed, and responsibilities are allocated to manage and regulate environmental issues (e.g., Gorris et al. 2019; Green and Hadden 2021; Gupta and van der Heijden 2019; Koch et al. 2021). They encompass legal, policy, and institutional frameworks guiding decision-making, resource allocation, and environmental management.

Processes involve the methods, procedures, mechanisms, and systematic approaches through which decisions are formulated, implemented, and evaluated within organizational structures (e.g., Asaduzzaman and Virtanen 2016; de Wit 2020; Vatn 2016; van Assche et al. 2020). They encompass activities such as data collection and analysis (e.g., tracking emissions, identifying climate vulnerabilities), stakeholder consultations (involving experts and communities), policy formulation (e.g., strategies for emission reduction and climate mitigation), policy execution (encompassing plans and measures), as well as monitoring to ensure effectiveness.

Environmental governance institutions hold a pivotal role in steering and overseeing environmental activities and outcomes (e.g., Azizi, Biermann and Kim 2019; de Wit 2020; Green and Hadden 2021; Simmons et al. 2018; Vatn 2016). Tasked with developing, enforcing, and administering environmental policies and regulations across diverse strata, these institutions collaborate closely with governmental bodies, NGOs, international entities, and other stakeholders. This cooperative approach is geared toward tackling environmental predicaments and fostering sustainable practices. In essence, governance structures, including institutions, delineate the organizational hierarchy and hence provide the organizational setup, while governance processes are the dynamic activities and workflows that outline how decisions are made and carried out within that hierarchy. Both are crucial for effective governance in various contexts.

Furthermore, implementing effective environmental governance strategies is crucial for achieving environmental objectives. Often integrated into comprehensive frameworks, these strategies encompass sustainable resource management, pollution control, biodiversity conservation, climate change mitigation and adaptation, circular economy initiatives, green infrastructure development, stakeholder engagement, policy coherence, science-based policies, monitoring and reporting, and capacity building (e.g., Bodin 2017; Beunen and Patterson 2019; de Wit 2020; Evans 2011; Nishant et al. 2020; Vatn 2016).

Enhancing the efficacy of environmental governance structures, processes, institutions, and strategies can be achieved through the integration of City Brain, SUM, and platform urbanism in the context of smarter eco-cities. This integration is justified by their collaborative functions, which synergistically contribute to addressing complex environmental challenges and achieving SDGs more effectively.

6.5.5.2 City Brain

City Brain plays an instrumental role in shaping the contours of contemporary environmental urban governance in terms of its processes. It employs AI technology directly in urban governance, significantly enhancing the informatization and intelligence of its processes (Zhu 2021). Through its collaborative functions, City Brain facilitates the coordination of various stakeholders, including government agencies, businesses, communities, and citizens, in addressing environmental issues. It equips city officials with the tools to implement measures that can effectively mitigate environmental impacts. This occurs through data-driven insights into key environmental factors and parameters, such as energy consumption, transport efficiency, traffic patterns, waste generation, and pollution control. These elements form the cornerstone of environmental governance strategies.

By harnessing advanced AI and AIoT technologies, it empowers cities to augment their capabilities for proactive decision-making, resource allocation, disaster management, and environmental stewardship. City officials can leverage data-driven insights to identify interventions, formulate effective and responsive policies, and make more informed decisions. This is at the core of environmental governance in terms of the

the methods, procedures, mechanisms, and practices through which decisions are formulated, implemented, and monitored (e.g., Asaduzzaman and Virtanen 2016; de Wit 2020). The dynamic relationship between City Brain and environmental governance fosters evidence-based policies that are more adaptive. Additionally, Xu et al. (2024) unveil the multi-scale policy framework of urban AI transitions, providing insights into the proactive potential of City Brain in addressing urban challenges. The study advances the understanding of anticipatory governance in the era of urban AI.

Drawing from diverse real-world implementations (e.g., Caprotti and Liu 2022; Kierans et al., 2023; Liu et al. 2022; Xie 2020; Xu 2022), City Brain integrates components of environmental governance by engaging a diverse array of stakeholders from both the public and private sectors. This collaborative endeavor seeks to enhance the management and coordination of a broad spectrum of urban systems and resources to foster sustainability and resilience. For example, Kierans, Jüngling, and Schütz

(2023) highlight the role played by City Brain in enhancing stakeholder collaboration among government agencies, businesses, communities, and citizens to ensure the adoption of environmental sustainability practices city-wide. Moreover, in the case of Haidian District, Liu et al. (2022) employ the concept of City Brain to analyze the avenues for enhancing and modernizing urban governance, emphasizing the importance of its conceptual advancement, multi-city approach, and process digitalization. Its implementation in Haidian District has shifted the governance mode from passive to active, thereby fostering knowledge in urban operation, evaluation, and development. According to the authors, City Brain drives innovation across transportation, urban management, ecological environmental protection, and smart energy. Furthermore, Xu (2022) examines the solutions and challenges surrounding the Hangzhou City Brain, exploring its construction and application through the lens of technology empowerment and multi-level governance. The author offers recommendations for enhancing City Brain initiatives to support the development of smart cities, emphasizing the crucial role of governance in shaping their advancement success.

Overall, City Brain's integration into contemporary environmental urban governance marks a significant advancement in urban management strategies. It enhances the intelligence and efficacy of decision-making by employing AI and AIoT in urban governance processes. Through its collaborative functions, it facilitates coordination among various stakeholders, enabling effective mitigation of environmental impacts. By harnessing advanced AI and AIoT technologies, smarter eco-cities can make proactive decisions and allocate resources efficiently. This dynamic relationship fosters evidence-based policies, contributing to more adaptive and resilient urban environments. The multi-scale policy framework of AI and AIoT transitions further emphasizes the proactive potential of City Brain in addressing urban challenges. Real-world implementations demonstrate its effectiveness in enhancing stakeholder collaboration and driving innovation across various urban sectors. As City Brain continues to evolve, its role in shaping the future of environmental governance remains paramount.

Theoretically, based on the underlying AIoT functionalities connected to the fields of environmental sustainability, climate change, and smart cities (Bibri et al. 2023a, b; El Himer et al. 2022; Gourisaria et al. 2022; Leal Filho et al. 2022; Mishra and Singh 2023; Puri et al. 2020; Samadi 2022; Seng et al. 2022; Zaidi et al. 2023; Zhang and Tao 2021), the key uses of City Brain in environmental governance include:

- Environmental monitoring: City Brain utilizes real-time data from various sources to continuously monitor environmental parameters, such as air quality, energy usage, water quality, waste management, and pollution and noise levels. This allows local authorities to quickly detect and respond to environmental issues.

- Resource optimization: City Brain optimizes resource allocation and utilization, leading to more efficient energy consumption, transport management, water management, and so on. It can help identify areas of resource inefficiency and suggest strategies for improvement.

- Waste management optimization: By providing comprehensive insights into waste management strategies, City Brain aids in developing holistic approaches to address waste-related challenges effectively. It can optimize waste collection routes, implement waste-to-energy initiatives, and enhance recycling programs to minimize the environmental impact of waste.

- Pollution control: By analyzing data on pollutant levels and their sources, City Brain aids in developing targeted pollution control measures. It can also support the implementation of emission reduction policies to combat air and water pollution.

- Sustainable urban planning: City Brain contributes to long-term urban planning by providing insights into sustainable development strategies. It aids in integrating renewable energy solutions, designing eco-friendly infrastructure, and promoting green spaces.

- Predictive analytics: City Brain analyzes historical data and real-time information to identify patterns, trends, and potential future outcomes in relation to transport, energy, water, and waste. By predicting future scenarios based on data analysis, City Brain can assist in making proactive and informed decisions, optimizing resource allocation, and implementing effective strategies for urban management and governance.

- Public engagement: City Brain can support citizen engagement initiatives by providing real-time data and visualizations. This fosters community involvement and cooperation in environmental protection efforts.

However, it is imperative that the design and deployment of City Brain prioritize critical ethical and social considerations, including privacy, security, fairness, algorithmic bias, accountability, and transparency. These considerations, which are associated with the utilization of AI and AIoT technologies (Ahmed et al. 2020; Bibri et al. 2023a; Parihar et al. 2023; Singha and Singha 2024; Yigitcanlar et al. 2020), are crucial for fostering trust among the stakeholders involved in environmental urban governance. They are also vital for ensuring the responsible deployment of City Brain in urban governance. Zhu (2021) explores the nexus between AI and urban governance, examining their relationship and significance through the lenses of human initiative, human experience, and future prospects. The author examines various aspects, including governance, ethics, experience, philosophy, and innovation, as well as the risks and conflicts arising from the integration of AI into urban governance.

6.5.5.3 Smart Urban Metabolism

SUM plays a crucial role in environmental governance as it provides a data-driven approach to understanding and managing the complex interactions between urban activities and their environmental impact. The essence of SUM lies in its ability to identify, control, utilize, and evaluate data concerning resource flows in urban areas, while also considering the social, political, and economic dimensions of urban development (Anttiroiko, Ari-Veikko 2023). Through this comprehensive approach, it has the potential to advance the sustainability and resilience of smarter eco-cities and improve the effectiveness of their governance practices. SUM enhances environmental governance processes by facilitating real-time monitoring and analysis of environmental data, thereby enabling policymakers to make informed decisions and implement effective interventions to mitigate environmental risks. Additionally, SUM fosters stakeholder collaboration, which facilitates the co-creation of environmental policies and initiatives that address the diverse needs and concerns of local communities.

The interplay between SUM and environmental governance, as exemplified in real-world urban settings and substantiated by several case studies (e.g., Anttiroiko, Ari-Veikko. 2023; Beck et al. 2023; Bibri et al. 2024a; Bibri and Krogstie 2020a; D'Amico et al. 2021; Peponi et al. 2022; Shahrokni et al. 2015; Wiedmann et al. 2021), is marked by interdependency and mutual reinforcement in various ways, including:

- Resource flows analysis: SUM enables a comprehensive analysis of resource flows within the urban environment, providing crucial insights into the consumption, distribution, and disposal of resources.

- Sustainability metrics: It contributes to the development and monitoring of sustainability metrics (e.g., GHG, energy consumption, air quality, waste generation and recycling, biodiversity and green space), enabling urban governance to evaluate the ecological impact of diverse activities and policies. These metrics provide quantitative data that informs decision-making processes and helps urban governance agencies prioritize interventions to promote sustainable development.

- Data-driven policy decisions: It provides comprehensive data-driven insights that can inform the formulation, implementation, and adjustment of environmental policies. This ensures that policies are not only responsive to current urban challenges but also align with SDGs. By leveraging real-time data and advanced analytics, policymakers can make informed decisions that promote sustainability, enhance urban resilience, and effectively address environmental issues.

- Pinpointing critical areas: By identifying areas with high resource consumption or environmental stress, it assists in pinpointing urban hotspots that require targeted governance interventions.

- Monitoring environmental impact: It facilitates continuous monitoring of the environmental impact of urban activities, aiding in the enforcement of regulations and the evaluation of their effectiveness.

- Integration with governance systems: It integrates seamlessly with urban governance systems, providing decision-makers with real-time information to enhance their capacity to respond promptly to environmental challenges.

- Scenario Planning: It allows for the simulation of different urban development scenarios, which enable governance to assess potential environmental outcomes and adopt strategies for optimal sustainability.

- Stakeholder engagement: By fostering transparency and data-driven decision-making, it encourages stakeholder engagement in environmental governance initiatives. This creates a more inclusive and collaborative approach to address environmental challenges and implement sustainable initiatives.

These mechanisms directly relate to governance processes by providing insights for policy decisions, informing resource management, monitoring environmental impact, and facilitating stakeholder engagement in sustainability initiatives. They highlight the innovative potential of SUM in enhancing environmental efficiency and resilience and its role in informing city officials about system inefficiencies in urban contexts. As a tool that integrates advanced technologies, data management, and analytics models, as well as environmental science and industrial ecology, SUM contributes to shaping the future of urban governance toward greater environmental well-being.

6.5.5.4 Platform Urbanism

The dynamic interaction between platform urbanism principles and urban governance dimensions has direct implications for advancing environmental sustainability in urban contexts. Platform urbanism, characterized by the role of digital platforms in facilitating various urban functions and services, plays a significant role in shaping environmental governance structures, processes, and institutions. These platforms act as intermediaries, connecting citizens, governments, and other stakeholders and enabling the exchange of information, resources, and services related to environmental management. By providing a digital infrastructure for collaboration and participation, platform urbanism enhances transparency, accountability, and inclusivity in environmental decision-making processes.

There is a discernible trend in the adoption of governance platforms by an increasing number of public institutions, constructed by diverse actors operating across various geographical scales and addressing a spectrum of issues, including sustainable development and environmental sustainability (Ansell and Miura 2019). According to Noori et al. (2020), the Amsterdam Smart City (ASC) platform illustrates a collaborative platform bringing together various stakeholders to develop smart solutions for urban sustainability. This platform integrates technology, data, and citizen participation, fostering a collective approach to advancing sustainability in Amsterdam City. Additionally, it enables the integration of diverse data sources and analytics tools, empowering policymakers to make evidence-based decisions and implement targeted interventions to address environmental challenges effectively.

Furthermore, governance platforms embody a new organizing logic that promotes distributed participation (Ansell and Miura 2019) and cross-cutting innovation (Hodson et al. 2021; Putra et al. 2018). Participation platforms can elevate citizen engagement in policy deliberation (Blasio and Selva 2019) and urban policy decisions (Bibri and Krogstie 2021), while multi-stakeholder platforms facilitate productive exchanges among diverse expert groups (ITU 2022). Haveri and Anttiroiko (2023) explore the role of platforms in local public governance in Finland's largest cities, classifying platforms as a distinct fourth mode of governance. The authors argue that these platforms have intrinsic features of networks and hierarchies, contributing to a novel approach to public governance. This hybrid model exhibits traits that defy complete attribution to conventional governance frameworks in terms of structures and processes (Ansell and Gash 2018). In a similar vein, Jiang et al. (2022) analyze the governance modes in Helsinki's smart city development, emphasizing the role of Forum Virium Helsinki as an integrative innovation platform. Platforms further support scalable and adaptive governance concerning public innovation and collaborative governance (Ansell and Gash 2018; Ansell and Miura 2019). Platform urbanism in urban governance involves leveraging digital platforms for citizen engagement, initiating open data initiatives, coordinating government agencies, and integrating policies (Brynskov et al. 2018; Hodson et al. 2021; Katmada et al. 2023; Repette et al. 2021).

Overall, platform urbanism transforms conventional governance models in smarter eco-cities by fostering decentralized, networked approaches to their environmental governance. As a rapidly evolving landscape, it introduces novel models that foster collaboration, innovation, and citizen participation, thereby contributing significantly to the pursuit of environmental sustainability goals in urban settings. Ultimately, digital platforms have ushered in transformative changes, redefining environmental urban governance paradigms comprehensively.

However, it is essential to acknowledge that the ethical and social issues related to City Brain as a platform extend to platform urbanism, as its functioning is enabled by AI and AIoT tools. Specifically, in relation to stakeholder engagement and citizen participation, algorithms are acknowledged to raise concerns regarding privacy, fairness, transparency, accountability, safety, equality, and other public and civic values across various domains. (e.g., König 2020; Noble 2028; Srivastava 2021; Van Dijck et al 2018), including platform urbanism. This relates to algorithmic governance in platform urbanism, which is associated with algocracy in terms of limiting citizen participation and curtailing public decision-making (Bibri 2023). This governance system is "organized and structured on the basis of computer-programmed algorithms…, a system in which algorithms are used to collect, collate, and organize the data upon which decisions are typically made and to assist in how that data is processed and communicated through the relevant governance system" (Danaher 2016, p. 247). Algorithmic mediation is increasingly being incorporated as a tool for governing urban societies—with both unintended and intended consequences (Bibri 2023). At its core, AI-driven or algorithmic governance is poised to extend its influence to various domains including laws, regulations, behaviors, and numerous facets of everyday life (e.g., König 2020; Hildebrandt 2018; Yeung 2017).

According to Graham (2020), urban platforms lack accountability, exert control over urban interactions, and typically operate as undemocratic entities that are distant and disinterested in promoting local voices. The author suggests three overarching strategies—regulate, replicate, and resist—that can be employed to construct alternative platform futures. Ethical and social considerations are crucial for fostering trust among stakeholders and citizens involved in environmental urban governance and ensuring the responsible deployment of governance platforms.

6.5.6 A Pioneering Framework for Data-Driven Environmental Governance in Smarter Eco-Cities

This section utilizes the insights derived from both the configurative and aggregative synthesis conducted in the preceding subsections to formulate the proposed novel conceptual framework. Configurative and aggregative synthesis techniques perform distinct yet complementary functions in shaping the proposed framework. Aggregative synthesis, through the consolidation and integration of diverse studies, distills essential concepts, theories, and models from existing literature. Accordingly, it identifies components pertinent to the framework and utilizes evidence to make assertions in conceptual categories. This process establishes a thematic overview crucial for the development of the proposed framework. Concurrently, configurative synthesis, by discerning common themes across synthesized studies, contributes to constructing an interconnected framework that captures contextual aspects and relationships among core components and elucidates the overall significance of research evidence. Together, aggregative and configurative synthesis approaches establish a robust foundation for the comprehensive conceptual framework. Therefore, they advance the understanding of the complex dynamics of environmental governance in smarter eco-cities in the era of digital evolution.

Considering the above, the study aims to harness the existing theoretical and empirical insights into AIoT, City Brain, SUM, and platform urbanism synergies. This involves a thorough examination of their foundations, functions, interconnections, and contributions toward enhancing environmental governance strategies for the advancement of smarter eco-cities. Through the integration of these data-driven environmental governance systems, the conceptual framework illustrated in Figure 6.6 establishes a holistic, cohesive, and interconnected ecosystem.

AIoT enhances the functionalities of City Brain, SUM, and platform urbanism through seamless integration. By enabling the harmonious

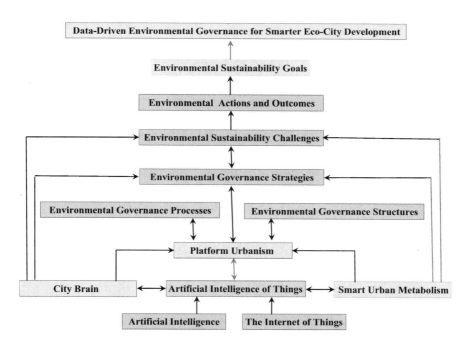

FIGURE 6.6 A pioneering framework for data-driven environmental governance in smarter eco-cities based on the integration of AIoT-powered City Brain, SUM, and platform urbanism.

coupling of AI algorithms with IoT devices, AIoT extends beyond mere data collection to enhance the processing and analysis of real-time data from diverse sources. This empowers City Brain, SUM, and platform urbanism to gather comprehensive insights into urban activities, environmental conditions, and infrastructure performance. Moreover, by leveraging AI capabilities, AIoT systems can perform advanced predictive analytics to forecast future trends, identify potential issues, and optimize resource allocation in urban environments. This aids City Brain, SUM, and platform urbanism in proactively addressing challenges and enhancing decision-making processes. Additionally, AIoT enables automated control systems that can respond to changing conditions in real-time, such as adjusting traffic signals, optimizing energy usage in buildings, managing waste disposal, and monitoring environmental conditions. This automation enhances the efficiency and effectiveness of City Brain and SUM in managing urban operations and resources. Furthermore, AIoT plays a crucial role in enhancing the functionality of digital platforms in urban environments by leveraging data-driven insights for effective stakeholder engagement, resource optimization, and collaborative approaches.

The convergence of City Brain, SUM, and platform urbanism empowered by AIoT represents a significant advancement in environmental urban governance processes and structures. City Brain optimizes urban management through data-driven insights, supporting informed decision-making and advanced automation processes. Utilizing its collaborative features, it streamlines the coordination among diverse stakeholders in tackling environmental challenges. SUM provides a comprehensive view of urban systems, aiding governance bodies in formulating policies for sustainable resource management, pollution control, and resilient urban development. Platform urbanism, facilitated by AIoT, promotes citizen participation and stakeholder coordination in environmental governance, fostering open data initiatives, policy deliberation, and cross-cutting innovation for a more inclusive and transparent governance model aligned with environmental sustainability goals.

Environmental governance strategies encompass a variety of organized and coordinated approaches and initiatives aimed at managing and protecting the environment. These strategies typically involve a combination of policy frameworks, regulatory measures, collaborative efforts, and technological innovations designed to promote sustainability and address environmental challenges. These include natural resource management, intensive energy consumption, transport inefficiency, infrastructure flaws, waste generation, pollution, biodiversity loss, ecosystem degradation, and climate change. However, key components of environmental governance strategies include (e.g., Azizi et al, 2019; de Wit 2020; Evans 2011; Gupta and van der Heijden 2019; Morrison et al. 2019; Nishant et al. 2020; van Assche et al. 2020; Vatn 2016):

- Implementing policies and regulations
- Engaging stakeholders in decision-making processes and collaborative initiatives
- Establishing mechanisms for monitoring and ensuring compliance
- Harnessing technological advancements and fostering innovations to develop sustainable solutions
- Implementing adaptive management approaches
- Promoting international cooperation and collaboration through agreements and partnerships

Environmental governance challenges, on the other hand, are the multifaceted obstacles and complexities faced in the process of implementing governance strategies to tackle environmental sustainability challenges. These obstacles and complexities underscore the need for integrative approaches to addressing these challenges effectively. Indeed, they include fragmented and siloed systems, lack of real-time data collection and analysis (Nishant et al. 2020), exclusionary decision-making processes (de Wit 2020), lack of citizen engagement ad participation (Fung 2006; Newig and Fritsch 2009; UN-Habitat 2023), ineffective multi-stakeholder collaboration (Evans 2011; Gupta and van der Heijden 2019), policy incoherence and fragmentation (Azizi et al, 2019), adaptive co-management approaches (Armitage 2009; Folk 2005; de Wit 2020), lack of evidence-based decision-making (Nishant et al. 2020), complexity and interconnectedness of ecosystems (de Wit 2022; Morrison et al. 2019), as well as the absence of collaborative data sharing infrastructure. These challenges pertain to smarter eco-cities as a paradigm of environmental urban governance.

Distilled based on the evidence from the configurative and aggregative synthesis, Table 6.4 outlines the collective contributions of City Brain, SUM, and platform urbanism to addressing a spectrum of environmental governance challenges. It serves as a succinct reference, illuminating

TABLE 6.4 Collaborative Contributions of City Brain, SUM, and Platform Urbanism to Addressing Environmental Governance Challenges

Environmental Governance Challenges	City Brain	SUM	Platform Urbanism
Lack of collaborative data sharing infrastructure	✓	✓	✓
Fragmented and siloed systems	✓	X	✓
Lack of real-time data collection and analysis	✓	✓	✓
Exclusionary decision-making processes	X	✓	✓
Lack of citizen engagement and participation	X	✓	✓
Ineffective multi-stakeholder collaboration	✓	✓	✓
Adaptive approaches to management	✓	✓	✓
Policy incoherence and fragmentation	X	✓	✓
Lack of evidence-based decisions (policies)	✓	✓	X
Complexity and interconnectedness of ecosystems	X	✓	✓

the nuanced roles each data-driven governance system plays in effectively addressing or mitigating these challenges. It also illuminates the significance of integrating these data-driven governance systems to tackle the challenges faced by smart eco-cities in the pursuit of environmental sustainability.

In addition, the connection between environmental governance strategies and actions is pivotal in driving positive outcomes. The effectiveness of these strategies becomes evident in tangible environmental outcomes. These include optimized resource allocation, improved air quality, reduced pollution levels, enhanced transportation systems, preserved natural resources, protected biodiversity and ecosystems, and mitigated climate change risks. Implementing concrete actions based on these strategies not only reinforces their impact on environmental sustainability goals but also translates intentions into measurable results. The actions carried out in established environmental governance processes and structures are pivotal, as they shape and define the overall success of efforts aimed at attaining environmental sustainability.

For example, effective climate action requires robust environmental governance strategies. Integrating AI and AIoT can enhance our capability to monitor, report, and verify climate actions, thereby strengthening the global response to climate change. Integrating robust environmental governance strategies with advanced AI and AIoT technologies is essential for effective climate action. These combined efforts can help cities navigate the complexities of climate change, ensuring responses are timely, equitable, and effective. AI and AIoT can support more informed and impactful decision-making at all levels of governance by enhancing our capacity to understand, predict, and respond to climate dynamics.

Building on AIoT's role as a pivotal enabler for integrating City Brain, SUM, and platform urbanism in the environmental governance of smarter eco-cities, its proven potential underscores its significance to advance environmental sustainability goals. This is evident in navigating environmental sustainability and climate change challenges (Bibri et al. 2023a, b; El Himer et al. 2022; Gourisaria et al. 2022; Leal Filho et al. 2022; Mishra and Singh 2023; Puri et al. 2020; Samadi 2022; Seng et al. 2022; Zaidi et al. 2023; Zhang and Tao 2021). Achieving environmental goals necessitates striking a delicate balance between fulfilling present societal demands while safeguarding natural resources, ecosystems, and biodiversity for posterity. These objectives are centered on curtailing environmental degradation, enhancing adaptability to climate change, and nurturing a symbiotic relationship between human endeavors and the planet. By leveraging the innovative and collaborative potential of AIoT, stakeholders, policymakers, and institutions are empowered to contribute significantly to the advancement of environmental governance in smarter eco-cities.

However, the integration of AI and AIoT into environmental strategies and actions is not without challenges. Issues such as energy consumption of these technologies, data availability, privacy and security, and potential biases in AI and AIoT systems need to be addressed to ensure equitable and effective use. Environmental governance structures and processes must therefore evolve to incorporate these considerations, promoting sustainable and inclusive AI applications for environmental actions.

6.6 DISCUSSION

In this discussion section, a comprehensive analysis and interpretation of the findings presented in the study are explored. The significance of these findings and their relevance to the research domain is highlighted. By comparing the findings with existing literature and theoretical frameworks, the study is situated within the broader scholarly landscape of smarter eco-cities, environmental governance, and emerging technologies. Additionally, the methodology employed in the study, combining the literature review and case studies, is critically evaluated, while acknowledging the limitations inherent in this mythological approach. This discussion provides valuable insights into the transformative potential of AIoT integration in data-driven environmental governance, offering implications for research, practice, and policymaking, along with suggestions for future research avenues.

6.6.1 Interpretation of Results

The findings reveal that the synergistic integration of City Brain, SUM, and platform urbanism through AIoT presents promising opportunities and prospects for advancing data-driven environmental governance in smarter eco-cities. Through an analysis of their functionalities, the study elucidates a symbiotic relationship among these data-driven governance systems. It highlights their collective capabilities to enable real-time monitoring, data-driven decision-making, and enhanced stakeholder engagement and collaboration. This synergy is pivotal in effectively addressing the environmental challenges in urban environments. It particularly underscores the significance of integrating AIoT in environmental governance strategies. By harnessing the complementary strengths of City Brain, SUM, and platform urbanism, local governments can optimize resource management, mitigate environmental impacts, and foster resilience in the face of complex ecological dynamics. They can further bolster environmental governance strategies by mitigating the complexities and obstacles surrounding the associated actions, thereby facilitating the attainment of better environmental outcomes. These findings underscore the transformative potential of AIoT-driven approaches in shaping the future of data-driven environmental governance, offering a pathway toward more efficient, resilient, and sustainable urban development practices.

6.6.2 Comparative Analysis

The findings of this study align with previous research highlighting the innovative role of AI and AIoT technologies in shaping urban governance and sustainability initiatives (Ajuriaguerra Escudero and Abdiu 2022; Ben and Peng 2020; Cugurello 2024; Liu et al. 2022; Nishant et al. 2020; Samuel et al. 2023; Xu et al. 2024; Zhu 2021). The existing literature also corroborates the significance of leveraging AI, IoT, big data science and analytics, and AIoT solutions to address the multifaceted environmental challenges confronting modern urban environments, including smart cities and smarter eco-cities (Bibri 2019; Bibri et al. 2023a, b; Cheng et al. 2023; El Himer et al. 2022; Gourisaria et al. 2022; Leal Filho et al. 2022; Mishra and Singh 2023; Puri et al. 2019; Samadi 2022; Seng et al. 2022; Zaidi et al. 2023). However, it predominantly centers on City Brain (e.g., Kierans et al.

2023; Liu et al. 2022; Xu 2024), Smart Urban Metabolism (SUM) (D'Amico et al. 2020, 2021; Shahrokni et al. 2015), and platform urbanism (e.g., Caprotti and Liu 2022; Repette et al. 2021; Komninos and Kakderi 2019; Haveri and Anttiroiko 2023) in isolation, rather than integrating them into comprehensive frameworks or integrated solutions to advance data-driven environmental urban governance practices. Moreover, only a limited subset of these studies, notably those focusing on City Brain, incorporate AI and AIoT technologies into the discourse of urban governance platforms.

Building upon this foundational and dynamic literature, the study introduces a novel framework that extends beyond traditional approaches by proposing a comprehensive integration of multiple AIoT-driven governance systems to advance environmental governance in smarter eco-cities. The notable gap in the current landscape of research and practice highlights the originality and significance of the study, which proposes a holistic approach to data-driven environmental governance by synergistically integrating City Brain, SUM, and platform urbanism through the lens of AIoT. By offering an integrated perspective, the proposed framework contributes to filling a critical void in the existing literature, paving the way for more effective environmental urban governance strategies in smarter eco-cities. This comparative analysis not only aligns with the existing theoretical frameworks but also introduces novel avenues for advancing environmental goals in urban contexts. Through this juxtaposition with prior literature, the study contributes to the ongoing discourse on the transformative impact of AI and AIoT technologies on future forms of urban governance paradigms, offering a nuanced understanding of their implications for environmental sustainability.

6.6.3 Methodological Evaluation and Limitations

While the methodology employed in this chapter, encompassing a thorough literature review and the analysis of case studies, laid a robust foundation for understanding the theoretical underpinnings of the synergistic integration of AIoT-driven governance systems, several limitations warrant critical examination. While the literature review provided valuable insights into existing research and theoretical frameworks, it is crucial to acknowledge the inherent biases and gaps that may exist within the selected literature. The case studies supplemented this theoretical framework by offering real-world insights into the practical application of these technologies. However, the limited availability of a comprehensive set of case studies poses a constraint on the depth of empirical evidence available for analysis. Despite efforts to include diverse case studies, the scarcity of comprehensive examples may undermine the breadth and applicability of these findings.

Furthermore, the evolving nature of AI and AIoT technologies presents a fundamental challenge to the generalizability of the findings. The rapid pace of technological advancement often outpaces scholarly inquiry, rendering certain aspects of the analysis potentially incomplete. Additionally, the dynamic landscape of technological innovation introduces uncertainties and complexities that may not be adequately captured within the confines of this study. As such, it is essential to approach the findings with a degree of caution and recognize the inherent limitations imposed by the evolving nature of technology and the availability of empirical evidence. Future research endeavors should strive to address these limitations by incorporating a more extensive range of case studies, embracing emerging methodologies, and continuously reassessing the implications of technological advancements for environmental governance practices.

6.6.4 Implications for Research, Practice, and Policymaking

The findings of this study carry significant implications for research, practice, and policymaking across multiple domains. This study offers implications for research by highlighting the importance of further investigation into the mechanisms and dynamics of AIoT-driven governance systems in urban contexts and the long-term impacts of AIoT integration on environmental governance practices. Researchers could explore the nuanced interactions between AIoT technologies, environmental conditions, and stakeholder behaviors. They could also explore the broader implications of AIoT integration for environmental governance frameworks and theoretical models. This can contribute to a deeper understanding of the role of emerging technologies in shaping environmental governance practices and policies.

Moreover, those practitioners operating in the realms of environmental urban planning and management stand to benefit significantly from the insights garnered from this study. The comprehensive framework proposed in this study offers policymakers a nuanced understanding of the synergistic effects of integrating AIoT-driven systems, enabling them to devise more effective and targeted interventions to address pressing environmental challenges. Practitioners are equipped with a diverse toolkit of technological solutions to tackle complex environmental issues in smarter eco-cities by leveraging the capabilities of AIoT integration, These tools include real-time monitoring systems, data-driven decision-making approaches, and participatory platforms that facilitate stakeholder engagement and collaboration. Equipped with this knowledge, practitioners can explore novel approaches to urban governance and sustainability, driving innovation and fostering resilience in urban environments.

The implications extend beyond the realms of research and practice to encompass policymaking. The integration of AIoT technologies presents policymakers with a promising opportunity to enact transformative policy measures that promote sustainability and resilience in urban environments. Policymakers can catalyze the adoption of innovative technologies, foster collaboration between public and private stakeholders, and create an enabling environment for sustainable urban development. This can be accomplished by aligning policy objectives with the guiding principles of AIoT integration. Ultimately, the findings of this study offer policymakers a roadmap for navigating the complexities of contemporary environmental urban governance, guiding them toward a future characterized by more efficient, inclusive, and environmentally sustainable cities.

6.6.5 Suggestions for Future Research Directions

Future research endeavors in the realm of AIoT integration and environmental governance should prioritize several key areas to advance the field and address emerging challenges. Firstly, there is a clear need for further research to explore and validate the proposed framework through empirical studies and field investigations. While the conceptual framework presented in this study offers a theoretical basis for understanding the potential benefits of AIoT integration, empirical validation is essential to ascertain its practical efficacy and applicability in real-world contexts. By conducting rigorous empirical studies, researchers can evaluate the performance of AIoT-driven systems in enhancing environmental governance outcomes and identify potential areas for optimization and improvement. This empirical validation will provide valuable insights into the feasibility and effectiveness of implementing the proposed framework in diverse urban environments, contributing to the development of more robust and scalable solutions for environmental governance in smarter eco-cities.

Additionally, future research should delve into the socioeconomic and ethical implications of AIoT integration in environmental governance. While the technological aspects of AIoT integration have been adequately studied, relatively little attention has been paid to the broader socioeconomic ramifications of deploying these technologies in urban environments. Investigating issues such as equity, fairness, and inclusivity in the context of AIoT-driven environmental governance initiatives is crucial for ensuring that this advanced technological framework contributes to more equitable and sustainable urban development. Researchers can identify opportunities to enhance the inclusiveness and effectiveness of environmental governance strategies in urban environments by exploring the social dimensions of AIoT integration. Future research should also give careful consideration to ethical implications. As AIoT becomes increasingly integrated into urban environments, it is crucial to address ethical concerns related to privacy, data security, transparency, and accountability. Researchers could develop frameworks for responsible and ethical adoption and implementation of AIoT in environmental urban governance.

Moreover, comparative studies across different urban contexts hold significant potential for enriching the understanding of the scalability and transferability of AIoT integration in environmental urban governance. Researchers can identify contextual factors that influence the adaptability of these technological and urban transformations by examining the implementation of AIoT-driven environmental governance solutions in diverse urban settings. Comparative studies can also provide valuable insights into best practices and lessons learned, enabling policymakers and practitioners to make more informed decisions when deploying AIoT-driven solutions for environmental governance in their respective contexts.

Finally, assessing the long-term sustainability of AIoT integration in environmental governance is paramount for ensuring the resilience and durability of this advanced technology. Longitudinal studies tracking the evolution of AIoT-driven governance systems over time can shed light on their durability and ability to adapt to changing environmental conditions. Researchers can inform future policy decisions and technological developments, ensuring that AIoT-driven environmental governance remains effective and sustainable in the face of evolving challenges. This entails monitoring the long-term impacts of AIoT integration on environmental outcomes. Of equal importance is the interdisciplinary research that bridges the gap between technology, evidence-based policy, and environmental science, which is essential for addressing complex urban challenges in an integrated and holistic manner. Overall, this study offers a roadmap for various future research endeavors, and by addressing these key research areas, researchers can contribute to the development of more effective approaches to environmental governance in smarter eco-cities.

6.7 CONCLUSION

This chapter explored the transformative potential of AIoT in seamlessly integrating City Brain, SUM, and platform urbanism to advance data-driven environmental governance in smarter eco-cities. In doing so, it proposed a pioneering framework that effectively leverages the synergies among these emerging AIoT-driven governance systems to enhance environmental sustainability practices in smarter eco-cities. This chapter synthesized empirical evidence and theoretical insights to demonstrate the benefits of leveraging these AIoT-driven systems to enhance environmental governance strategies and tackle complex environmental challenges. AIoT integration offers promising avenues for enhancing sustainability and resilience in urban environments by facilitating real-time monitoring, data-driven decision-making, and enhanced stakeholder engagement and collaboration. The findings revealed that the synergistic and collaborative integration of City Brain, SUM, and platform urbanism through AIoT presents promising opportunities and prospects for advancing data-driven environmental governance in smarter eco-cities.

In response to RQ1, the co-evolution of AI and IoT into AIoT is driven by various factors, including advancements in computing power, the proliferation of sensor technology, the exponential growth of IoT-generated data, and the increasing demand for interconnected and intelligent systems in smarter eco-cities. The architecture of AIoT consists of specific technical components, including IoT sensors for data collection, AI algorithms for data analysis, cloud computing for data storage and processing, and communication protocols for seamless connectivity.

Regarding RQ2, a comparative analysis reveals both commonalities and differences among City Brain, SUM, and platform urbanism within the frameworks of AIoT and environmental governance in smarter eco-cities. While these data-driven systems utilize AIoT technologies for real-time monitoring and data-driven decision-making, they differ in terms of their specific functionalities, governance processes, and environmental outcomes.

Moving to RQ3, City Brain, SUM, and platform urbanism collectively contribute to data-driven environmental governance within the framework of smarter eco-cities by providing valuable insights, facilitating informed decision-making, and enabling proactive interventions to address environmental challenges. These AIoT-driven governance systems leverage real-time data analytics, predictive modeling, and stakeholder engagement to optimize resource management and enhance environmental sustainability.

Transitioning to RQ4, a pioneering framework should effectively utilize AIoT to synergize the functions of City Brain, SUM, and platform urbanism to advance data-driven environmental governance in emerging smarter eco-cities. By integrating these AIoT-driven systems, the framework can enable holistic and data-driven approaches to environmental governance, fostering innovation, resilience, and sustainability in smarter eco-cities.

Furthermore, the comparison with the existing literature highlighted the originality and significance of the proposed framework, which extends beyond traditional approaches to propose an integrated solution for data-driven environmental governance in smarter eco-cities. While the literature primarily focuses on City Brain, SUM, and platform urbanism in isolation, this chapter fills this critical gap by offering a holistic perspective that leverages the collective capabilities of these AIoT-driven governance systems.

Despite methodological limitations, the study provides valuable insights that can inform policy decisions, guide practice, and inspire further research in the pursuit of environmental sustainability transitions. It offers a roadmap for advancing environmental sustainability goals in emerging smarter eco-cities by providing policymakers, practitioners, and researchers with fertile insights.

Looking ahead, future research endeavors should prioritize refining and validating the proposed framework through rigorous empirical studies and field investigations. Given the dynamic nature of smarter eco-cities as experimental sites for innovation, it is essential to conduct in-depth analyses in real-world urban environments to assess the effectiveness and scalability of the integrated AIoT-driven governance systems. Researchers can gain valuable insights into the practical implementation and impact of the proposed framework on environmental governance practices by conducting comprehensive field investigations and gathering empirical data from diverse urban contexts. Additionally, conducting comparative studies across different urban contexts is essential for enhancing the understanding of the scalability and transferability of AIoT-driven governance solutions.

This chapter contributes to the growing body of literature on smarter eco-cities, environmental governance, environmental sustainability, and urban AIoT by proposing a pioneering framework for synergizing City

Brain, SUM, and platform urbanism. This framework holds promise for transforming environmental governance practices in smarter eco-cities. By leveraging AIoT technologies, local governments can enhance their environmental strategies by monitoring, managing, and mitigating environmental challenges, thereby promoting sustainability and resilience in urban environments.

REFERENCES

Ahmad, M. A., Teredesai, A., & Eckert, C. (2020). Fairness, accountability, transparency in ai at scale: Lessons from national programs. In: Proceedings of the 2020 Conference on Fairness, Accountability, and Transparency (pp. 690–690), Barcelona, Spain.

Ajuriaguerra Escudero, M. A., & Abdiu, M. (2022). Artificial intelligence in European urban governance. In: J. Saura, & F. Debasa (Eds.), *Handbook of Research on Artificial Intelligence in Government Practices and Processes* (pp. 88–104). IGI Global. https://doi.org/10.4018/978-1-7998-9609-8.ch006

Alahakoon, D., Nawaratne, R., Xu, Y., De Silva, D., Sivarajah, U., & Gupta, B. (2020). Self-building artificial intelligence and machine learning to empower big data analytics in smart cities. *Information Systems Frontiers, 25*(1), 221–240. https://doi.org/10.1007/s10796-020-10056-x.

Alahi, M. E. E., Sukkuea, A., Tina, F. W., Nag, A., Kurdthongmee, W., Suwannarat, K., & Mukhopadhyay, S. C. (2023). Integration of IoT-enabled technologies and artificial intelligence (AI) for smart city scenario: Recent advancements and future trends. *Sensors, 23*(11), 5206.

Ansell, C., & Gash, A. (2018). Collaborative platforms as a governance strategy. *Journal of Public Administration Research and Theory, 28*(1), 16–32.

Ansell, C., & Miura, S. (2019). Can the power of platforms be harnessed for governance? *Public Administration.* https://doi.org/10.1111/padm.12636

Anttiroiko, A.-V. (2016). City-as-a-platform: The rise of participatory innovation platforms in Finnish cities. *Sustainability, 8*(9), 922. https://doi.org/10.3390/su8090922

Anttiroiko, A.-V. (2023). Smart circular cities: Governing the relationality, spatiality, and digitality in the promotion of circular economy in an urban region. *Sustainability, 15*(17), 12680. https://doi.org/10.3390/su151712680

Armitage, D. R., Plummer, R., Berkes, F., Arthur, R. I., Charles, A. T., Davidson-Hunt, I. J., Diduck, A. P., Doubleday, N. C., Johnson, D. S., Marschke, M., McConney, P., Pinkerton, E. W., & Wollenberg, E. K. (2009) Adaptive co-management for social-ecological complexity. *Frontiers in Ecology and the Environment, 7,* 95–102. https://doi.org/10.1890/070089

Asaduzzaman, M., & Virtanen, P. (2016). Governance theories and models. In: *Global Encyclopedia of Public Administration, Public Policy, and Governance* (pp. 1–13). Springer.

Azizi, D., Biermann, F., & Kim, R. E. (2019). Policy integration for sustainable development through multilateral environmental agreements. *Global Governance, 25*, 445–475. https://doi.org/10.1163/19426720-02503005

Baccini, P., & Brunner, P. H. (2012). Metabolism of the Anthroposphere: Analysis, Evaluation, Design (2nd ed.). MIT Press.

Baker, S., & Xiang, W. (2023). Artificial intelligence of things for smarter healthcare: A survey of advancements, challenges, and opportunities. *IEEE Communications Surveys & Tutorials.* https://doi.org/10.1109/COMST.2023.3256323

Barns, S. (2019) Negotiating the platform pivot: From participatory digital ecosystems to infrastructures of everyday life. *Geography Compass, 13,* e12464

Barns, S. (2020). Re-engineering the city: Platform ecosystems and the capture of urban big data. *Frontiers in Sustainable Cities, 2.* https://doi.org/10.3389/frsc.2020.00032

Ben, Q., & Peng, X. (2020). Artificial intelligence application embedded in government governance: Practice, mechanism and risk architecture: Taking Hangzhou City Brain as an example. *Journal of Gansu Administration College, 3,* 29–42 + 125.

Benbya, H., Davenport, T. H., & Pachidi, S. (2020). Artificial intelligence in organizations: Current state and future opportunities. *SSRN Electronic Journal, 19*(4). https://doi.org/10.2139/ssrn.3741983

Beunen, R., & Patterson, J. J. (2019). Analyzing institutional change in environmental governance: Exploring the concept of 'institutional work'. *Journal of Environmental Planning and Management, 62*(1), 12–29. https://doi.org/10.1080/09640568.2016.1257423

Bibri, S. E. (2019). Smart sustainable urbanism: Paradigmatic, scientific, scholarly, epistemic, and discursive shifts in light of big data science and analytics. In: *Big Data Science and Analytics for Smart Sustainable Urbanism: Advances in Science, Technology & Innovation.* Springer. https://doi.org/10.1007/978-3-030-17312-8_6

Bibri, S. E. (2020). Data-driven environmental solutions for smart sustainable cities: Strategies and pathways for energy efficiency and pollution reduction. *Euro-Mediterranean Journal for Environmental Integration, 5*(66). https://doi.org/10.1007/s41207-020-00211-w

Bibri, S. E. (2021a). Data-driven smart eco-cities and sustainable integrated districts: A best-evidence synthesis approach to an extensive literature review. *European Journal of Futures Research, 9,* 1–43. https://doi.org/10.1186/s40309-021-00181-4

Bibri, S. E. (2021b). The underlying components of data-driven smart sustainable cities of the future: A case study approach to an applied theoretical framework. *European Journal of Futures Research, 9*(13). https://doi.org/10.1186/s40309-021-00182-3

Bibri, S. E. (2021c). Data-driven smart sustainable cities of the future: Urban computing and intelligence for strategic, short-term, and joined-up planning. *Computational Urban Science, 1,* 1–29.

Bibri, S. E. (2023). The metaverse as a virtual model of platform urbanism: Its converging AIoT, XReality, neurotech, and nanobiotech and their applications, challenges, and risks. *Smart Cities, 6*(3), 1345–1384. https://doi.org/10.3390/smartcities6030065

Bibri, S. E., & Krogstie, J. (2020a). Environmentally data-driven smart sustainable cities: Applied innovative solutions for energy efficiency, pollution reduction, and urban metabolism. *Energy Informatics, 3*, 29. https://doi.org/10.1186/s42162-020-00130-8

Bibri, S. E., & Krogstie, J. (2020b). Data-driven smart sustainable cities of the future: A novel model of urbanism and its core dimensions, strategies, and solutions. *Journal of Futures Studies, 25*(2), 77–94. https://doi.org/10.6531/JFS.202012_25(2).0009

Bibri, S. E., Alexandre, A., Sharifi, A., & Krogstie, J. (2023b). Environmentally sustainable smart cities and their converging AI, IoT, and big data technologies and solutions: An integrated approach to an extensive literature review. *Energy Informatics, 6*, 9.

Bibri, S. E., Huang, J., & Krogstie, J. (2024a). Artificial intelligence of things for synergizing smarter eco-city brain, metabolism, and platform: Pioneering data-driven environmental governance. *Sustainable Cities and Society*, 105516. https://doi.org/10.1016/j.scs.2024.105516

Bibri, S. E., Huang, J., Jagatheesaperumal, S. K., & Krogstie, J. (2024b). The synergistic interplay of artificial intelligence and digital twin in environmentally planning sustainable smart cities: A comprehensive systematic review. *Environmental Science and Ecotechnology, 20*, 100433. https://doi.org/10.1016/j.ese.2024.100433

Bibri, S. E., & Huang, J. (2025). Artificial intelligence of sustainable smart city brain and digital twin: A pioneering framework for advancing environmental sustainability. *Environmental Technology and Innovation* (in press).

Bibri, S. E., Krogstie, J., Kaboli, A., & Alahi, A. (2023a). Smarter eco-cities and their leading-edge artificial intelligence of things solutions for environmental sustainability: A comprehensive systemic review. *Environmental Science and Ecotechnlogy, 19*. https://doi.org/10.1016/j.ese.2023.100330

Blasio, E., & Selva, D. (2019). Implementing open government: A qualitative comparative analysis of digital platforms in France, Italy and United Kingdom. *Quality & Quantity: International Journal of Methodology, 53*(2), 871–896. https://doi.org/10.1007/s11135-018-0803-9

Bodin, Ö. (2017). Collaborative environmental governance: Achieving collective action in social-ecological systems. *Science, 357*(6352), eaan1114. https://doi.org/10.1126/science.aan1114

Borghys, K., Van Der Graaf, Sh., Walravens, N., & Van Compernolle, M. (2020). Multi-stakeholder innovation in smart city discourse: Quadruple helix thinking in the age of "platforms". *Frontiers in Sustainable Cities, 2*, 1–6.

Brand, D. (2022). Responsible artificial intelligence in government: Development of a legal framework for South Africa. *eJournal of eDemocracy and Open Government, 14*(1), 1. https://doi.org/10.29379/jedem.v14i1.678

Brynskov, M., Calvillo, N., & Borup, M. (2018). The urban data platform as city orchestrator: Participatory urbanism in Copenhagen. *European Journal of Cultural Studies, 21*(4), 474–490.

Bulkeley, H, Castán Broto V (2013) Government by experiment? Global cities and the governing of climate change. *Transactions of the Instit Brit Geographers, 38*(3):361–375

Buyya, R., Netto, M., Toosi, A., Rodriguez, M., Llorente, I., Vimercati, S., Samarati, P., Milojicic, D., Varela, C., Bahsoon, R., Assuncao, M., Srirama, S., Rana, O., Zhou, W., Jin, H., Gentzsch, W., Zomaya, A., Shen, H., Casale, G., & Calheiros, R. (2018). A manifesto for future generation cloud computing. *ACM Computing Surveys, 51*(5), 1–38.

Calvo, P. (2019). The ethics of Smart City (EoSC): Moral implications of hyperconnectivity, algorithmization and the datafication of urban digital society. *Ethics and Information Technology, 22*, 141–149.

Caprotti, F., & Cowley, R. (2017) Interrogating urban experiments. *Urban Geography, 38*(9):1441–1450

Caprotti, F., & Liu, D. (2022). Platform urbanism and the Chinese smart city: The co-production and territorialisation of Hangzhou City Brain. *GeoJournal, 87*, 1559–1573. https://doi.org/10.1007/s10708-020-10320-2

Caprotti, F., Chang, I.-C. C., & Joss, S. (2022). Beyond the smart city: A typology of platform urbanism. *Urban Transformations, 4*(4). https://doi.org/10.1186/s42854-022-00033-9

Caprotti, F., Cowley, R., Bailey, I., Joss, S., Sengers, F., Raven, R., Spaeth, P., Jolivet, E., Tan-Mullins, M., & Cheshmehzangi, A. (2017). *Smart Eco-City Development in Europe and China: Policy Directions*. University of Exeter (SMART-ECO Project.

Cooper, H. M. (2017). *Research Synthesis and Meta-Analysis: A Step-by-Step Approach*. SAGE.

Cugurullo, F., Caprotti, F., Cook, M., Karvonen, A., McGuirk, P., & Marvin, S. (2024). *Artificial Intelligence and the City: Urbanistic Perspectives on AI*. Taylor & Francis.

D'Amico, G., Arbolino, R., Shi, L., Yigitcanlar, T., & Ioppolo, G. (2021). Digital technologies for urban metabolism efficiency: Lessons from urban agenda partnership on circular economy. *Sustainability, 13*(11), 6043. https://doi.org/10.3390/su13116043

D'Amico, G., Taddeo, R., Shi, L., Yigitcanlar, T., & Ioppolo, G. (2020). Ecological indicators of smart urban metabolism: A review of the literature on international standards. Ecological Indicators, 118, 106808.

Danaher, J. (2016). The threat of algocracy: Reality, resistance and accommodation. *Philosophy & Technology, 29*, 245–268.

de Wit, M. P. (2020). Environmental governance: Complexity and cooperation in the implementation of the SDGs. In: W. Leal Filho, A. Azul, L. Brandli, A. Lange Salvia, & T. Wall (Eds.), *Affordable and Clean Energy. Encyclopedia of the UN Sustainable Development Goals*. Springer. https://doi.org/10.1007/978-3-319-71057-0_25-1

Deng, W., Cheshmehzangi, A., Ma, Y., & Peng, Z. (2021). Promoting sustainability through governance of eco-city indicators: A multi-spatial perspective. *International Journal of Low-Carbon Technologies*, 16(1), 61–72.

El Himer, S., Ouaissa, M., & Boulouard, Z. (2022). Artificial intelligence of things (AIoT) for renewable energies systems. In: S. El Himer, M. Ouaissa, A. A. A. Emhemed, M. Ouaissa, & Z. Boulouard (Eds.), *Artificial Intelligence of Things for Smart Green Energy Management (Studies in Systems, Decision and Control* (vol. 446, pp. 1–19). Springer. https://doi.org/10.1007/978-3-031-04851-7_1

Evans, J. (2011). *Environmental Governance*. Routledge.

Evans, J., Karvonen, A., & Raven, R. (2016). *The Experimental City*. Routledge. https://doi.org/10.4324/9781315719825

Flynn, A., Yu, L., Feindt, P., & Chen, C. (2016). Eco-cities, governance and sustainable lifestyles: The case of the Sino-Singapore Tianjin Eco-City. *Habitat International*, 53, 78–86.

Folke, C., Hahn, T., Olsson, P., & Norberg, J. (2005). Adaptive governance of social-ecological systems. *Annual Review of Environment and Resources*, 30, 441–473.

Fragkos, G., Tsiropoulou, E. E., & Papavassiliou, S. (2020). Artificial intelligence enabled distributed edge computing for Internet of Things applications. In: Proceedings of the 2020 16th International Conference on Distributed Computing in Sensor Systems (DCOSS) (pp. 450–457). IEEE.

Fung, A. (2006). Varieties of participation in complex governance. *Public Administration Review*, 66(s1), 66–75.

Ghosh, R., & Sengupta, D. (2023). Smart urban metabolism: A big-data and machine learning perspective. In: R. Bhadouria, S. Tripathi, P. Singh, P. K. Joshi, & R. Singh (Eds.), *Urban Metabolism and Climate Change*. Springer. https://doi.org/10.1007/978-3-031-29422-8_16

Gorris, M., Glaser, R., Idrus, A., & Yusuf, A. (2019). The role of social structure for governing natural resources in decentralized political systems: Insights from governing a fishery in Indonesia. *Public Administration*, 97(3), 654–670. https://doi.org/10.1111/padm.12586

Gough, D., Oliver, S., & Thomas, J. (2017). *An Introduction to Systematic Reviews*. SAGE

Gourisaria, M. K., Jee, G., Harshvardhan, G., Konar, D., & Singh, P. K. (2022). Artificially intelligent and sustainable smart cities. In: P. K. Singh, M. Paprzycki, M. Essaaidi, & S. Rahimi (Eds.), *Sustainable smart cities: Theoretical foundations and practical considerations* (pp. 237–268). Springer, Cham. https://doi.org/10.1007/978-3-031-08815-5_14.

Gracias, J. S., Parnell, G. S., Specking, E., Pohl, E. A., & Buchanan, R. (2023). Smart cities: A structured literature review. *Smart Cities*, 6(4), 1719–1743. https://doi.org/10.3390/smartcities6040080

Graham, M. (2020). Regulate, replicate, and resist: The conjunctural geographies of platform urbanism. *Urban Geography*, 41, 453–457. https://doi.org/10.1080/02723638.2020.1717028

Green, J. F., & Hadden, J. (2021). How did environmental governance become complex? Understanding mutualism between environmental NGOs and international organizations. *International Studies Review, 23*(4), 1–21

Gupta, J., & van der Heijden, J. (2019). Introduction: The co-production of knowledge and governance in collaborative approaches to environmental management. *Current Opinion in Environmental Sustainability, 39*, 45–51.

Habbal, A., Ali, M. K., & Abuzaraida, M. A. (2024). Artificial intelligence trust, risk and security management (AI TRiSM): Frameworks, applications, challenges and future research directions. *Expert Systems with Applications, 240*, 122442. https://doi.org/10.1016/j.eswa.2023.122442

Haveri, A., & Anttiroiko, A.-V. (2023). Urban platforms as a mode of governance. *International Review of Administrative Sciences, 89*(1), 3–20. https://doi.org/10.1177/00208523211005855

Hildebrandt, M. (2018). Algorithmic regulation and the rule of law. *Philosophical Transactions of the Royal Society A: Mathematical, Physical and Engineering Sciences, 376*, 20170355.

Hoang, T. V. (2024). Impact of integrated artificial intelligence and internet of things technologies on smart city transformation. *Journal of Technical Education Science, 19*(1), 64–73. https://doi.org/10.54644/jte.2024.1532

Hodson, M., Kasmire, J., McMeekin, A., Stehlin, J. G., & Ward, K. (2021). *Urban Platforms and the Future City: Transformations in Infrastructure, Governance, Knowledge and Everyday Life*. Routledge.

Huynh-The, T., Pham, Q. V., Pham, X. Q., Nguyen, T. T., Han, Z., Kim, & D.-S. (2023). Artificial intelligence for the metaverse: A survey. *Engineering Applications of Artificial Intelligence, 117*, 105581.

Ishengoma, F. R., Shao, D., Alexopoulos, C., Saxena, S., & Nikiforova, A. (2022). Integration of artificial intelligence of things (AIoT) in the public sector: Drivers, barriers and future research agenda. *Digital Policy, Regulation and Governance, 24*(5), 449–462.

Ismagilova, E., Hughes, L., Rana, N. P., & Dwivedi, Y. K. (2020). Security, privacy and risks within smart cities: Literature review and development of a smart city interaction framework. *Information Systems Frontiers, 24*, 393–414.

ITU (2022). The United for Smart Sustainable Cities (U4SSC). The International Telecommunication Union. https://u4ssc.itu.int (accessed 10 December 2022).

Jagatheesaperumal, S. K., Bibri, S. E., Ganesan, S., & Jeyaraman, P. (2023). Artificial Intelligence for road quality assessment in smart cities: A machine learning approach to acoustic data analysis. *Computers in Urban Science, 3*, 28. https://doi.org/10.1007/s43762-023-00104-y

Jagatheesaperumal, S. K., Bibri, S. E., Huang, J Ganesan, S., & Jeyaraman, P. (2024). Artificial intelligence of things for smart cities: Advanced solutions for enhancing transportation safety. *Computers, Environment and Urban Systems, 4*, 10. https://doi.org/10.1007/s43762-024-00120-6

Jan, Z., Ahamed, F., Mayer, W., Patel, N., Grossmann, G., Stumptner, M., & Kuusk, A. (2023). Artificial intelligence for industry 4.0: Systematic review of applications, challenges, and opportunities. *Expert Systems with Applications, 216*, 119456. https://doi.org/10.1016/j.eswa.2022.119456

Jiang, H., Geertman, S., & Witte, P. (2022). Smart urban governance: An alternative to technocratic "smartness". *GeoJournal, 87*(7), 1639–1655. https://doi.org/10.1007/s10708-020-10326-w

Joss, S. (2011). Eco-city governance: A case study of treasure island and sonoma mountain village. *Journal of Environmental Policy & Planning, 13*(4), 331–348.

Katmada, A., Katsavounidou, G., & Kakderi, C. (2023). Platform urbanism for sustainability. In: N. A. Streitz, & S. Konomi (Eds.), *Distributed, Ambient and Pervasive Interactions. HCII 2023*, Lecture Notes in Computer Science (vol. 14037). Springer. https://doi.org/10.1007/978-3-031-34609-5_3

Kennedy, C., & Hoornweg, D. (2012). Mainstreaming urban metabolism. *Journal of Industrial Ecology, 16*, 780–782. https://doi.org/10.1111/j.1530-9290.2012.00548.x.

Kennedy, C., Baker, L., Dhakal, S., & Ramaswami, A. (2012). Sustainable urban systems. *Journal of Industrial Ecology, 16*(6), 775–779.

Kennedy, C., Cuddihy, J., & Engel-Yan, J. (2007). The changing metabolism of cities. *Journal of Industrial Ecology, 11*(2), 43–59.

Kennedy, C., Pincetl, S., & Bunje, P. (2011). The study of urban metabolism and its applications to urban planning and design. *Environmental Pollution, 159*(8–9), 1965–1973.

Kierans, G., Jüngling, S., & Schütz, D. (2023). Society 5.0 2023. *EPiC Series in Computing 93*, 82–96

Koch, L., Gorris, P., & Pahl, W. C. (2021). Narratives, narration and social structure in environmental governance. *Global Environmental Change, 69*, 102317. https://doi.org/10.1016/j.gloenvcha.2021.102317 (accessed 23 July 2021).

Komninos, N., & Kakderi, C. (Eds.). (2019). *Smart Cities in the Post-algorithmic Era: Integrating Technologies, Platforms and Governance*. Edward Elgar.

König, P. D. (2019). Dissecting the algorithmic leviathan: On the socio-political anatomy of algorithmic governance. *Philosophy & Technology, 33*, 467–485.

Koumetio Tekouabou, S. C., Diop, E. B., Azmi, R., & Chenal, J. (2023). Artificial intelligence based methods for smart and sustainable urban planning: A systematic survey. *Archives of Computational Methods in Engineering, 30*(5), 1421–1438. https://doi.org/10.1007/s11831-022-09844-2

Kramers, A., Wangel, J., & Höjer, M. (2016). Governing the smart sustainable city: The case of the Stockholm Royal Seaport. *ICT for Sustainability* 2016, 46, 99–108.

Kuguoglu, B. K., van der Voort, H., & Janssen, M. (2021). The giant leap for smart cities: Scaling up smart city artificial intelligence of things (AIoT) initiatives. *Sustainability, 13*(21), 12295. https://doi.org/10.3390/su132112295

Leal Filho, W., Wall, T., Mucova, S. A. R., Nagy, G. J., Balogun, A.-L., Luetz, J. M., Ng, A. W., Kovaleva, M., Azam, F. M. S., & Alves, F. (2022). Deploying artificial intelligence for climate change adaptation. *Technological Forecasting and Social Change, 180*, 121662.

Lemos, M. C., & Agrawal, A. (2006). Environmental governance. *Annual Review of Environment and Resources, 31*, 297–325. https://doi.org/10.1146/annurev.energy.31.042605.135621

Liu, F, Ying, L., & Yunqin, Z. (2021). Discussion on the definition and construction principles of city brain. In: 2021 IEEE 2nd International Conference on Big Data, Artificial Intelligence and Internet of Things Engineering (ICBAIE), Nanchang, China. https://doi.org/10.1109/ICBAIE52039.2021.9390064

Liu, F., Liu, F. Y., & Shi, Y. (2018). City brain, a new architecture of smart city based on the internet brain. In: IEEE 22nd International Conference on Computer Supported Cooperative Work in Design, Nanjing (pp. 9–11). https://doi.org/10.1109/CSCWD.2018.8465164

Liu, W., Cui, P., Nurminen, J. K., & Wang, J. (2017). Special issue on intelligent urban computing with big data. *Machine Vision and Applications, 28*, 675–677. https://doi.org/10.1007/s00138-017-0877-8

Liu, W., Mei, Y., Ma, Y., Wang, W., Hu, F., & Xu, D. (2022). City brain: A new model of urban governance. In: M. Li, G. Bohács, A. Huang, D. Chang, & X. Shang (Eds.), *IEIS* 2021. Lecture Notes in Operations Research. Springer. https://doi.org/10.1007/978-981-16-8660-3_12

Madan, R., & Ashok, M. (2022). AI adoption and diffusion in public administration: A systematic literature review and future research agenda. *Government Information Quarterly, 40*(1), 101774. https://doi.org/10.1016/j.giq.2022.101774

Mahdavinejad, M. S., Rezvan, M., Barekatain, M., Adibi, P., Barnaghi, P., & Sheth, A. P. (2018). Machine learning for internet of things data analysis: A survey. *Digital Communications and Networks, 4*(3), 161–175. https://doi.org/10.1016/j.dcan.2017.10.002

Marasinghe, R., Yigitcanlar, T., Mayere, S., Washington, T., & Limb, M. (2024). Computer vision applications for urban planning: A systematic review of opportunities and constraints. *Sustainable Cities and Society, 100*, 105047. https://doi.org/10.1016/j.scs.2023.105047

Mastorakis, G., Mavromoustakis, C. X., Batalla, J. M., & Pallis, E. (Eds.). (2020). *Convergence of Artificial Intelligence and the Internet of Things*. Springer.

Meuleman, L., & Niestroy, I. (2015). Common but differentiated governance: A metagovernance approach to make the SDGs work. *Sustainability, 7*(9), 12295–12321. https://doi.org/10.3390/su70912295

Mishra, P., & Singh, G. (2023). Artificial intelligence for sustainable smart cities. In: *Sustainable Smart Cities*. Springer. https://doi.org/10.1007/978-3-031-33354-5_6

Mohamed, K. S. (2023). Deep learning for IoT "artificial intelligence of things (AIoT)". In: *Deep Learning-Powered Technologies. Synthesis Lectures on Engineering*, Science, and Technology. Springer. https://doi.org/10.1007/978-3-031-35737-4_3

Morrison, T. H., Adger, W. N., Brown, K., Lemos, M. C., Huitema, D., Phelps, J., Evans, L., Cohen, P., Song, A. M., Turner, R., Quinn, T., & Hughes, T. P. (2019). The black box of power in polycentric environmental governance. *Global Environmental Change, 57*, 101934. https://doi.org/10.1016/j.gloenvcha.2019.101934

Newig, J., & Fritsch, O. (2009). Environmental governance: Participatory, multi-level – and effective? *Environmental Policy and Governance, 19*, 197–214. https://doi.org/10.1002/eet.509

Nishant, R., Kennedy, M., & Corbett, J. (2020). Artificial intelligence for sustainability: Challenges, opportunities, and a research agenda. *International Journal of Information Management, 53*, 102104.

Niza, S., Rosado, L., & Ferrão, P. (2009). Urban metabolism: Methodological advances in urban material flow accounting based on the Lisbon case study. *Journal of Industrial Ecology, 13*(3), 384–405.

Noble, S. U. (2018). *Algorithms of Oppression: How Search Engines Reinforce Racism*. New York University Press.

Noori, N., Hoppe, T., & de Jong, M. (2020). Classifying pathways for smart city development: Comparing design, governance and implementation in Amsterdam, Barcelona, Dubai, and Abu Dhabi. *Sustainability, 12*, 4030.

Page, M. J., McKenzie, J. E., Bossuyt, P. M., Boutron, I., Hoffmann, T. C., Mulrow, C. D., Shamseer, L., Tetzlaff, J. M., Akl, E. A., Brennan, S. E., Chou, R., Glanville, J., Grimshaw, J. M., Hróbjartsson, A., Lalu, M. M., Li, T., Loder, E. W., Mayo-Wilson, E., McDonald, S., McGuinness, L. A., Stewart, L. A., Thomas, J., Tricco, A. C., Welch, V. A., Whiting, P., & Moher, D (2021). The PRISMA 2020 statement: An updated guideline for reporting systematic reviews. *International Journal of Surgery, 88*, 105906. https://doi.org/10.1016/j.ijsu.2021.105906

Parihar, V., Malik, A., Bhawna, Bhushan, B., & Chaganti, R. (2023). From smart devices to smarter systems: The evolution of artificial intelligence of things (AIoT) with characteristics, architecture, use cases and challenges. In: *AI Models for Blockchain-Based Intelligent Networks in IoT Systems: Concepts, Methodologies, Tools, and Applications* (pp. 1–28). Springer. https://doi.org/10.1007/978-3-031-31952-5_1

Peponi, A., Morgado, P., & Kumble, P. (2022). Life cycle thinking and machine learning for urban metabolism assessment and prediction. *Sustainable Cities and Society, 80*, 103754. https://doi.org/10.1016/j.scs.2022.103754

Poell, T., Nieborg, D., & Van Dijck, J. (2019). Platformisation. *Internet Policy Review, 8*, 1–13.

Princetl, S., & Bunje, P. (2009). *Potential Targets and Benefits for Sustainable Communities Research, Development, and Demonstration*. Los Angeles.

Puri, V., Jha, S., Kumar, R., Priyadarshini, I., Son, L. H., Abdel-Basset, M., Elhoseny, M., & Long, H. V. (2019). A hybrid artificial intelligence and Internet of Things model for generation of renewable resource of energy. *IEEE Access, 7*, 111181–111191.

Putra, W., Wahidayat, Z. D., & van der Knaap, W. (2018). Urban innovation system and the role of an open web-based platform: The case of Amsterdam smart city. *Journal of Regional and City Planning, 29*(3), 234–249.

Repette, P., Sabatini-Marques, J., Yigitcanlar, T., Sell, D., & Costa, E. (2021). The evolution of city-as-a-platform: Smart urban development governance with collective knowledge-based platform urbanism. *Land, 10*, 33. https://doi.org/10.3390/land10010033

Sadowski, J. (2020). Cyberspace and cityscapes: On the emergence of platform urbanism. *Urban Geography, 41*(3), 448–452.

Samadi, S. (2022). The convergence of AI, IoT, and big data for advancing flood analytics research. *Frontiers in Water, 4*, 786040. https://doi.org/10.3389/frwa.2022.786040

Samuel, P., Jayashree, K., Babu, R., & Vijay, K. (2023). Artificial intelligence, machine learning, and IoT architecture to support smart governance. In: K. Saini, A. Mummoorthy, R. Chandrika, & N. Gowri Ganesh (Eds.), *AI, IoT, and Blockchain Breakthroughs in E-Governance* (pp. 95–113). IGI Global. https://doi.org/10.4018/978-1-6684-7697-0.ch007

Sanchez, T. W. (2023). Planning on the verge of AI, or AI on the verge of planning. *Urban Science, 7*, 70. https://doi.org/10.3390/urbansci7030070

Sarker, I. (2021). Machine learning: Algorithms, real-world applications and research directions. *SN Computer Science, 2*(3), 160. https://doi.org/10.1007/s42979-021-00592-x

Seng, K. P., Ang, L. M., & Ngharamike, E. (2022). Artificial intelligence Internet of Things: A new paradigm of distributed sensor networks. *International Journal of Distributed Sensor Networks.* https://doi.org/10.1177/15501477211062835

Shahrokni, H., Årman, L., Lazarevic, D., Nilsson, A., & Brandt, N. (2015). Implementing smart urban metabolism in the Stockholm Royal Seaport: Smart city SRS. *Journal of Industrial Ecology, 19*(5), 917–929.

Shi, F., Ning, H., Huangfu, W., Zhang, F., Wei, D., Hong, T., & Daneshmand, M. (2020). Recent progress on the convergence of the Internet of Things and artificial intelligence. *IEEE Network, 34*(5), 8–15.

Simboli, A., Taddeo, R., & Raggi, R. (2019). The multiple dimensions of urban contexts in an industrial ecology perspective: An integrative framework. *International Journal of Life Cycle Assessment, 24*(7), 1285–1296.

Simmons, G., Giraldo, J. E. D., Truong, Y., & Palmer, M. (2018). Uncovering the link between governance as an innovation process and socio-economic regime transition in cities. *Research Policy, 47*(1), 241–251. https://doi.org/10.1016/j.respol.2017.11.002

Singha, R., & Singha, S. (2024). AIoT concepts and integration: Exploring customer interaction, ethics, policy, and privacy. In: *Artificial Intelligence of Things (AIoT) for Productivity and Organizational Transition* (pp. 1–26). https://doi.org/10.4018/979-8-3693-0993-3.ch002

Soe, R.-M., Ruohomäki, T., & Patzig, H. (2022). Urban open platform for borderless smart cities. *Applied Sciences, 12*(2), 700. https://doi.org/10.3390/app12020700

Son, T. H., Weedon, Z., Yigitcanlar, T., Sanchez, T., Corchado, J. M., & Mehmood, R. (2023). Algorithmic urban planning for smart and sustainable development: Systematic review of the literature. *Sustainable Cities and Society, 94*, 104562.

Song, H., Bai, J., Yi, Y., Wu, J., & Liu, L. (2020). Artificial intelligence enabled internet of things: Network architecture and spectrum access. *IEEE Computational Intelligence Magazine, 15*(1), 44–51. https://doi.org/10.1109/MCI.2019.2954643

Späth, P., & Rohracher, H. (2011). The "eco-cities" Freiburg and Graz: The social dynamics of pioneering urban energy and climate governance. In: H. Bulkeley, V. Castan-Broto, M. Hodson, & S. Marvin (Eds.), *Cities and Low Carbon Transitions. Routledge Studies in Human Geography* (pp. 88–106). Routledge.

Späth, P., Hawxwell, T., John, R., Li, S., Löffler, E., Riener, V., & Utkarsh, S. (2017). *Smart-Eco Cities in Germany: Trends and City Profiles.* University of Exeter Press.

Srivastava, S. (2021). Algorithmic governance and the international politics of big tech. In: *Perspectives on Politics* (pp. 1–12). Cambridge University Press.

Tan Mullins, M., Cheshmehzangi, A., Chien, S., & Xie, L. (2017). *Smart-Eco Cities in China: Trends and City Profiles 2016.* University of Exeter.

Tarek, S. (2023). Smart eco-cities conceptual framework to achieve UN-SDGs: A case study application in Egypt. *Civil Engineering and Architecture, 11*(3), 1383-1406. https://doi.org/10.13189/cea.2023.110322

Trantopoulos, K., Schläpfer, M., & Helbing, D. (2011). Toward sustainability of complex urban systems through techno: Social reality mining. *Environmental Science & Technology, 45*, 6231–6236.

Underdal, A. (2010). Complexity and challenges of long-term environmental governance. *Global Environmental Change, 20*(3), 386–393.

UN-Habitat. (2023). Urban Governance. https://unhabitat.org/topic/urban-governance (accessed 12 March 2023).

Vadiati, N. (2022). Alternatives to smart cities: A call for consideration of grassroots digital urbanism. *Digital Geography and Society, 3*, 100030.

Van Assche, K., Beunen, R., Gruezmacher, M., & Duineveld, M. (2020) Rethinking strategy in environmental governance. *Journal of Environmental Policy & Planning, 22*(5), 695–708, https://doi.org/10.1080/1523908X.2020.1768834

van der Heijden, J. (2016). Experimental governance for low-carbon buildings and cities: Value and limits of local action networks. *Cities, 53*, 1–7. https://doi.org/10.1016/j.cities.2015.12.008

Van Dijck, J., Poell, T., & De Waal, M. (2018). *The Platform Society: Public Values in a Connective World.* Oxford University Press.

van Noordt, C., & Misuraca, G. (2022). Artificial intelligence for the public sector: Results of landscaping the use of AI in government across the European Union. *Government Information Quarterly, 39*(3), 101714. https://doi.org/10.1016/j.giq.2022.101714

Vatn, A. (2016). *Environmental Governance: Institutions, Policies and Actions.* Edward Elgar Publishing.

Vogl, T. M. (2021). Artificial intelligence in local government: Enabling artificial intelligence for good governance in UK local authorities. *SSRN Electronic Journal*. https://doi.org/10.2139/ssrn.3840222

Webb, R., Bai, X., Smith, M. S., Costanza, R., Griggs, D., Moglia, M., Neuman, M., Newman, P., Newton, P., Norman, B., & Ryan, C. (2018). Sustainable urban systems: Co-design and framing for transformation. *Ambio*, *47*, 57–77.

Weissenrieder, D. (2023). Approaching platform urbanism. In: *Exploring Platform Urbanism Using Counter-Mapping: BestMasters*. Springer. https://doi.org/10.1007/978-3-658-40648-6_3

Wiedmann, N. S., Athanassiadis, A., & Binder, C. R. (2021). Data mining in the context of urban metabolism: A case study of Geneva and Lausanne, Switzerland. *Journal of Physics: Conference Series*, 2042, 012020. https://doi.org/10.1088/1742-6596/2042/1/012020

Xie, J. (2020). Exploration and Reference of Urban Brain Construction in Xiacheng District of Hangzhou. *Communications World*, *31*, 26–27.

Xu, R. (2022). Deeply understand, analyze and draw lessons from Hangzhou's urban brain construction paradigm. *BCP Business & Management*, *29*, 441–446. https://doi.org/10.54691/bcpbm.v29i.2310

Xu, Y., Cugurullo, F., Zhang, H., Gaio, A., & Zhang, W. (2024). The emergence of artificial intelligence in anticipatory urban governance: Multi-scalar evidence of China's transition to city brains. *Journal of Urban Technology*. https://doi.org/10.1080/10630732.2023.2292823

Yang, L., Chen, X., Perlaza, S. M. et al. (2020). Special issue on artificial-intelligence-powered edge computing for Internet of Things. *IEEE IoT Journal*, *7*(10), 9224–9226.

Yang, S., Xu, K., Cui, L., Ming, Z., Chen, Z., & Ming, Z. (2021). EBI-PAI: Towards an efficient edge-based IoT platform for artificial intelligence. *IEEE IoT Journal*, *8*, 9580–9593. https://doi.org/10.1109/JIOT.2020.3019008.

Yeung, K. (2017). Algorithmic regulation: A critical interrogation. *Regulation & Governance*, *12*, 505–523.

Yigitcanlar, T., Agdas, D., & Degirmenci, K. (2022). Artificial intelligence in local governments: Perceptions of city managers on prospects, constraints and choices. *AI & Society*. https://doi.org/10.1007/s00146-022-01450-x

Yigitcanlar, T., Agdas, D., & Degirmenci, K. (2023a). Artificial intelligence in local governments: Perceptions of city managers on prospects, constraints and choices. *AI & Society*, *38*(3), 1135–1150. https://doi.org/10.1007/s00146-022-01450-x

Yigitcanlar, T., Corchado, J., Mehmood, R., Li, R., Mossberger, K., & Desouza, K. (2021). responsible urban innovation with local government artificial intelligence (AI): A conceptual framework and research agenda. *Journal of Open Innovation: Technology, Market, and Complexity*, *7*(1), 1. https://doi.org/10.3390/joitmc7010071

Yigitcanlar, T., David, A., Wenda, L., Bibri, S. E., Fooks, C., & Ye, X. (2024) Unlocking artificial intelligence adoption in local governments: Best practice lessons from real-world implementations. *Smart Cities*, *7*(4), 1576-1625. https://doi.org/10.3390/smartcities7040064

Yigitcanlar, T., Desouza, K., Butler, L., & Roozkhosh, F. (2020). Contributions and risks of artificial intelligence (AI) in building smarter cities: Insights from a systematic review of the literature. *Energies, 13*(6), 1473.

Yigitcanlar, T., Li, R., Beeramoole, P., & Paz, A. (2023b). Artificial intelligence in local government services: Public perceptions from Australia and Hong Kong. *Government Information Quarterly, 40*(3), 101833. https://doi.org/10.1016/j.giq.2023.101833

Yigitcanlar, T., Senadheera, S., Marasinghe, R., Bibri, S. E., Sanchez, T., Cugurullo, F., & Sieber, R. (2024). Artificial intelligence and the local government: A five-decade scientometric analysis on the evolution, state-of-the-art, and emerging trends. *Cities, 152*, 105151. https://doi.org/10.1016/j.cities.2024.105151.

Zaidi, A., Ajibade, S.-S. M., Musa, M., & Bekun, F. V. (2023). New insights into the research landscape on the application of artificial intelligence in sustainable smart cities: A bibliometric mapping and network analysis approach. *International Journal of Energy Economics and Policy, 13*(4), 287–299. https://doi.org/10.32479/ijeep.14683

Zaidi, A., Ajibade, S.-S. M., Musa, M., & Bekun, F. V. (2023). New insights into the research landscape on the application of artificial intelligence in sustainable smart cities: A bibliometric mapping and network analysis approach. *International Journal of Energy Economics and Policy, 13*(4), 287–299. https://doi.org/10.32479/ijeep.14683

Zhang, J., & Tao, D. (2021). Empowering things with intelligence: A survey of the progress, challenges, and opportunities in artificial intelligence of things. *IEEE Internet of Things Journal, 8*(10), 7789–7817. https://doi.org/10.1109/JIOT.2020.3039359

Zhang, J., Hua, X. S., Huang, J., Shen, X., Chen, J., Zhou, Q., Fu, Z., & Zhao, Y. (2019). City brain: Practice of large-scale artificial intelligence in the real world. *IET Smart Cities, 1*, 28–37.

Zhu, W. (2021) Artificial intelligence and urban governance: Risk conflict and strategy choice. *Open Journal of Social Sciences, 9*, 250–261. https://doi.org/10.4236/jss.2021.94019

This article was adapted from Sustainable Cities and Society, Vol. 108, by Simon Elias Bibri, Jeffrey Huang, John Krogstie, titled "Artificial intelligence of things for synergizing smarter eco-city brain, metabolism, and platform: Pioneering data-driven environmental governance," on pages 105516, Copyright Elsevier (Year 2024).

CHAPTER 7

Artificial Intelligence and Its Generative Power for Advancing the Sustainable Smart City Digital Twin

A Novel Framework for Data-Driven Environmental Planning and Design

Abstract

Recent advancements in Artificial Intelligence (AI) and its subfields or domains, notably Machine Learning (ML), Deep Learning (DL), Computer Vision (CV), Natural Language Processing (NLP), and Generative AI (GenAI), present promising opportunities for advancing city planning and design. However, there is a notable absence of comprehensive studies on their combined applications in the realm of sustainable smart city planning and design. Furthermore, the existing body of literature provides limited insights into the integration of GenAI into UDT as a data-driven planning and design system, primarily due to the emerging and evolving nature of these technologies. Therefore, this chapter aims to fill these gaps by exploring the transformative potential of AI and its five subfields—ML, DL, CV, NLP, and GenAI—in reshaping sustainable smart city planning and design through UDT. To guide this investigation, four objectives are formulated: (1) analyze and synthesize an extensive body of literature on AI and its five subfields, alongside UDT, with a specific focus on their foundational underpinnings and practical applications in the context of sustainable smart city planning and design; (2) investigate a case study of Lausanne City on Generative Spatial AI (GSAI), a specialized branch of GenAI tailored for UDT; (3) develop a novel framework for GenAI-driven sustainable smart city planning and design; and (4) provide a comprehensive discussion encompassing a comparative analysis, implications, limitations, and suggestions for future research directions. The findings highlight the pivotal role of AI and its five subfields in enhancing data-driven planning and design processes through UDT. They also reveal the untapped potential of GenAI and GSAI in propelling the advancement of sustainable smart cities. These insights offer valuable guidance for researchers, practitioners, and policymakers, catalyzing groundbreaking research endeavors, facilitating practical implementations, and informing policy decisions in the realm of emerging GenAI-driven sustainable urban development.

7.1 INTRODUCTION

In recent years, the pace of urbanization has accelerated dramatically, escalating ecological degradation and leading to fundamental transformations in urban landscapes. This phenomenon has been further fueled by the widespread integration and adoption of advanced technologies, which have exerted profound effects on urban environments. This transformation poses both unprecedented challenges and promising opportunities for urban planners, designers, and policymakers. The growing complexity of urban development, coupled with the urgent imperative to bolster sustainability and resilience, underscores the critical need for innovative solutions in urban planning and design. As a response to this pressing need, the convergence of Artificial Intelligence (AI) and Urban Digital Twin (UDT), a data-driven urban planning and design system, has emerged as a promising frontier, offering novel pathways to shape the future of sustainable smart cities (Bibri et al. 2024a; Bibri and Huang 2024).

As subfields of AI, Machine Learning (ML), Deep Learning (DL), Computer Vision (CV), and Natural Language Processing (NLP) have shown significant potential in addressing the multifaceted challenges faced by sustainable smart cities, particularly those related to environmental sustainability and climate change. This potential has been evidenced by numerous studies (e.g., Bibri et al. 2023a, b; Gourisaria et al. 2022; Efthymiou and Egleton 2023; Mishra 2023; Tyagi and Bhushan 2023; Szpilko et al. 2023; Ullah et al. 2020, 2023; Zaidi et al. 2023), which demonstrate the capacity of these technologies to analyze vast datasets, identify patterns, and generate actionable insights for decision-making processes in various domains of sustainable smart cities. One of these domains where these technologies have seen a steady integration is in urban planning and design processes (Koumetio et al. 2023; Sanchez 2023; Son et al. 2022; Marasinghe et al. 2024). These technologies have also found their way into the functionalities of UDT, enhancing decision-making processes and enriching data-driven planning and design initiatives in sustainable smart cities (e.g., Almusaed and Yitmen 2023; Beckett 2022; Bibri et al. 2024a; Efthymiou and Egleton 2023; GKontzis et al. 2024; Wu et al. 2022; Ye et al. 2023; Zayed et al. 2023). By integrating ML, DL, CV, and NLP, planners and designers can effectively manage vast datasets, identify patterns, and discern trends, either through AI-driven analytical tools or UDT systems, thereby facilitating more informed decision-making across various domains, including transportation, traffic, energy, water, waste, and environmental management through automation, optimization, and prediction.

Within this dynamic landscape, Generative AI (GenAI), another specialized subset of AI, has garnered significant attention among urban planners, designers, and policymakers in more recent years. As a set of innovative computational methodologies for tackling the multifaceted challenges associated with sustainable urban development, GenAI models have shown significant potential to transform many disciplines and application domains (Castelli and Manzoni 2022; Ooi et al. 2023), including urban planning (e.g. Ahn et al. 2023; Huang et al. 2022; Pérez-Martínez et al. 2023; Zhang et al. 2023) and design (e.g., Jiang et al. 2023; Sun and Dogan 2022, 2023; Yang et al. 2023a). These models harness the generative power of AI to advance sustainable smart city development practices. Among the most commonly used ML, DL, CV, and NLP models in GenAI for urban planning and design include Generative Adversarial Networks (GANs), Variational Autoencoders (VAEs), Convolutional Neural Networks (CNNs), and Transformers (e.g., GPT (Generative Pre-trained Transformer) and BERT (Bidirectional Encoder Representations from Transformers)).

Furthermore, GenAI models are increasingly being integrated into the functionalities of UDT, driving advancements in city modeling and simulation (e.g., Liu et al. 2023; Somanath et al. 2023; Xu et al. 2024) and hence data-driven planning and design practices in sustainable smart cities. These advancements mark a significant stride toward fostering more efficient, resilient, environmentally conscious, and intelligently structured urban environments. Indeed, the incorporation of GenAI into the current UDT frameworks signifies a strategic approach to addressing critical challenges encountered by its current simulation models and computational processes. These challenges include its inability to model complex urban systems, quantify uncertainty, adapt in real-time, integrate data effectively, and support decision-making (Batty et al. 2018; Lei et al. 2022; Weil et al. 2023). Furthermore, AI models in UDT systems encounter difficulties due to inadequate data availability (Wang et al. 2022b; Ballouch et al. 2022), data inconsistency and inaccuracy (e.g., Ferré-Bigorra et al. 2022; Lenfers et al. 2021; Wang and Yin 2022b; White et al. 2021), computational constraints requiring multilevel integrated models (Marcucci et al. 2020; Wu et al. 2021), as well as issues related to simulation realism, dynamic modeling, scenario exploration and optimization, prediction accuracy, and computational intensity (e.g., Beckett et al. 2022; Dembski et al. 2020; Lei et al. 2022; Lenfers et al. 2021; Papyshev and Yarime 2021; Topping et al. 2021; Weil et al. 2023).

These challenges hinder the development and implementation of UDT (Lei et al. 2022), including in sustainable smart cities (Weil et al. 2023), prompting a proactive response to tackle these obstacles and alleviate the complexities faced by urban planners and designers. As urban ecosystems undergo continuous evolution, there is a growing demand for innovative approaches to enhance UDT's modeling and simulation capabilities. In contrast, UDT, as a computational model replicating the physical and functional aspects of urban environments for planning, design, and policy purposes, has generated much criticism for its inherent limitations. Acknowledging and addressing these limitations, while harnessing the potential of GenAI, offers unparalleled opportunities to surmount pivotal challenges encountered by UDT, thus bolstering its efficacy in city modeling and simulation, especially in the realm of sustainable smart city planning and design.

Developing comprehensive, innovative, and insightful solutions to urban challenges from a holistic perspective is becoming increasingly costly and demanding. The intricacy of modern urban environments requires multidisciplinary approaches, extensive data collection, and advanced analytical tools, all of which contribute to the growing expense and effort required. As cities expand and their needs become more diverse, the challenge of integrating various systems and addressing multifaceted problems further complicates the process, necessitating significant investments in both financial and human resources in the context of UDT. GenAI can address the high costs, time consumption, and labor-intensive efforts required to gather scenario-specific data and generate relevant scenarios using processed data and simulations for predictive analytics and decision-making (Xu et al. 2024). Additionally, it can streamline the creation of comprehensive 3D city models and enhance multi-scale urban design processes, making them more efficient and less resource-intensive (Xu et al. 2024).

The current research landscape provides only a narrow glimpse into the integration of GenAI and GSAI into UDT, primarily due to the emerging and evolving nature of these advanced technologies. This reflects the early stages of exploration and experimentation with GenAI within the framework of UDT. Furthermore, there is a notable absence of comprehensive studies examining the combined application of GenAI alongside other subfields of AI in the realm of sustainable smart city planning and design. Therefore, this study aims to fill these gaps by exploring the transformative potential of AI and its five subfields—ML, DL, CV, NLP, and GenAI—alongside UDT in reshaping sustainable smart city planning and design. In alignment with the overarching aim, we delineate four key objectives to guide this investigation:

1. Analyze and synthesize an extensive body of literature on AI and its five subfields, alongside UDT, with a specific focus on their foundational underpinnings and practical applications in the context of sustainable smart city planning and design.

2. Examine a case study on GSAI tailored for UDT and its functionalities to provide empirical insights into its relevance and significance in urban planning and design.

3. Develop a novel framework for GSAI-driven sustainable smart city planning and design, drawing upon insights derived from the synthesized literature and the GSAI case study.

4. Engage in a comprehensive discussion encompassing a comparative analysis, implications, limitations, and recommendations for future research avenues.

The motivation of the study stems from the recognition of the growing importance of AI technologies in shaping the future of cities. As cities worldwide face increasing challenges, there is a pressing need for innovative solutions that can enhance sustainability and resilience. By harnessing the power of AI and its subfields, the study seeks to uncover new possibilities for data-driven planning and design processes in sustainable smart cities. Overall, the study is motivated by the desire to contribute to the development of more sustainable, efficient, and livable cities through cutting-edge research and innovation.

This study contributes significantly to the field of sustainable smart city planning and design by offering novel insights and advancements in knowledge. Firstly, it sheds light on the foundational principles and practical applications of AI and its subfields while highlighting their synergistic and collaborative potential in advancing sustainable urban development practices. Secondly, it introduces an innovative framework for GenAI-driven sustainable smart city planning and design. This framework presents an advanced approach to urban planners, policymakers, and researchers, offering both theoretical insights and practical methodologies.

The remainder of this study is structured as follows: Section 7.2 provides a survey of related work. Section 7.3 describes and justifies the research design. Section 7.4 defines and establishes connections between the fundamental concepts that underpin sustainable smart city planning and design. Section 7.5 analyzes and synthesizes a large body of literature on the applications of AI and its five subfields as well as UDT in the realm of city planning and design. Section 7.6 examines the case study of GSAI for UDT. Section 7.7 derives a novel framework for GenAI-driven sustainable smart city planning and design based on the insights gleaned from the synthesized studies and the GSAI case study. Section 7.8 provides a comprehensive discussion of the analysis and results. This study concludes in Section 7.9, by providing a summary of findings and presenting concluding thoughts and takeaways.

7.2 RELATED WORK

Urban planning and design are increasingly influenced by AI, reflecting the growing role of ML, DL, CV, NLP, and GenAI in shaping the development trajectory of urban landscapes, with its expanding adoption and academic significance highlighting this trend. Indeed, scholars have explored various aspects of this intersection, ranging from the impact of these models on city design and planning to their application in data-driven planning and design systems. Kamrowska-Załuska (2021) explores the emergence of urban big data analytics driven by AI tools, focusing on their potential to revolutionize the design and planning of cities. The author offers a conceptual framework to assess the impact of these tools on urban design and planning practices, providing insights for urban planners and theorists. In a similar vein, Koumetio et al. (2023) provide a systematic survey of AI-based methods for smart and sustainable urban planning, identifying common urban planning issues addressed and the predominant data and ML techniques employed. By synthesizing existing literature, the authors identify key areas of focus, data sources, and geographic regions, offering valuable insights for urban planning researchers and practitioners.

Sanchez (2023) discusses the transformative potential of AI in urban planning, emphasizing its role in enhancing decision-making processes. The author provides insights into how AI applications, such as ML and NLP, can streamline planning procedures and address urban challenges. Sanchez's work offers an overview of AI's implications for planning practice and community development. Additionally, Marasinghe et al. (2024) conduct a systematic review of CV applications in urban planning, highlighting the opportunities offered by CV to support various planning tasks while acknowledging existing challenges. The findings inform urban policy and decision-makers about the benefits and limitations of integrating CV into planning practices.

Furthermore, Son et al. (2023) provide a systematic review of algorithmic urban planning for smart and sustainable development, focusing on the application of AI technologies in urban contexts. Their study highlights the role of AI, especially ML and DL, in addressing economic, social, and environmental challenges in cities, emphasizing the importance of collaboration and data-driven approaches. The findings underscore the need for integrating AI into urban planning processes to achieve smarter and more sustainable cities. Jiang et al. (2023) contribute to the discourse on Generative Urban Design (GUD) by conducting a systematic review of existing literature. Their study explores the use of AI-driven generative methods in urban design, with an emphasis on problem formulation, design generation, and decision-making. By synthesizing current trends and challenges, the authors provide a comprehensive understanding of GUD's potential and limitations, offering valuable insights for researchers and practitioners in the field.

Despite these contributions, there remains a need to synthesize and bridge the gaps in existing literature, particularly regarding the integration of the commonly applied AI subfields, namely ML, DL, CV, NLP, and GenAI, within the framework of UDT. This study aims to address this gap by offering a thorough analysis and synthesis of a large body of literature on AI and its five subfields, alongside UDT, emphasizing their foundational underpinnings and practical applications in the context of sustainable smart city planning and design. Additionally, it offers a novel framework for GenAI-driven sustainable smart city planning and design based on the outcome of the synthesized literature and the GSAI case study. This research anticipates to enrich and broaden the understanding of the potential benefits and limitations of GenAI-driven sustainable urban development and to provide guidance for future research and practice in this field.

7.3 METHODOLOGY

The primary aim of this study is to explore the transformative potential of AI and its five subfields—ML, DL, CV, NLP, and GenAI—in reshaping sustainable smart city planning and design through UDT. The methodological approach adopted was designed to achieve the four specified objectives outlined in the introduction. This exploration process followed a structured approach (Figure 7.1), which included a comprehensive literature review and a detailed case study analysis. The latter entailed an investigation into the development and application of GSAI in relation to UDT, offering empirical evidence related to GSAI's functionalities and practical implications for urban planning and design. The focus was on GSAI's relevance and significance in enhancing decision-making processes and promoting sustainable urban development initiatives in the realm of AI-driven UDT.

To establish solid groundwork for developing the synthesized framework, an extensive review of existing literature was conducted, focusing on the foundational principles and practical applications of its essential components. We selected academic databases such as Scopus, Web of Science

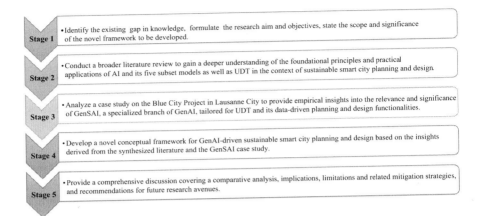

FIGURE 7.1 A flow diagram outlining the step-by-step process of developing the conceptual framework.

(WoS), and ScienceDirect for their comprehensive coverage of high-quality, peer-reviewed studies on the topic. A set of chosen keywords related to the five subfields of AI and UDT in the context of sustainable smart city planning and design ensured specificity in the search and retrieval process. These were used to search against the title, abstract, and keywords of articles to produce initial insights. Inclusion criteria were applied to filter studies based on their relevance, reliability, publication date, publication type, language, and alignment with the interlinkages between the topics under study. Conversely, exclusion criteria were employed to eliminate studies that were not pertinent to the intended analysis and synthesis.

Data necessary for our analysis and synthesis were extracted from the abstracts and relevant sections of the selected studies. The extracted insights and findings facilitated the identification of key themes, patterns, and variations, ensuring a comprehensive analysis and synthesis. Configurative and aggregative synthesis methods were employed as complementary approaches to develop the conceptual framework. Aggregative synthesis involved summarizing and combining multiple studies to distill key concepts, theories, models, techniques, platforms, and applications from the existing body of literature. This process identified components relevant to the conceptual framework and leveraged evidence to make statements based on conceptual categories, establishing a thematic summary crucial for framework development. Simultaneously, configurative synthesis identified common themes across synthesized studies, contributing to the development of an interconnected framework. It captured relationships between core components and constructed the overall meaning of research evidence. Together, aggregative synthesis identified key components, while configurative synthesis focused on relationships and contexts. These processes formed a robust foundation for the comprehensive conceptual framework, advancing the understanding of the complex phenomenon of GenAI-driven sustainable smart city planning and design.

7.4 CONCEPTUAL BACKGROUND

This section examines the fundamental concepts that underpin sustainable smart city planning and design. These concepts include AI and its subfields, including ML, DL, CV, NLP, and GenAI, as well as UDT, given its relevance to urban planning and design and its synergistic interplay with these models. The primary aim is to establish a comprehensive understanding of the intricate relationships between these technologies and their implications for shaping sustainable smart urban development.

7.4.1 Sustainable Smart Cities and Their Planning and Design Processes

A smart city integrates advanced technologies and data-driven solutions to optimize city operations, enhance resilience, improve citizens' quality of life, and promote sustainability. Silva et al. (2018) offer a comprehensive review of these urban environments, covering trends, architectures, components, and open challenges. However, smart city projects often prioritize economic gains over environmental and social considerations (Ahvenniemi et al. 2017; Evans 2019; Toli and Murtagh 2020). This imbalance has led to a growing focus on research and practical efforts dedicated to the environmental objectives of sustainable development (Al-Dabbagh et al. 2022; Saravanan and Sakthinathan 2021; Shruti et al. 2022; Tripathi et al. 2022) or Sustainable Development Goals (SDG) (Mishra et al. 2022; Sharifi et al. 2024; Visvizi and del Hoyo 2021). This shift toward sustainable smart cities represents a significant leap beyond traditional smart cities, aiming to harmonize environmental, social, and economic dimensions of sustainability. Specifically, sustainable smart cities strive to advance environmental conservation, reduce ecological footprints, promote social inclusivity, and drive economic prosperity while embracing cutting-edge technologies and innovative approaches.

While transitioning from smart to sustainable smart cities requires a holistic approach, environmental sustainability has become a key focus in recent years. In this evolving landscape, sustainable smart cities are increasingly embracing and leveraging AI models and techniques for addressing complex environmental and climate change challenges (Bibri et al. 2023a; Gourisaria et al. 2022; Efthymiou and Egleton 2023; Fang et al. 2023; Mishra 2023; Szpilko et al. 2023; Zaidi et al. 2023), positioning AI as a pivotal force in shaping the trajectory of urban planning practices.

Bibri (2019) examines the transformative shifts occurring in sustainable smart urbanism, propelled by the integration of big data science and analytics. Through a comprehensive exploration, the author elucidates the paradigmatic, scientific, scholarly, epistemic, and discursive changes shaping the landscape of sustainable smart urbanism. The study provides insightful analysis that equips researchers, practitioners, and policy makers with the knowledge needed to navigate the complexities of modern urban planning and design. This work sheds light on the transformative shifts driven by big data science and analytics and also underscores their profound implications for future technologies. It serves as a vital resource for understanding how the integration of big data science and analytics paves the way for the advancement of data-driven solutions in shaping the future of sustainable urban development.

Big Data, IoT, AI, and AIoT play crucial roles in smart sustainable urbanism, working synergistically to enhance urban environments' efficiency, sustainability, and resilience. Big Data provides the foundation by collecting vast amounts of information from various urban sources, enabling comprehensive analysis and insights. IoT enhances this by connecting physical devices and systems, facilitating real-time data exchange and monitoring. AI processes and analyzes this data, generating actionable insights and automating decision-making processes to optimize urban operations. AIoT further amplifies these capabilities by creating intelligent, responsive systems that can adapt to changing conditions and demands. Together, these technologies enable smarter, more sustainable urban development.

In the context of sustainable smart cities, urban planning emerges as a multifaceted and systematic process crucial for designing, regulating, and managing land use and built environments, as well as their interconnected infrastructure. This process spans various conceptual domains, including strategic, sustainable, spatial, land use, transportation, local, regional, infrastructure, and environmental planning, operating across diverse scales from district and neighborhood to urban and regional levels (Bibri 2021a). Urban design, closely linked with urban planning, focuses on shaping the physical forms of cities, emphasizing both aesthetic and functional aspects. It entails arranging and organizing buildings, streets, public spaces, and landscapes (Jiang et al. 2023) to create visually appealing, livable, and functional environments, while also ensuring their performance aligns with sustainable development and smart growth initiatives.

Bibri and Krogstie (2020) provide an overview of key design strategies aimed at enhancing the sustainability of smart cities in terms of their physical form. These strategies focus on integrating environmentally responsible urban planning principles with advanced technological solutions to create cities that are not only smart but also sustainable. By emphasizing the importance of energy efficiency, green infrastructure, density, diversity, mixed land use, and data-driven approaches to resource management, the authors highlight how sustainable design can play a pivotal role in shaping urban environments. These design approaches promote long-term sustainability, resilience, and improved quality of life for city inhabitants, ensuring that future cities are adaptable to both environmental and social needs.

Based on a case study analysis, Bibri (2021b) develops an applied theoretical framework for strategic sustainable urban development planning. This framework focuses on identifying and integrating the core components of data-driven sustainable smart cities, considering the dimensions, strategies, and solutions drawn from the leading global paradigms of smart urbanism and sustainable urbanism. The proposed framework synthesizes compact urban design strategies, eco-city design strategies, and technology solutions to promote sustainability. It further incorporates data-driven smart city technologies, competencies, and solutions that align with sustainability goals, alongside environmentally data-driven sustainable smart city solutions and strategies.

The relationship between urban planning and urban design is symbiotic, as they both aim to shape the physical form and function of cities. The former provides the strategic framework and policies for guiding development and growth, while the latter translates these plans into spatial layouts and architectural designs. Together, they contribute to the creation of cohesive, sustainable, and vibrant urban environments.

7.4.2 Artificial Intelligence

In theory, AI involves simulating human intelligence processes by developing computer systems capable of mimicking human-like behaviors and capabilities. These cognitive processes include perception, learning, reasoning, problem-solving, language understanding, and decision-making (Bibri 2015), and "implementing aspects of human intelligence in computer systems is one of the main practical goals of AI within data-driven sustainable smart cities of the future" (Bibri 2021c). By enabling machines to perform tasks traditionally requiring human intelligence, such as visual perception, object detection, speech recognition, decision-making, and language generation, AI has become integral across various fields. AI encompasses diverse subfields, including ML, DL, CV, and NLP. In practice, AI often involves a combination of these subfields. ML and DL are practical applications of AI, where algorithms are trained on data to perform specific tasks. CV focuses specifically on visual data, while NLP deals with textual data. Each of these subfields has its own set of models, algorithms, and methodologies tailored to their specific domain. But they can also be combined to solve more complex problems in various domains—including urban planning and design (e.g., Marasinghe et al. 2024; Sanchez 2023; Son et al. 2022), smart cities (Nishant et al. 2020; Yigitcanlar et al. 2020; Yigitcanlar et al. 2021), and sustainable smart cities (Bibri et al. 2023a, 2024a).

It was not until recently that AI gained prominence in urban planning and design (e.g., Koumetio et al. 2023; Sanchez 2023; Son et al. 2022; Marasinghe et al. 2024). This emergence underscores the increasing recognition of AI's potential to revolutionize urban development practices. Especially, the evolving landscape of sustainable smart cities are increasingly embracing and harnessing AI to address complex challenges (e.g., Bibri et al. 2023a; Gourisaria et al. 2022; Efthymiou and Egleton 2023; Mishra 2023; Szpilko et al. 2023; Zaidi et al. 2023) aligns with this trend, highlighting the innovative role of AI in shaping the future of urban planning and design. The intelligence of AI lies in its capacity to learn from analysis results and outputs, a process that mirrors the fundamental logic of planning practice: gathering information, analyzing it, and producing plans or policies to address challenges while improving the overall situation (Sanchez 2023). In this context, the integration of ML, DL, CV, NLP, and GenAI further enriches the potential applications and capabilities within urban planning and design.

7.4.2.1 Machine Learning
ML, a subset of AI, involves various techniques and algorithms enabling computers to learn patterns from large datasets and make predictions or decisions based on unseen data without explicit programming. Mitchell (1997) defines ML as "a computer program learning from experience 'E'

with respect to some class of tasks 'T' and performance measure 'P,' if its performance at tasks in 'T' as measured by 'P,' improves with experience E." This definition underscores the iterative nature of machine learning algorithms, which continuously refine their performance through exposure to data and feedback. The accuracy and confidence of ML models' decisions and predictions depend on how well algorithms are trained during the learning process. However, a plethora of studies have comprehensively explored various aspects of ML, including techniques, algorithms, architectures, applications, opportunities, and challenges (e.g., Azevedo et al. 2024; Naeem et al. 2023; Sharma et al. 2021; Sharifani and Amini 2023; Shinde and Shah 2018; Ullah 2020; Verma et al. 2024). Overall, the evolving landscape of ML continues to drive advancements in AI, shaping its applications across diverse fields and domains. As ML techniques become increasingly sophisticated and versatile, they hold immense promise for revolutionizing decision-making processes and driving innovation in various domains of sustainable smart cities.

7.4.2.2 Deep Learning

DL, a subfield of ML, utilizes artificial neural networks with multiple layers (deep neural networks) to learn complex hierarchical representations of data. It can automatically discern intricate features from raw data, minimizing the need for manual feature engineering. Multiple studies have explored DL concepts, architectures, applications, challenges, and future directions (e.g., Alhijaj and Khudeyer 2023; Alom et all. 2019; Alzubaidi et al. 2021; Dong et al. 2021; Joshi et al. 2023; Mathew et al. 2021). Compared to ML, DL has significantly advanced various AI tasks, such as image recognition, speech recognition, and language processing. Deep neural networks, including GANs for generative tasks, CNNs for image processing, and RNNs for sequence modeling, have demonstrated remarkable performance in learning hierarchical representations of data across a range of fields. For example, Chen et al. (2019) provide a comprehensive examination of GANs for image synthesis, highlighting their versatility and effectiveness in generating realistic images across various domains and applications. Their review encompasses the theoretical foundations, architectural advancements, and practical implementations of GANs. Moreover, recent research has explored novel DL architectures and methodologies, expanding the capabilities of DL models and driving innovation in AI applications across diverse domains of sustainable smart cities.

7.4.2.3 Computer Vision

CV involves developing techniques and algorithms to extract meaningful insights, recognize patterns, and make decisions based on visual data from digital images or videos. Sharma et al. (2021) define CV as a versatile field in ML, which is concerned with the ability of machines to process and analyze visual data by employing techniques from both ML and its subset DL. CV encompasses a wide range of tasks, including image classification, object detection, segmentation, and pose estimation. While CV has some of its own algorithms and methods (e.g., Marasinghe et al. 2024), it is primarily considered a multidisciplinary field and a subfield of AI and ML due to its heavy reliance on algorithms from these domains (Mahadevkar et al. 2022; Ranjana et al. 2022). It is worth noting that DL techniques, particularly CNNs, have revolutionized the field of CV by allowing for end-to-end learning directly from raw pixel data. This significant advancement has led to state-of-the-art performance in various CV tasks. Moreover, ongoing research continues to explore innovative approaches and advancements in CV, further enhancing its capabilities and applicability across diverse domains of sustainable smart cities.

7.4.2.4 Natural Language Processing

NLP, a subfield of AI, focuses on enabling computers to understand, interpret, and generate human language in a manner that is both meaningful and contextually appropriate. It encompasses a wide range of tasks, including text classification, sentiment analysis, machine translation, topic modeling, named entity recognition, and text generation. NLP has garnered considerable interest following the introduction of ChatGPT, which demonstrates rapidly evolving capabilities to learn, analyze, and interpret human language content (Khurana et al. 2023). Numerous review studies (e.g., Bowden 2023; Brown et al. 2020; Fu 2024; Ghazizadeh and Zhu 2020; Khurana et al. 2023; Sharma et al. (2022); Tyagi and Bhushan 2023; Wolf et al. 2020; Zeroual and Lakhouaja 2018) have extensively researched and discussed NLP techniques, algorithms, applications, advances, challenges, and future research directions from both computer science and data science perspectives.

NLP heavily relies on ML and DL techniques to process and analyze language data, leveraging large data volumes effectively for various NLP tasks. In particular, DL models have demonstrated remarkable performance in NLP tasks by learning complex patterns and representations from raw text data. Architectures like GANs, CNNs, RNNs, Long Short-term Memory

Networks (LSTMs), and transformers have become standard models for machine translation, text summarization, and language understanding (e.g., Li and Mao 2019; Otter et al. 2020; Torfi et al. 2020; Wolf et al. 2020). For instance, Cao et al (2023) provide a comprehensive review of the history of generative models, particularly GANs, and recent advances in Artificial Intelligence Generated Content (AIGC), focusing on the transition from unimodal to multimodal interaction, and discussing existing open problems and future challenges in the field.

Furthermore, integrating CV techniques with NLP enables systems to analyze textual and visual data simultaneously, enhancing their ability to understand and describe images or answer questions about visual content. For example, Contrastive Language-Image Pre-training (CLIP), a neural network model developed by OpenAI, can understand both images and text in a unified manner (Radford et al. 2021). CLIP is trained to understand images and text bidirectionally. This means it can generate textual descriptions of images and predict the most relevant images for given textual prompts. Moreover, CLIP is pre-trained on a large dataset of images and text pairs, learning to associate images with corresponding textual descriptions in a contrastive manner. This contrastive learning approach encourages the model to embed similar images and text close to each other in the feature space. Additionally, CLIP can generalize to unseen tasks and concepts by leveraging its understanding of both images and text. It can perform zero-shot learning, meaning it can make predictions on tasks it has never been explicitly trained on. CLIP's ability to understand both images and text has led to its application in various domains, including NLP, CV, and multimodal AI. Overall, CLIP is a powerful multimodal model that bridges the gap between images and text, enabling it to perform a wide range of tasks that require understanding and reasoning about both modalities. In sum, NLP intersects with ML, DL, and CV, utilizing methodologies from these fields to develop advanced systems for processing, understanding, and generating human language.

7.4.3 Generative Artificial Intelligence, Large Language Models, and Pre-Trained Foundation Models

GenAI is a subset of DL, which is a broad field of ML that involves algorithms and models inspired by the structure and function of the human brain's neural networks. It focuses on creating or generating new content rather than just recognizing patterns or making predictions based on existing data. It involves various generative models that possess the capability to generate novel data instances by leveraging learned patterns from existing datasets. Through pattern analysis of training data, GenAI models grasp the underlying distribution of data and can subsequently produce new samples that closely resemble the training data, facilitating their diverse application across various disciplines and domains (Castelli and Manzoni 2022; Ooi et al. 2023). They have the ability to generate a diverse array of outputs, including new images, text, audio, code, video, and other forms of data that mimic the characteristics of the training data they were exposed to. They are particularly noteworthy for their ability to generate realistic, novel, and often indistinguishable data from human-generated content (Eigenschink et al. 2021).

Additionally, GenAI techniques and services powered by Large Language Models (LLMs) have shown exceptional capabilities in interpreting natural human language, automating code generation, and processing information (Brown et al. 2020; Radford et al. 2021; Wolf et al. 2020). These capabilities have been applied to develop autonomous systems across various research fields (Li and Ning 2023). They are associated primarily with four types of GenAI models: GANs, VAEs, Transformer-Based Models, and Generative Diffusion Models. The latter, which have gained prominence in recent years, offer a novel approach to synthesizing high-quality data in machine learning (Ho et al. 2020). These models function by progressively converting random noise into structured data, emulating the physical process of diffusion. This innovative mechanism has challenged the longstanding dominance of GANs in the field (Yang et al. 2023b).

DL has significantly advanced ML generative models, such as GANs, VAEs, transformers, flow-based, and bigGAN and Large-scale models. GANs, introduced by Goodfellow et al. (2014), have been a groundbreaking development in generative modeling. GANs consist of two neural networks, a generator and a discriminator, trained simultaneously in a competitive setting. The generator tries to create data that is indistinguishable from real data, while the discriminator aims to differentiate between real and generated data. Through this adversarial process, GANs can produce remarkably realistic samples across various domains, including images and texts. VAEs are another DL architecture for generative modeling, proposed by Kingma and Welling (2013). VAEs are based on variational inference techniques and consist of an encoder and a decoder. The encoder maps input data to a latent space representation, while the decoder reconstructs the input data from the latent space. VAEs are capable of generating new samples by sampling from the learned latent space distribution. They have been widely used for tasks like image generation, molecular design, and natural language processing.

Additionally, Transformers, introduced by Vaswani et al. (2017), have revolutionized many areas of DL, including generative modeling. Transformers rely on self-attention mechanisms to capture dependencies between input and output sequences, making them particularly effective for tasks involving sequential data. Models like GPT have achieved remarkable success in generating coherent and contextually relevant text. Transformers have also been applied to other generative tasks such as image captioning and language translation. They fall under the broader category of DL because they utilize neural networks with multiple layers to learn representations of data. While they are a subset of DL models, they have unique characteristics that set them apart and are often associated with advances in NLP tasks.

Flow-based models are a class of generative models that directly model the data distribution through a series of invertible transformations (Kingma and Dhariwal 2018; Dinh et al. 2016). These models offer tractable likelihood estimation and efficient sampling. Flow-based models have demonstrated excellent performance in generating high-fidelity images and have been applied to tasks like image generation, super-resolution, and style transfer. Finally, the scale of DL models has also played a crucial role in advancing generative modeling. Models like BigGAN, introduced by Brock et al. (2018), leverage large-scale architectures and massive amounts of data to generate high-resolution and diverse images. These models are trained on powerful computational infrastructure and benefit from techniques like distributed training and mixed-precision arithmetic to handle the computational demands.

In the realm of DL, GenAI has become a transformative field, showcasing its remarkable ability to create innovative content from realistic data. This capability is increasingly vital for the development of sustainable smart city applications. By tackling challenges related to data availability and quality, generating accurate scenarios, and facilitating advanced 3D city modeling and urban design, GenAI models have the potential to significantly enhance the functionality, reliability, and efficiency of UDT. The integration of these models can streamline urban planning processes, improve predictive analytics, and foster more resilient and adaptive urban environments, ultimately contributing to the creation of smarter, more sustainable cities.

Furthermore, LLMs, a subset of GenAI and thus DL, are specifically designed to understand and generate human language. They are commonly used for tasks such as text generation, text completion, dialogue generation, language translation, summarization, sentiment analysis, and named entity recognition. GenAI models learn from extensive datasets to generate original content that mimics the patterns and structures found in the training data (e.g., Dinh et al. 2016; Brock et al. 2018; Goodfellow et al. 2014; Kingma and Dhariwal 2018). Examples of GenAI models, which include LLMs, are GPT-3 (Generative Pre-trained Transformer 3) (Brown et al. 2020), BERT (Bidirectional Encoder Representations from Transformers) (Wolf et al. 2020), Text-To-Text Transfer Transformer (T5) (Raffel et al. 2020), and CLIP (Contrastive Language-Image Pre-training) (Radford et al. 2021).

Pre-trained foundation models serve as the backbone for these LLMs, providing a robust framework that can be fine-tuned for specific applications. These models have undergone extensive training on large and diverse datasets, enabling them to capture a wide array of linguistic patterns and contextual nuances. In the context of LLMs, pre-trained foundation models, such as BERT and GPT-3, leverage their comprehensive training to perform a variety of natural language processing tasks with high accuracy and efficiency, making them essential tools in the development of advanced AI applications. Recent advancements in generative models have significantly enhanced the capabilities of LLMs, fostering innovation and value creation across various domains, including urban planning and design and environmental sustainability and climate change. These advancements hold immense potential for transforming data-driven solutions through GenAI applications, offering powerful tools for addressing pressing and complex urban challenges.

Other notable examples of GenAI include generalist geospatial AI (Jakubik et al. 2023), geo-foundation models (Janowicz 2023; Jakubik et al. 2023; Xie et al. 2023), joint foundation models (Xu et al. 2023), and geographically diverse foundation models (Liu et al. 2024). These can be considered as GenAI and pre-trained foundation models. Generalist geospatial AI models are designed to understand and generate data specific to geographic information systems (GIS). They can analyze spatial data, generate maps, and make predictions based on geographic patterns. They leverage large datasets of geospatial information to generate new maps, predict urban growth, and simulate environmental changes, contributing to urban planning and environmental sustainability. Geo-foundation models are pre-trained on vast amounts of geospatial data, capturing a broad understanding of spatial relationships and geographic features. They serve as a base for various geospatial applications. They can be fine-tuned to generate synthetic geographic data, predict spatial trends, and simulate urban dynamics, aiding in the development of smart cities and environmental planning. Joint foundation models are designed to handle multiple types of data or tasks simultaneously, integrating information from different sources to improve performance on complex tasks. They can combine textual, visual, and spatial data to generate comprehensive outputs. For instance, they can be used to create detailed urban models that integrate building designs, traffic patterns, and environmental data. Geographically diverse models are trained on data from various geographic regions, making them robust and adaptable to different spatial contexts. They can generate realistic scenarios for different urban environments, predict the impact of urban policies in diverse settings, and create tailored solutions for region-specific challenges.

These models are all examples of pre-trained foundation models tailored for specific generative tasks related to geospatial and urban planning applications. They leverage the principles of GenAI to analyze, predict, and generate new data, providing valuable tools for urban planners and researchers working on sustainable smart city projects. In essence, they enhance the capabilities of GenAI by providing a robust base of geographic and spatial knowledge, enabling the generation of high-quality, contextually relevant data and simulations.

Furthermore, pre-trained foundation models form the cornerstone of many GenAI systems, providing a powerful basis for a wide range of applications. These models, having been trained on extensive datasets, possess a deep understanding of data structures and patterns, allowing them to produce highly realistic and contextually appropriate content. Here are some key aspects of how pre-trained foundation models contribute to the field of GenAI (Bommasani et al. 2022; Jakubik et al. 2023; Janowicz 2023; Li et al. 2024; Yuan et al. 2021; Zhou et al. 2023):

- *Transfer learning*: Pre-trained models enable efficient transfer learning by allowing knowledge gained from one task or domain to be applied to another related task with minimal extra training.
- *Resource efficiency*: These models significantly cut down on the computational resources and time needed to train new models from the ground up, particularly for tasks with scarce labeled data.

- *Generalization*: By capturing general patterns and knowledge from a wide variety of datasets, pre-trained models enhance generalization and performance on new, unseen data.
- *Versatility*: These models can be fine-tuned for numerous applications, making them highly versatile tools for various specific AI tasks. For instance, a pre-trained language model can be fine-tuned to generate poetry, write code, or simulate conversation, making them versatile tools in the GenAI domain.

In summary, pre-trained foundation models are essential to the functionality and success of GenAI systems, providing the necessary foundation for creating diverse and high-quality generative outputs. Their capabilities drive innovation in GenAI, enabling the development of sophisticated applications such as advanced simulation models for urban planning and design and hence UDT.

7.4.4 Urban Digital Twin and Its Synergistic Integration with Artificial Intelligence

UDT is a computational model that replicates the physical, spatial, and functional aspects of a city. By integrating diverse data sources, such as sensor data, historical data, geospatial data, spatiotemporal data, satellite imagery, environmental data, and socioeconomic data, UDT creates a dynamic virtual representation of buildings, infrastructures, systems, dynamics, and natural features. This virtual replica, continuously updated with real-time data from IoT sensors and devices and processed using AI techniques, enables city planners and stakeholders to monitor, analyze, model, simulate, and visualize complex urban systems, understand behavioral patterns, and test scenarios for planning and decision-making purposes (Almusaed and Yitmen 2023; Beckett et al. 2022; Bibri et al. 2024a; Zayed et al. 2023). UDT supports decision-making by facilitating the optimization of urban operations, the prediction of future trends, and the development of innovative solutions to urban challenges such as environmental sustainability. In this context, it enhances the performance, efficiency, and resilience of sustainable smart cities (Weill et al. 2023). It also allows city designers to explore different development projections, assess their potential impacts, and optimize spatial layouts and other design solutions (e.g., Caprari et al. 2022; Ferré-Bigorra et al. 2022; Martella et al. 2023). For a detailed account of the functionalities of UDT for planning and design, the interested readers might be directed to Bibri and Huang (2024). Additionally, for a comprehensive examination of UDT technical and computational features, Weil et al. (2023) provide extensive insights. Overall, UDTs play a crucial role in fostering smart, resilient, and sustainable cities for the future.

The synergistic integration of AI with UDT has sparked a transformative wave in the realm of sustainable smart cities, bringing about significant advancements in how these urban environments can be planned and designed (Bibri et al. 2024a). In this context, Zvarikova et al. (2022) propose UDT algorithms leveraging 3D spatiotemporal simulations, ML, and DL for accurate urban modeling and simulation. Austin et al. (2020) integrate semantic knowledge representation and ML in their UDT architecture, showcasing the practical impact of AI in smart city contexts. Beckett (2022) underscores the significant potential of integrating AI and UDT in enhancing urban design and planning strategies through the integration of 3D modeling, visualization tools, and spatial cognition algorithms.

7.5 APPLICATIONS OF ARTIFICIAL INTELLIGENCE, GENERATIVE ARTIFICIAL INTELLIGENCE, AND URBAN DIGITAL TWIN IN URBAN PLANNING AND DESIGN

In this section, we embark on an in-depth analysis and synthesis of a diverse body of literature exploring the applications of AI and its subfields, namely ML, DL, CV, NLP, and GenAI, in the domain of city planning and design. Additionally, we review the pivotal role of UDT in this context. By scrutinizing the synergistic and collaborative integration of these technologies, we endeavor to illuminate their collective impact on shaping urban landscapes, highlighting emerging opportunities for city planning and design.

7.5.1 Artificial Intelligence and Its Five Subfields

In recent years, models and techniques from ML, DL, CV, NLP, and GenAI subfields of AI have emerged as transformative forces reshaping the landscape of urban development, planning, and design. These advanced technologies collectively empower urban policymakers, planners, and designers to analyze vast amounts of data, extract valuable insights from various forms of data, and make informed decisions, ultimately fostering the creation of more sustainable, efficient, resilient, and livable cities. ML continues to be extensively applied in smart cities (e.g., Alahakoon et al. 2020; Alahi et al. 2023; Bibri et al. 2023a, b; Din et al.

2022; Mahamuni et al. 2022; Nishant et al. 2020; Ullah et al. 2020, 2023) and sustainable smart cities (Bibri et al. 2023a; Gourisaria et al. 2022; Efthymiou and Egleton 2023; Mishra 2023; Szpilko et al. 2023; Zaidi et al. 2023)—compared to DL, CV, NLP, and GenAI. This applies by extension to the planning and design of these urban environments. Looking ahead, there are prospects for these comparative models to become increasingly prevalent alongside ML in the years to come, especially considering the significance of their computational capabilities and engineering applications in the urban context.

7.5.1.1 Machine Learning

In urban planning and design, ML techniques are pivotal in harnessing vast amounts of urban data to inform decision-making, optimize systems, and enhance the overall quality of urban environments. Numerous studies have explored and demonstrated the application of ML techniques and algorithms in this domain (e.g., Kamrowska-Załuska et al. 2021; Koumetio et al. 2023; Li and Ma 2022; Marasinghe et al. 2024; Sanches 2023; Son et al. 2022; Yin et al. 2021), highlighting their level of accuracy and efficiency across diverse prediction tasks. These applications span the utilization of supervised learning algorithms, such as Decision Trees (DT), Random Forests (DF), and Support Vector Machines (SVM), for classification and regression tasks, along with unsupervised learning methods, such as k-means clustering and hierarchical clustering, for segmenting urban areas based on socioeconomic characteristics. Reinforcement learning algorithms, including Q-learning and policy gradient methods, offer insights for optimizing resource allocation and decision-making in dynamic urban environments, with potential for optimal decision-making in complex environments (Kamrowska-Załuska 2021). For instance, in urban transportation planning, reinforcement learning optimizes traffic signal control strategies, public transit systems management, and adaptive routing algorithms for autonomous vehicles (Peng et al. 2021; Priya and Saranya 2023).

Furthermore, ML techniques are used for traffic prediction and optimization, energy management and consumption prediction, waste management and recycling optimization, air quality monitoring and pollution prediction, and public safety prediction. They facilitate predictive modeling and scenario analysis, allowing stakeholders to anticipate future urban trends, simulate alternative development scenarios, and evaluate the potential impacts of policy interventions. This capability is closely associated with the integration of AI and ML with UDT in urban planning and design (Beckett 2022; Bibri et al. 2024a), which will be explored separately. Overall, the emergence of advanced ML techniques presents unparalleled opportunities for modeling intricate urban processes and dynamics, enabling urban planners and designers to analyze complex urban systems, detect trends and patterns, and devise evidence-based strategies for sustainable urban development.

7.5.1.2 Deep Learning

DL has significantly advanced ML techniques in urban planning and design by introducing more sophisticated algorithms capable of learning intricate patterns and relationships from complex urban datasets, thereby enabling more accurate predictions and insights. One key aspect where DL has enhanced ML in urban planning and design is its ability to handle large, high-dimensional, and unstructured datasets, offering advanced capabilities for predictive modeling. Traditional ML techniques often struggle to process and extract meaningful information from such datasets due to their complexity and variability. By integrating advanced AI models, especially DL, into planning practices, urban planners can significantly enhance their analytical capabilities, recognizing patterns and trends and make informed predictions through modeling and simulation (Son et al. 2023). Leveraging DL techniques, researchers and practitioners in urban planning can develop models that accurately predict various urban phenomena, including land use changes, mobility patterns, pedestrian movement, traffic patterns (Bibri et al. 2024a), and air quality (Janarthanan et al. 2021). DL methods are more commonly employed for addressing land use/cover, buildings, climate-related issues, the natural environment (Koumetio et al. 2023) and public transport networks (Aqib et al. 2019). DL models contribute to more informed decision-making processes and help urban planners design smarter and more sustainable cities by providing a comprehensive understanding of urban dynamics and informing policy-making and infrastructure development.

Furthermore, DL models and neural networks are increasingly being used in conjunction with CV (Luo et al. 2022a; Suel et al. 2023; Sanchez 2023) to analyze and visually assess built environments as well as address complex urban challenges. In recent years, the utilization of DL algorithms for common CV tasks, such as object detection, image segmentation, and classification, has become widespread, demonstrating outstanding performance (Doiron et al. 2022; Liu et al. 2023; Zhao et al. 2023). The expansion of CV technologies, driven by DL advancements, has enabled their broad application in transportation management, public space analysis, and policy formulation (Vanky and Le 2023). This proliferation provides planners and designers with new opportunities to document, analyze, and strategize urban development (Vanky and Le 2023), thanks to the abundance of visual data from cameras, satellites, and street-view imagery (Liu et al. 2023). DL applications in CV tasks encompass examining the spatial distribution and temporal dynamics of the green view index (Li 2021), conducting multi-scale analysis to assess the impact of a street-built environment on crime occurrence using street-view images (He et al. 2022), comprehending images of urban public spaces through advanced analysis techniques (Sun and Dogan 2023), and forecasting outdoor comfort levels by leveraging human-centric digital twins (Liu et al. 2023). In addition, Ibrahim et al. (2021) introduce a framework for distinguishing planned and unplanned urban areas, emphasizing that DL and CV enable the understanding and analysis of urban environment dynamics through automated processes. Recent advances in DL and CV have the potential to enhance the precision of measuring the human-scale urban environment and are effective in extracting information from various visual datasets (Zhang et al. 2021, 2022).

Similarly, there are also many applications of ML for CV tasks in the domain of urban planning and design, including exploring associations between people's urban density and urban characteristics (Garrido-Valenzuela, Cats and van Cranenburgh 2023), analyzing associations between eye-level urban design quality and on-street crime density (Su, Li and Qiu 2023), predicting outdoor comfort using human-centric DT (Liu et al. 2023), and retrieving urban land-use information at the building block level (Li et al. 2017).

In sum, DL has significantly advanced ML and CV in urban planning and design by enabling the analysis of high-dimensional and unstructured urban data, improving predictive modeling accuracy, facilitating data integration from diverse sources, and enhancing urban simulation and visualization capabilities. With the rise in processing and computational capabilities, coupled with broader access to pre-trained Artificial Neural Network (ANN) algorithms, DL models have the potential to become a widely adopted tool for urban analytics. These advancements and prospects pave the way for more effective and sustainable urban development practices, ultimately culminating in the creation of smarter and more sustainable cities.

7.5.1.3 Computer Vision

In the context of urban planning and design, CV plays a crucial role in analyzing and understanding urban environments through diverse forms of quantitative and qualitative information from visual data. In a recent systematic literature review conducted by Marasinghe et al. (2024), exploring the integration of CV in the process of urban planning, the authors provide several key insights: CV's multifaceted potential spans data collection and analysis; issue identification and prioritization; public engagement; and plan design, implementation, and evaluation, while enhancing decision-making through diverse visual information and promoting endeavors for sustainable urban development. CV techniques are increasingly being used to analyze satellite imagery and aerial photographs to assess land-use patterns, identify urban features such as buildings and roads, monitor changes in the built environment over time, and so on. Visual data are of critical importance in the process of urban planning and design, thereby the significance of CV (Liang et al. 2023; Guzder-Williams et al. 2023; Yuan et al. 2022) as a valuable tool for decision-making and planning processes (Guzder-Williams et al. 2023).

By leveraging advanced visual data analytics, urban planners and designers can gain valuable insights into the spatial characteristics of cities, inform decision-making processes, and optimize urban development strategies. Based on a comprehensive systematic review conducted by Bibri et al. (2024a), it is evident that CV techniques hold great potential for advancing urban planning and design. The following key insights highlight the significant opportunities and applications of CV techniques in this field:

- CV methods collect diverse environmental and socioeconomic data from images and videos.
- Understanding visual environments includes identifying physical objects, environmental attributes, and human experiences and perceptions.
- CV is crucial for enhancing AI in urban planning by investigating user experiences and providing empirical evidence for evaluating planning and design interventions.
- CV allows professionals to analyze different locations using fine-grained information on microscale urban features and large datasets, predicting people's perceptions of urban environments.

- ML applications for CV in urban planning include examining associations between urban density and characteristics and analyzing the link between urban design quality and on-street crime density.
- Utilizing CV techniques helps urban planners and policymakers investigate spatial correlations and factors contributing to urban challenges and prospects.
- The rise of spatial-explicit GeoAI, integrating spatial attributes into AI computations, offers diverse applications for sustainable urban planning.
- AI empowers urban researchers and planners to use CV for designing, observing, and modeling urban environments, enhancing planning and evaluation processes, and facilitating stakeholder participation.
- CV offers accessible and cost-effective tools for urban assessment and modeling.
- CV helps evaluate scenarios and their impacts on city sustainability and resilience throughout the planning cycle, tracking progress and outcomes of interventions.
- CV is pivotal for smart and sustainable urbanism initiatives, creating environmentally friendly urban environments through data-driven methodologies.

Overall, the comprehensive review by Bibri et al. (2024a) underscores the transformative potential of CV techniques in urban planning and design. By enabling the collection and analysis of extensive environmental and socioeconomic data, CV facilitates a deeper understanding of urban environments. The integration of CV with AI and IoT technologies offers scalable and cost-effective solutions for urban assessment, modeling, and planning. These advancements not only enhance the decision-making processes for urban planners and policymakers but also contribute to creating resilient, sustainable, and environmentally friendly urban spaces. As CV continues to evolve, its role in shaping the future of sustainable smart urban development will undoubtedly become even more pivotal.

Indeed, CV enhances our understanding of the physical and spatial spaces. Its applications span various domains, including satellite image mapping, autonomous driving, and studies of the built environment and urban dynamics (Wael et al. 2022; Sanchez 2023; Vanky and Le 2023;

Garrido-Valenzuela et al. 2023; Salazar-Miranda et al. 2023). Moreover, CV contributes to sustainable urban policy and planning development, aiding in the implementation of sustainable urban development goals (Jaad and Abdelghany 2021; Liu et al. 2023). Also, CV enables stakeholders to visualize ground changes and prioritize human-centric aspects of the planning process (Yuan et al. 2023; Andrews et al. 2022). By offering opportunities to evaluate built environments and analyze urban data and human activities from high-resolution visual images, CV supports detailed spatial analysis for understanding urban patterns and comparisons between areas (Garrido-Valenzuela et al. 2023; Salazar-Miranda et al. 2023; Zhang et al. 2023). CV holds promise in supporting the design of sustainable and resilient urban built environments by addressing data gaps in planning and enabling comprehensive spatial analysis (Hudson and Sedlackova 2021; Hosseini et al. 2022). Additionally, CV technologies can facilitate the creation of DT and 3D models of urban areas (e.g., Beckett 2022; Liu et al. 2023), enabling planners and policymakers to visualize and simulate proposed design interventions and assess their potential impacts on the built environment. Overall, CV, along with ML and DL, offers powerful tools for enhancing the efficiency, effectiveness, and sustainability of urban planning and design processes.

7.5.1.4 Natural Language Processing
NLP presents a notable opportunity for planners to enhance the efficiency and effectiveness of processing colossal amounts of textual data. Urban planners have dealt with extensive textual content comprising plans (e.g., land use, transportation, and hazard mitigation plans), policies (e.g., zoning ordinances and building codes), reports (e.g., housing and traffic analysis), as well as stakeholder engagement and community feedback obtained from public engagement initiatives (Fu 2024). NLP techniques can be used primarily to analyze these forms of textual data. They can extract insights from these data, including sentiment analysis, topic modeling, and named entity recognition, to understand public opinions, identify trends, and inform decision-making processes in urban planning and design. For example, sentiment analysis identifies and categorizes opinions expressed in textual data to determine the positive, negative, or neutral sentients toward urban development projects and policies. Planning scholars have utilized NLP methods to analyze social media content for capturing public sentiment and opinions regarding various subjects across different temporal and spatial contexts (Kong et al. 2022; Zhai et al. 2020),

perceptions of urban parks based on environmental features (Huai and Van de Voorde 2022), and analyzing and visualizing users' sentiment pertaining to the built environment (Chang et al. 2022). Additionally, NLP models have been employed in planning research to monitor research themes and trends (Fang and Ewing 2020) and to extract pertinent information from planning documents, eliminating the need for manual reading (Brinkley and Wagner 2022; Fu et al. 2023; Fu et al. 2023), as well as integrating spatial development plans (Kaczmarek et al. 2022).

In addition, Zhou et al. (2024) introduce a novel urban planning approach integrating Large Language Models (LLMs) into the participatory process. Employing a crafted LLM agent, the framework facilitates role-play, collaborative generation, and feedback iteration to address community-level land-use tasks. Empirical experiments showcase the adaptability and effectiveness of LLMs across diverse planning scenarios. Evaluation metrics reveal LLMs' superiority over human experts in satisfaction and inclusion through natural language reasoning and robust scalability and comparable performance to state-of-the-art reinforcement learning methods in service and ecology. Additionally, NLP models can be applied to generate natural language reports, summarize complex urban data, or facilitate communication between stakeholders involved in urban projects. Fu (2024) provides an overview of contemporary planning studies utilizing Natural Language Processing (NLP), aiming to identify major topics and obstacles while proposing a unified direction for future investigations. Findings indicate that current research predominantly explores NLP applications with a disjointed scholarly domain. Future endeavors should prioritize data sharing, standardize NLP methods, promote collaborative research specific to planning, and address ethical concerns to fully exploit NLP's capabilities in planning endeavors.

7.5.1.5 Generative Models for Urban Planning and Design

GenAI models are being increasingly utilized in the domain of urban planning and design to generate various types of content, including images, text, and even entire urban layouts. In recent years, a surge in studies has explored various aspects of GenAI in this domain, employing innovative computational methodologies to address the multifaceted challenges of sustainable urban development. Pérez-Martínez et al. (2023) propose a novel methodology for automating urban planning tasks through generative design principles, aiming to enhance planning processes and decision-making efficiency. Yang et al. (2023a) focus on generative urban space design, particularly emphasizing the conservation and reuse of historical blocks. Their study demonstrates the effectiveness of their method in quickly generating urban space design schemes while preserving historical fabric. Sun and Dogan (2022) introduce an accelerated environmental performance-driven design approach for parametric urban blocks using GANs. Their method accelerates optimization while considering environmental performance, significantly reducing computation time. Oyama et al. (2023) contributed with a deep generative model tailored for super-resolving spatially correlated multiregional climate data, crucial for accurately downscaling global climate simulations. Their approach preserves spatial correlations, contributing to accurate downscaling and addressing systems with spatial expanse. Tian et al. (2023) focus on generative parallel transformer development for vehicles with intelligent systems, specifically targeting transport automation.

In addition, Sun and Dogan (2023) introduce generative methods for urban design and rapid solution space exploration. Their study presents a generative urban modeling toolkit for rapid design space exploration and multi-objective optimization. Calderaa et al. (2021) explore computational urban planning utilizing ML and generative design for assessing multiple planning scenarios, contributing to a better understanding of planning outcomes. Zheng et al. (2023) introduce an AI urban-planning model leveraging graph-based techniques to generate spatial urban plans and address complex urban geography issues. Their deep reinforcement learning model, based on graph neural networks, surpasses human-designed plans, efficiently generating adaptive spatial plans. This proposed human–AI collaborative workflow enhances productivity in urban planning, demonstrating the potential of computational methods in tackling real-world urban challenges. Furthermore, Pérez et al. (2023) address data augmentation through multivariate scenario forecasting using GAN, and Rane (2023a) investigates the potential roles of ChatGPT and similar GenAI in architectural engineering.

These studies employ various GenAI techniques, including GANs, VAEs, RNNs, and autoregressive models. These techniques enable the exploration of design alternatives, optimization of urban layouts, and simulation of urban scenarios to facilitate decision-making in city planning and design. Collectively, they represent significant advancements in the emerging field of GenAI-driven planning and design, each offering unique insights and methodologies to address the complexities of sustainable urban development. However, more specialized generative models are emerging for innovative approaches to urban planning and design.

However, based on a comprehensive study conducted by Bibri et al. (2024a) on the synergistic interplay of urban AI and UDT in planning and designing sustainable smart cities, the following key insights and conclusions can be drawn:

The adoption of AI and its subfields in urban planning and design is expected to be a gradual process, requiring considerable time and resources. Despite the promising prospects and opportunities AI offers for advancing environmental goals in sustainable smart cities, challenges remain in translating research into practical applications. Urban planners and designers express concerns about adopting AI in the field, which hampers the integration of AI-driven urban planning and design into real-world scenarios. Nonetheless, given the anticipated pivotal role of AI in strategic city planning and design aimed at enhancing environmental sustainability and resilience, planners and designers must proactively prepare for the transformative changes these technologies herald for sustainable urban development. This preparation includes reimagining how future cities will be planned, designed, and managed.

To equip planners for the AI revolution, several challenges must be addressed. Overcoming fear and uncertainty surrounding AI adoption is paramount. Urban planners need to acquire new skills necessary for effectively leveraging AI technologies. This entails continuous education and training to stay updated with the latest advancements in AI. Moreover, adapting to changing data needs and considerations inherent in AI-driven planning and design processes is crucial. Establishing clear goals to guide AI implementation strategies will help align these technologies with urban planning and design objectives. Ensuring transparency, accountability, and explainability in AI systems is essential to building trust among stakeholders and the public. Mitigating bias in AI algorithms and decision-making processes is another critical challenge, as biased AI systems can lead to unfair and ineffective urban planning and design outcomes. Addressing ethical concerns arising from the use of new methods and data in AI-driven planning and design endeavors is also vital to ensure responsible use of AI technologies. Indeed, urban planners must ensure the effective integration and ethical use of AI technologies in such endeavors. The expanding role of AI in the planning and design profession offers significant potential implications for communities. Planners must ensure fairness and inclusivity while integrating AI into their practices.

Furthermore, the implications for GenAI in urban planning and design are profound. GenAI can revolutionize urban planning and design by automating complex tasks, generating realistic urban scenarios, and enhancing decision-making processes. However, the adoption of GenAI will require overcoming significant challenges, such as ensuring data privacy, maintaining transparency in AI-generated outcomes, and addressing ethical concerns. Urban planners must navigate these challenges to harness the full potential of GenAI for sustainable and resilient urban development.

In conclusion, while the adoption of AI and its subfields in urban planning and design is anticipated to be gradual and resource-intensive, its potential for advancing sustainable smart cities is immense. Overcoming the outlined challenges and equipping urban planners and designers with the necessary skills and tools will be crucial for realizing the transformative potential of AI and GenAI in urban planning and design.

7.5.1.6 Generative Artificial Intelligence for Architectural and Urban Design

GenAI is transforming the fields of architecture and urban design, offering new tools and methodologies to address the growing complexities of modern cities. From urban mobility systems to building structures, GenAI has introduced novel approaches to planning, designing, and managing spaces. As AI continues to evolve, it offers architects and urban planners enhanced creative potential and efficiency, pushing the boundaries of design. This synthesis explores a series of studies that examine how AI, particularly GenAI, is being applied in architectural and urban contexts, highlighting its role in reshaping urban form, transportation, and design processes.

Bratton (2021) provides a conceptual foundation for understanding AI's impact on urban design and governance by discussing how embedded AI technologies influence urban form and governance models. The study emphasizes the need to move beyond anthropocentric AI models and instead focus on AI as a distributed, material process that reshapes urban systems through zoning, data governance, and platform cognition. The framework lays the groundwork for a more holistic understanding of AI in urban environments and opens the door for practical applications in urban governance and form.

Moving from urban-scale applications to the architectural design process, Fitriawijaya and Jeng (2024) integrate multimodal GenAI and blockchain technologies to enhance the early stages of architectural design. Their work expands on AI's potential by demonstrating how it can not only generate innovative designs but also ensure the security and authenticity of those designs through blockchain. The study's framework incorporates blockchain for data management, ensuring transparency and traceability in the architectural design process. This highlights AI's role in both enhancing creativity and ensuring the integrity of the design workflow.

Li et al. (2024) extend the application of GenAI to more advanced design technologies, providing a comprehensive review of how deep generative models like GANs and VAEs have advanced architectural design innovation. The authors emphasize the expanding role of state-of-the-art AI models, such as diffusion models and 3D generative models, in architectural design. Their work builds on Fitriawijaya and Jeng's (2024) integration of AI in design processes by showing how emerging AI models offer architects a broader creative toolkit, increasing both the diversity and efficiency of architectural design.

Further deepening the discussion on AI's role in the construction industry, Liao et al. (2023) examine how GenAI enhances building structure design by learning from previous design data and generating efficient, optimized structural models. Their focus on AI's ability to integrate complex structural data complements Li et al.'s (2024) exploration of AI in architectural design by highlighting how AI not only improves design creativity but also optimizes structural integrity and functionality. The study demonstrates how AI's capacity to analyze and generate structural designs streamlines processes and addresses key challenges in building construction.

To bridge architectural design with construction management, Ma et al. (2021) explore the integration of Generative Design (GD) with Building Information Modelling (BIM), focusing on the constructability of automated design solutions in the building industry. Their work complements both Liao et al. (2024) and Li et al. (2024) by emphasizing the practical applications of AI-generated designs in BIM, ensuring that design solutions are not only innovative but also feasible for construction. Ma et al.'s (2021) study pushes the boundaries of AI-driven design by integrating it with BIM, thus offering a comprehensive approach to both design and construction processes.

Finally, Michelle and Gemilang (2022) provide a broader bibliometric analysis that tracks trends in generative, algorithmic, and parametric design in architecture. Their work highlights the increasing integration of AI and computational design tools in the architectural community, mapping the rapid growth of AI technologies in architecture and urban design.

This study builds on the foundational and applied research by showing the broader trajectory of AI adoption and its impact on the discipline.

These studies demonstrate the innovative potential of GenAI in reshaping both urban and architectural design. From urban governance and transportation systems to architectural design processes and construction management, AI is being leveraged to address the complex challenges of modern cities. By integrating advanced AI techniques like GANs, VAEs, and blockchain, these studies showcase the creative potential of AI but also emphasize its ability to optimize efficiency, security, and sustainability in design. As AI continues to evolve, its application in architecture and urban planning will likely expand, offering new opportunities for innovation, efficiency, and more intelligent, sustainable urban environments.

The integration of GenAI in architectural and urban design, as demonstrated in these studies, aligns with the development of the sustainable smart city digital twin as a virtual replica that uses real-time data, simulation, and advanced AI to optimize urban systems. By leveraging GenAI technologies, this replica can model complex urban dynamics. GenAI-driven design methodologies enhance the creative and operational aspects of urban planning and contribute to the sustainable management of resources. Integrating blockchain technology, as outlined by Fitriawijaya and Jeng (2024), further ensures data security and transparency in managing the sustainable smart city's digital infrastructure. These advancements position GenAI as a critical component in creating intelligent, sustainable urban environments that adapt to real-time changes, enabling the efficient management of resources and supporting long-term sustainability goals.

7.5.2 Urban Digital Twin

7.5.2.1 Artificial Intelligence

AI models and techniques, especially ML and DL, are employed within UDT frameworks to analyze vast amounts of urban data, extract meaningful insights, and predict future trends. The applications of AI-driven UDT have demonstrated their efficacy across various domains. In urban mobility, AI-driven UDT plays a key role in improving energy efficiency, predicting traffic patterns, enhancing sustainable transportation, optimizing traffic management, improving transportation infrastructure, and strengthening resilience.

Shen, Saini, and Zhang (2021) review the application of DT in Positive Energy Districts (PEDs) to enhance livability and sustainability.

The authors highlight DT's role in integrating energy, ICT, and mobility systems through AI and big data analytics, identifying key concepts, tools, and applications while discussing challenges and opportunities in optimizing PEDs. Deena et al. (2022) explore using DT and AI to optimize energy management in residential buildings. A case study in Rome's Rinascimento III neighborhood showed how DT, connected to IoT and AI, could model and simulate energy-efficient scenarios to maintain comfort levels and meet near zero energy building (NZEB) criteria. Strielkowski, Rausser, and Kuzmin (2022) review the impact of DT in the energy sector, focusing on its application in smart grids. The authors discuss various types of DTs and their roles in enhancing safety, reliability, and efficiency in energy networks through real-time monitoring and predictive maintenance. Almusaed and Yitmen (2023) examined the role of AI simulation models and DT in designing smart buildings. Their research verified that AI-driven DT could simulate building performance, improving user comfort and efficiency by evaluating different design scenarios. The authors focus on its potential in optimizing resource management, efficiently monitoring and allocating resources, such as energy, water, and waste.

Wu et al. (2022) analyze the application of DT and AI in transportation infrastructure. The authors categorize intelligent development in transportation, highlighting how DTs and AI can optimize infrastructure management and integration with new media, providing references for future smart city developments. Sabri et al. (2023) discussed designing UDT for smart water infrastructure and flood management. The authors use case studies from Orange County and Victoria to illustrate the importance of accurate GIS data and GeoAI in managing water resources and flood impacts in smart cities. Kamal et al. (2024) propose a DT-based deep reinforcement learning approach for adaptive traffic signal control. Their study demonstrates that DT could optimize traffic signals to reduce CO_2 emissions and fuel consumption in urban areas, using simulations from a neighborhood in Amman, Jordan.

Manocha et al. (2023) introduce a DT framework for flood prediction using AI and blockchain for secure data management. Their case study demonstrates the framework's efficacy in improving situational analysis and decision-making for flood management, achieving high training and testing accuracy. Fan et al. (2021) introduce the Disaster City DT concept to improve disaster management by integrating AI for enhanced situational assessment and decision-making. The authors propose a framework combining multi-data sensing, data integration, game-theoretic decision-making, and dynamic network analysis to improve crisis response and humanitarian actions.

Agostinelli et al. (2020) discuss the potential of integrating a DT model with AI systems, focusing on its application in optimizing energy management and efficiency in residential districts. The authors present a case study of Rinascimento III, a residential area in Rome, which comprises 16 eight-floor buildings with a total of 216 apartment units. The buildings have an impressive 70% self-renewable energy production, classifying them as Near Zero Energy Buildings (NZEBs). The implemented DT model is designed as a comprehensive three-dimensional data system capable of intelligent optimization and automation of building systems. It allows for real-time monitoring and analysis of energy consumption, production, and distribution in the residential district. By integrating AI systems, it can predict energy demand patterns, optimize energy usage, and automate energy-related processes to enhance overall efficiency. The study highlights the importance of NZEBs in achieving sustainable energy goals and reducing environmental impact. It emphasizes the significance of on-site renewable energy production, which contributes to the overall self-sufficiency of the building complex. The integration of DT with AI offers a promising approach to further improve the performance of NZEBs by enabling proactive and data-driven energy management strategies. Overall, the study demonstrates the potential of AI-driven DT models in optimizing energy systems and promoting sustainability in residential areas. It underscores the importance of leveraging advanced technologies to address the challenges of urban energy management and achieve energy efficiency targets.

The contribution of the study by Gkontzis et al. (2024) lies in its innovative method by integrating Extract, Transform, Load (ETL) processes with AI and DT techniques to process and interpret urban data streams. This integration facilitates the development of dynamic simulations that provide actionable insights for urban planners and decision-makers. The study aims to strengthen urban resilience by employing smart city data analyses, forecasts, and DT modeling and simulation techniques at the district level. It utilizes these methodologies to monitor and replicate urban systems in real-time, with a focus on proactive planning, intervention strategies, and bottleneck identification to optimize the efficiency of smart cities. The findings highlight the crucial role of real-time monitoring, simulation, and analysis in supporting test scenarios, leading to improvements in urban functionality, resilience, and resident quality of life. Integrating advanced techniques enables the visualization, analysis, and prediction of urban system behaviors and responses, thereby informing more effective urban planning and decision-making processes. Ultimately, the study contributes to the advancement of urban resilience strategies by offering a comprehensive framework for leveraging smart city data at the neighborhood level.

Bibri et al. (2024a) conduct a thorough analysis and synthesis of a large body of literature to uncover the intricate interactions among AI, AIoT, UDT, data-driven planning, and environmental sustainability, revealing the nuanced dynamics and untapped synergies in the complex ecosystem of sustainable smart cities. The integration of these technologies and models is revolutionizing data-driven planning strategies and practices and hence supporting the achievement of environmental sustainability goals. The study has several key objectives: to identify and synthesize the theoretical and practical foundations that support the convergence of AI, AIoT, UDT, data-driven planning, and environmental sustainability into a comprehensive framework; to investigate how the integration of AI and AIoT transforms data-driven planning to enhance the environmental performance of sustainable smart cities; to explore how AI and AIoT can augment UDT capabilities to improve data-driven environmental planning processes in sustainable smart cities; and to identify the challenges and barriers in integrating and implementing AI, AIoT, and UDT in data-driven environmental urban planning, and develop strategies to overcome or mitigate these challenges. The integration of AI, AIoT, and UDT technologies significantly enhances data-driven environmental planning processes, providing innovative solutions to complex environmental challenges and promoting sustainable urban development. However, this integration also introduces significant challenges and complexities that must be carefully managed to achieve successful outcomes. Beyond theoretical enrichment, the study offers valuable insights into the transformative potential of integrating AI, AIoT, and UDT technologies to advance sustainable urban development practices. It serves as a comprehensive reference guide, stimulating groundbreaking research, practical implementations, strategic initiatives, and policy formulations in sustainable urban development.

These studies underscore the transformative potential of AI in advancing UDT functionalities, thereby shaping sustainable, resilient, efficient, and ecologically oriented urban landscapes and profoundly impacting the trajectory of sustainable urban development. By integrating AI and real-time data analytics, UDT provides comprehensive tools for optimizing urban systems, enhancing sustainability, and improving the quality of life in sustainable smart cities. These contributions underscore the critical role of UDT in driving forward urban innovation and resilience in the face of growing urbanization challenges.

7.5.2.2 Generative Artificial Intelligence

In the context of sustainable smart cities, strategic approaches to envisioning and analyzing various potential future scenarios are crucial for enhancing urban planning and design. They necessitate foresight into future needs and challenges, which is essential for developing resilient urban environments. However, implementing these strategies often involves costly, time-intensive, and laborious efforts to collect data specific to each scenario. Furthermore, producing relevant scenarios using processed data and simulations to aid predictive analytics and decision-making is equally complex (Dong et al. 2021; Argota Sánchez-Vaquerizo 2022). The challenge of acquiring real-world data that accurately represents future scenarios adds another layer of complexity to urban planning and design. Moreover, creating a comprehensive 3D city model to digitize and computerize large urban areas can be a repetitive and labor-intensive endeavor (Botín-Sanabria et al. 2022; Singh et al. 2021). This task involves the detailed modeling of urban systems as 3D models with various levels of detail, requiring significant financial and human resources.

The planning of sustainable smart cities often involves hierarchical and multi-scale urban design processes based on previous data analytics and city modeling and simulation (UDT) applications. These processes are crucial for configuring land use and land cover, transportation networks, buildings, and infrastructure (Van Nes and Yamu 2021; Pérez-Martínez et al. 2023). The repetitive, costly, and time-consuming nature of these processes, which involve numerous design alternatives and complex decision-making for optimization, presents significant challenges. Scaling up these design processes to encompass large urban areas with unique design requirements remains a formidable task. Thus, innovative solutions and efficiencies in planning and design are critical for the sustainable development of smart cities or sustainable smart cities.

In light of the above, UDT has recently started to integrate GenAI into its functionalities to advance decision-making processes. By leveraging UDT as a platform, GenAI contributes to the advancement of more data-driven, efficient, and innovative approaches to urban planning and design. It has proven more effective in generating various outputs of practical relevance to UDT, including synthetic data for scenario testing, optimization, and simulation processes; design proposals for urban spaces; adaptive spatial plans; scenario simulations for visualizing potential outcomes; virtual environments representing proposed urban developments; predictive models for urban phenomena; customized solutions for specific urban challenges; and optimized designs for efficiency, sustainability, and resilience. UDT, as a "cyber-physical systems of systems" due to its unique characteristics in terms of system size, complexity, and requirements (Bibri and Huang 2024; Mylonas et al. 2021), allows for the seamless integration of these diverse outputs into its multifaceted functions, enhancing decision-making capabilities, fostering innovation, and empowering stakeholders to create smarter and more sustainable urban environments.

The integration of UDT and GenAI in sustainable smart cities signifies a transformative shift in urban planning and design methodologies. Xu et al. (2024) investigate the innovative use of GenAI techniques and UDT to tackle data-related challenges in smart cities. In their structured review of how recent advancements in GenAI are revolutionizing key areas of sustainable smart city applications, the authors provide a detailed examination of its numerous applications. They primarily focus on the roles of various GenAI models, including GANs, VAE), transformer-based models, and generative diffusion models. Among the specific application areas of GenAI models highlighting their transformative ability to address challenges in UDT applications, include:

1. Data Augmentation:

    - *Mobility and transportation management:* Enhancing the diversity and volume of data used for training models that manage traffic flow and public transportation systems.

    - *Urban energy systems:* Improving the predictive models for energy consumption and generation by creating varied synthetic datasets.

    - *Building and infrastructure management:* Generating comprehensive data sets to optimize the maintenance and management of urban infrastructure.

2. Data Synthesis and Scenario Generation:

    - *Mobility and transportation management:* Creating realistic simulations of traffic patterns and public transport scenarios to improve planning and response strategies.

    - *Urban analytics:* Producing diverse data sets for better analysis of urban phenomena and trends.

    - *Urban energy systems:* Synthesizing data for predictive modeling in energy usage and distribution.

    - *Urban water management:* Simulating various water management scenarios to optimize resource allocation and disaster preparedness.

    - *Urban disaster management:* Generating possible disaster scenarios to enhance emergency response and resilience planning.

3. 3D City Modeling:

    Utilizing GenAI models to create detailed, three-dimensional representations of urban environments, aiding in urban planning and development.

4. 3D Building Generation: Designing realistic and structurally sound building models to assist architects and planners in visualizing and optimizing urban spaces.

5. 3D Street Environment Construction: Creating immersive street-level models that help in planning pedestrian pathways, vehicular routes, and public spaces.

6. Generative Urban Design and Optimization: Employing GenAI to explore innovative urban design solutions and optimize city layouts for sustainability and efficiency.

The key findings reveal that GenAI can significantly aid in data augmentation, synthetic data and scenario generation, automated 3D city modeling, and generative urban design and optimization, thereby contributing to more sustainable and resilient smart city developments. These advancements promise to improve urban monitoring, predictive analytics, and the overall reliability, scalability, and automation of smart city management.

Indeed, there is a growing demand for scalable and autonomous solutions that leverage GenAI techniques to enhance the availability and quality of urban data. These solutions also automate the creation of various scenarios, aid in the development of 3D city models, and streamline complex urban design and optimization processes (Wang et al. 2023; Kirwan and Zhiyong 2020).

Additionally, Somanath et al. (2023) focus on procedural UDT generation and visualization of large-scale data. The authors develop a semi-automated workflow for generating detailed 3D models of buildings and their context from aerial imagery, LiDAR, or footprints. Their findings highlight the challenges in achieving an automated end-to-end workflow and propose a workflow that extends building reconstruction to include procedural context generation using Unreal Engine. Xu et al. (2023) address the integration of GenAI and DT into intelligent process planning to overcome the limitations of current intelligent process planning methods by leveraging GenAI techniques. The authors propose a conceptual framework called GenAI and DT-enabling intelligent process planning, which includes methodologies such as ProcessGPT modeling and DT-based process verification. The contributions include a prototype system for milling a specific thin-walled component.

Xu et al. (2024) investigate the integration of GenAI models with emerging UDT applications, such as web-based platforms, cyber-physical systems, and cyberinfrastructure. The goal is to enhance the current capabilities of these systems for more efficient and sustainable smart city management. The authors also explore the potential technical and methodological challenges associated with developing generative UDT, highlighting the complexities and innovations required for this forward-thinking approach. They cover cognitive DT opportunities, GenAI-DT integration, and the associated challenges and limitations.

Overall, through the collaborative and synergistic integration of UDT (or advanced city modeling and simulation) and GenAI, urban planners and designers can explore diverse scenarios, evaluate their potential impacts, and devise informed strategies to enhance the resilience, livability, and environmental sustainability of urban landscapes.

7.5.3 Generative Artificial Intelligence for Environmental Sustainability and Climate Change

The integration of generative models and GenAI models in various fields is paving the way for groundbreaking advancements. These models can analyze large datasets related to environmental factors, such as air quality, water usage, and energy consumption, to identify trends and predict future conditions. For example, they can simulate the effects of various environmental policies or climate scenarios on urban areas, providing valuable insights for policymakers. This aids in crafting strategies that mitigate adverse environmental impacts and promote sustainability. Furthermore, LLMs can aid in disseminating complex environmental data to the public, fostering greater awareness and engagement in sustainability initiatives. The following synthesized studies highlight the application of GenAI models in oceanography, disaster management, climate disaster preparedness, extreme weather forecasting, and nature-based solutions. They also underscore the application of generative models in detecting specific disasters and simulating their potential impacts, thereby enhancing preparedness strategies thanks to their adeptness to synthesize realistic images and scenarios.

Zhou et al. (2023) employ GANs for predicting urban morphology related to local climate zones (LCZs). They integrate LCZ and urban morphology data from six cities using GANs, particularly excelling with the Pix2pix model. This approach allows for rapid and responsive 3D urban morphology prediction, providing significant insights for improving local climates in urban areas through 3D generative design techniques.

Chen et al. (2023) explore the creation of DT for the ocean using data science and AI. Their goal is to merge extensive marine data with AI to develop a prototype DT of the ocean. The research presents a structured architecture for the Generative Digital Twin (GenDT) of the ocean and sets forth future research directions. The focus is on leveraging AI and oceanography to enhance the analysis and simulation of marine data. This method aims to tackle marine scientific challenges by utilizing advanced technologies such as big data, cloud computing, 5G communication, virtual reality, supercomputing, and AI, creating a detailed and interactive digital model of the ocean.

Ahn et al. (2023) focus on generating virtual earthquake scenarios for disaster management in smart cities using Auxiliary Classifier GAN (AC-GAN). The authors address the challenge of constructing earthquake response models in regions with limited training data by generating earthquake scenarios for response training. Their findings demonstrate the effectiveness of the proposed AC-GAN model in generating earthquake scenarios with similar characteristics to actual data. The integration of UDT and GenAI into sustainable smart city planning and design enhances creativity and efficiency in addressing urgent environmental challenges and advancing long-term sustainability goals.

Delacruz (2020) introduces an innovative method utilizing GANs for the swift classification of earthquake-induced damage to railways and roads. The study highlights the application of GANs to generate synthetic damage images, addressing the issue of limited real-world data for training deep learning models. This approach showcases the potential of GANs to significantly improve the efficiency and accuracy of structural damage assessment, which is vital for emergency response and infrastructure recovery efforts. In earthquake and flood management, GANs produce detailed images of damaged infrastructure, facilitating rapid damage assessment and efficient resource allocation (Tilon et al. 2020). This technology greatly enhances response times and decision-making accuracy, playing a crucial role in mitigating the impacts of urban disasters and strengthening resilience in city planning and emergency responses.

Hofmann and Schüttrumpf (2021) introduce a novel deep learning technique for real-time pluvial flood prediction. Their research unveils floodGAN, a deep GAN that significantly enhances the speed and accuracy of flood prediction. This model leverages the power of deep learning, particularly GANs, to interpret and predict intricate flood dynamics driven by spatially distributed rainfall. The innovative aspect of this approach lies in its ability to deliver rapid, precise flood predictions, crucial for real-time applications such as early warning systems, marking a substantial advancement in flood modeling and risk management.

McCormack and Grierson (2024) delve into the application of GenAI technologies to develop realistic simulations aimed at climate disaster preparedness. They emphasize the importance of immersive, narrative-driven simulations that enable users to safely experience the impacts of extreme climate events, thereby improving planning and preparedness strategies. Their study reviews the current capabilities of GenAI models but also identifies future advancements necessary to enhance the realism and effectiveness of these simulations. By leveraging the strengths of GenAI, they highlight the potential for experiential learning to prepare stakeholders for climate-related disasters, allowing for more effective planning and response before actual events occur. This approach underscores the critical role of advanced simulations in enhancing resilience and readiness in the face of climate change.

Sha et al. (2024) introduce an innovative method that combines a 3-D Vision Transformer (ViT) with a Latent Diffusion Model (LDM) to enhance probabilistic forecasts of extreme precipitation events across the conterminous United States. This approach enhances the accuracy of 6-hourly precipitation ensemble forecasts by generating spatiotemporally consistent precipitation trajectories. Verified against Climate-Calibrated Precipitation Analysis (CCPA) data, these improved forecasts demonstrate better skill in predicting extreme precipitation events, as evidenced by higher Continuous Ranked Probabilistic Skill Scores (CRPSSs) and Brier Skill Scores (BSSs).

Yadav et al. (2024) present an innovative unsupervised ML approach for flood mapping using Synthetic Aperture Radar (SAR) time series data. The authors utilize a VAE trained with reconstruction and contrastive learning techniques. This model excels at detecting changes by analyzing the latent feature distributions between pre-flood and post-flood data. Evaluated across nine different flood events, it shows superior accuracy compared to existing unsupervised models, representing a significant advancement in applying DL to environmental monitoring and enhancing flood detection capabilities using satellite data.

Salih et al. (2022) investigate the application of GANs and Mask Region-Based Convolutional Neural Networks (Mask-R-CNN) for detecting and segmenting floodwater in images. The study demonstrates the ability of GANs to generate segmented images of flooded areas, highlighting their potential for quick and accurate floodwater identification. This research underscores the effectiveness of GANs in image-based flood detection, offering a novel approach to improving flood response and management strategies.

Lago et al. (2023) introduce cGAN-Flood, a conditional generative adversarial network aimed at enhancing flood prediction in urban catchments. Unlike traditional ANN models, cGAN-Flood generalizes flood predictions to areas outside its training dataset, addressing varying boundary conditions. It uses two cGAN models to identify wet cells and estimate water depths, trained with HEC-RAS outputs. When tested on different urban catchments and compared with HEC-RAS and the rapid flood model WCA2D, cGAN-Flood successfully predicted water depths, though it underestimated channels during low-intensity events. This model offers a promising, efficient alternative for large-scale flood forecasting, being significantly faster than existing models.

Richards et al. (2024) explore the potential of GenAI to automate and scale up the communication and implementation of nature-based solutions. They present three case studies: reporting on ecosystem services and land use options for farms, providing interactive guidance for biodiversity-friendly garden design, and visualizing future landscape scenarios that integrate nature-based and technological solutions. These studies illustrate how GenAI can facilitate the dissemination of nature-based design strategies, reaching a broader audience and promoting sustainability.

GenAI is proving to be invaluable in enhancing resilience, improving extreme weather forecasts, simulating climate events, supporting nature-based solutions, and promoting environmental sustainability. The studies reviewed demonstrate the diverse applications and transformative potential of GenAI across various domains. Harnessing the power of GenAI alongside human expertise can tackle complex challenges and build a more resilient and sustainable future. Furthermore, fenerative models are increasingly utilized to detect specific disasters and simulate potential impacts, thus enhancing preparedness strategies. The summarized studies showcase the diverse applications of generative models, ranging from real-time flood prediction and SAR data flood mapping to floodwater detection in images and urban morphology prediction. These advancements not only accelerate prediction processes but also improve accuracy, offering innovative solutions for managing and mitigating the impacts of natural disasters.

7.5.4 Generative Artificial Intelligence for Urban Planning and Design

GenAI models, including LLMs, have substantial applications in the realm of urban planning and design. By leveraging these advanced models, urban planners can simulate and analyze various scenarios to make informed decisions about city development. For instance, LLMs can process vast amounts of textual data, such as zoning laws, building codes, and community feedback, to generate comprehensive urban plans that balance regulatory compliance with community needs. Additionally, GenAI models can create realistic 3D models of urban environments, enabling planners to visualize potential developments and their impacts on the urban landscape. This capability is particularly useful for optimizing land use, transportation networks, and infrastructure placement, ensuring that urban growth is sustainable and efficient.

The advent of GenAI models, including LLMs such as ChatGPT, has revolutionized various fields by enhancing human-machine interaction and computational capabilities. Several studies have explored the roles, challenges, and potential of these AI models in the contexts of urban planning, design, and sustainability. This synthesis aims to summarize and integrate the key findings from these studies, focusing on how GenAI can advance SDGs and enhance UDT capabilities in urban planning and design.

Wang et al. (2022b) examine the integration of generative and ChatGPT-like AI into urban planning. It highlights the significant contributions of AI to urban planning from various perspectives, including sustainability, living conditions, economic factors, disaster management, and environmental impacts. The authors discuss fundamental concepts of urban planning and relate them to key problems in ML, such as adversarial learning and GNNs. The central problem identified is automated land-use configuration, which involves generating land uses and building configurations based on surrounding geospatial, human mobility, social media, environmental, and economic data. The study concludes by outlining the implications of AI for urban planning and proposing key research areas at the intersection of AI and urban planning.

Jiang et al. (2023) focus on GUD, which uses AI and computational methods to enhance the efficiency and effectiveness of urban design processes. The authors cover three key stages: design problem formulation, design option generation, and decision-making. They highlight how GUD can address the inefficiencies of traditional urban design, which relies heavily on manual work and expert experience. By employing computer-aided generative methods such as evolutionary optimization and deep generative models, GUD can efficiently explore complex solution spaces and generate design options that satisfy various constraints and conflicting objectives. The study identifies current trends, findings, and limitations in GUD research and suggests future directions and potential challenges.

Ćirić (2024) explores new generative and AI design methods for transportation systems and urban mobility. It focuses on the application of AI in urban computing, architectural studies, and spatial planning, particularly in the context of urban mobility and transportation systems. The author provides a theoretical review of recent technological developments in digital architectural and urban design, including computational design methods and available software tools. The study uses the Grand Paris transportation network as a case study to illustrate how generative design problem modeling and solution methodologies can be applied to urban mobility and transportation planning. It emphasizes the importance of a systematic cross-disciplinary approach to addressing design problems in urban mobility.

Rane et al. (2023a) investigate the impact of GenAI systems like ChatGPT on various aspects of architectural engineering, including structural engineering, HVAC systems, electrical engineering, and sustainability. The study highlights the potential of ChatGPT to expedite design processes, optimize materials and costs, and enhance energy efficiency. It also addresses challenges such as ethical concerns, data security, and the need for skilled professionals to interpret AI-generated insights. The authors propose that GenAI can significantly contribute to sustainable building design and project management by facilitating seamless collaboration and automating model generation.

Rane et al. (2023b) delves into the integration of advanced GenAI models like ChatGPT and Bard within architectural design and engineering. The authors examine the impact of AI on various aspects of architectural theory, design processes, representation and visualization, interior design, and urban design and planning. The study highlights the potential of AI to foster creativity, streamline ideation and collaboration, and generate realistic visualizations. It also explores the role of AI in structural engineering, building systems, construction management, compliance with building codes, and sustainability.

These studies collectively highlight the transformative potential of GenAI and LLMs in urban planning and design. Wang et al. (2022b) emphasize the importance of automated land-use configuration, which is crucial for developing realistic and responsive UDTs. Jiang et al. (2023) focus on GUD, which can enhance UDT capabilities by efficiently exploring design solutions and optimizing urban environments. Ćirić (2024) expands on the application of AI in urban mobility, a critical component of UDTs, by demonstrating how AI can address transportation and infrastructure planning challenges. Rane et al. (2023a, b) provide insights into the broader applications of GenAI in architectural engineering, which can be integrated into UDTs to improve building design, energy efficiency, and project management.

By integrating GenAI and LLMs into UDTs, cities can improve their planning and design processes but also enhance their resilience to climate change. These technologies enable the creation of dynamic, responsive urban environments that can adapt to changing conditions and support long-term sustainability goals. Through predictive analytics and scenario modeling, urban planners and policymakers can develop more effective strategies for reducing carbon footprints, managing natural resources, and enhancing overall urban resilience. As a result, the application of GenAI in UDTs holds immense potential for transforming urban environments into more sustainable, livable, and climate-resilient spaces.

As regards SDGs, the study by Rane (2023b) aims to explore the roles and challenges of GenAI models, such as ChatGPT, in enhancing SDGs. It seeks to understand how these AI models can contribute to various SDGs and the obstacles they face in achieving sustainable development. The study highlights the transformative impact of ChatGPT on human-machine interactions and their potential to advance SDGs. It concludes that addressing SDG challenges requires global collaboration and robust policy frameworks aligned with the SDGs. It emphasizes the need for innovative approaches to leverage ChatGPT's capabilities effectively, ensuring they align with sustainable development goals. By overcoming ethical and technical obstacles and fostering stakeholder collaboration, GenAI can significantly enhance the global effort toward sustainable development, creating a more inclusive, informed, and interconnected world.

In summary, the integration of GenAI models and LLMs into urban planning, design, and sustainability efforts holds immense potential for transforming urban environments. These studies demonstrate how AI can address complex challenges, enhance design processes, and contribute to achieving SDGs. By leveraging AI's capabilities in data analysis, simulation, and generative design, urban planners and engineers can develop more efficient, sustainable, and resilient cities. Future research should focus on overcoming ethical and technical challenges, fostering interdisciplinary collaboration, and developing robust policy frameworks to fully harness the potential of GenAI in shaping the future of urban planning and design.

7.6 BLUE CITY PROJECT CASE STUDY: GENERATIVE SPATIAL ARTIFICIAL INTELLIGENCE FOR LAUSANNE CITY DIGITAL TWIN

The Blue City Project in Lausanne, Switzerland, is an innovative initiative aimed at transforming the city into a test bed and living lab. At the core of the project lies the development of a responsive AI-powered UDT. This platform is designed to empower citizens and stakeholders by fostering collective, evidence-based decision-making processes aimed at enhancing sustainability, resilience, quality of life, and ecological value (Huang et al. 2025). Central to the project's mission is the creation of an integrated, open-source platform, serving as the foundation for examining emerging patterns and facilitating informed, proactive decisions in urban planning and design. A transdisciplinary consortium collaborates to map the multi-layered, interconnected network of flows in the city, including people (human mobility), energy, materials, goods, biodiversity, and waste. For example, to understand human mobility within urban environments, the research team employs advanced techniques to quantify urban form with high spatial resolution across various dimensions. The aim of establishing this detailed digital representation of the built environment is to correlate it with observed behaviors and the demand for urban mobility. This approach allows to assess different archetypes of urban form and design, evaluating their respective impacts on energy consumption and emissions in the transportation sector.

The integrated platform will enable: (1) the detection of correlations and evidence for causalities across flow categories over time and space; (2) seeing emerging patterns not discernible with traditional analytical tools, (3) testing intervention hypotheses and scenario prediction; and (4) leveraging engineering and AI tools for planning a more resilient city (ETH 2024). This holistic approach, merging AI, architecture, urban planning, engineering, and environmental science, underpins a comprehensive analysis and modeling of flows and infrastructures (ETH 2024).

The Blue City Project signifies an innovative endeavor toward developing a smarter and more sustainable future for Lausanne City. As part of this ambitious project, the research team embarked on an exploratory journey to explore GSAI, marking a significant milestone as the first to envision a transformative approach to completing, estimating, and predicting various spatial phenomena associated with city structures and flows. These phenomena encompass the organization, utilization, and interactions of physical space within urban environments. Examples include pedestrian traffic patterns, ridership for public transportation systems, land use allocation, energy infrastructure distribution, waste collection logistics, retail and commercial center distribution, air quality variations, and the accessibility of parks and green spaces within the city. The development of the GSAI model is part of the Blue City Innosuisse Flagship Project.

The GSAI model will play a crucial role in advancing the capabilities of the AI-powered UDT platform. By integrating GSAI into this open platform, stakeholders will gain access to advanced spatial modeling and simulation tools, enabling them to generate adaptive spatial plans specific to urban environments, simulate urban scenarios, and visualize alternative design options. This integration empowers stakeholders to tackle urban planning and design challenges with a high level of accuracy and efficiency. The GSAI model aligns with the project's overarching goal of developing a responsive and proactive AI-powered platform to improve the environmental, social, and economic performances of Lausanne City. Indeed, the GSAI-driven UDT holds significant computational, analytical, and simulation relevance to the four objectives outlined for the integrated platform in the Blue City Project, namely:

Detection of correlations and evidence for causalities: GSAI can contribute to this objective by generating spatial representations and scenarios that allow stakeholders to analyze correlations and causal relationships across different flow categories over time and space. GSAI can aid in identifying hidden correlations and causalities that may not be readily apparent with traditional analytical tools by means of generating synthetic data and simulations based on learned patterns. For example, it can simulate the effects of changes in land use patterns on transportation demand or the impact of green spaces on air quality. This capability aids in understanding the complex dynamics of urban systems and supports evidence-based decision-making in urban planning and design.

Seeing emerging patterns: GSAI's ability to generate spatial layouts, configurations, and scenarios based on learned patterns enables stakeholders to visualize and understand emerging patterns within urban environments. For example, in a city experiencing rapid population growth, GSAI can generate spatial layouts and configurations based on historical data and projected trends. GSAI can aid stakeholders in identifying emerging trends and patterns that may not be obvious with current analytical methods by providing advanced visualization tools and simulations. This supports informed decision-making in urban planning and design.

Testing intervention hypotheses and predicting scenarios: GSAI's capacity to generate alternative spatial plans and configurations enables stakeholders to simulate the impacts of diverse interventions and policies. This functionality empowers urban planners to assess potential outcomes, facilitating informed decision-making in urban development strategies. For instance, urban planners can use GSAI to model the effects of implementing green infrastructure projects, such as urban parks or green roofs, on reducing heat island effects or improving air quality. GSAI facilitates evidence-based decision-making by offering a realistic simulation environment, thus contributing to the development of more resilient and sustainable urban environments.

Deploying engineering and AI tools for planning a more resilient city: By providing sophisticated spatial representations of urban environments, capturing the dynamic interplay between different elements such as buildings, infrastructure, green spaces, and transportation networks, GSAI enables stakeholders to optimize urban layouts, simulate various scenarios, and test resilience strategies, thereby supporting the development of more resilient urban environments in the face of climate change, natural disasters, and socioeconomic disruptions. These representations serve as the foundation for conducting comprehensive simulations that emulate real-world conditions and dynamics. GSAI contributes to the development of urban environments that are better prepared to withstand and adapt to a wide range of uncertainties and disruptions.

Overall, the capabilities and functionalities of GSAI make it a valuable asset for advancing the Blue City Project's goals of sustainable urban development and resilience. GSAI offers an innovative approach to addressing the complexities of urban planning and design, equipping stakeholders with advanced spatial modeling, simulation, and prediction tools. The project is poised to advance decision-making processes, leading to more informed, proactive, adaptive, and effective strategies for building smarter and more resilient and sustainable cities.

7.7 A SYNTHESIZED FRAMEWORK FOR GENAI-DRIVEN SUSTAINABLE SMART CITY PLANNING AND DESIGN

This section presents a comprehensive conceptual framework that outlines the synergistic and collaborative integration of AI and its five subfields with UDT as a data-driven planning and design system in the context of sustainable smart cities (Figure 7.2). By synthesizing the characteristics, applications, and interconnections of these models, this framework aims to provide a holistic understanding of their collective role in reshaping the future of the urban landscape and advancing sustainable urban development practices. This synthesized framework entails a cohesive and integrated structure that combines various elements, concepts, theories, and practices into a unified system, drawing from different disciplines, to

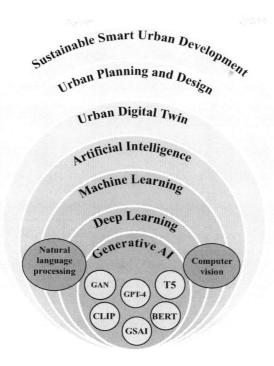

FIGURE 7.2 A novel synthesized framework for GenAI-driven sustainable smart city planning and design.

provide a lens through which complex issues can be examined, challenges can be identified, and solutions can be formulated.

The synthesized framework elucidates five distinct subfields of AI and their interactions at the center of UDT as a platform of data-driven planning and design for sustainable smart cities. AI envelopes all five subfields, symbolizing its foundational role in the family of its models. In this encompassing circle, ML and DL stand out, emphasizing their specialized functions in learning from data and extracting complex features. Alongside ML and DL, CV and NLP intersect with AI by relying on ML and DL techniques for the analysis of visual data and the processing of human language, respectively. Lastly, GenAI interfaces with ML, DL, CV, and NLP, epitomizing their capability in generating novel content, thereby enriching a range of AI-driven UDT applications in relation to planning and design. In more detail, GenAI encompasses a wide range of models and techniques based on these four subfields of AI. While not every GenAI model utilizes all these subfields simultaneously, various GenAI models may leverage elements from one or more of ML, DL, CV, and NLP, depending on the specific task or application. For example, GANs primarily fall under the realm of DL and are widely used for generating realistic images. On the other hand, RNNs and transformer-based architectures, often used in NLP tasks, can be employed for generating text or sequences. Similarly, CNNs can be adapted for generating visual content. Therefore, while not all GenAI models incorporate every aspect, they draw upon a diverse set of models and techniques from those subfields of AI to achieve their objectives.

Furthermore, drawing upon the previously reviewed studies in the field, CV, like NLP, intersects with various AI models and techniques. Although CNNs predominantly serve in CV tasks, they have been adapted for NLP tasks such as sentiment analysis, owing to their capability to capture local patterns in textual data efficiently. CNNs serve as the cornerstone of contemporary CV systems, continually pushing the boundaries of image recognition accuracy. RNNs, predominantly employed for sequential data in NLP tasks, also find applicability in CV tasks requiring sequential processing, such as video analysis and captioning. Popular variants of RNNs, such as LSTM and Gated Recurrent Unit (GRU), mitigate the limitations of existing RNNs by integrating gating mechanisms that regulate the information flow in the network. NLP intersects with transformer models, exemplified by BERT (e.g., Tian et al. 2023; Wolf et al. 2020), for tasks like text classification, named entity recognition, and question answering. Through these exemplars of commonly employed DL models in NLP and CV, both fields witness rapid advancements in DL architectures and techniques, catalyzing progress across an extensive spectrum of applications, including urban planning and design. It is important to note that the techniques mentioned in Figure 7.2 are only for illustrative and descriptive purposes.

Within the framework of GSAI, GANs and VAEs can be utilized, albeit with distinct applications and adaptations compared to GenAI. GANs and VAEs are frequently tailored to tackle spatial data challenges and address complexities inherent in UDT. As evidenced by the synthesized studies, GANs exhibit utility in generating synthetic urban images or simulating urban scenarios pertinent to city planning and design. Conversely, VAEs are employed to discern underlying patterns from existing spatial data and subsequently generate novel outputs or new samples with similar characteristics. These and other DL generative models utilized in GenAI in urban planning and design encompass Spatial GANs (sGANs), Conditional GANs (cGANs), and Conditional VAEs (cVAEs) (e.g., Eltahan et al. 2021;

Oyama et al. 2023; Sun and Dogan 2023; Xu et al. 2024). For example, cGANs is a significant advancement of the original GAN model, developed to incorporate a directed feature into the generative process (Mirza and Osindero 2014).

This nuanced utilization reflects the unique requirements and complexities inherent in spatial data and urban contexts. In other words, while GSAI may capitalize on similar DL models as GenAI, its emphasis and applications are specialized for spatial data and urban contexts. This underscores the nuanced approach required in leveraging GenAI techniques (Castelli and Manzoni 2022; Ooi et al. 2023) in the domain of urban planning and design (Xu et al. 2024), where considerations of spatiality and urban dynamics necessitate tailored solutions and adaptations of existing models.

The synthesized framework accentuates the dynamic relationships and overlaps among the different subfields of AI, reflecting their collaborative essence in propelling technological innovation and augmenting problem-solving capabilities. The small circle at the bottom of GenAI represents GSAI, a specialized DL model designed to complete, estimate, and predict various spatial phenomena related to the structures and flows of Lausanne City. This ensures the provision of spatially explicit representations within the framework of the Lausanne City Digital Twin. In addition, the synthesized framework encompasses additional circles representing UDT, urban planning and design, and sustainable smart urban development. This illustrates how UDT harnesses the identified subfields of AI to advance sustainable smart urban development through planning and design interventions based on data-driven decision-making processes enabled by AI and its subfields within the framework of UDT.

7.8 DISCUSSION

The discussion section serves as a critical component of this study, providing an opportunity to highlight the significance of the findings and their relevance to the broader context of sustainable smart city planning and design. Specifically, it interprets the findings presented in the results section, compares them with existing literature, and explores the implications of the findings for research, practice, and policymaking. Additionally, it acknowledges any limitations of the study and suggests avenues for future research to address these limitations and beyond.

7.8.1 Interpretation of Findings

The findings of this study highlight the transformative potential of AI and its subfields, particularly GenAI, within the UDT framework. Through a comprehensive analysis and synthesis of the existing literature on AI and its subfields, the study reveals how these technologies can augment the capabilities of UDT in terms of data-driven planning and design processes. Specifically, the results elucidate how ML, DL, CV, NLP, and GenAI contribute to enhancing the computational and analytical processes of UDT, enabling more accurate and efficient urban modeling and simulation. This integration empowers UDT to conduct sophisticated analysis and visualization of urban data, thereby facilitating more informed decision-making processes in urban planning and design. Furthermore, the case study on GSAI for UDT provides empirical evidence of the feasibility and efficacy of integrating GenAI into UDT frameworks. These outcomes highlight the significance of AI technologies in advancing data-driven approaches to urban planning and design, offering new opportunities for optimizing resource allocation, energy efficiency, environmental sustainability, as well as spatial organizations in sustainable smart cities. Through the interpretation of these results, the study contributes to a deeper understanding of the potential applications of AI in urban contexts, paving the way for future research and innovation in the field.

7.8.2 Comparative Analysis

The study's findings are situated within the broader context of existing literature on AI-driven urban planning and design. Several studies have explored the application of AI and its subfields, especially ML, DL, and CV, in urban planning and design (Bibri et al. 2024a; Kamrowska-Załuska 2021; Koumetio et al. 2023; Sanchez 2023; Son et al. 2022; Marasinghe et al. 2024), highlighting their potential to enhance decision-making processes from a general perspective. The findings of this study align with these studies, demonstrating the efficacy of integrating AI technologies into urban planning and design frameworks, with a specific focus on sustainable smart cities. Moreover, the limited literature review in this field is characterized by a predominant focus on individual AI models in urban planning. For instance, Marasinghe et al. (2024) focus on CV tasks, Son et al. (2022) on ML/DL techniques, and Jiang et al. (2023) on generative models for urban design. This study stands out for its comprehensive approach, encompassing all five subfields of AI in the realm of urban planning and design. This holistic perspective allows us to explore the synergies and interactions among these models, providing a more nuanced understanding of their collective impact on sustainable smart city development. To the best of our knowledge, no previous study has addressed the combination of applications of AI and its five subfields in the context of urban planning and design. By filling this gap, this research contributes significantly to the field, offering novel insights into and interactional knowledge about the potential of integrating multiple AI technologies for enhancing city planning and design practices.

The conceptual framework derived from the synthesized studies and the case study represents a key advancement in the field of sustainable smart city planning and design. Its novelty lies in its comprehensive integration of AI and its five subfields within the framework of UDT. This framework not only provides theoretical insights into AI and its five models but also offers practical foundations for implementing AI-driven solutions and strategies in real-world contexts. By combining theoretical foundation with practical applicability, the framework serves as a valuable guide for urban planners, policymakers, and researchers seeking to harness the transformative power of AI in sustainable urban development initiatives.

7.8.3 Implication for Research, Practice, and Policymaking

As the integration of AI and its subfields continues to reshape the landscape of urban planning and design, understanding the implications of these advancements is paramount for research, practice, and policymaking. The findings of this study hold significant implications for further research in the field of sustainable smart city planning and design. By highlighting the transformative potential of integrating AI and its subfields, particularly GenAI, into the UDT framework, this study opens avenues for future investigations into innovative approaches for sustainable urban development. Researchers can build upon the synthesized framework to explore the synergistic interactions between AI technologies and urban planning and design methodologies, exploring novel strategies and solutions for optimizing city planning processes and enhancing sustainability outcomes. Furthermore, researchers can investigate the integration of GenAI and GSAI techniques within UDT frameworks to augment their analytical and predictive capabilities. By leveraging GenAI algorithms, UDT can generate synthetic urban data that complements existing datasets, enabling more comprehensive simulations and scenario analyses.

From a practical perspective, the study offers valuable insights for urban planners, architects, and designers involved in sustainable smart city projects and initiatives. The conceptual framework derived from the synthesized literature and the GSAI case study provides a roadmap for incorporating advanced AI and GenAI technologies into real-world planning and design initiatives and practical implementations of UDT. Practitioners can leverage these insights to develop more efficient, resilient, and sustainable urban environments by harnessing the power of AI and GenAI to address complex challenges. By integrating these cutting-edge technologies into their workflows, practitioners can streamline design iterations, optimize spatial configurations, and explore alternative scenarios with unprecedented efficiency and precision.

The utilization of GenAI enables practitioners to generate diverse design options that respond to various urban complexities. Additionally, GSAI can enhance the spatial understanding and representation of urban environments within UDT frameworks, facilitating more accurate modeling and visualization of complex urban systems. The integration of GSAI with UDT platforms further enhances their capabilities, allowing practitioners to simulate and visualize urban scenarios in a virtual environment. However, the successful adoption of GeAI technologies may require overcoming various challenges, including data availability, technical expertise, and institutional support. Therefore, collaboration with experts in the field is essential to navigate these obstacles and ensure the effective implementation of GenAI solutions in practice.

This study informs policymakers and city officials entrusted with formulating sustainable urban development policies. It emphasizes the significance of integrating innovative strategies into urban policymaking procedures by illuminating the transformative capabilities of GenAI-driven methods within the UDT framework. It is of strategic value for policymakers to adopt forward-looking policies that acknowledge AI and GenAI technologies as catalysts for sustainable urban progress. This involves fostering partnerships between the public and private sectors to harness AI and GenAI advancements efficiently and promote equitable and inclusive urban development based on the responsible and ethical use of AI and GenAI. These implications underscore the broader significance of this study for advancing knowledge, informing practice, and guiding policy-making efforts in the realm of sustainable smart city planning and design.

Moreover, policymakers have a unique opportunity to leverage GSAI to inform and enhance urban planning policies and strategies, ultimately promoting resilience in cities. By integrating GSAI into policy frameworks, policymakers can facilitate data-driven decision-making processes that address key urban challenges and promote equitable outcomes. Investments in GenAI, including its spatial branch, and capacity-building initiatives are critical to support the widespread adoption of this technology. Additionally, fostering stakeholder engagement and community involvement is paramount to ensure that policy decisions reflect the diverse needs and aspirations of urban populations. Embracing GenAI and its spatial branch and nurturing collaboration between policymakers, practitioners, and researchers, cities can unlock the innovative potential of this advanced technology and pave the way for a more sustainable urban future.

7.8.4 Limitations and Mitigation Strategies

While this study contributes to advancing knowledge about the integration of AI and its subfields into the UDT framework for sustainable smart city planning and design, it is essential to acknowledge its limitations. One of these limitations is the reliance on existing literature and a single case study for analysis, which may restrict the generalizability of the findings to diverse urban contexts. Additionally, the study focuses primarily on the technical and functional aspects of AI integration, overlooking broader socioeconomic and political factors that may influence the adoption of GenAI-driven approaches in urban planning and design. To address these limitations, future research could explore a more extensive range of case studies on GenAI and its specialized branches across various urban settings and incorporate interdisciplinary perspectives to provide a more comprehensive understanding of the implications of GenAI for sustainable urban development. Moreover, longitudinal studies could track the implementation and impact of GenAI-driven interventions over time, offering insights into their long-term effectiveness and sustainability. This can further advance knowledge in the field of GenAI-driven sustainable urban development and inform evidence-based policy and practice.

Furthermore, the study's scope may have been narrow, concentrating on specific aspects of AI integration into UDT while neglecting other relevant factors. This could impact the depth of the analysis and hence limit the comprehensiveness of the findings and their applicability to broader contexts. To address this limitation, future research could adopt a more holistic approach, considering a wider range of factors that influence AI integration into UDT and its adoption in general. This would provide a deeper understanding of the complexities involved and enhance the applicability of the findings to diverse urban contexts.

In addition, a limitation inherent in the conceptual framework for GenAI-driven sustainable smart urban planning and design is the potential lack of empirical validation. Existing models lack proper validation (e.g., Papyshev and Yarime 2021; Luo et al. 2022a) and are considered inadequate in handling the complexity of urban systems (Dembski et al. 2020; Wang et al. 2022a; Topping et al. 2021). While the conceptual framework is based on synthesized literature and a case study on GSAI, empirical validation in real-world urban contexts is necessary to assess its effectiveness and practical applicability. To address this limitation, future research could conduct empirical studies or pilot projects to validate the framework's efficacy in diverse urban settings. This would provide empirical evidence of its utility and enhance its credibility for further development and adoption in practice. Rigorous evaluation and validation of generative GenAI models and UDT platforms are essential to ensure their effectiveness and reliability in real-world applications.

7.8.5 Key Challenges and Risks of Generative Artificial Intelligence

GenAI and LLMs are technological marvels, widely celebrated for their capabilities in NLP and multimodal content generation. These technologies promise to revolutionize industries, including the development of sustainable smart cities, by enhancing urban planning and design, optimizing energy efficiency, and improving resource management. In these cities, GenAI and LLMs can drive a future that is not only innovative but also environmentally sustainable. However, as with any powerful tool, they come with inherent risks and challenges. Alongside their potential to spur innovation, there is a growing concern about their misuse, highlighting the need for ethical considerations and robust governance to prevent nefarious applications, particularly in the sensitive infrastructures of sustainable smart cities.

GenAI presents several significant challenges that need to be addressed to ensure its ethical and effective deployment. Rane et al. (2023b) address ethical considerations, bias mitigation, and user adaptability and outline future directions for interdisciplinary collaboration and ongoing research. Richards et al. (2024) highlight the risks associated with GenAI, such as bias, data privacy issues, mistrust, and high energy consumption. The authors emphasize the need for integrated social research into ethics, public acceptability, and user experience in order to maximize the benefits of GenAI and mitigate its risks.

Furthermore, Ferrara (2024) examines the profound risks associated with GenAI and LLMs. Despite their transformative capabilities, the author cautions about the dangers that arise when these technologies are misused. The study outlines various ways in which GenAI can be exploited, including the production of indistinguishable deepfakes, synthetic identities for misinformation campaigns, and the creation of highly targeted scams. It also raises concerns about fabricated realities and the development of sophisticated malware, stressing that the line between the virtual and real worlds is increasingly blurring. This work synthesizes the potential risks posed by GenAI and serves as a call to action, urging society to prepare for and mitigate the harmful applications of these powerful technologies. Expanding on this, Chu-Ke and Dong (2024) explore the global issue of misinformation in the context of GenAI, examining both theoretical advancements and empirical findings. The authors highlight how AI-driven misinformation complicates the information landscape and advocate for ethical AI development through stricter regulations and improved AI literacy. They propose a framework for AI literacy that emphasizes the importance of understanding the cultural and ethical implications of AI, critically assessing AI-generated content, and employing feedback mechanisms to manage AI's influence at the institutional level.

Data privacy is also a major concern; GenAI systems often require large amounts of data, which raises questions about the security and privacy of sensitive information. Mistrust in AI systems can stem from a lack of transparency and understanding of how these models work, which is exacerbated by the opaque nature of many GenAI algorithms. Security risks are another critical issue, as GenAI systems can be vulnerable to adversarial attacks, where malicious inputs are used to deceive the AI into making incorrect predictions.

One of the primary concerns with GenAI systems is their potential to perpetuate and amplify existing biases present in the training data, which can result in unfair outcomes, particularly for marginalized communities. Ensuring fairness in these systems is equally crucial, as achieving equitable treatment and outcomes across different user groups continues to be a significant challenge. GenAI models, such as GANs and GPT, are trained on vast datasets that often contain real-world biases (Grossman et al. 2023). Other models, like VAEs and Diffusion Models, also face similar challenges, as they rely on large-scale data that may reflect societal inequalities. These models, while highly innovative in generating creative outputs, risk perpetuating existing biases if not carefully managed during the data curation and training processes. These biases can manifest in AI outputs that undermine the trust and acceptance of AI technologies. Training data may embed societal biases related to factors such as race, gender, and age, which AI models can then perpetuate and even amplify. The consequences of these biases are significant, resulting in discriminatory outcomes where certain groups are disproportionately favored, ultimately resulting in unfair outcomes (Ferrara 2023).

Addressing these biases requires a multi-faceted approach. One key strategy is to use more diverse and representative datasets that include a wide range of demographics, experiences, and perspectives to minimize inherent biases (Nazer et al. 2023). Moreover, developing fairness-aware algorithms that incorporate bias detection and mitigation during the model training process can help reduce biased outcomes. Regular auditing and monitoring of GenAI systems for biased behavior, along with implementing transparency mechanisms, can further ensure that biases are identified and addressed over time. Lastly, involving interdisciplinary teams in the GenAI development process, including ethicists and social scientists, can provide valuable insights into the social impacts of GenAI, helping to create more equitable models. These strategies are essential for fostering trust in GenAI technologies and ensuring their ethical and fair deployment.

The high energy consumption of GenAI models, due to the extensive computational resources required for training and running these models, poses significant environmental and sustainability concerns. DL models require enormous amounts of data, which must be acquired, transferred, stored, and processed. They demand significant equipment and energy, leading to substantial environmental impacts (Ligozat et al. 2021). The process of training these models is resource-intensive, necessitating considerable computation time and power as the model develops comprehensive data representations. Strubell et al. (2019) highlight that training a typical AI model in NLP can emit over 284 tonnes of CO_2 equivalent, illustrating the significant environmental footprint of AI training. Recent GenAI models, LLMs and other pre-trained foundation models can worsen environmental challenges due to their high computational requirements for training large models, which results in increased energy consumption and carbon emissions.

Despite these challenges, the field of green AI is dedicated to creating GenAI technologies that are environmentally sustainable. This involves efforts to reduce the carbon footprint and energy consumption of AI development and deployment (Verdecchia 2022; Yokoyama et al. 2023). Research endeavors in this field focus on optimizing algorithms, improving hardware efficiency, and enhancing data center operations to mitigate the environmental impact of GenAI systems. Green AI initiatives also work on developing metrics and standards to assess and promote the sustainability of AI technologies, ensuring that advancements in AI do not compromise ecological health (Schwartz et al. 2020; Raman et al. 2024).

Addressing these challenges is crucial for the responsible development and deployment of GenAI technologies. While GenAI offers remarkable potential for advancements in urban planning and design, it also presents significant challenges that need to be meticulously addressed to ensure its ethical and sustainable deployment. The issues of bias, fairness, data privacy, and mistrust underscore the need for transparency and robust security measures in the development of GenAI systems. To maximize the benefits and mitigate these risks, While the environmental impact of GenAI technologies, particularly their high energy consumption and substantial carbon footprint, further complicates their deployment, the burgeoning field of green AI shows promise in mitigating these environmental concerns. Addressing these multifaceted challenges is critical for harnessing the full potential of GenAI while safeguarding ethical standards and environmental sustainability.

7.8.6 Suggestions for Future Research Directions

In addition to the future research directions suggested earlier, there remain other areas in the domain of GenAI-driven sustainable urban planning and design that warrant further investigation. One such area pertains to the long-term impacts and implications of integrating GenAI into UDT frameworks. Future studies could assess the environmental costs and risks of GenAI-driven planning and design initiatives over extended periods, providing insights into the sustainability of these approaches. Additionally, there is a need to explore the scalability and replicability of GenAI-driven solutions across diverse urban contexts. Understanding how these approaches can be adapted and implemented in different cities worldwide will be crucial for facilitating widespread adoption and maximizing their potential benefits. Also, future research could explore the ethical and governance dimensions associated with the use of AI technologies in urban planning and design to address concerns related to data privacy, data security, algorithmic bias, explainability, fairness, accountability, and social equality. Examining these multifaceted aspects enable researchers to contribute to the development of robust frameworks and guidelines for responsible AI deployment in urban contexts, ensuring that technology-driven interventions align with societal values. Research could explore participatory and community-based approaches to AI governance and decision-making, empowering local communities to shape the future of their cities collaboratively.

Furthermore, future research efforts should focus on advancing GenAI techniques to address specific challenges in sustainable smart city planning and design. This includes exploring novel algorithms and models that can better simulate urban dynamics, optimize spatial configurations, and generate alternative design scenarios. Additionally, research could investigate the integration of GenAI with other emerging technologies, such as blockchain and Internet of Things (IoT), Artificial Intelligence of Things (AIoT) to further enhance its capabilities and applications in urban environments.

Given the complex and multifaceted nature of urban planning and design challenges, future research should prioritize interdisciplinary collaboration. This includes fostering partnerships between urban planners, architects, data scientists, experimentalists, policymakers, and community stakeholders to co-create innovative solutions that address diverse urban needs and priorities. Interdisciplinary research teams can leverage their collective expertise to develop holistic and context-specific approaches to sustainable smart city development.

Finally, future research should consider the long-term sustainability of GenAI models and UDT platforms in urban environments. This includes investigating the environmental impact of GenAI and UDT technologies, such as their energy consumption and carbon footprint, and exploring strategies to mitigate potential negative effects. Furthermore, research could explore innovative financing mechanisms and business models for sustaining the development and maintenance of GenAI and UDT infrastructures over time.

7.9 CONCLUSION

This study explored the transformative potential of AI and its five subfields, alongside UDT, in reshaping sustainable smart city planning and design. Through a thorough analysis and synthesis of the existing literature and a detailed examination of a case study on GSAI, this study underscored the critical and innovative role of AI and its five subfields in augmenting the capabilities of UDT in regard to data-driven planning and design processes. Additionally, it reveals the untapped potential of GenAI and GSAI in propelling the advancement of sustainable smart cities.

The development of a novel framework for GenAI-driven sustainable smart city planning and design represents a significant contribution of this study to the field. This framework offers innovative methodologies for optimizing resource utilization, enhancing energy efficiency, mitigating environmental impacts, and improving spatial organizations in urban environments. It provides a systematic approach to leveraging AI technologies for enhancing sustainable urban development strategies and practices. It can benefit urban planners, designers, and policymakers through utilizing GenAI-driven solutions to address complex challenges associated with UDT and hence data-driven planning and design. The solution it embodies is poised to enable cities to enhance their data-driven decision-making processes in urban planning and design and ultimately move toward more sustainable urban development trajectories.

However, it is essential to acknowledge the limitations of this study and the conceptual framework, including the narrow scope of analysis and the potential biases inherent in the research process. Future research endeavors should aim to address these limitations by adopting a more holistic approach and incorporating diverse perspectives and methodologies.

In light of the findings and insights generated by this study, it is evident that AI and its five subfields present new opportunities and prospects to advance sustainable urban development practices and pave the way for more efficient, resilient, and environmentally conscious cities. By embracing AI technologies and fostering interdisciplinary collaborations, cities can strive toward achieving greater efficiency, resilience, and inclusivity in urban development practices.

Lastly, this study represents a significant step forward in understanding the role of AI in shaping the future of urban environments. It sets the stage for further exploration and innovation in the field of sustainable smart city planning and design, with the ultimate goal of creating more livable, equitable, and sustainable cities for future generations. Overall, this study contributes to advancing knowledge in the field and provides fertile insights for urban scholars, planners, and policymaker.

REFERENCES

Agostinelli, S., Cumo, F., Guidi, G., & Tomazzoli, C. (2021). Cyber-physical systems improving building energy management: Digital twin and artificial intelligence. *Energies*, *14*(8), 2338.

Ahn, J.-K., Kim, B., Ku, B., & Hwang, E.-H. (2023). Virtual scenarios of earthquake early warning to disaster management in smart cities based on auxiliary classifier generative adversarial networks. *Sensors, 23*(22), 9209.

Ahvenniemi, H., Huovila, A., Pinto-Seppä, I., Airaksinen, M., & Valvon-takonsultit Oy, R. (2017). What are the differences between sustainable and smart cities? *Cities, 60*, 234–245. https://doi.org/10.1016/j.cities.2016.09.009

Alahi, M. E. E., Sukkuea, A., Tina, F. W., Nag, A., Kurdthongmee, W., Su-wannarat, K., & Mukhopadhyay, S. C. (2023). Integration of IoT-enabled technologies and artificial intelligence (AI) for smart city scenario: Recent advancements and future trends. *Sensors, 23*(11), 5206.

Al-Dabbagh, R. H. (2022). Dubai, the sustainable, smart city. *Renewable Energy and Environmental Sustainability, 7*(3), 12.

Alhijaj, J. A., & Khudeyer, R. S. (2023). Techniques and applications for deep learning: A review. *Journal of Al-Qadisiyah for Computer Science and Mathematics, 15*(2), 114–126. https://doi.org/10.29304/jqcm.2023.15.2.1236

Almusaed, A., & Yitmen, I. (2023). Architectural reply for smart building design concepts based on artificial intelligence simulation models and digital twins. *Sustainability, 15*(6), 4955.

Alom, M. Z., Taha, T. M., Yakopcic, C., Westberg, S., Sidike, P., Nasrin, M. S., Hasan, M., Van Essen, B. C., Awwal, A. A., & Asari, V. K. (2019). A state-of-the-art survey on deep learning theory and architectures. *Electronics, 8*(3), 292.

Alzubaidi, L., Zhang, J., Humaidi, A. J., Al-Dujaili, A., Duan, Y., Al-Shamma, O., Santamaría, J., Fadhel, M. A., Al-Amidie, M., & Farhan, L. (2021). Re-view of deep learning: Concepts, CNN architectures, challenges, applications, future directions. *Journal of Big Data, 8*(1). https://doi.org/10.1186/s40537-021-00444-8

Andrews, C., Cooke, K., Gomez, A., Hurtado, P., Sanchez, T., Shah, S., & Wright, N. (2022). AI in planning: Opportunities and challenges and how to prepare. *American Planning Association*. https://www.planning.org/publications/document/9255930/

Aqib, M., Mehmood, R., Alzahrani, A., Katib, I., Albeshri, A., & Altowaijri, S. (2019). Rapid transit systems: Smarter urban planning using big data, in-memory computing, deep learning, and GPUs. *Sustainability, 11*(10), 2736. https://doi.org/10.3390/su11102736

Argota Sánchez-Vaquerizo, J. (2022). Getting real: The challenge of building and validating a large-scale digital twin of Barcelona's traffic with empirical data. *ISPRS International Journal of Geo-Information, 11*(1), 24. https://doi.org/10.3390/ijgi11010024

Austin, M., Delgoshaei, P., Coelho, M., & Heidarinejad, M. (2020). Architecting smart city digital twins: Combined semantic model and machine learning approach. *Journal of Management in Engineering, 36*(4), 04020026.

Azevedo, B. F., Rocha, A. M. A. C., & Pereira, A. I. (2024). Hybrid approaches to optimization and machine learning methods: A systematic literature review. *Machine Learning*. https://doi.org/10.1007/s10994-023-06467-x

Ballouch, Z., Hajji, R., Poux, F., Kharroubi, A., & Billen, R. (2022). A prior level fusion approach for the semantic segmentation of 3d point clouds using deep learning. *Remote Sensing, 14*, 3415. https://doi.org/10.3390/rs14143415

Batty, M. (2018). Digital twins. *Environment and Planning B: Urban Analytics and City Science, 45*(5), 817–820. https://doi.org/10.1177/2399808318796416

Beckett, S. (2022). Smart city digital twins, 3D modeling and visualization tools, and spatial cognition algorithms in artificial intelligence-based urban design and planning. *Geopolitics, History, and International Relations, 14*(1), 123–138. https://www.jstor.org/stable/48679657

Bibri, S. E. (2015). Ambient intelligence: A new computing paradigm and a vision of a next wave in ICT. In: *The Human Face of Ambient Intelligence: Atlantis Ambient and Pervasive Intelligence* (vol. 9). Atlantis Press. https://doi.org/10.2991/978-94-6239-130-7_2

Bibri, S. E. (2019). Smart sustainable urbanism: Paradigmatic, scientific, scholarly, epistemic, and discursive shifts in light of big data science and analytics. In: Big Data Science and Analytics for Smart Sustainable Urbanism. Advances in Science, Technology & Innovation. Springer. https://doi.org/10.1007/978-3-030-17312-8_6

Bibri, S. E. (2020). Data-driven environmental solutions for smart sustainable cities: Strategies and pathways for energy efficiency and pollution reduction. *Euro-Mediterranean Journal for Environmental Integration, 5*(66). https://doi.org/10.1007/s41207-020-00211-w

Bibri, S. E. (2021a). Data-driven smart sustainable cities of the future: Urban computing and intelligence for strategic, short-term, and joined-up plan-ning. *Computational Urban Science, 1*, 1–29.

Bibri, S. E. (2021b). The underlying components of data-driven smart sustainable cities of the future: A case study approach to an applied theoretical framework. *European Journal of Futures Research, 9*(13). https://doi.org/10.1186/s40309-021-00182-3

Bibri, S. E. (2021c). The core academic and scientific disciplines underlying data-driven smart sustainable urbanism: An interdisciplinary and transdisciplinary framework. *Computational Urban Science, 1*(1), 1. https://doi.org/10.1007/s43762-021-00001-2

Bibri, S. E., & Krogstie, J. (2020). Data-driven smart sustainable cities of the future: A novel model of urbanism and its core dimensions, strategies, and solutions. *Journal of Futures Studies, 25*(2), 77–94. https://doi.org/10.6531/JFS.202012_25(2).0009

Bibri, S. E., Alexandre, A., Sharifi, A., & Krogstie, J. (2023a). Environmentally sustainable smart cities and their converging AI, IoT, and big data tech-nologies and solutions: An integrated approach to an extensive literature review. *Energy Informatics, 6*, 9.

Bibri, S. E., Huang, J., Jagatheesaperumal, S. K., & Krogstie, J. (2024a). The synergistic interplay of artificial intelligence and digital twin in environmentally planning sustainable smart cities: A comprehensive systematic review. *Environmental Science and Ecotechnology, 20*, 100433. https://doi.org/10.1016/j.ese.2024.100433

Bibri, S. E., Huang, J., & Krogstie, J. (2024b). Artificial intelligence of things for synergizing smarter eco-city brain, metabolism, and platform: Pioneering data-driven environmental governance. *Sustainable Cities and Society*, 105516. https://doi.org/10.1016/j.scs.2024.105516

Bibri, S. E., & Huang, J. (2025). Artificial intelligence of sustainable smart city brain and digital twin: A pioneering framework for advancing environmental sustainability. *Environmental Technology and Innovation* (in press).

Bibri, S., Krogstie, J., Kaboli, A., & Alahi, A. (2023b). Smarter eco-cities and their leading-edge artificial intelligence of things solutions for environmental sustainability: A comprehensive systemic review. *Environmental Science and Ecotechnlogy, 19*, https://doi.org/10.1016/j.ese.2023.100330

Bommasani, R., Hudson, D. A., Adeli, E., Altman, R., Arora, S., von Arx, S., & Wang, W. (2022). On the opportunities and risks of foundation models. https://doi.org/10.48550/arXiv.2108.07258

Botín-Sanabria, D. M., Mihaita, A. S., Peimbert-García, R. E., Ramírez-Moreno, M. A., Ramírez-Mendoza, R. A., & Lozoya-Santos, J. D. J. (2022). Digital twin technology challenges and applications: A comprehensive review. *Remote Sensing, 14*, 1335

Bowden, M. (2023). A review of textual and voice processing algorithms in the field of natural language processing. *Journal of Computing and Natural Science. 3*(4), 194–203.

Bratton, B. (2021). AI urbanism: A design framework for governance, program, and platform cognition. *AI & Society, 36*(4), 1307–1312. https://doi.org/10.1007/s00146-020-01121-9

Brock, A., Donahue, J., & Simonyan, K. (2018). Large scale GAN training for high fidelity natural image synthesis. In: *International Conference on Learning Representations*.

Brown, T. B., Mann, B., Ryder, N., Subbiah, M., Kaplan, J., Dhariwal, P., Neelakantan, A., Shyam, P., Sastry, G., Askell, A., Agarwal, S., Herbert-Voss, A., Krueger, G., Henighan, T., Child, R., Ramesh, A., Ziegler, D. M., Wu, J., Winter, C., Hesse, C., Chen, M., Sigler, E., Litwin, M., Gray, S., Chess, B., Clark, J., Berner, C., McCandlish, S., Radford, A., Sutskever, I., & Amodei, D. (2020). Language models are few-shot learners. In: *Advances in Neural Information Processing Systems* (pp. 1877–1901). https://papers.nips.cc/paper/2020/hash/1457c0d6bfcb4967418bfb8ac142f64a-Abstract.html

Calderaa, C., Ostorerob, C., Mannic, V., Gallid, A., & Valzanoe, L. S. (2021). Cities in transformation. Computational urban planning through big data analytics. *Techne, 2*, 76. https://doi.org/10.13128/techne-10686

Cao, Y., Li, S., Liu, Y., Yan, Z., Dai, Y., Yu, P. S., & Sun, L. (2023). A comprehensive survey of ai-generated content (AIGC): A history of generative AI from GAN to ChatGPT. https://doi.org/10.48550/arXiv.2303.04226

Caprari, G., Castelli, G., Montuori, M., Camardelli, M., & Malvezzi, R. (2022). Digital twin for urban planning in the green deal era: A state of the art and future perspectives. *Sustainability, 14*(10), 6263

Castelli, M., & Manzoni, L. (2022). Generative models in artificial intelligence and their applications. *Applied Sciences, 12*(9), 4127.

Chang, S. W., Rhee, D. Y., & Jun, H. J. (2022). A natural language processing framework for collecting, analyzing, and visualizing users' sentiment on the built environment: Case implementation of New York City and seoul residences. *Architectural Science Review, 65*(4), 278–294.

Chen, G., Yang, J., Huang, B. X., Ma, C. Y., Tian, F. L., Ge, L. Y., Xia, L., & Li, J. (2023). Toward digital twin of the ocean: From digitalization to cloning. *Intelligent Marine Technology and Systems, 1*. https://doi.org/10.1007/s44295-023-00003-2

Chen, L., Zhang, Z., Xu, X., Cai, Z., & He, S. (2019). A review of generative adversarial networks for image synthesis. *Multimedia Tools and Applications, 78*(10), 12539–12571.

Chu-Ke, C., & Dong, Y. (2024). Misinformation and literacies in the era of generative artificial intelligence: A brief overview and a call for future research. *Emerging Media, 2*(1), 70–85. https://doi.org/10.1177/27523543241240285

Ćirić, D. (2024). New generative and AI design methods for transportation systems and urban mobility design, planning, operation, and analysis: Contribution to urban computing theory and methodology. In: Keeping Up with Technologies to Imagine and Build Together Sustainable, Inclusive, and Beautiful Cities (vol. 8, pp. 490–503). International Academic Conference on Places and Technologies. https://doi.org/10.18485/arh_pt.2024.8.ch59

Deena, G., Gulati, K., Jha, R., Bajjuri, U. R., Sahithullah, M., & Singh, M. (2022, July). Artificial intelligence and a digital twin are effecting building energy management. In: 2022 International Conference on Innovative Computing, Intelligent Communication and Smart Electrical Systems *(ICSES)* (pp. 1–8). IEEE.

Delacruz, G. P. (2020). Using generative adversarial networks to classify structural damage caused by earthquakes. Ph.D. thesis. California Polytechnic State University.

Dembski, F., Wössner, U., Letzgus, M., Ruddat, M., & Yamu, C. (2020). Urban digital twins for smart cities and citizens: The case study of Herrenberg, Germany. *Sustainability, 12*(6), 2307. https://doi.org/10.3390/su12062307

Din, I. U., Guizani, M., Rodrigues, J. J. P. C., Hassan, S., & Korotaev, V. V. (2019). Machine learning in the internet of things: Designed techniques for smart cities. *Future Generation Computer Systems, 100*, 826–843.

Dinh, L., Sohl-Dickstein, J., & Bengio, S. (2016). Density estimation using Real NVP. https://doi.org/10.48550/arXiv.1605.08803

Doiron, D., Setton, E., Brook, J., Kestens, Y., McCormack, G., Winters, M., Shooshtari, M., Azami, S., & Fuller, D. (2022). Predicting walking-to-work using street-level imagery and deep learning in seven Canadian cities. *Scientific Reports, 12*(1), 1. https://doi.org/10.1038/s41598-022-22630-1

Dong, S., Wang, P., & Abbas, K. (2021). A survey on deep learning and its applications. *Computer Science Review, 40*. https://doi.org/10.1016/j.cosrev.2021.100379

Efthymiou, I.-P., & Egleton, T. E. (2023). Artificial intelligence for sustainable smart cities. In: *Handbook of Research on Applications of AI, Digital Twin, and Internet of Things for Sustainable Development* (pp. 1–11). IGI Global.

Eltahan, M., Daoud, N., & Moharm, K. (2021). Generative adversarial networks (GANs) for spatial upward fluxes radiation estimation. In: 2021 International Conference on Advances in Electrical, Computing, Communication and Sustainable Technologies (ICAECT), Bhilai, India (pp. 1–5). https://doi.org/10.1109/ICAECT49130.2021

ETH. (2024). The Blue City Project. https://esd.ifu.ethz.ch/research/research-projects/research-and-theses/bluecity.html (accessed 12 October 2023).

Evans, J., Karvonen, A., Luque-Ayala, A., Martin, C., McCormick, K., Raven, R., & Palgan, Y. V. (2019). Smart and sustainable cities? Pipedreams, practicalities and possibilities. *Local Environment, 24*(7), 557–564. https://doi.org/10.1080/13549839.2019.1624701

Fan, C., Zhang, C., Yahja, A., & Mostafavi, A. (2021). Disaster city digital twin: A vision for integrating artificial and human intelligence for disaster management. *International Journal of Information Management, 56*, 102049

Fang, B., Yu, J., Chen, Z., Osman, A. I., Farghali, M., Ihara, I., Hamza, E. H., Rooney, D. W., & Yap, P.-S. (2023). Artificial intelligence for waste management in smart cities: A review. *Environmental Chemistry Letters, 21*(6), 1959–1989. https://doi.org/10.1007/s10311-023-01604-3

Fang, L., & Ewing, R. (2020). Tracking our footsteps: Thirty years of publication in JAPA, JPER, and JPL. *Journal of the American Planning Association, 86*(4), 470–80.

Ferrara, E. (2024). GenAI against humanity: Nefarious applications of generative artificial intelligence and large language models. *Journal of Computational Social Science, 7*, 549–569. https://doi.org/10.1007/s42001-024-00250-1

Ferré-Bigorra, J., Casals, M., & Gangolells, M. (2022). The adoption of urban digital twins. *Cities, 131*, 10390.

Fitriawijaya, A., & Jeng, T. (2024). Integrating multimodal generative AI and blockchain for enhancing generative design in the early phase of architectural design process. *Buildings, 14*(8), 2533. https://doi.org/10.3390/buildings14082533

Fu, X., Li, C., & Zhai, W. (2023). Using natural language processing to read plans: A study of 78 resilience plans from the 100 resilient cities network. *Journal of the American Planning Association, 89*(1), 107–19.

Fu, X., Wang, R., & Li, C. (2023). Can ChatGPT evaluate plans? *Journal of the American Planning Association*, 1–12. https://doi.org/10.1080/01944363.2023.2271893

Garrido-Valenzuela, F., Cats, O., & van Cranenburgh, S. (2023). Where are the people? Counting people in millions of street-level images to explore associations between people's urban density and urban characteristics. *Computers, Environment and Urban Systems, 102*, 101971. https://doi.org/10.1016/j.compenvurbsys.2023.101971

Ghazizadeh, E., & Zhu, P. (2020). A systematic literature review of natural language processing: Current state, challenges and risks. In: *Proceedings of the Future Technologies Conference* (pp. 634–647). Springer.

Gkontzis, A. F., Kotsiantis, S., Feretzakis, G., & Verykios, V. S. (2024). Enhancing urban resilience: Smart city data analyses, forecasts, and digital twin techniques at the neighborhood level. *Future Internet, 16*(2), 47. https://doi.org/10.3390/fi16020047

Goodfellow, I., Pouget-Abadie, J., Mirza, M., Xu, B., Warde-Farley, D., Ozair, S., Courville, A., & Bengio, Y. (2014). Generative adversarial nets. In: *Proceedings of the 27th International Conference on Neural Information Processing Systems.* (Vol. 2, pp. 2672–2680).

Gourisaria, M. K., Jee, G., Harshvardhan, G., Konar, D., & Singh, P. K. (2022). Artificially intelligent and sustainable smart cities. In: P. K. Singh, M. Paprzycki, M. Essaaidi, & S. Rahimi (Eds.), *Sustainable Smart Cities: Theoretical Foundations and Practical Considerations* (pp. 237–268). Springer, Cham. https://doi.org/10.1007/978-3-031-08815-5_14

Guzder-Williams, B., Mackres, E., Angel, S., Blei, A., & Lamson-Hall, P. (2023). Intra-urban land use maps for a global sample of cities from Sentinel-2 satellite imagery and computer vision. *Computers, Environment and Urban Systems, 100*, 101917. https://doi.org/10.1016/j.compenvurbsys.2022.101917

He, Z., Wang, Z., Xie, Z., Wu, L., & Chen, Z. (2022). Multiscale analysis of the influence of street built environment on crime occurrence using street-view images. *Computers, Environment and Urban Systems, 97*, 101865. https://doi.org/10.1016/j.compenvurbsys.2022.101865

Ho, J., Jain, A., & Abbeel, P. (2020). Denoising diffusion probabilistic models. *Advances in Neural Information Processing Systems, 33*, 6840–6851.

Hofmann, J., & Schüttrumpf, H. (2021). FloodGAN: Using deep adversarial learning to predict pluvial flooding in real time. *Water, 13*, 2255.

Hosseini, M., Miranda, F., Lin, J., & Silva, C. (2022). CitySurfaces: City-scale semantic segmentation of sidewalk materials. *Sustainable Cities and Society, 79*, 103630. https://doi.org/10.1016/j.scs.2021.103630

Huai, S., & Van de Voorde, T. (2022). Which environmental features contribute to positive and negative perceptions of urban parks? A cross-cultural comparison using online reviews and natural language processing methods. *Landscape and Urban Planning, 218*, 104307.

Huang, C., Zhang, G., Yao, J., Wang, X., Calautit, J. K., Zhao, C., An, N., & Peng, X. (2022). Generative models for synthetic urban mobility data: A systematic literature review. *ACM Computing Surveys*. https://doi.org/10.1145/3610224

Huang, J. Bibri, S. E., & Keel, P. (2025). Generative spatial artificial intelligence for sustainable smart cities: A pioneering large flow foundation model for urban digital twin. *Environmental Science and Ecotechnology* (in press).

Hudson, L., & Sedlackova, A. (2021). Urban sensing technologies and geospatial big data analytics in internet of things-enabled smart cities. *Geopolitics, History, and International Relations, 13*(2), 37–50.

Ibrahim, M., Haworth, J., & Cheng, T. (2021). URBAN-i: From urban scenes to mapping slums, transport modes, and pedestrians in cities using deep learning and computer vision. *Environment and Planning B: Urban Analytics and City Science, 48*(1), 76–93. https://doi.org/10.1177/2399808319846517

Jaad, A., & Abdelghany, K. (2021). The story of five MENA cities: Urban growth prediction modeling using remote sensing and video analytics. *Cities, 118*, 103393. https://doi.org/10.1016/j.cities.2021.103393

Jakubik, J., Roy, S., Phillips, C. E., Fraccaro, P., Godwin, D., Zadrozny, B., Szwarcman, D., Gomes, C., Nyirjesy, G., Edwards, B., Kimura, D., Simumba, N., Chu, L., Mukkavilli, S., Lambhate, D., Das, K., Bangalore, R., Oliveira, D., Muszynski, M., Ankur, K., Ramasubramanian, M., Gurung, I., Khallaghi, S., Li, H., Cecil, M., Ahmadi, M., Kordi, F., Alemohammad, H., Maskey, M.,

Ganti, R., Weldemariam, K., & Ramachandran, R. (2023). Foundation models for generalist geospatial artificial intelligence. https://doi.org/10.48550/arXiv.2310.18660

Janarthanan, R., Partheeban, P., Somasundaram, K., & Elamparithi, P. N. (2021). A deep learning approach for prediction of air quality index in a metropolitan city. *Sustainable Cities and Society*, 67, 102720.

Janowicz, K. (2023). Philosophical foundations of GeoAI: Exploring sustainability, diversity, and bias in GeoAI and spatial data science. In: *Handbook of Geospatial Artificial Intelligence* (pp. 26–42). CRC Press.

Jiang, F., Ma, J., Webster, C. J., Chiaradia, A. J. F., Zhou, Y., Zhao, Z., & Zhang, X. (2023). Generative urban design: A systematic review on problem formulation, design generation, and decision-making. Progress in Planning, 180, 100795. https://doi.org/10.1016/j.progress.2023.100795.

Joshi, K., Kumar, V., Anandaram, H., Kumar, R., Gupta, A., & Krishna, K. H. (2023). A review approach on deep learning algorithms in computer vision. In: Nitin Mittal, Amit Kant Pandit, Mohamed Abouhawwash, Shubham Mahajan (Eds.) *Intelligent Systems and Applications in Computer Vision* (1st ed., pp. 15). CRC Press.

Kaczmarek, I., Iwaniak, A., Świetlicka, A., Piwowarczyk, M., & Nadolny, A. (2022). A machine learning approach for integration of spatial development plans based on natural language processing. *Sustainable Cities and Society*, 76, 103479.

Kamal, H., Yánez, W., Hassan, S., & Sobhy, D. (January 2024). Digital-twin-based deep reinforcement learning approach for adaptive traffic signal control. *IEEE Internet of Things Journal*, 9, 1–1. https://doi.org/10.1109/JIOT.2024.3377600

Kamrowska-Załuska, D. (2021). Impact of AI-based tools and urban big data analytics on the design and planning of cities. *Land*, 10, 1209. https://doi.org/10.3390/land10111209

Khurana, D., Koli, A., Khatter, K., & Singh, S. (2023). Natural language processing: State of the art, current trends and challenges. *Multimedia Tools and Applications*, 82(3), 3713–3744.

Kingma, D. P., & Dhariwal, P. (2018). Glow: Generative flow with invertible 1x1 convolutions. In Advances in Neural Information Processing Systems (Vol. 31). https://doi.org/10.48550/arXiv.1807.03039

Kingma, D. P., & Welling, M. (2022). Auto-encoding variational Bayes. arXiv. https://doi.org/10.48550/arXiv.1312.6114

Kirwan, C. G., & Zhiyong, F. (2020). *Smart Cities and Artificial Intelligence: Convergent Systems for Planning, Design, and Operations*. Elsevier, Amsterdam.

Komninos, N., & Kakderi, C. (Eds.). (2019). *Smart Cities in the Post-algorithmic Era: Integrating Technologies, Platforms and Governance*. Edward El-gar Publishing, Cheltenham

Kong, L., Liu, Z., Pan, X., Wang, Y., Guo, X., & Wu, J. (2022). How do different types and landscape attributes of urban parks affect visitors' positive emotions? *Landscape and Urban Planning*, 226, 104482.

Koumetio, T. S. C., Diop, E. B., Azmi, R., & Chenal, J. (2023). Artificial intelligence-based methods for smart and sustainable urban planning: A systematic survey. *Archives of Computational Methods in Engineering, 30*, 1421–1438. https://doi.org/10.1007/s11831-022-09844-2

Lago, C. A. F. D., Giacomoni, M. H., Bentivoglio, R., Taormina, R., Gomes Junior, M. N., & Mendiondo, E. M. (2023). Generalizing rapid flood predictions to unseen urban catchments with conditional generative adversarial networks. *Journal of Hydrology, 618*, 1–15.

Lei, B., Janssen, P., Stoter, J., & Biljecki, F. (2023). Challenges of urban digital twins: A systematic review and a Delphi expert survey. *Automation in Construction, 147*, 104716. https://doi.org/10.1016/j.autcon.2022.104716

Lenfers, U. A., Ahmady-Moghaddam, N., Glake, D., Ocker, F., Osterholz, D., Ströbele, J., & Clemen, T. (2021). Improving model predictions—integration of real-time sensor data into a running simulation of an agent-based model. *Sustainability, 13*, 7000. https://doi.org/10.3390/su13137000

Li, C., Gan, Z., Yang, Z., Yang, J., Li, L., Wang, L., & Gao, J. (2024). Multimodal foundation models: From specialists to general-purpose assistants. *Foundations and Trends® in Computer Graphics and Vision, 16*(1–2), 1–214. https://doi.org/10.1561/0600000110

Li, C., Zhang, T., Du, X., Zhang, Y., & Xie, H. (2024). Generative AI for architectural design: A literature review. arXiv (Cornell University). https://doi.org/10.48550/arxiv.2404.01335

Li, F., Yigitcanlar, T., Nepal, M., Nguyen, K., & Dur, F. (2023). Machine learning and remote sensing integration for leveraging urban sustainability: A review and framework. *Sustainable Cities and Society, 96*, 104653. https://doi.org/10.1016/j.scs.2023.104653

Li, P., & Mao, K. (2019). Knowledge-oriented convolutional neural network for causal relation extraction from natural language texts. *Expert Systems with Applications, 115*, 512–523.

Li, X. (2021). Examining the spatial distribution and temporal change of the green view index in New York City using Google Street View images and deep learning. *Environment and Planning B: Urban Analytics and City Science, 48*(7), 2039–2054. https://doi.org/10.1177/2399808320962511

Li, X., Zhang, C., & Li, W. (2017). Building block level urban land-use information retrieval based on Google Street View images. *GIScience & Remote Sensing, 54*(6), 819–835. https://doi.org/10.1080/15481603.2017.1338389

Li, Z., & Ma, J. (2022). Discussing street tree planning based on pedestrian volume using machine learning and computer vision. *Building and Environment, 219*, 109178. https://doi.org/10.1016/j.buildenv.2022.109178

Li, Z., & Ning, H. (2023). Autonomous GIS: The next-generation AI-powered GIS.

Liang, X., Zhao, T., & Biljecki, F. (2023). Revealing spatio-temporal evolution of urban visual environments with street view imagery. *Landscape and Urban Planning, 237*, 104802. https://doi.org/10.1016/j.landurbplan.2023.104802

Liao, W., Lu, X., Fei, Y., Gu, Y., & Huang, Y. (2023). Generative AI design for building structures. *Automation in Construction, 157*, 105187. https://doi.org/10.1016/j.autcon.2023.105187

Ligozat, A.-L., Lefevre, J., Bugeau, A., & Combaz, J. (2022). Unraveling the hidden environmental impacts of AI solutions for environment: Life cycle assessment of AI solutions. *Sustainability*, *14*(9), 5172. https://doi.org/10.3390/su14095172

Liu, P., Zhao, T., Luo, J., Lei, B., Frei, M., Miller, C., & Biljecki, F. (2023). Towards human-centric digital twins: Leveraging computer vision and graph models to predict outdoor comfort. *Sustainable Cities and Society*, *93*, 104480. https://doi.org/10.1016/j.scs.2023.104480

Liu, Z., Janowicz, K., Currier, K., & Shi, M. (2024). Measuring geographic diversity of foundation models with a natural language–based geo-guessing experiment on GPT-4. AGILE: *GIScience Series*, *5*(38). https://doi.org/10.5194/agile-giss-5-38-2024

Luo, J., Liu, P., & Cao, L. (2022a). Coupling a physical replica with a digital twin: A comparison of participatory decision-making methods in an urban park environment. *ISPRS International Journal of Geo-Information*, *11*, 452. https://doi.org/10.3390/ijgi11080452

Luo, J., Zhao, T., Cao, L., & Biljecki, F. (2022b). Semantic riverscapes: Perception and evaluation of linear landscapes from oblique imagery using computer vision. *Landscape and Urban Planning*, *228*, 104569. https://doi.org/10.1016/j.landurbplan.2022.104569

Mahadevkar, S., Khemani, B., Patil, S., Kotecha, K., Vora, D., Abraham, A., & Gabralla, L. (2022). A review on machine learning styles in computer vision: Techniques and future directions. *IEEE Access*, *10*, 107293–107329. https://doi.org/10.1109/ACCESS.2022.3209825

Mahamuni, C. V., Sayyed, Z., & Mishra, A. (2022). Machine learning for smart cities: A survey. In: 2022 IEEE International Power and Renewable Energy Conference (IPRECON), Kollam, India (pp. 1–8). https://doi.org/10.1109/IPRECON55716.2022.10059521

Ma, W., Wang, X., Wang, J., Xiang, X., & Sun, J. (2021). Generative design in building information modelling (BIM): Approaches and requirements. *Sensors*, *21*(16), 5439. https://doi.org/10.3390/s21165439

Mai, G., Cundy, C., Choi, K., Hu, Y., Lao, N., & Ermon, S. (2022). Towards a foundation model for geospatial artificial intelligence (vision paper). In: Proceedings of the 30th International Conference on Advances in Geographic Information Systems, SIGSPATIAL '22 (pp. 1–4).

Manocha, A., Sood, S. K., & Bhatia, M. (2023). Digital twin-assisted fuzzy logic-inspired intelligent approach for flood prediction. *IEEE Sensors Journal*. https://doi.org/10.1109/JSEN.2023.3322535

Marasinghe, R., Yigitcanlar, T., Mayere, S., Washington, T., & Limb, M. (2024). Computer vision applications for urban planning: A systematic review of opportunities and constraints. *Sustainable Cities and Society*, *100*, 105047. https://doi.org/10.1016/j.scs.2023.105047

Marcucci, E., Gatta, V., Le Pira, M., Hansson, L., & Bråthen, S. (2020). Digital twins: A critical discussion on their potential for supporting policy-making and planning in urban logistics. *Sustainability*, *12*, 10623. https://doi.org/10.3390/su122410623

Martella, A., Ramadan, A. I. H. A., Martella, C., Patano, M., & Longo, A. (2023). State of the art of urban digital twin platforms. In: L. T. De Paolis, P. Arpaia, & M. Sacco (Eds.), *Extended Reality: XR Salento 2023*. Lecture Notes in Computer Science (vol. 14218). Springer. https://doi.org/10.1007/978-3-031-43401-3_20

Mathew, A., Amudha, P., & Sivakumari, S. (2021). Deep learning techniques: An overview. In: *Advances in Intelligent Systems and Computing* (pp. 599–608). https://doi.org/10.1007/978-981-15-3383-9_54

McCormack, J., & Grierson, M. (2024). Building simulations with generative artificial intelligence. In: D. Del Favero, S. Thurow, M. J. Ostwald, & U. Frohne (Eds.), *Climate Disaster Preparedness: Arts, Research, Innovation and Society*. Springer. https://doi.org/10.1007/978-3-031-56114-6_11

Michelle, B., & Gemilang, M. P. (2022). Bibliometric analysis of generative design, algorithmic design, and parametric design in architecture. *Journal of Artificial Intelligence in Architecture*, 1(1), 30–40. https://doi.org/10.24002/jarina.v1i1.4921

Mirza, M., & Osindero, S. (2014). Conditional generative adversarial nets. arXiv. https://doi.org/10.48550/arXiv.1411.1784

Mishra, P., & Singh, G. (2023). Artificial intelligence for sustainable smart cities. In: *Sustainable Smart Cities*. Springer. https://doi.org/10.1007/978-3-031-33354-5_6

Mohamed, K. S. (2023). Deep learning for IoT "artificial intelligence of things (AIoT)". In: *Deep Learning-Powered Technologies*. Synthesis Lectures on Engineering, Science, and Technology. https://doi.org/10.1007/978-3-031-35737-4_3

Mylonas, G., Kalogeras, A., Kalogeras, G., Anagnostopoulos, C., Alexakos, C., & Muñoz, L. (2021). Digital twins from smart manufacturing to smart cities: A survey. *IEEE Access*, 9, 143222–143249.

Naeem, S., Ali, A., Anam, S., & Ahmed, M. M. (2023). An unsupervised machine learning algorithms: Comprehensive review. *International Journal of Computer and Digital Systems*, 13, 911–921.

Naik, N., Kominers, S. D., Raskar, R., & Glaeser, E. L. (2017). Computer vision uncovers predictors of physical urban change. *Proceedings of the National Academy of Sciences*, 114(29), 7571–7576.

Nishant, R., Kennedy, M., & Corbett, J. (2020). Artificial intelligence for sustain-ability: Challenges, opportunities, and a research agenda. *International Journal of Information Management*, 53, 102104.

Niu, T., Qing, L., Han, L., Long, Y., Hou, J., Li, L., Tang, W., & Teng, Q. (2022). Small public space vitality analysis and evaluation based on human trajectory modeling using video data. *Building and Environment*, 225, 109563. https://doi.org/10.1016/j.buildenv.2022.109563

Ooi, K. B., Tan, G. W. H., Al-Emran, M., Al-Sharafi, M. A., Capatina, A., Chakraborty, A., Wong, L. W., Dwivedi, Y. K., Huang, T.-L., Kar, A. K., Lee, V.-H., Loh, X.-M., Micu, A., Mikalef, P., Mogaji, E., Pandey, N., Raman, R., Rana, N. P., Sarker, P., Sharma, A., & Teng, C.-I. (2023). The potential of generative artificial intelligence across disciplines: Perspectives and future directions. *Journal of Computer Information Systems*, 1–32. https://doi.org/10.1080/08874417.2023.2261010

Otter, D. W., Medina, J. R., & Kalita, J. K. (2020). A survey of the usages of deep learning for natural language processing. *IEEE Transactions on Neural Networks and Learning Systems*, 32(2), 604–624.

Oyama, N., Ishizaki, N. N., Koide, S., & et al. (2023). Deep generative model super-resolves spatially correlated multiregional climate data. *Scientific Reports*, 13, 5992. https://doi.org/10.1038/s41598-023-32947-0

Page, M. J., McKenzie, J. E., Bossuyt, P. M., Boutron, I., Hoffmann, T. C., Mulrow, C. D., Shamseer, L., Tetzlaff, J. M., Akl, E. A., Brennan, S. E., Chou, R., Glanville, J., Grimshaw, J. M., Hróbjartsson, A., Lalu, M. M., Li, T., Loder, E. W., Mayo-Wilson, E., McDonald, S., McGuinness, L. A., Stewart, L. A., Thomas, J., Tricco, A. C., Welch, V. A., & Moher, D. (2021). The PRISMA 2020 statement: An updated guideline for reporting systematic reviews. *Systematic Reviews*, 10, 89. https://doi.org/10.1186/s13643-021-01626-4

Papyshev, G., & Yarime, M. (2021). Exploring city digital twins as policy tools: A task-based approach to generating synthetic data on urban mobility. *Data & Policy*, 3, e16. https://doi.org/10.1017/dap.2021.17

Paramesha, M., Rane, N., & Rane, J. (2024). Enhancing resilience through generative artificial intelligence such as ChatGPT. *SSRN*: https://ssrn.com/abstract=4832533, https://doi.org/10.2139/ssrn.4832533

Peng, N., Xi, Y., Rao, J., Ma, X., & Ren, F. (2021). Urban multiple route planning model using dynamic programming in reinforcement learning. *IEEE Transactions on Intelligent Transportation Systems*, 23(7), 8037–8047.

Pérez, J., Arroba, P., & Moya, J. M. (2023). Data augmentation through multivariate scenario forecasting in data centers using generative adversarial networks. *Applied Intelligence*, 53(2), 1469–1486.

Pérez-Martínez, I., Martínez-Rojas, M., & Soto-Hidalgo, J. M. (2023). A methodology for urban planning generation: A novel approach based on generative design. *Engineering Applications of Artificial Intelligence*, 124, 106609.

Priya, S., & Saranya, K. (2023). Significance of artificial intelligence in the development of sustainable transportation. *The Scientific Temper*, 14(02), 418–425.

Radford, A., Kim, J. W., Hallacy, C., Ramesh, A., Goh, G., Agarwal, S., Sastry, G., Askell, A., Mishkin, P., Clark, J., Krueger, G., & Sutskever, I. (2021). Learning transferable visual models from natural language supervision. arXiv. https://doi.org/10.48550/arXiv.2103.00020

Raffel, C., Shazeer, N., Roberts, A., Lee, K., Narang, S., Matena, M., Zhou, Y., Li, W., & Liu, P. J. (2020). Exploring the limits of transfer learning with a unified text-to-text transformer. arXiv. https://doi.org/10.48550/arXiv.1910.10683

Raman, R., Pattnaik, D., Lathabai, H. H., Kumar, C., Govindan, K., & Nedungadi, P. (2024). Green and sustainable AI research: An integrated thematic and topic modeling analysis. Journal of Big Data, 11(55). https://doi.org/10.1186/s40537-024-00920-x

Rane, N. (2023a). Potential role and challenges of ChatGPT and similar generative artificial intelligence in architectural engineering. *SSRN*, http://dx.doi.org/10.2139/ssrn.4607767.

Rane, N. (2023b). Roles and challenges of ChatGPT and similar generative artificial intelligence for achieving the sustainable development goals (SDGs). *SSRN*, 4603244. https://doi.org/10.2139/ssrn.4603244

Rane, N., Choudhary, S., & Rane, J. (2023). Integrating ChatGPT, Bard, and leading-edge generative artificial intelligence in architectural design and engineering: Applications, framework, and challenges. https://doi.org/10.2139/ssrn.4645595

Sharma, R., Agarwal, P., & Arya, A. (2022). Natural language processing and big data: A strapping combination. In R. Sharma & D. Sharma (Eds.), New trends and applications in Internet of Things (IoT) and big data analytics (Vol. 221). Springer, Cham. https://doi.org/10.1007/978-3-030-99329-0_16

Ranjana, R., Subha, T., Varsha, J., & Kumar, P. (2022). Applications and implications of artificial intelligence and deep learning in computer vision. In: *Applications and Implications of Artificial Intelligence and Deep Learning in Computer Vision* (pp. 35–52). De Gruyter. https://doi.org/10.1515/9783110750584-003

Richards, D., Worden, D., Song, X. P., & Lavorel, S. (2024). Harnessing generative artificial intelligence to support nature-based solutions. *People and Nature*, 6(2), 882–893. https://doi.org/10.1002/pan3.10622

Sabri, S., Alexandridis, K., Koohikamali, M., Zhang, S., & Ozkaya, H. E. (2023, November). Designing a spatially-explicit urban digital twin framework for smart water infrastructure and flood management. In: 2023 IEEE 3rd International Conference on Digital Twins and Parallel Intelligence *(DTPI)* (pp. 1–9). IEEE.

Salazar-Miranda, A., Zhang, F., Sun, M., Leoni, P., Duarte, F., & Ratti, C. (2023). Smart curbs: Measuring street activities in real-time using computer vision. *Landscape and Urban Planning*, 234, 104715. https://doi.org/10.1016/j.landurbplan.2023.104715

Salunke, A. A. (2023). Reinforcement learning empowered digital twins: Pioneering smart cities towards optimal urban dynamics. *EPRA International Journal of Research & Development (IJRD)*. https://doi.org/10.36713/epra13959

Sanchez, T. (2023). *Planning With Artificial Intelligence*. American Planning Association. https://www.planning.org/publications/report/9270237/.

Sanchez, T. W. (2023). Planning on the verge of AI, or AI on the verge of planning. *Urban Science*, 7, 70. https://doi.org/10.3390/urbansci7030070

Saravanan, K., & Sakthinathan, G. (2021). *Handbook of Green Engineering Technologies for Sustainable Smart Cities*. CRC Press

Sarp, S., Kuzlu, M., Zhao, Y., Cetin, M., & Guler, O. (2022). A comparison of deep learning algorithms on image data for detecting floodwater on roadways. *Computer Science and Information Systems*, 19(1), 397-414. https://doi.org/10.2298/CSIS210313058S

Schwartz, R., Dodge, J., Smith, N. A., & Etzioni, O. (2020). Green AI. *Communications of the ACM*, 63(12), 54–63. https://doi.org/10.1145/3381831

Sha, Y., Sobash, R. A., & Gagne, D. J. II. (2024). Improving ensemble extreme precipitation forecasts using generative artificial intelligence. https://doi.org/10.48550/arXiv.2407.04882

Sharifani, K., & Amini, M. (2023). Machine learning and deep learning: A review of methods and applications. *World Information Technology and Engineering Journal, 10*(7), 3897–3904. https://ssrn.com/abstract=4458723

Sharifi, A., Allam, Z., Bibri, S. E., & Khavarian-Garmsir, A. R. (2024). Smart cities and sustainable development goals (SDGs): A systematic literature review of co-benefits and trade-offs. *Cities, 146*, 104659. https://doi.org/10.1016/j.cities.2023.104659

Sharma, N., Sharma, R., & Jindal, N. (2021). Machine learning and deep learning applications: A vision. *Global Transitions Proceedings, 2*(1), 24–28. https://doi.org/10.1016/j.gltp.2021.01.004

Shen, J., Saini, P. K., & Zhang, X. (2021). Machine learning and artificial intelligence for digital twin to accelerate sustainability in positive energy districts. In: Data-driven Analytics for Sustainable Buildings and Cities: From Theory to Application (pp. 411–422).

Shi, W. (2021). Introduction to urban sensing. In: W. Shi, M. F. Goodchild, M. Batty, M.-P. Kwan, & A. Zhang (Eds.), *Urban Informatics* (pp. 311–314). Springer. https://doi.org/10.1007/978-981-15-8983-6_19

Shinde, P. P., & Shah, S. (2018). A review of machine learning and deep learning applications. In: 2018 Fourth International Conference on Computing Communication Control and Automation (ICCUBEA), Pune, India (pp. 1–6). https://doi.org/10.1109/ICCUBEA.2018.8697857

Shruti, S., Singh, P. K., Ohri, A., & Singh, R. S. (2022). Development of environmental decision support system for sustainable smart cities in India. *Environmental Progress & Sustainable Energy, 41*(5), e13817.

Singh, M., Fuenmayor, E., Hinchy, E. P., Qiao, Y., Murray, N., & Devine, D. (2021). Digital twin: Origin to future. *Applied System Innovation, 4*(2), 36. pp.1-19. https://doi.org/10.3390/asi4020036

Somanath, S., Naserentin, V., Eleftheriou, O., Sjölie, D., Wästberg, B. S., & Logg, A. (2023). On procedural urban digital twin generation and visualization of large scale data.

Son, T. H., Weedon, Z., Yigitcanlar, T., Sanchez, T., Corchado, J. M., & Mehmood, R. (2023). Algorithmic urban planning for smart and sus-tainable development: Systematic review of the literature. *Sustainable Cities and Society, 94*, 104562.

Son, T., Weedon, Z., Yigitcanlar, T., Sanchez, T., Corchado, J., & Mehmood, R. (2023). Algorithmic urban planning for smart and sustainable development: Systematic review of the literature. *Sustainable Cities and Society, 94*, 104562. https://doi.org/10.1016/j.scs.2023.104562

Strielkowski, W., Rausser, G., & Kuzmin, E. (2022). Digital revolution in the energy sector: Effects of using digital twin technology. In: *Digital Transformation in Industry: Digital Twins and New Business Models* (pp. 43–55). Springer.

Su, N., Li, W., & Qiu, W. (2023). Measuring the associations between eye-level urban design quality and on-street crime density around New York subway entrances. *Habitat International, 131*, 102728. https://doi.org/10.1016/j.habitatint.2022.102728

Suel, E., Muller, E., Bennett, J., Blakely, T., Doyle, Y., Lynch, J., Mackenbach, J., Middel, A., Mizdrak, A., Nathvani, R., Brauer, M., & Ezzati, M. (2023). Do poverty and wealth look the same the world over? A comparative study of 12 cities from five high-income countries using street images. *EPJournal of Data Science, 12*(1), 1. https://doi.org/10.1140/epjds/s13688-023-00394-6

Sun, Y., & Dogan, T. (2022). Accelerated environmental performance-driven design of the parametric urban block with generative adversarial network. *Building and Environment*. https://doi.org/10.1016/j.buildenv.2022

Sun, Y., & Dogan, T. (2023). Generative methods for Urban design and rapid solution space exploration. *Environment and Planning B: Urban Analytics and City Science, 50*(6), 1577–1590. https://doi.org/10.1177/23998083221142191

Szpilko, D., Naharro, F. J., Lăzăroiu, G., Nica, E., & Gallegos, A. D. L. T. (2023). Artificial intelligence in the smart city: A literature review. *Engineering Management in Production and Services, 15*(4), 53–75. https://doi.org/10.2478/emj-2023-0028

Tian, Y., Li, X., Zhang, H., Zhao, C., Li, B., Wang, X., Wang, X., & Wang, F.-Y. (2023). VistaGPT: Generative parallel transformers for vehicles with intelligent systems for transport automation. *IEEE Transactions on Intelligent Vehicles, 8*(9), 4198–4207. https://doi.org/10.1109/TIV.2023.3307012

Tilon, S., Nex, F., Kerle, N., & Vosselman, G. (2020). Post-disaster building damage detection from Earth observation imagery using unsupervised and transferable anomaly detecting generative adversarial networks. *Remote Sensing, 12*(24), 4193. https://doi.org/10.3390/rs12244193

Toli, A. M., & Murtagh, N. (2020). The concept of sustainability in smart city definitions. *Frontiers in Built Environment, 6*, 77. https://doi.org/10.3389/fbuil.2020.00077

Topping, D., Bannan, T. J., Coe, H., Evans, J., Jay, C., Murabito, E., & Robinson, N. (2021). Digital twins of urban air quality: Opportunities and challenges. *Frontiers in Sustainable Cities, 3*. https://doi.org/10.3389/frsc.2021.786563

Torfi, A., Shirvani, R. A., Keneshloo, Y., Tavaf, N., & Fox, E. A. (2020). Natural language processing advancements by deep learning: A survey.

Tripathi, S. L., Ganguli, S., Kumar, A., & Magradze, T. (2022). *Intelligent Green Technologies for Sustainable Smart Cities*. John Wiley & Sons, New York.

Tyagi, N., & Bhushan, B. (2023). Demystifying the role of Natural Language Processing (NLP) in smart city applications: Background, motivation, recent advances, and future research directions. *Wireless Personal Communications, 130*, 857–908. https://doi.org/10.1007/s11277-023-10312-8

Ullah, A., Anwar, S. M., Li, J., Nadeem, L., Mahmood, T., Rehman, A., & Saba, T. (2023). Smart cities: The role of Internet of Things and machine learning in realizing a data-centric smart environment. *Complex Intelligent Systems, 10*, 1607–1637. https://doi.org/10.1007/s40747-023-01175-4

Ullah, Z., Al-Turjman, F., Mostarda, L., & Gagliardi, R. (2020). Applications of artificial intelligence and machine learning in smart cities. *Computer Communications*, *154*, 313–323.

Van Nes, A., & Yamu, C. (2021). Introduction *to Space Syntax in Urban Studies*. Springer, New York.

Vanky, A., & Le, R. (2023). Urban-semantic computer vision: A framework for contextual understanding of people in urban spaces. *AI & Society*, *38*(3), 1193–1207. https://doi.org/10.1007/s00146-022-01625-6

Vaswani, A., Shazeer, N., Parmar, N., Uszkoreit, J., Jones, L., Gomez, A. N., Kaiser, Ł., & Polosukhin, I. (2017). Attention is all you need. In: *Advances in Neural Information Processing Systems* (pp. 5998–6008). https://doi.org/10.48550/arXiv.1706.03762

Verdecchia, R., Cruz, L., Sallou, J., Lin, M., Wickenden, J., & Hotellier, E. (2022). Data-centric green AI: An exploratory empirical study. In C. Calero, A. Karvonen, E. Somova, J. P. Fernandes, A.-K. Peters, & J. Cunha (Eds.), 2022 International Conference on ICT for Sustainability (ICT4S) (pp. 35–45). Institute of Electrical and Electronics Engineers Inc.. https://doi.org/10.1109/ICT4S55073.2022.00015

Verma, T., Kishore Kumar, B., Rajendar, J., & Kumara Swamy, B. (2024). A review on quantum machine learning. In: A. Kumar & S. Mozar (Eds.), Proceedings of the 6th International Conference on Communications and Cyber Physical Engineering (ICCCE 2024), Lecture Notes in Electrical Engineering (vol. 1096). Springer. https://doi.org/10.1007/978-981-99-7137-4_39

Visvizi, A., & del Hoyo, R. P. (2021). *Smart Cities and the Un Sdgs*. Elsevier.

Wael, S., Elshater, A., & Afifi, S. (2022). Mapping user experiences around transit stops using computer vision technology: Action priorities from Cairo. *Sustainability*, *14* (17), 17. https://doi.org/10.3390/su141711008

Wang, B., Wang, Q., Cheng, J. C. P., & Yin, C. (2022a). Object verification based on deep learning point feature comparison for scan-to-BIM. *Automation in Construction*, *142*, Article 104515. https://doi.org/10.1016/j.autcon.2022.104515

Wang, D., Lu, C. T., & Fu, Y. (2023). Towards automated urban planning: When generative and chatgpt-like AI meets urban planning. https://doi.org/10.48550/arXiv.2304.03892

Wang, D., Lu, C.-T., & Fu, Y. (2022b). Towards automated urban planning: When generative and ChatGPT-like AI meets urban planning. *ACM Transactions on Spatial Algorithms and Systems*, *1*(1), *1*, 19. https://doi.org/10.1145/3524302

Wang, X., Song, Y., & Tang, P. (2020). Generative urban design using shape grammar and block morphological analysis. *Frontiers of Architectural Research*, *9*(4), 914–924. https://doi.org/10.1016/j.foar.2020.09.001

Weil, C., Bibri, S. E., Longchamp, R., Golay, F., & Alahi, A. (2023). Urban digital twin challenges: A systematic review and perspectives for sustainable smart cities. *Sustainable Cities and Society*, *99*, 104862.

White, G., Zink, A., Codecá, L., & Clarke, S. (2021). A digital twin smart city for citizen feedback. *Cities*, *110*, 103064. https://doi.org/10.1016/j.cities.2020.103064

Wolf, T., Debut, L., Sanh, V., Chaumond, J., Delangue, C., Moi, A., Cistac, P., Rault, T., Louf, R., Funtowicz, M., Davison, J., Shleifer, S., von Platen, P., Ma, C., Jernite, Y., Plu, J., Xu, C., Le Scao, T., Gugger, S., Drame, M., Lhoest, Q., & Rush, A. M. (2020). Transformers: State-of-the-art natural language processing. In: Proceedings of the 2020 Conference on Empirical Methods in Natural Language Processing: System Demonstrations (pp. 38–45). Association for Computational Linguistics (ACL). https://doi.org/10.18653/V1/2020.EMNLP-DEMOS.6

Wu, J., Wang, X., Dang, Y., & Lv, Z. (2022). Digital twins and artificial intelligence in transportation infrastructure: Classification, application, and future research directions. *Computers and Electrical Engineering, 101*, 107983. https://doi.org/10.1016/j.compeleceng.2022.107983

Wu, Y., Zhang, K., & Zhang, Y. (2021). Digital twin networks: A survey. *IEEE Internet of Things Journal, 8*, 13789–13804. https://doi.org/10.1109/JIOT.2021.3079510

Xie, Y., Wang, Z., Mai, G., Li, Y., Jia, X., Gao, S., & Wang, S. (2023). Geo-foundation models: Reality, gaps and opportunities. In: SIGSPATIAL '23: Proceedings of the 31st ACM International Conference on Advances in Geographic Information Systems (pp. 1–4). https://doi.org/10.1145/3589132.3625616

Xu, H., Omitaomu, F., Sabri, S., Li, X., & Song, Y. (2024). Leveraging generative AI for smart city digital twins: A survey on the autonomous generation of data, scenarios, 3D city models, and urban designs. https://doi.org/10.48550/arXiv.2405.19464

Xu, M., Niyato, D., Zhang, H., Kang, J., Xiong, Z., Mao, S., & Han, Z. (2023). Joint foundation model caching and inference of generative AI services for edge intelligence. https://doi.org/10.48550/arXiv.2305.12130

Yadav, R., Nascetti, A., Azizpour, H., & Ban, Y. (2024). Unsupervised flood detection on SAR time series using variational autoencoder. *International Journal of Applied Earth Observation and Geoinformation, 126*, 103635.

Yang, L., Li, J., Chang, H.-T., Zhao, Z., Ma, H., & Zhou, L. (2023a). A generative urban space design method based on shape grammar and urban induction patterns. *Land, 12*(6), 1167. https://doi.org/10.3390/land12061167

Yang, L., Zhang, Z., Song, Y., Hong, S., Xu, R., Zhao, Y., Zhang, W., Cui, B., & Yang, M. H. (2023b). Diffusion models: A comprehensive survey of methods and applications. *ACM Computing Surveys, 56*, 1–39.

Ye, X., Du, J., Han, Y., Newman, G., Retchless, D., Zou, L., Ham, Y., & Cai, Z. (2023). Developing human-centered urban digital twins for community infra-structure resilience: A research agenda. *Journal of Planning Literature, 38*(2), 187–199.

Yigitcanlar, T., Desouza, K. C., Butler, L., & Roozkhosh, F. (2020). Contributions and risks of artificial intelligence (AI) in building smarter cities: Insights from a systematic review of the literature. *Energies*. https://doi.org/10.3390/en13061473

Yigitcanlar, T., Mehmood, R., & Corchado, J. (2021). Green artificial intelligence: Towards an efficient, sustainable and equitable technology for smart cities and futures. *Sustainability, 13*(16), 8952. https://doi.org/10.3390/su13168952

Yin, X., Li, J., Kadry, S. N., & Sanz-Prieto, I. (2021). Artificial intelligence assisted intelligent planning framework for environmental restoration of terrestrial ecosystems. *Environmental Impact Assessment Review, 86*, 106493.

Yokoyama, A. M., Ferro, M., de Paula, F. B., Vieira, V. G., & Schulze, B. (2023). Investigating hardware and software aspects in the energy consumption of machine learning: A green AI-centric analysis. *Concurrency and Computation: Practice and Experience, 35*(24), e7825.

Yuan, J., Zhang, L., & Kim, C. (2023). Multimodal interaction of MU plant landscape design in marine urban based on computer vision technology. *Plants, 12*(7), 7.https://doi.org/10.3390/plants12071431

Yuan, L., Chen, D., Chen, Y.-L., Codella, N., Dai, X., Gao, J., Hu, H., Huang, X., Li, B., Li, C., Liu, C., Liu, M., Liu, Z., Lu, Y., Shi, Y., Wang, L., Wang, J., Xiao, B., Xiao, Z., Yang, J., Zeng, M., Zhou, L., & Zhang, P. (2021). Florence: A new foundation model for computer vision. arXiv. https://doi.org/10.48550/arXiv.2111.11432

Zaidi, A., Ajibade, S.-S. M., Musa, M., & Bekun, F. V. (2023). New insights into the research landscape on the application of artificial intelligence in sustainable smart cities: A bibliometric mapping and network analysis approach. *International Journal of Energy Economics and Policy*(4), 287.

Zayed, S. M., Attiya, G. M., El-Sayed, A., & Hemdan, E. E.-D. (2023). A review study on digital twins with artificial intelligence and internet of things: Concepts, opportunities, challenges, tools and future scope. *Multimedia Tools and Applications, 82*(47081–47107). https://doi.org/10.1007/s11042-023-15611-7

Zeroual, I., & Lakhouaja, A. (2018). Data science in light of natural language processing: An overview. *Procedia Computer Science, 127*, 82–91.

Zhai, W., Peng, Z. R., & Yuan, F. (2020). Examine the effects of neighborhood equity on disaster situational awareness: Harness machine learning and geotagged Twitter Data. *International Journal of Disaster Risk Reduction, 48*, 101611.

Zhang, K., Zhang, J., Li, C., Jiao, Y., & Wang, Y. (2021). Tourists' perceptions of urban space: A computer vision approach. *Tourism Review, 77*(4), 1203–1218. https://doi.org/10.1108/TR-06-2020-0287

Zhang, L., Han, X., Wu, J., & Wang, L. (2023). Mechanisms influencing the factors of urban built environments and coronavirus disease 2019 at macroscopic and microscopic scales: The role of cities. *Frontiers in Public Health, 11*, 1137489.

Zhang, Y., Ong, G. X., Jin, Z., Seah, C. M., & Chua, T. (2022). The effects of urban greenway environment on recreational activities in tropical high-density Singapore: A computer vision approach. *Urban Forestry & Urban Greening, 75*, 127678. https://doi.org/10.1016/j.ufug.2022.127678

Zhao, C., Ogawa, Y., Chen, S., Oki, T., & Sekimoto, Y. (2023). Quantitative land price analysis via computer vision from street view images. *Engineering Applications of Artificial Intelligence, 123*, 106294. https://doi.org/10.1016/j.engappai.2023.106294

Zheng, H., & Yuan, P. F. (2021). A generative architectural and urban design method through artificial neural networks. *Building and Environment, 205*, 108178. https://doi.org/10.1016/j.buildenv.2021.108178

Zheng, Y., Lin, Y., Zhao, L., Wu, T., Jin, D., & Li, Y. (2023). Spatial planning of urban communities via deep reinforcement learning. *Nature Computational Science*, *3*, 748–762. https://doi.org/10.1038/s43588-023-00503-5

Zhong, Z., Rempe, D., Xu, D., Chen, Y., Veer, S., Che, T., Ray, B., & Pavone, M. (2023). Guided conditional diffusion for controllable traffic simulation. In: 2023 IEEE International Conference on Robotics and Automation (ICRA) (pp. 3560–3566). IEEE

Zhou, C., Li, Q., Li, C., Yu, J., Liu, Y., Wang, G., Zhang, K., Ji, C., Yan, Q., He, L., Peng, H., Li, J., Wu, J., Liu, Z., Xie, P., Xiong, C., Pei, J., & Yu, P. S. (2023). A comprehensive survey on pretrained foundation models: A history from BERT to ChatGPT. https://arxiv.org/abs/2302.09419.

Zhou, Z., Lin, Y., & Li, Y. (2024). Large language model empowered participatory urban planning. https://doi.org/10.48550/arXiv.2402.01698

Zou, J., Kim, B., Kim, H., & Al-Hussein, M. (2012). An automated system for the creation of an urban infrastructure 3D model using image processing techniques. *KSCE Journal of Civil Engineering*, *16*(1), 9–17. https://doi.org/10.1007/s12205-012-1272-7

Zvarikova, K., Horak, J., & Downs, S. (2022). Digital twin algorithms, smart city technologies, and 3d spatio-temporal simulations in virtual urban environments. *Geopolitics, History and International Relations*, *14*(1), 139–154.

CHAPTER 8

Transforming Smarter Eco-Cities with Artificial Intelligence

Harnessing the Synergies of Circular Economy, Metabolic Circularity, and Tripartite Sustainability

Abstract

The urban landscape is experiencing a revolutionary transformation, fueled by the powerful convergence of Artificial Intelligence (AI) and the Internet of Things (IoT) into AIoT. This groundbreaking fusion is redefining how cities operate, paving the way for smarter, more sustainable urban environments. It promises innovative solutions to urban challenges, fostering the development of "smarter eco-cities" characterized by enhanced efficiency, resilience, and sustainability. However, there is a limited understanding of how AI and AIoT can leverage the synergistic potential of circular economy, metabolic circularity, and tripartite sustainability to advance smarter eco-cities. While these technologies are recognized as pivotal in this dynamic landscape, comprehensive

DOI: 10.1201/9781003536420-8

studies exploring their integrative role in enhancing these practices to boost the contribution of smarter eco-cities to sustainability goals are scant. Therefore, this study aims to explore the transformative role of AI and AIoT technologies in harnessing the synergies of circular economy, metabolic circularity, and tripartite sustainability to advance the development of smarter eco-cities toward achieving Sustainable Development Goals (SDGs). The findings indicate that AI and AIoT can significantly enhance data-driven decision-making, enabling smarter eco-cities to align with these goals. AI and AIoT technologies facilitate real-time data collection and analysis, supporting the circular economy's principles of resource regeneration and waste minimization. Additionally, they bolster urban metabolic circularity by optimizing the flow, reuse, and closed-loop systems of materials and energy within smarter eco-cities. Furthermore, they promote tripartite sustainability by balancing environmental, economic, and social dimensions, ensuring a holistic approach to urban development. The contributions of this study lie in providing a comprehensive examination of AI and AIoT's capabilities in fostering sustainable urban development and providing actionable insights and valuable guidance for researchers, practitioners, and policymakers. The implications are profound, suggesting that the strategic deployment of AI and AIoT technologies can lead to more sustainable, efficient, resilient, and inclusive urban environments. This work underscores the pivotal role of these advanced technologies in achieving urban sustainability goals and provides a framework for future research endeavors and practical implementations in the realm of smarter eco-cities.

8.1 INTRODUCTION

Cities account for about 75% of natural resource consumption, 75% of Greenhouse Gas (GHG) emissions, and over 50% of global waste production. They face numerous challenges due to rapid urbanization, escalating ecological degradation, and increasing resource scarcity. These challenges encompass energy consumption, material usage, waste generation, transport inefficiency, traffic congestion, infrastructure flaws, air pollution, biodiversity loss, and climate change risks. Therefore, the urban landscape is undergoing significant transformation, driven by the rapid advancement and convergence of data-driven technologies and approaches, which can provide the innovative solutions needed to tackle these urban challenges effectively. This surge of innovation has engendered a drastic change in urban development, giving rise to the visionary concept of "smarter eco-cities," (Bibri et al. 2023b, 2024a), poised to shape the urban landscape of the future. This change is propelled by the dynamic and seamless integration of two transformative forces: Artificial Intelligence (AI) and the Internet of Things (IoT), synergistically brought together under Artificial Intelligence of Things (AIoT). This has "reshaped the way we interact with technology, leading us into a future where everyday objects, devices, and systems are not only interconnected but also imbued with intelligence, autonomy, and unprecedented capabilities" (Koffka 2023).

With the synergistic potential of AIoT, the seeds of new approaches to sustainable urban development are sown. This promises profound implications for circular economy and Sustainable Development Goals (SDGs) (e.g., Agrawal et al. 2021; Ali et al. 2023; Akinode and Oloruntoba 2020; Lampropoulos et al. 2024; Rathore and Malawalia 2021; Siakas et al. 2023). This innovative approach emphasizes the integration of intelligent technologies to create more efficient, resilient, and sustainable urban environments. At its core, AIoT embodies a combination of AI's cognitive capabilities with IoT's pervasive connectivity and real-time data generation, creating a dynamic ecosystem where devices, sensors, and systems communicate intelligently, learn autonomously, and adapt purposefully. This ecosystem enables both smart cities and smarter eco-cities to optimize resource utilization, streamline waste management, reduce environmental impacts, enhance service delivery, and improve the quality of life (e.g., Bibri et al. 2023a, b, 2024b; El Himer et al. 2022; Fang et al. 2023; Olawade et al. 2024; Popescu et al. 2024; Rane et al. 2024; Thamik et al. 2024). AIoT has shown significant potential to harness the collection of real-time data from myriad IoT devices and sensors, which are fed into AI systems to enhance data-driven decision-making processes pertaining to numerous application urban domains.

In the realm of emerging smarter eco-cities, powerful AI models and techniques are used to derive valuable insights from data. These insights become the bedrock upon which the circular economy and tripartite sustainability practices stand. These two models are seen as systems solution frameworks that tackle global challenges. The circular economy model, often referred to as the "cradle-to-cradle" model, extends beyond the mere transformation of waste into valuable resources. It espouses a vision where resources are utilized, regenerated, and repurposed in a closed-loop system within urban environments (urban circularity), minimizing waste, reducing resource extraction, and ultimately contributing to environmental conservation. As such, it aligns well with the broader framework of tripartite sustainability, which seeks to harmonize environmental, economic, and social dimensions in the pursuit of holistic urban development (e.g., Dincă et al. 2022; Dong et al. 2021; Rodríguez-Anton et al. 2019). It embodies a sustainable and efficient economic model that demonstrates tripartite value and viability. It yields significant advantages across the three pillars of sustainability known as the Triple Bottom Line (TBL): economic, environmental, and societal benefits (Ma et al. 2020; Husain et al. 2021). In addition, it presents a structured strategy and an innovative solution for tackling urban sustainability challenges, which play a key role in advancing the United Nations' SDGs (Dong et al. 2021; Rosa et al. 2023; Schroeder et al. 2019).

As both smart cities and smarter eco-cities rapidly burgeon due to rapid urbanization, they consume vast amounts of resources, generate copious amounts of waste, and emit GHG at unprecedented rates. This unsustainable urban trajectory poses significant challenges to environmental equilibrium, social fabric, and economic stability. To mitigate these challenges, urban planners, policymakers, technologists, and scientists alike have been diligently seeking innovative strategies and solutions that can pivot modern cities toward a sustainable future. In this light, circular economy emerges as a potent approach to the linear "use-make-dispose" or "take-make-waste," revolving around resource renewal and ecosystem restoration. Indeed, the linear model, which dominates contemporary resource consumption, is characterized by heavy wastage, improper disposal, and high GHG emissions (Riesener et al. 2019). The circular model introduces a regenerative approach to urban resource management, encouraging the reuse and recycling of materials, prolonging product lifecycles, and minimizing waste generation. Unlike the linear model, the circular model is rooted in a systemic and holistic approach that seeks to emulate the open, diverse, adaptive, complex, resilient, and optimized characteristics of natural systems.

The circular model has garnered growing attention and acceptance as a viable means to reconcile urban development with environmental conservation toward achieving prosperity while recognizing the limits imposed by ecological and social factors. More importantly, deploying emerging technologies, especially AI, IoT, and big data analytics, offers promising solutions to overcome the practical challenges of implementing the circular model (e.g., Agrawal et al. 2021; Akinode and Oloruntoba 2020; Cîmpeanu et al. 2022; Lampropoulos et al. 2024; Ramadoss et al. 2018). Similarly, tripartite sustainability, a guiding principle for urban planning and development, posits that urban sustainability can only be achieved when its environmental, economic, and social dimensions are balanced and this process is supported by these advanced technologies, creating a synergy that fosters vibrant, resilient, and equitable smart(er) eco-cities (e.g., Bibri et al. 2023b, 2024a; Tarek 2023).

Furthermore, the United Nations has recognized the circularity of urban metabolism, driven by advanced technologies, as a crucial tool for achieving the SDGs (D'Amico et al. 2022). This approach is closely related to Smart Urban Metabolism (SUM) (Bibri and Krgostie 2020a; D'Amico et al. 2020; Shahrokni 2015), which involves the comprehensive analysis and optimization of energy and material flows within cities. It aims to understand how resources are consumed, transformed, and disposed of in urban areas, leveraging data-driven technologies to enhance efficiency, reduce waste, and improve sustainability and resilience. Urban metabolic circularity, on the other hand, focuses on creating closed-loop systems where waste and byproducts are minimized and repurposed. This concept emphasizes the recycling, reuse, and regeneration of materials and energy within cities, aiming for sustainability by reducing reliance on external resources and minimizing environmental impact. In essence, SUM is about analyzing and optimizing urban resource flows using advanced technologies, while urban metabolic circularity is about designing these flows to be circular, minimizing waste and maximizing resource reuse.

While SUM focuses on understanding and quantifying the flows of resources and waste within cities, typically in a linear fashion, urban metabolic circularity emphasizes managing a city's resources by applying this understanding to create more sustainable, efficient, resilient, and integrated urban systems, modeled after natural metabolic and circular processes. This approach is increasingly being enhanced by AI and AIoT technologies within the SUM framework (Bibri et al. 2024a; Peponi et al. 2022; Ghosh and Sengupta 2023). Moreover, the digitalization of circularity is advocated as a foundational strategy to steer SUM toward more sustainable and automated pathways. This entails optimizing the circulation and enhancement of materials, waste, byproducts, emissions, energy, knowledge, information, and data (D'Amico et al. 2022; Levoso et al. 2020; Venkata Mohan et al. 2020; Yao et al. 2020) in the context of emerging smarter eco-cities (Bibri et al. 2024a). AI and AIoT entail efficient data collection, adaptive algorithm development, and meticulous data engineering, resulting in rapid problem-solving. This approach minimizes the chances of errors, delays, and resource commingling, thus facilitating the transition to a circular economy (Rusch et al. 2023) and hence urban metabolic circularity.

However, there is a limited understanding of how AI and AIoT can leverage the synergistic potential of circular economy, metabolic circularity, and tripartite sustainability to advance smarter eco-cities. While these technologies are recognized as pivotal in this dynamic landscape, comprehensive studies exploring their integrative role in enhancing these practices to boost the contribution of smarter eco-cities to sustainability goals are scant. This gap underscores the need for research on the strengths and capabilities of AI and AIoT in addressing sustainability challenges in smarter eco-cities. Therefore, this study aims to explore the transformative role of AI and AIoT technologies in harnessing the synergies of circular economy, metabolic circularity, and tripartite sustainability to advance the development of smarter eco-cities toward achieving SDGs.

The study makes several significant contributions to the field of sustainable urban development:

- Provides a comprehensive examination of AI and AIoT's capabilities in fostering sustainable urban development.
- Highlights the practical applications and benefits of AI and AIoT technologies in enhancing urban sustainability.
- Emphasizes the importance of integrating circular economy principles, metabolic circularity, and tripartite sustainability frameworks in urban development.
- Demonstrates how AI and AIoT technologies can optimize resource utilization, waste management, and environmental impacts.
- Underscores the potential of AI and AIoT to support the principles of circular economy, including resource regeneration and waste minimization.
- Showcases the ability of AI and AIoT to enhance urban metabolic circularity by optimizing the flow and reuse of materials and energy.

This chapter is structured as follows: Section 8.2 introduces key conceptual definitions and discusses their relationships. Section 8.3 outlines and describes the research design. Section 8.4 presents the results. Section 8.5 provides a detailed discussion, covering various types of challenges, limitations, and future research directions. Section 8.6 introduces an innovative framework for integrating AI and AIoT with circular economy, metabolic circularity, and tripartite sustainability. This chapter ends, in Section 8.6, with a summary of key findings, contributions, implications, and final thoughts.

8.2 CONCEPTUAL DEFINITIONS AND DISCUSSIONS

Understanding the core concepts and their interrelationships is crucial for advancing the discourse on sustainable urban development. This section delves into the fundamental ideas underpinning urban circularity and circular economy, exploring their commonalities and differences. Examining the integration of circularity and circular economy principles within the framework of SUM can provide a deeper understanding of how these concepts converge to support sustainable transformations in the context of smarter eco-cities. Furthermore, this section highlights the interplay between smart urban metabolic circularity and tripartite sustainability, elucidating how technological advancements and innovative practices can drive cities toward more resilient, efficient, and environmentally balanced futures. Tripartite sustainability is the foundation of smart sustainable cities, which serve as an overarching umbrella term encompassing smart eco-cities.

8.2.1 Urban Circularity and Circular Economy: Commonalities and Differences

Circularity and circular economy form a symbiotic relationship in the pursuit of sustainability. This suggests that these two concepts are closely bound together and have profound interdependence, emphasizing the significance of their interplay in advancing the goals of sustainability. They advocate for sustainable practices that prioritize minimizing waste and maximizing the recycling or regeneration of resources. Circularity encapsulates the broader ethos of minimizing waste and optimizing resource use. In contrast, the circular economy operationalizes this concept within economic systems, fostering practices that prioritize the continual use, recycling, and regeneration of materials. Essentially, circularity provides the guiding principle beyond the economic sector of society, while the circular economy offers the actionable framework for integrating sustainable practices into various industries and business domains. Together, they form a cohesive approach toward creating more resilient and environmentally responsible systems.

Circularity is a broader concept that encompasses the principles of circular economy. It refers to the idea of creating closed-loop sustainable systems where products, materials, and resources are constantly reused and recycled rather than disposed of as waste. It involves designing processes and systems in a way that minimizes waste and maximizes sustainability. It can be applied in various contexts and domains, including urban planning, urban development, and environmental management. The application of circularity to the urban setting is aimed at fostering sustainability through the efficient use of resources at different spatial scales, including household, building, district, community, city, and region.

Circular economy is a model of production and consumption, in which products, materials, and resources are used efficiently, shared, leased, reused, repaired, refurbished, recycled, and recovered as long as possible in a closed-loop system. It is defined as "a strategy which aims at reducing both inputs of virgin materials and output of wastes by closing economic and ecological loops of resource flows" (Husain et al. 2021). This results in reducing the need for new resource extraction, contributing to environmental conservation, and minimizing ecological degradation.

Circular economy focuses on maintaining the value of products and materials for as long as possible in the economic system. It is a systematic economic approach designed to minimize waste generation and resource depletion, preserving the value of resources throughout their lifecycle, with the overarching goal of fostering sustainable production and consumption patterns. It centers on several key principles and strategies to achieve these objectives (e.g., Dincă et al. 2022; Dong et al. 2021; Kirchherr et al. 2017; Lantto et al. 2019; Merli et al. 2018; Rodríguez-Anton 2019; Rosa et al. 2023; Schroeder et al. 2019; Trica et al. 2019; Velte et al. 2018), including:

- Minimizing raw material use: Instead of relying heavily on raw materials, the circular economy model emphasizes the importance of reducing the consumption of primary resources. This involves finding ways to use fewer natural resources in the production process.

- Extending product lifespan: To reduce waste, products are designed to have longer lifespans. This includes the use of durable materials, modular designs, and repairable components to extend the useful life of products.
- Resource efficiency: Maximizing the efficiency of resource utilization is a central tenet of the circular economy model. This entails optimizing production processes to minimize waste and inefficiencies while maximizing the utilization of resources.
- Recycling: Recycling plays a crucial role in the circular economy by reintroducing materials back into the production cycle. This reduces the need for new raw materials and minimizes waste disposal.
- Reusing: Encouraging the reuse of products and components whenever possible involves refurbishing, repairing, or repurposing items to extend their use rather than discarding them.
- Remanufacturing: Remanufacturing involves disassembling and refurbishing products to like-new condition. This not only extends the life of products but also reduces the environmental impact compared to manufacturing entirely new items.
- Waste reduction: The circular economy model prioritizes the reduction of waste at every stage of a product's lifecycle. This includes minimizing waste during production, use, and disposal.
- Closed-loop systems: In circular economy, products are designed with the end in mind. This means considering how products will be disassembled, reused, or recycled at the end of their life, creating closed-loop systems where resources are continually cycled.
- Consumer awareness: Education and awareness campaigns are essential components of the circular economy model. Consumers are encouraged to make sustainable choices by selecting products with longer lifespans and considering the environmental impact of their consumption habits.
- Policy and industry support: Governments and industries play a pivotal role in facilitating the transition to a circular economy through regulations, incentives, and investments in research and development.

Overall, circular economy seeks to transform traditional linear consumption patterns (take, make, dispose) into a circular and regenerative system that minimizes waste, conserves resources, and promotes sustainable economic growth.

Various researchers have formulated the circular economy concept based on diverse principles, all with the shared objective of crafting a regenerative economic system aimed at attaining sustainable development through the promotion of environmental quality, economic prosperity, and social equity. In this context, there are various frameworks applied in circular economy, known as R-principles (D'Amico et al. 2021; Kirchherr et al. 2017; Potting et al. 2017; Sihvonen and Ritola 2015). These include 3Rs (recycle, reuse, reduce), 6Rs (reuse, reduce, redesign, remanufacture, recycle, recover), 4R (Reduce, Reuse, Recycle, and Recover), and 9/10Rs (reduce, recycle, reuse, remanufacture, rethink, refuse, refurbish, repair, repurpose, and recover). This has created confusion in the field, leading to numerous definitions of the circular economy concept. Kirchherr et al. (2017) highlight the lack of consensus and clarity in the understanding of this concept among scholars and practitioners, emphasizing the need for greater transparency. The authors identify and analyze 114 circular economy definitions, revealing that it is most commonly associated with "reduce, reuse, and recycle" activities, often neglecting the need for a systemic shift. Additionally, there are limited explicit connections between the circular economy and sustainable development, with economic prosperity being the primary focus, followed by environmental quality. Social equity and future generations receive little attention, and business models and consumers are rarely seen as contributors to the circular economy. This work contributes to a better understanding of the circular economy concept and its various interpretations. It underscores the importance of achieving coherence in these definitions, as significant variations may undermine the concept's integrity.

8.2.2 Circularity, Circular Economy, and Smart Urban Metabolism

The principles of circularly and circular economy, when applied practically, guide the development of eco-friendly and sustainable urban economies. In the urban context, urban circularity is typically the implementation of circular economy principles in urban environments, focusing on resource efficiency, waste reduction, and the closing of material loops. Urban circularity initiatives often leverage digital technologies and data-driven approaches to optimize resource use, monitor waste streams, and facilitate circular practices. It aims to holistically connect the built, natural, and digital environment (D'Amico et al. 2022).

A strong foundation is crucial for promoting practices such as waste reduction, recycling, and the repurposing of resources, leading to the creation of practical, circular economic systems in smarter eco-cities. Unlike the traditional linear economy, where resources are extracted, used, and then discarded as waste in cities, a circular economy approach aims to minimize waste and make the most of resources in urban environments. Urban metabolism and circularity are closely linked concepts in sustainable urban development. These concepts aim to create resilient cities that efficiently manage resources and minimize waste, thereby enhancing sustainability and reducing ecological footprint. The focus on reusing and maximizing resources aligns closely with the goals of SUM: improve the management of resources by understanding where inefficiencies and waste occur and providing data and insights to inform urban planning and policy decisions aimed at making cities more sustainable (Bibri et al. 2024a).

SUM marks a notable leap forward in urban metabolism by comprehending resource consumption, waste generation, and environmental impact using AI and AIoT technologies (e.g., Ghosh and Sengupta 2023; Peponi et al. 2022) in smarter eco-cities (Bibri et al. 2024a). Building on the foundations of urban metabolism (Kennedy et al. 2007; 2012), SUM incorporates advanced technologies and data analytics to refine resource management and boost urban sustainability and resilience. This advancement allows cities to more effectively monitor and assess their metabolic functions, facilitating informed decisions, precise interventions, and the development of policies aimed at resource efficiency, waste minimization, and environmental protection.

By leveraging advanced technologies to monitor and manage urban flows, SUM represents a practical application of circular economy principles at the urban scale. This synergy enhances the ability of cities to not only implement sustainable practices but also improve overall urban efficiency and resilience. Expanding upon these circular economy principles, the innovative SUM framework integrates AI and AIoT technologies, enabling a more dynamic approach to urban development. SUM provides policymakers, urban planners, urban managers, and municipal utilities with the essential tools to gather, monitor, analyze, and assess the circularity of environmental, social, and economic resources in smarter eco-cities, leading to more effective urban planning and governance.

These stakeholders acknowledge the significance of advanced digital technologies and infrastructures in optimizing the efficiency of circularity across several dimensions of SUM, including governance, environment, energy, waste, water and wastewater, and health (D'Amico et al. 2020, 2021; Shahrokni et al. 2015).

D'Amico et al. (2021) examine the circularity of SUM in terms of digitalization practices using the "Circular Resource Efficiency Management Framework." The authors analyze initiatives implemented by various municipalities within the Partnership on Circular Economy of the Urban Agenda of the European Union. The findings reveal diverse practices such as real-time monitoring for water and energy, digital cameras for traffic control, web platforms for resource sharing, and tracking sensors for public transport. These initiatives collectively aim to enhance the efficiency of urban metabolic flows. Hence, the fusion of the circular economy with advanced technologies presents numerous benefits for enhancing the efficiency of digitalized UM or SUM.

Furthermore, as cities pursue more sustainable futures, two key concepts—metabolic circularity and circular metabolism—have emerged to guide this transformation (e.g., D'Amico et al. 2022; Lucertini, & Musco 2020; Cui 2020). Both approaches are integrally used to create sustainable, closed-loop systems where resources are efficiently used, reused, and recycled, minimizing waste. Metabolic circularity draws on the analogy of cities functioning like living organisms, where energy, water, and material flows are optimized much like natural ecosystems. It focuses on fostering holistic urban systems that mimic nature's efficient cycles, integrating not only infrastructure but also community and ecological aspects. Circular metabolism, on the other hand, is more focused on the technical management of these resource flows within urban systems, applying circular economy principles—such as waste-to-resource cycles, energy efficiency, and material recovery—to make cities more sustainable. Cui (2022) proposes a Circular Urban Metabolism (CUM) framework to assess resource use patterns and identify circular development potential in urban areas. The study highlights the significant resource and environmental challenges posed by large populations and economic activity, making the circular economy critical for sustainability. The CUM framework, applied to Shanghai, analyzes sectoral material flows, resource use, and circularity potential, revealing a dramatic increase in material inputs and outputs from 2000 to 2019. The study identifies construction-related non-metallic mineral use and energy-induced pollution as major obstacles to circularity and recommended sector-specific policies to improve resource efficiency at the city level.

The relationship between the two is complementary: metabolic circularity provides the overarching ecological framework, while circular metabolism implements the technical solutions to achieve those goals within urban infrastructure and planning. The current study explores aspects of both, highlighting how AI and AIoT technologies can integrate ecological models with circular resource management to create smarter eco-cities. In these cities, real-time monitoring, smart grids, and data-driven platforms enhance the efficiency of urban metabolic flows, fostering resilience and sustainability through both natural-inspired cycles and technical innovation.

8.2.3 Smart Urban Metabolic Circularity and Tripartite Sustainability

The circularity of urban metabolism builds on the framework of urban metabolism and applies the principles of circular economy by emphasizing the reuse, recycling, and regeneration of resources within an urban ecosystem. It aims to minimize waste and maximize resource efficiency by creating closed-loop systems. This approach focuses on circularity, systems integration, and a holistic approach (e.g., integrating water, energy, and waste systems) to create synergies and optimize overall resource flow. Its continuous cycle mirrors the natural metabolic processes found in living organisms, where waste products are reused for new growth.

In the urban context, metabolic circularity aims to minimize waste and maximize resource efficiency by creating systems where outputs from one process can be inputs for another, effectively reducing the overall environmental footprint. This approach fosters sustainable urban development by integrating ecological considerations into economic and social activities, ensuring that cities can support their citizens without depleting natural resources or causing ecological damage. Eco-cities are urban systems designed so that the inflows of materials and energy and the disposal of wastes do not exceed the capacity of their surrounding environments (Kennedy et al. 2007). Digital technologies and circularity principles play a crucial role in achieving this balance (D'Amico et al. 2022) in the context of smarter eco-cities.

Smart urban metabolic circularity builds on the concept of metabolic circularity by incorporating advanced technologies such as AI, IoT, and data analytics—in short, AIoT—to enhance the efficiency and effectiveness of circular economy practices within urban ecosystems. This approach utilizes smart technologies to monitor, control, and optimize the flows of energy, materials, and resources, ensuring that they are reused and recycled at their highest utility and value. In smart urban metabolic circularity, technology plays a critical role in:

- Monitoring resource flows in real-time to detect inefficiencies and opportunities for reuse.
- Analyzing data to identify patterns and optimize systems for better resource allocation.
- Automating processes to ensure the continuous and efficient reuse and recycling of materials and energy.

SUM circularity offers a valuable resource for policymakers, urban managers, planners, and administrators, enabling them to identify, manage, and assess a broad spectrum of data related to the flows of social, environmental, and economic resources (D'Amico et al. 2022). Its goals include enhancing urban sustainability, increasing the resilience of urban systems by creating more self-sufficient and adaptive resource flows, and promoting innovative solutions and technologies that enable the circular use of resources in cities. This entails intelligently integrating economic, environmental, and social dynamics into urban operational management, planning, and governance.

The broader tripartite sustainability framework encompasses three key dimensions of sustainability—environmental, economic, and social. It seeks to achieve a balance and harmony between these three dimensions in the context of urban development. In essence, it aims to create cities that are environmentally friendly, economically viable, and socially inclusive. AI and AIoT significantly contribute to urban metabolic circularity within the framework of tripartite sustainability. Regarding environmental sustainability, AI and AIoT can significantly enhance the environmental aspect of urban metabolism by optimizing resource use and reducing waste. AI algorithms can predict patterns in energy consumption and waste production, while IoT devices can monitor and control the flow of resources in real-time (Bibri et al. 2024a; Ghosh and Sengupta 2023). AI can also forecast and mitigate potential environmental impacts by analyzing vast datasets from various urban activities, helping cities to maintain ecological balances and reduce their carbon footprints.

Economically, AI and AIoT drive efficiency and foster economic growth by streamlining urban operations and reducing costs. AI can optimize supply chains and logistics for urban businesses, reducing overhead costs and increasing profitability. Smart grids and AI-managed energy systems can reduce energy expenditures for cities by dynamically adjusting energy production and distribution based on actual demand. This saves costs and attracts businesses looking for sustainable and technologically advanced locales. From a social perspective, urban metabolic circularity supported by AI and AIoT ensures that urban environments are livable and equitable. AI technologies can help in designing cities that are more accessible and inclusive, using data to optimize public transportation routes, improve public safety, and ensure that all citizens have access to essential services. AIoT enhances community engagement by facilitating more responsive governance systems where feedback from citizens is quickly integrated into urban planning through real-time data collection (Bibri et al. 2024a).

Overall, AI and AIoT are pivotal in advancing urban metabolic circularity by creating systems that are not only self-sustaining and efficient but also adaptable to the needs of their citizens and the pressures of global environmental challenges. This technology-driven approach aligns with the principles of tripartite sustainability, promising a balanced and forward-thinking development model for cities of the future.

8.2.4 Smart Sustainable Cities: A Tripartite Sustainability Perspective

In the evolving landscape of urban development, smart eco-cities and smart sustainable cities have emerged as central paradigms of urbanism in recent years. These city models represent a shift toward integrating advanced technologies with sustainable practices to create cities that are more efficient, resilient, inclusive, and environmentally friendly. They pave the way for a future where urban areas can thrive sustainably and intelligently. They are deeply interlinked, each striving to redefine the conventional paradigms of eco-cities and sustainable cities in pursuit of a holistic approach to achieving urban sustainability (Bibri 2021a, b). At the heart of this evolution lies the convergence of three dimensions of sustainability—where cities are not merely hubs of technological innovation but also stewards of resources and well-being. In this context, the concept of smart sustainable cities takes center stage, serving as the bridge that connects the ideals of smart cities, eco-cities, and sustainable cities while fostering a symbiotic relationship between sustainable development and technological development. In this tapestry of urban transformation, the interplay between these concepts paints a promising picture of urbanization that is intelligently and environmentally regenerative, prioritizing the well-being of people and the responsible use of resources. Regenerative urbanization aims to create urban environments that contribute positively to the natural ecosystem, promote biodiversity, improve air and water quality, and support sustainable living practices.

The relationship between sustainable cities (eco-cities) and smart sustainable cities lies in their shared goal of achieving sustainability through the integration of advanced technologies and data-driven solutions. In essence, smart sustainable cities represent an evolution of sustainable cities, leveraging technological innovations to tackle urban challenges more effectively in an increasingly connected and urbanized world (Bibri 2019, 2021b). Smart sustainable cities combine the principles of sustainable cities (eco-cities) and smart cities. These urban areas utilize digital and green technologies, integrated approaches, and innovative strategies to address a range of environmental, economic, and social challenges. By merging these elements, they aim to create urban environments that are not only technologically advanced but also environmentally sustainable, socially inclusive, and economically viable. They have emerged in response to the growing critique of the fragmented development of sustainable cities and smart cities, particularly in light of the emergence and widespread diffusion of the UN SDGs. This fragmentation has been identified as a significant barrier to progress toward achieving the targets outlined in SDG 11, which focuses on making cities inclusive, safe, resilient, and sustainable with support of advanced ICT.

The distinction and demarcation between sustainable cities and smart cities has sparked extensive research, prompting suggestions to embrace the concept of smart sustainable cities as a more suitable concept (Ahvenniemi et al. 2017). It is noteworthy that sustainable cities and smart cities exhibit notable disparities across several dimensions and therefore do not necessarily converge in their approaches, priorities, goals, and visions (Ahvenniemi et al. 2017; Bibri 2019, 2021a, b; Martin et al. 2018; Yigitcanlar et al. 2019). A substantial body of research underscores the prevailing emphasis of existing smart cities on economic imperatives, often at the expense of addressing environmental and social issues (Cugurullo 2018; Evans et al. 2019; Trencher 2019; Verrest and Pfeffer 2019). In response to these challenges, recent research has increasingly interwoven the concepts of smartness and sustainability, developing novel strategies for smart city development that address these concerns (Mishra et al. 2022; Shariffi et al. 2024; Visvizi and del Hoyo 2021).

In view of the above, the concept of smart sustainable cities embodies a tripartite perspective of sustainability, integrating its environmental, economic, and social dimensions (Table 8.1). Serving to gain a comprehensive and holistic understanding of sustainability, this approach to urban development recognizes that complex issues often have multiple dimensions that need to be integrated in order to arrive at a more nuanced form of knowledge and a holistic solution. By adopting a tripartite perspective, researchers, practitioners, policymakers, and decision-makers can avoid oversimplification and consider the intricate and dynamic interplay of multiple dimensions that contribute to a more balanced assessment and integration of sustainability dimensions. Achieving sustainability means finding a balance between these dimensions, where progress in one area does not come at the expense of the others. The tripartite approach entails that these dimensions are interconnected and mutually reinforcing, and that addressing one dimension in isolation may lead to unintended consequences in the others—as documented by numerous studies since the widespread diffusion of sustainable development in the early 1990s (see Bibri 2021a, b for a detailed discussion).

In the context of advancing smart sustainable cities with a strong emphasis on environmental sustainability, smart eco-cities come to the

TABLE 8.1 Tripartite Sustainability in the Development of Smart Sustainable Cities

Environmental Sustainability	Social Sustainability	Economic Sustainability
This dimension focuses on the impact of human activities on the natural environment. It involves strategies that aim to minimize negative environmental impacts and promote the conservation of ecosystems, biodiversity, and natural resources. It emphasizes mitigating and adapting to climate change impacts and promoting sustainable practices to maintain ecological balance.	This dimension centers on the well-being and quality of life of individuals and communities. It involves promoting equity, social justice, inclusivity, and addressing issues such as inequality and social disparities, and hence ensuring equal access to resources, services, and opportunities for all members of the community without compromising the well-being of future generations.	This dimension focuses on the long-term viability of economic systems. It involves practices that support economic growth, innovation, and prosperity while also considering the efficient use of resources, minimizing waste, and avoiding overexploitation. Economic sustainability seeks to ensure that economic activities are balanced with the well-being of society and the environment.

forefront. These two emerging paradigms of urbanism coexist and complement each other in the pursuit of urban development that prioritizes ecological well-being alongside economic and social progress. Building on eco-cities, smart eco-cities aspire to establish a comprehensive and interconnected ecosystem that harmonizes environmental preservation, social justice, and economic growth (e.g., Caprotti 2020; Tarek 2023). These aspirations draw inspiration from the principles of urban ecology (Roseland 1997), ecological sustainability (Holmstedt et al. 2017), or urban sustainability (Kenworthy 2006; Register 2006) that underlie the eco-city model.

Both paradigms share a fundamental goal: leveraging digital technologies and intelligent systems to optimize resource management, improve urban infrastructure, and enhance the quality of life for residents. However, smart sustainable cities take this a step further by aiming to develop urban ecosystems that are resilient, resource-efficient, and capable of addressing the challenges of urbanization while ensuring the well-being of current and future generations. Central to SDGs is the integration of environmental considerations into economic decision-making, the promotion of social inclusivity within environmental and economic initiatives, and the equitable distribution of benefits and opportunities. The strategies and technologies used to achieve these goals can vary significantly from one smart sustainable city to another. These variations are influenced by the specific challenges, resources, technological capabilities, and objectives unique to each urban area. Despite these differences, the primary objective remains the same: to create urban environments that are technologically advanced, environmentally sustainable, socially inclusive, and economically viable (see Figure 8.2). The "United for Smart Sustainable Cities (U4SSC)" initiative aims to provide guidance for cities worldwide. This initiative helps cities become smarter and more sustainable while accelerating their digital transformation in alignment with the SDGs (ITU 2022).

As depicted in Figure 8.2, the tripartite sustainability framework hinges upon the convergence and intersection of three pivotal dimensions, each playing a crucial role in shaping a sustainable future through intentional actions, decisions, and outcomes. Within this framework, at the core of the convergence, lie three interconnected and compound forms of sustainability:

Socioeconomic sustainability constitutes a multifaceted concept that encompasses the intricate interplay or the inextricable link between social well-being and economic prosperity. It underscores the imperative to cultivate societies and economies in ways that prioritize the welfare of individuals, communities, and future generations while simultaneously fostering economic growth and stability. Socioeconomic sustainability acknowledges that economic growth should be a means to enhance the lives of people and communities while respecting the constraints and boundaries of the environment. To attain socioeconomic sustainability, it is necessary to take into account the seamless integration of social and economic factors in decision-making and policy formulation (Figure 8.1).

Enviro-economic sustainability adopts a comprehensive approach, aiming to strike a balance between economic prosperity and environmental preservation. This perspective underscores the necessity of developing and managing economies with a long-term vision of economic growth and stability while safeguarding our invaluable natural resources, ecosystems, and the overall health of our planet. Enviro-economic sustainability recognizes that the vitality of economic systems is inherently intertwined with the health of the environment. It places significant emphasis on the integration of environmental considerations into economic decision-making processes and policy development, ensuring that economic advancement and environmental protection are pursued hand in hand. To achieve

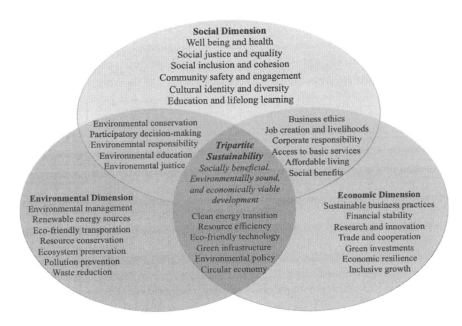

FIGURE 8.1 A framework for balancing the three dimensions of sustainability within smart sustainable cities.

enviro-economic sustainability, it is indispensable to carefully balance economic and environmental objectives.

Socio-environmental sustainability embodies a holistic approach that combines social well-being and environmental conservation as integral pillars of a sustainable future. It is dedicated to the notion of developing and managing societies and economies in ways that elevate human welfare, promote equity, and enhance overall quality of life while preserving and protecting our natural environment. It recognizes the dynamic interconnectedness between human societies and the natural world, emphasizing that the well-being of individuals is inextricably linked to the health and resilience of the environment. To realize socio-environmental sustainability, it is imperative to take into consideration the interdependency of social and environmental factors in decision-making and policy development.

8.2.5 From Eco-Cities to Smarter Eco-Cities: An Environmental and Technological Trajectory Perspective

In the realm of urban development, the concept of eco-cities has emerged as a response to the imperative of minimizing resource input and waste output in urban environments (Register 1987). Eco-cities are based on the idea of a living organism that monitors the inflow of resources and the outflow of products and waste (Bai 2007). They embody the principles of urban metabolism, a framework for understanding and analyzing the flows of energy, materials, water, nutrients, waste, and emissions within cities. They encompass a diverse array of strategies and solutions or urban-ecological approaches striving for achieving urban sustainability. These approaches entail a variety of policies targeting environmental, social, and institutional aspects, all aimed at effectively managing urban spaces to achieve sustainable outcomes. However, eco-cities prioritize environmental sustainability and ecological balance (Holmstedt et al. 2017; Mostafavi and Doherty 2010; Rapoport and Verney 2021), seeking to reduce their carbon footprint while fostering harmony between human activities and the natural environment. Principles such as energy efficiency, renewable energy adoption, waste reduction, green spaces, sustainable transportation, and resource preservation constitute the foundation of eco-cities (Bibri 2020, 2021a). Their overarching goal is to establish resilient, low-carbon, and environmentally friendly and community-oriented urban spaces that support the well-being of ecosystems and people alike (De Jong et al. 2015).

Eco-city initiatives are progressively embracing data-driven technologies to enact environmental improvements. This is achieved through the fusion of the advantages offered by smart cities and eco-cities, leveraging the synergies of their solutions and strategies under smart eco-cities (Bibri 2021a), thereby empowering eco-cities to enhance their environmental sustainability performance. In the trajectory of urban development, the paradigm of smart eco-cities has risen as a transformative model for eco-urbanism, blending the principles of urban ecology and technological advancement. Urban ecology, which is concerned with the interactions between humans and their environment within urban areas, is key to achieving eco-cities (Roseland 1997). It focuses on understanding how natural ecosystems function and adapt in urban environments, as well as the impact of human activities on these ecosystems. It aims to develop strategies for sustainable development, biodiversity conservation, and well-being improvement in cities while minimizing negative environmental impacts.

Furthermore, smart eco-cities harness the potential of IoT and Big Data technologies in conjunction with environmental strategies to achieve urban sustainability (Bibri and Krogstie 2020a; Caprotti et al. 2017; Shahrokni 2015; Späth 2017). They represent urban environments where data-driven and environmental technologies synergize to advance energy efficiency, renewable energy deployment, waste management, metabolic flow analysis, and sustainable planning, thereby reducing environmental impact, conserving resources, and promoting sustainable urban living (see Bibri and Krogstie 2020a for illustrative case studies). Their holistic approach extends to the incorporation of nature-based solutions and community engagement in environmental stewardship, fostering urban environments that prioritize nature conservation and promote long-term sustainability and resilience. This approach underscores a commitment to creating urban spaces that embrace the coexistence of natural ecosystems and advanced technologies.

The concept of "smarter eco-cities" has recently emerged as an evolution of smart eco-cities, leveraging the potential of AI and AIoT technologies to maximize sustainable systems' performance and amalgamating them with smart systems (Bibri et al. 2023b). This integration aims to yield amplified effects, enhancing the benefits of environmental sustainability through the synergistic integration of urban ecology and urban computing paradigms. Smarter eco-cities draw inspiration from both smart city systems and eco-city systems, utilizing AI and AIoT technologies to enhance their environmental performance and tackle the growing complex challenges they are facing. As such, they are characterized by their innovative sustainability practices, improved sustainability outcomes, synergistic integration of smart city and eco-city systems, optimized system performance, and enhanced urban management, planning, and governance practices (Bibri et al. 2023b).

In essence, the journey from eco-cities to smart eco-cities and then to smarter eco-cities represents a continuum of urban development models that intertwine environmental sustainability with technological innovation. This evolution underscores the role of technology in reshaping urban environments to be both ecologically balanced and technologically advanced, serving as an experimental ground for innovative urban approaches in the pursuit of sustainability transitions.

In summary, the exploration of these conceptual definitions provides a comprehensive understanding of the theoretical foundations necessary for implementing sustainable development strategies in smarter eco-cities as an approach to smart sustainable urbanism. Distinguishing between urban circularity and circular economy and examining their integration with SUM uncovers the potential for creating more sustainable and resilient cities. The focus on smart urban metabolic circularity and its alignment with tripartite sustainability as a foundation of smart sustainable cities further emphasizes the importance of a holistic approach in addressing the multifaceted challenges of urbanization, resource scarcity, and ecological degradation. This foundational knowledge sets the stage for exploring innovative solutions and practical applications in the pursuit of smarter, eco-friendly, and environmentally conscious urban environments.

8.3 RESEARCH DESIGN

The aim of this study is to explore the transformative role of emerging AI and AIoT in harnessing the synergies among circular economy, metabolic circularity, and tripartite sustainability to advance the development of smarter eco-cities toward achieving SDGs. To achieve this aim, an overarching methodology rooted in a thematic review approach was designed and employed. This methodology integrates aggregative and configurative synthesis techniques to develop a conceptual framework (Figure 8.2). This framework aims to illustrate how AI and AIoT technologies can effectively contribute to urban environments by optimizing resource efficiency, reducing waste, and improving the overall resilience and sustainability in smarter co-cities.

FIGURE 8.2 A flow chart outlining the process of conducting the thematic review using aggregative and configurative synthesis techniques to develop the conceptual framework.

Thematic review is designed to organize and synthesize extant research around key themes that emerge from the existing body of literature, rather than offering a comprehensive review of the existing literature on a particular topic. It allows to gain a deeper understanding of the concepts, theories, models, and challenges within a specific interdisciplinary field by examining the contributions of multiple studies to these themes. Accordingly, it facilitated the organization and synthesis of a broad spectrum of the existing literature, research findings, and empirical evidence related to smarter eco-cities, smart sustainable cities, AI/AIoT, circularity, circular economy, and tripartite sustainability. Analyzing and structuring the identified studies around themes serves to effectively identify the intersections, interlinkages, patterns, and gaps within these domains in the research landscape.

A comprehensive search of academic articles, conference papers, reports, case studies, and other relevant material from diverse sources was conducted. Subsequently, data extraction techniques were employed to gather pertinent information from the selected sources. Through rigorous analysis, the findings were synthesized to identify commonalities, implications, gaps, and critical intersections between the key topics in focus. Thus, the diverse findings were not only consolidated but also meaningful insights and conclusions were derived from the large combined body of work.

Aggregative synthesis is a method that entails the systematic collection and integration of numerous research studies to create a cohesive and holistic summary of their outcomes. The interpretation stage follows the synthesis process, culminating in a thematic summary that frames the findings of the synthesized studies (Cooper 2017). In essence, aggregative synthesis accumulates and integrates this evidence to formulate statements that align with specific conceptual categories or frameworks (Gough et al. 2017). As an initial phase in the thematic review, this approach involves the analysis of a wide range of systematically garnered evidence, allowing to identify themes that shed light on the intersection of the key topics involved in the study.

Configurative synthesis extends beyond the mere aggregation of research findings. In essence, it involves a comprehensive process of interpretation during synthesis to unveil the underlying patterns, arrangements, and relationships among the research findings (Dixon-Woods et al. 2005). Thematic configuration seeks to uncover the broader context and establish an overarching meaning, i.e., thematic synthesis (Gough et al. 2017). This is to construct a conceptual framework that explains how the findings of the synthesized studies are interconnected and where variations exist. The main objective is to gain a deeper understanding of the research evidence by exploring the contextual relationships that influence the findings. The configurative synthesis phase involved an in-depth examination of the synthesized studies to understand the nuanced relationships, interdependencies, and potential causal links between the main topics of the study.

Drawing upon the insights derived from both the aggregative and configurative synthesis phases, a comprehensive conceptual framework was developed. By employing this integrated approach of thematic review, aggregative synthesis, and configurative synthesis, a holistic understanding was sought of how AI and AIoT technologies can serve as transformative forces, advancing the development of emerging smarter eco-cities by harnessing the synergies between metabolic circularity and tripartite sustainability.

8.4 RESULTS

This section endeavors to unravel the trajectories, nuances, and complexities surrounding the paradigm of smarter eco-cities in relation to AIoT, circular economy, and tripartite sustainability. By examining the evolution from traditional eco-cities to smarter eco-cities, this discussion highlights the integration of advanced technologies and sustainable practices aimed at enhancing urban intelligence. The focus is on understanding how AIoT-powered innovations and principles of industrial ecology can drive circular futures, ensuring that cities not only meet environmental sustainability goals but also foster technological innovation and resilient systems. This analysis aims to provide insights into the dynamic interplay of environmental stewardship, technological advancement, and sustainable urban development.

8.4.1 Artificial Intelligence and Artificial Intelligence of Things for Metabolic and Circular Futures in Smarter Eco-Cities

Eco-cities aim to reduce resource consumption, minimize waste, alleviate pollution, and enhance the overall quality of life for residents. Urban metabolism represents the dynamic flow of energy, materials, and resources within cities. It is defined by Kennedy et al. (2007, p. 44) as "the total sum of the technical and socioeconomic processes that occur in cities, resulting in growth, production of energy, and elimination of waste." It serves as a fundamental concept in the development of eco-cities (Niza et al. 2009; Kennedy et al. 2011), which are urban systems designed to ensure that the inflows of materials and energy, as well as the disposal of wastes, remain within the capacity of their hinterlands (Kennedy et al. 2007). However, a comprehensive analysis of urban metabolism necessitates substantial data on urban metabolic flows (Kennedy et al. 2007; Weisz and Steinberger 2010; Shahrokni et al. 2015). The need for detailed data analysis and advanced computational tools has given rise to SUM, an AI- or AIoT-driven platform or model capable of collecting and integrating vast amounts of high-spatial and temporal resolution data from diverse urban systems (Bibri et al. 2024a; Ghosh and Sengupta 2023).

To effectively manage and analyze urban metabolism, it is essential to identify and utilize various data sources. Table 8.2 lists potential data sources, data points, and relevant data owners, along with the methods, frequency, accessibility, resolution, and applications. It expands on the work of Shahrokni et al. (2015) and incorporates insights from D'Amico et al. (2021, 2022), Bibri et al. (2024a), and Ghosh and Sengupta (2023). By accurately capturing local, regional, and global flows of energy and materials, SUM provides a more comprehensive understanding of metabolic flows. Understanding these elements allows eco-cities to develop smarter, more sustainable practices that enhance urban living while minimizing environmental impact.

By integrating these diverse data sources, smarter eco-cities can gain a holistic view of their metabolism. This enables more informed decision-making, supports sustainability initiatives, and fosters resilience against environmental challenges. Effective data management is crucial for leveraging these insights to improve urban living conditions and promote long-term sustainability. Moreover, these data sources contribute to the circularity of SUM. By continuously monitoring resource flows and waste generation, smarter eco-cities can identify opportunities to reuse,

TABLE 8.2 Comprehensive Data Sources and Management for Developing Smart Urban Metabolism

Flows	Data Points	Data Owners	Data Collection Methods	Frequency of Data Collection	Data Resolution
I. Local					
Electricity, Heating, Cooling, Water Use, On-Site Generation	Billing meters, submeters	Energy utilities, building owners, homeowners	Automated meters, manual reporting	Monthly	Building level
Transportation (Car, Public, Goods, Fuel Use)	GPS, GSM, road tolls, vehicle registrations, taxi logistics, traveler card swipes, package tracking numbers, road numbers, gasoline station billing records	Car owners, telecommunications companies, transportation authorities, taxi companies, app developers, statistics bureaus, package tracking services, road tolls, gasoline stations	Sensors, GPS tracking, surveys	Real-time, daily	Individual, road segment level
Waste Generation	Weight of waste fractions, weight of organic waste	Municipalities, waste management companies, customers (homeowners, building owners)	Weighing systems, manual reporting	Weekly, monthly	Household, neighborhood level
Food and Goods Consumption, Service Use	Credit card statements, e-bills, digitized receipts, sales statistics, economic input-output LCA	Citizens (card owners), finance institutions, shops, scientific journals	Electronic transactions, surveys	Daily, monthly	Individual, store level
Local Agriculture and Urban Gardening	Harvest volumes, water usage, fertilizer usage, pesticide usage	Local farms, urban gardeners, community garden organizations	Manual reporting, sensors	Seasonal, monthly	Plot, farm level

(Continued)

TABLE 8.2 (Continued) Comprehensive Data Sources and Management for Developing Smart Urban Metabolism

Flows	Data Points	Data Owners	Data Collection Methods	Frequency of Data Collection	Data Resolution
Air Quality and Pollution	Air quality sensor data, emissions data	Environmental protection agencies, research institutions, municipalities	Sensors, manual reporting	Real-time, daily	City, regional level
Green spaces and Biodiversity	Tree inventories, biodiversity surveys, park usage statistics	Municipalities, environmental NGOs, research institutions	Surveys, remote sensing	Yearly, seasonal	Park, city level
On-Site Carbon Sequestration	Municipal tree inventory, carbon sequestration volume	Municipality	Surveys, remote sensing	Yearly, seasonal	Plot, city level
Construction and Maintenance	Construction materials purchased, reused, recycled content, shipping distance, fuel use for construction	Construction developers, subcontractors, logistic centers	Automated reporting, manual tracking	Monthly, project-based	Project, site level
Water Treatment and Distribution	Water use by building and neighborhood, sewer water generation, sludge water to biogas generation, water treatment and distribution electricity, water quality	Water utilities	Automated meters, sensors	Real-time, monthly	Building, neighborhood level
Waste Management	Nutrients in treated water, heat recovery from sewer water, recycling rates of waste, weight of waste fractions, waste stream assessment	Waste management companies, municipalities, refurbishing, and second-hand organizations	Weighing systems, sensors	Weekly, monthly	City, regional level

(Continued)

TABLE 8.2 (Continued) Comprehensive Data Sources and Management for Developing Smart Urban Metabolism

Flows	Data Points	Data Owners	Data Collection Methods	Frequency of Data Collection	Data Resolution
Global					
Mining, Goods Manufacturing	Open data and LCA databases	Public data, local companies, and residents	Automated systems, manual reporting	Yearly, monthly	National, global level
Energy Production	Public electricity use (e.g., parks), public transportation electricity use, EV charging electricity and location, electricity generation fuel mix, grid failures, peak load reduction, local heat/cooling production to the grid	Energy utilities	Automated meters, sensors	Real-time, monthly	City, regional level
Food Production	Crop yields, livestock numbers, water usage, fertilizer, and pesticide usage	Agricultural businesses, research institutions, government agencies	Remote sensing, surveys, sensors	Seasonal, monthly	Farm, regional, national level
Goods Shipping	Shipping volumes, shipping routes, transportation modes, fuel usage	Shipping companies, logistics firms, trade associations	Automated systems, manual reporting	Real-time, monthly	National, global level
Flights	Airport flight data (real-time), airline carbon disclosures, company travel expenditures, resident travel expenses	Airports, airlines, local companies, and residents	Automated systems, manual reporting	Real-time, monthly	National, global level

(Continued)

TABLE 8.2 (Continued) Comprehensive Data Sources and Management for Developing Smart Urban Metabolism

Flows	Data Points	Data Owners	Data Collection Methods	Frequency of Data Collection	Data Resolution
Climate Data	Temperature, precipitation, extreme weather events	Meteorological agencies, climate research institutions	Sensors, remote sensing	Real-time, daily	City, regional, global level
Global Supply Chain Logistics	Shipping volumes, shipping routes, transportation modes, fuel usage	Shipping companies, logistics firms, trade associations	Automated systems, manual reporting	Real-time, monthly	National, global level
International Trade & Economics	Import/export data, trade balances, international economic indicators	Government trade agencies, international economic organizations, research institutions	Automated systems, manual reporting	Monthly, yearly	National, global level

recycle, and reduce materials, effectively closing the loop on resource use. For instance, data on waste fractions and recycling rates can inform waste management strategies that prioritize recycling and composting, reducing the need for landfills.

SUM enables real-time processing, analysis, and visualization of data on metabolic flows, leading to a more profound understanding of how cities consume resources, generate waste, and impact the environment. Through comprehensive data collection and analysis, smarter eco-cities can enhance their circular economy practices, minimizing waste and maximizing resource efficiency. This holistic approach supports environmental sustainability and fosters economic resilience and social well-being. Toward this end, to enhance its analytical capabilities for real-time monitoring and resource management, SUM is increasingly integrating ML/DL algorithms (Ghosh and Sengupta 2023; Peponi et al. 2022) with IoT and Big Data technologies (Bibri and Krogstie 2020a; Shahrokni et al. 2015)—in short, AIoT. AIoT overcomes temporal and spatial communication constraints, enabling real-time monitoring and rapid response to system conditions. This leads to the creation of advanced models, yielding actionable insights on metabolic flow patterns. Indeed, a key strength of AIoT in SUM is its capacity to develop predictive models and optimization algorithms. Data-driven AI models are central to SUM, efficiently managing complex datasets, autonomously learning, and providing advanced predictive analysis (Ghosh and Sengupta 2023). Further, AIoT sensors and data analytics provide real-time insights into resource flows, enabling more precise and dynamic resource management (Bibri and Huang 2024). For example, AIoT can optimize energy usage, water distribution, waste management, and transportation systems in real-time. The AIoT-powered, data-driven approach enhances our understanding of urban metabolism dynamics and identifies opportunities to foster efficient resource use and environmental performance.

Ghosh and Sengupta (2023) emphasize the crucial role that AI, ML, and Big Data technologies play in advancing SUM. They highlight how self-learning ML models and predictive analytics can effectively manage and analyze complex, multidimensional datasets, providing novel insights. According to their study, SUM, powered by these advanced technologies, serves as an invaluable tool for urban planners to identify and address intricate issues related to the flow of energy and materials within urban environments.

Peponi et al. (2022) propose an innovative, evidence-based methodology to manage urban metabolic complexities through the concept of smart and regenerative urban metabolism. They combine Life Cycle Thinking and ML to assess the metabolic processes of a city's urban core using multidimensional indicators, including urban ecosystem service dynamics. The authors develop a multilayer perceptron network to identify key metabolic drivers and predict changes by 2025, with validation through prediction error standard deviations and training graphs. The study identifies significant drivers of urban metabolic changes such as employment rates, energy systems, waste management, demography, cultural assets, and air pollution. This research framework provides a knowledge-based tool to support sustainable and resilient urban development policies.

Bibri et al. (2024a) explore the integration of AI and AIoT into SUM as a means to advance the development of smarter eco-cities. This fusion enhances sustainability, resilience, and governance effectiveness by enabling data-driven decision-making and targeted interventions. They stress the pivotal role of AI and AIoT in developing predictive models and optimization algorithms that transform urban metabolic processes, thus making smart eco-cities more resource-efficient and sustainable. The study underscores the significance of these technologies in facilitating real-time city functions and informed decision-making, which are crucial for the successful implementation of environmental policies in smarter eco-cities.

All three studies emphasize the critical role of advanced technologies—AI, ML, AIoT, and Big Data—in enhancing SUM. They collectively demonstrate the transformative potential of integrating advanced technologies into SUM to enhance urban sustainability, resilience, and governance. Each study provides a unique perspective and methodology, contributing valuable insights into the development of smarter and more sustainable urban environments.

Furthermore, the relationship among eco-cities, urban metabolism, and SUM extends to industrial ecology and circular economy, reflecting the evolution of sustainable urban development. In this regard, urban metabolism draws on ecology in terms of the flow of resources (e.g., energy, water, and materials) into, within, and out of a city. Ecology is concerned with the interactions between organisms and their environment. It encompasses the relationships between living organisms, their habitats, and the surrounding ecosystems. Ecological principles guide practices that promote the harmonious coexistence of humans and the natural environment in eco-cities (Rosalind 1997). Industrial ecology is a multidisciplinary field that studies the interactions between industrial systems, ecosystems, and human activities. It aims to optimize the use of resources, minimize environmental impact, and create more sustainable industrial processes by adopting principles inspired by natural ecosystems. It applies ecological principles to industrial systems. It emphasizes resource efficiency, waste reduction, and the circular flow of materials. In the context of cities, industrial ecology focuses on the interplay of urban activities, industries, and the environment to minimize environmental impact. Circular economy emphasizes product design for reusability, recycling, and remanufacturing, thereby focusing on the resource efficiency and waste elimination aspects of industrial ecology. Moreover, its principles align closely with sustainability goals, as they aim to reduce environmental impact while promoting economic growth.

The circular city is described as a complex ecosystem that leverages advanced technologies in circular economy practices to address social, economic, and environmental aspects simultaneously (D'Amico et al. 2022). In this context, the urban metabolism of circular cities involves a multitude of actors, including authorities; municipal, regional, and national utilities; citizens, communities, universities, entrepreneurial ventures, research centers, innovation labs, businesses, and non-profit organizations (D'Amico et al. 2021). These actors interact continuously and systematically to plan, monitor, develop, and evaluate increasingly sustainable and digitalized urban processes and operations (Panori et al. 2020). To sustain urban circularity through digital technologies (Campbell-Johnston et al. 2019), it is essential to thoroughly understand the social, economic, environmental, and technological flows that define urban processes and operations (Stapper and Duyvendak 2020) and to reorient their metabolism accordingly (Maranghi et al. 2020). Moreover, the digitalization of circularity necessitates comprehensive and integrated frameworks designed to manage, monitor, and quantify the social, economic, and environmental metabolic flows (Alcayaga et al. 2019; D'Amico et al. 2020; 2022; Shahrokni et al. 2015).

Overall, urban metabolism and SUM provide frameworks for understanding and managing resource flows within cities, with AI and AIoT enhancing the precision and real-time adaptability of resource management. Industrial ecology extends the principles of resource efficiency and waste reduction to urban and industrial systems, aligning with circular economy principles. Circular economy principles support eco-cities and smarter eco-cities by promoting resource efficiency, waste reduction, and sustainable product design. In essence, these approaches are interconnected, and their integration contributes to more sustainable and efficient urban environments.

8.4.2 Smart Eco-Cities and Symbiotic Cities as Prominent Approaches to Environmentally Smart Sustainable Urbanism

Environmentally smart sustainable urbanism is an umbrella term for various models for urban development that harness advanced technologies and nature-based solutions to advance the goals of sustainability goals, with a focus on its environmental dimension. Among these models are eco-cities, symbiotic cities, circular cities, green cities, low-carbon cities, zero-waste cities, and net-zero energy cities (Bibri in 2021a). Each of these models seeks to improve and sustain its contributions to specific facets of environmental sustainability. The foundation of these models rests upon the application of a cohesive set of principles related to environmental sustainability, ecological sustainability, urban ecology, urban sustainability, industrial ecology, industrial symbiosis, industrial metabolism, green economy, and/or circular economy—with overlaps between them.

As regards these principles, for example, smart eco-cities in the green economy prioritize environmental sustainability, alongside social and economic considerations, and involve adopting practices and technologies that minimize environmental impact, promote resource efficiency, and support sustainable development (Caprotti 2020). Environmental sustainability entails the ability to maintain the health and integrity of the natural environment through responsible and green approaches to economic development over the long term, as well as using resources efficiently and sustainably for better societal outcomes.

As another example, symbiotic cities, also known as eco-industrial or circular cities, apply the concept of industrial symbiosis to promote sustainable urban development. Industrial symbiosis is a framework where different industries and sectors within a city collaborate to optimize resource efficiency, reduce waste, promote sustainable practices, and enhance environmental sustainability. The SymbioCity model places the well-being, comfort, safety, and overall quality of life at the center of urban development, emphasizing a people-centric approach (Backmann et al. 2014). It aims to enhance urban integration, sustainability, resilience, and the overall quality of life for both current and future urban residents, and its objectives are to (Backmann et al. 2014):

- Foster collaboration across multiple sectors and disciplines among stakeholders.

- Build capacity by facilitating the sharing of knowledge and experiences among stakeholders.

- Promote cooperation between local, regional, and national stakeholders.

- Guide sustainability assessments and planning processes across various scales.

- Contribute to the development of strategies for the short, medium, and long-term enhancement of urban areas and various aspects of sustainability.

- Assist in identifying practical and integrated system solutions and synergies that support sustainable urban development.

- Enhance existing urban development policies, plans, processes, and practices.

Symbiotic cities promote circular economy principles, emphasizing product design that prioritizes reusability, remanufacturing, and recycling. Industries within symbiotic cities are encouraged to design products and processes with a focus on minimizing waste and extending the life of products. Accordingly, symbiotic cities and circular cities overlap in many aspects, including resource sharing, waste exchange, cluster development, urban planning, stakeholder engagement, environmental benefits, and data and technology integration. In more detail, in symbiotic cities, industries and businesses collaborate to share resources such as water, energy, materials, and waste products. Moreover, industrial symbiosis fosters the exchange of waste and byproducts between different businesses. One company's waste can become another's raw material. For instance, a food processing plant might provide organic waste to a nearby biogas facility, which can convert it into renewable energy.

Stuiver and O'Hara (2022) explore two visions for symbiotic cities in 2050, focusing on the integration of nature into urban areas and their social, built, and geographic characteristics. The first vision is for a City Region Food System in Washington, DC, emphasizing food as the cornerstone of a nature-based and socially inclusive urban economy within the larger metropolitan region. This vision values natural resources, social and cultural diversity, and innovative, nature-based initiatives. The second vision is for a region in the Netherlands, comprising eight neighborhoods that each embodies different aspects of nature-based thinking, valuing natural resources like water, soils, and biodiversity. Both visions illustrate the potential for symbiotic cities by 2050, though further specification and implementation strategies are needed. The study also raises the importance of various actors in the urban innovation ecosystem. Indeed, symbiotic cities encourage the clustering of related industries and businesses to facilitate resource exchange and collaboration. By co-locating industries with complementary needs and outputs, cities can create industrial ecosystems that are more resource-efficient and environmentally friendly.

Moreover, symbiotic cities incorporate industrial symbiosis principles into urban planning and zoning regulations. This involves creating designated areas or zones where industries with synergistic relationships can locate, making resource sharing more feasible. In addition, engaging various stakeholders, including local businesses, government agencies, research institutions, and community members, is essential in symbiotic cities. Collaboration and communication among these groups are vital to identify opportunities for resource optimization and environmental improvement. Additionally, symbiotic cities reduce their ecological footprint, conserve natural resources, and lower GHG emissions. Lastly, modern technologies, including IoT, Big Data, and AI, play a pivotal role in symbiotic cities and hence circular cities. They enable real-time monitoring of resource flows, predictive analytics, and efficient resource allocation, helping businesses and industries make data-driven decisions to optimize resource use and minimize waste within cities. Indeed, symbiotic cities rely heavily on interconnected data networks, which function as their nervous system to enable constant communication between various AI and AIoT systems, smart materials, and biotechnological processes.

Furthermore, data-driven technologies play a pivotal role as essential enablers, facilitating the seamless integration of the aforementioned principles. This enhances the overall effectiveness of smart eco-cities due to the synergistic potential provided by the convergence of AI, IoT, and Big Data. These advanced technologies allow for real-time data collection, analysis, and application, optimizing resource management and decision-making processes. Table 8.3 presents a curated selection of smart eco-cities, illustrating how they incorporate different sets of principles into their development strategies.

TABLE 8.3 Smart Eco-City Models and Their Underlying Principles

Urbanism Models	Domain Principles	Citations
Smart Eco-cities	Industrial ecology, economy, and environmental sustainability	D'Amico et al. (2020)
Smart eco-cities	Green economy	Caprotti et al. (2020)
Smart eco-cities	Environmental sustainability and urban metabolism	Bibri and Krgostie (2020a)
Smart eco-cities	Industrial ecology/metabolism/symbiosis	Shahrokni et al. (2015)
Smart eco-cities	Environmental sustainability and industrial symbioses	Bibri and Krogstie (2020b)
Smart eco-cities	SDGs and environmental sustainability	Tarek (2023)
Sustainable-smart-resilient low-carbon eco-knowledge cities	Environmental sustainability and urban ecology	De Jong et al. (2015)

In addition, Bibri et al. (2024a) explore the crucial role of AIoT in integrating SUM, City Brain, and platform urbanism to advance data-driven environmental governance in smarter eco-cities. Specifically, the authors introduce a pioneering framework that leverages the synergies among these AIoT-powered governance systems to enhance environmental sustainability practices. A comparative analysis reveals commonalities and differences among SUM, City Brain, and platform urbanism in the contexts of AIoT and environmental governance. These data-driven systems collectively contribute to environmental governance by leveraging real-time data analytics, predictive modeling, and stakeholder engagement. The proposed framework emphasizes the importance of data-driven decision-making, optimization of resource management, reduction of environmental impact, collaboration among stakeholders, engagement of citizens, and formulation of evidence-based policies. The findings reveal that the synergistic and collaborative integration of SUM, City Brain, and platform urbanism through AIoT presents promising opportunities for advancing environmental governance in smarter eco-cities.

The study by Bibri (2024a) provides valuable insights for informing symbiotic cities. One of its key contributions is its emphasis on enhanced environmental governance. By utilizing AIoT-powered systems, symbiotic cities can improve their management of natural resources through real-time data analytics, predictive modeling, and stakeholder engagement. This approach allows for more effective monitoring of environmental impacts and better decision-making processes aimed at reducing waste and preserving resources. Additionally, symbiotic cities can adopt the integrated framework to harmonize various urban functions and services, ensuring a balanced relationship between human activities and natural ecosystems. Moreover, by leveraging data-driven insights, symbiotic cities can implement more sustainable urban development strategies and develop adaptive measures to address ecological challenges effectively. They can also create a cooperative environment that supports sustainable practices and innovation by involving various actors. This collaborative approach ensures that diverse perspectives are considered in urban planning and development, leading to more robust and effective solutions. Lastly, the efficient utilization of resources such as water, energy, and materials aligns with the principles of the circular city, promoting the reuse and recycling of resources to create closed-loop systems. Overall, by applying the gained insights, symbiotic cities can develop more sustainable, efficient, and resilient urban systems that integrate human and natural elements harmoniously.

8.4.3 The Role of Artificial Intelligence of Things in Advancing Metabolic Circularity in Smarter Eco-Cities: Integrating Eco-Cities, Smart Cities, and Circular Cities

A recent surge in research has centered on the integration of various models for urban development under different labels, and a noteworthy example is the amalgamation of smart eco-cities and circular smart cities. This convergence is particularly compelling due to their shared utilization of AI, IoT, and Big Data technologies to enhance the circularity of existing eco-cities under smarter eco-cities. Indeed, eco-cities and circular economy models share fundamental principles and strategies, including waste reduction, material circulation (back into use without waste), and pollution prevention (waste reduction at its source), as previously discussed. This integration is often referred to as a "circular economy eco-city" (Li et al. 2021).

Moreover, the fusion of circular economy with digital technologies presents numerous benefits for enhancing the efficiency of SUM (D'Amico et al. 2020, 2021). This involves optimizing the circulation and improving materials, waste, byproducts, emissions, energy, knowledge, information, and data (D'Amico et al. 2022; Levoso et al. 2020; Venkata

Mohan et al. 2020; Yao et al. 2020) in smart eco-cities (Bibri and Krogstie 2020a, Shahrokni et al. 2015). Utilizing a data-driven approach in urban computing and intelligence has advantages for analyzing resource flows within urban environments, especially when assessing the development of sustainable eco-districts or cities. This approach enhances decision-making by providing precise insights into resource allocation, consumption patterns, and environmental impacts, ultimately supporting more efficient and sustainable urban planning. Big Data, AI, and IoT technologies are key to advancing SUM. ML models and predictive analytics can efficiently manage and analyze complex, multidimensional datasets, generating new insights (Ghosh and Sengupta 2023). This aligns with the evidence-driven approach proposed by Peponi et al. (2022) to manage urban metabolic complexities through smart and regenerative urban metabolism, driven by changes in energy systems, waste management, air pollution, and demography, among others.

The integration of eco-cities and circular cities through smart cities is steadily gaining traction, resulting in the emergence of circular smart sustainable cities as an umbrella term for circular smart eco-cities, all made possible by the capabilities of AI, IoT, and Big Data technologies. The confluence of smart eco-cities and circular economy is exemplified through the integrated paradigm of circular smart sustainable cities. Similar to smarter eco-cities in the context of AIoT (Bibri et al. 2024a), this emerging paradigm acts as testing grounds for innovative solutions, data-driven approaches, and enhanced strategies in urban development. IoT facilitates the creation of urban environments aligned with the principles of circular economy, with Big Data and AI technologies making it possible to achieve the goal of self-reliance and environmental well-being (Rudskaya et al. 2019). Dincă et al. (2022) explore the drivers behind smart environment development in smart cities, particularly in mitigating air pollution. Their study utilizes AI-driven regression models to evaluate how circular economy practices, fiscal policies, and environmental initiatives influence air pollution reduction. Their research underscores the critical role of AI, renewable energy adoption, environmental education, circular economy strategies, and effective policy frameworks in achieving significant reductions in air pollution.

With respect to smart cities, Kierans et al. (2023) propose an AIoT-based smart city for circular economy framework. This centralized platform utilizes data-driven technologies to manage resources and waste within a closed-loop system. It enables real-time resource tracking, where sensors and devices monitor resource, energy, and waste flows in real-time, ensuring efficient and sustainable management. It also involves AI-driven resource optimization, where AI and ML are leveraged to optimize resource allocation, reduce waste, and forecast future requirements. The study incorporates product material data, a broader approach that supports circularity in cities. Sastry (2020) delves into the opportunities and challenges inherent in implementing circular economy principles to manage and govern smart cities, especially in the context of energy conservation and reduction. The author explores a range of strategies, including waste-to-energy production, recycling, reuse, sharing, and demand-side management. Additionally, a novel approach to visualizing key environmental performance indicators through dashboards is proposed. It is worth noting that many of those strategies align closely with the objectives of smart eco-cities (Bibri and Krogstie 2020a, b).

Furthermore, Sertyesilisik (2023) investigates the role of IoT, Big Data, and advanced ICT in enhancing environmental sustainability, resilience, and circularity in the context of smart, sustainable, circular, and resilient cities. The author offers detailed frameworks and strategies for this purpose, supported by an extensive review of existing literature. The study underscores the essential nature of these technologies in urban transformation, improving both sustainability and resilience. Núñez-Cacho et al. (2022) present a smart management model for circular smart cities, incorporating various factors that influence air quality. The authors employ AI techniques for data analytics to monitor current air quality and predict its future trends, thereby enhancing decision-making processes in circular sustainable cities. Abou Baker et al. (2021) direct their attention to the automated recycling of e-waste, a rapidly growing waste stream globally, driven by AI technologies in the realm of circular smart cities. Their work showcases the potential of Transfer Learning techniques in facilitating automated recycling, significantly reducing the size of training sets and computation time while enhancing recycling processes.

Building upon the idea that new technologies are central to the vision of circular cities, Vishkaei (2022) explores the potential role of the Metaverse in smart cities to improve transportation, environmental quality, and overall quality of life. The author advocates for smarter actions in smart cities to support the implementation of the circular economy. In line with this perspective, a study conducted by Allam et al. (2022) investigates the potential of the Metaverse as a virtual model of smart cities, highlighting its capacity to enhance resource management, climate change mitigation and adaptation, virtual mobility, workplace dynamics, and overall quality of life. The authors emphasize the role of AI, IoT, and Big Data in providing valuable datasets and computational insights into human behavior, which can reshape city design and service delivery, boosting urban efficiencies and performance.

8.4.4 Advanced Artificial Intelligence of Things Solutions for Waste Management in Smarter Eco-Cities

AI and AIoT have played a key role in the advancement of sophisticated waste collection systems, effectively optimizing multiple parameters to maximize overall efficiency. This is increasingly being implemented in emerging smarter eco-cities. Further, Fang et al. (2023) delve into diverse AI applications in waste management, including:

- Waste-to-energy processes
- Intelligent waste bins
- Automated waste-sorting robots
- Predictive waste generation models
- Real-time waste monitoring and tracking systems
- Optimized waste logistics
- Efficient waste disposal methods
- Innovative waste resource recovery techniques
- Enhanced waste process efficiency
- Cost savings in waste management

This research underscores how AI-driven solutions in waste logistics yield substantial reductions in transportation distances and time, as well as enhanced waste pyrolysis, accurate carbon emission estimations, and efficient energy conversion. Additionally, it stresses AI's pivotal role in enhancing precision and reducing costs within waste identification and sorting processes in smart urban contexts.

Focusing on the application of AI in improving waste management practices, Olawade et al. (2024) examine how AI aids in collection, sorting, recycling, and monitoring waste, highlighting both the advantages and challenges, such as data quality, privacy, and costs. The review also considers future AI integration with IoT and ML, stressing the need for collaborative frameworks and policy initiatives. In summary, while AI offers considerable benefits for waste management, addressing its associated challenges is crucial for maximizing its potential for sustainability and efficiency.

In the context of circular economy, the integration of AI and AIoT in urban waste management facilitates a transformative approach to sustainability. The adoption of these technologies contributes significantly to the circular economy by closing the loop of product lifecycles through advanced waste handling and by turning waste into valuable resources, ultimately promoting sustainable urban development. Lanzalonga et al. (2024) explore how AI) can improve decision-making in the utility sector to support circular economy practices. It focuses on a case study of Alia Servizi Ambientali Spa, an Italian utility company, to analyze AI's use in waste management. The findings reveal that user engagement is crucial, economic incentives are necessary to enhance technology adoption, and AI plays a transformative role in managing specific waste types. The study contributes to the circular economy by highlighting AI's impact on waste management and providing practical insights for managers on using AI-driven data algorithms for decision-making.

Furthermore, Rathore and Malawalia (2021) discuss the deployment of AI in smart waste management. The authors utilize Convolutional Neural Networks (CNN) for waste estimation and classification. Their work demonstrates the use of AI, particularly CNN, in improving waste management efficiency. Nañez Alonso et al. (2021) validate the use of AI, specifically CNN, for automatic recycling of materials. They achieve a high reliability rate in classifying recyclable materials. Their research emphasizes the role of AI in waste recycling, supporting circular economy practices. The work of Nasir and Aziz Al-Talib (2023) explores the challenges inherent in waste classification and highlights the potential of AI coupled with image processing techniques to tackle these issues. The authors acknowledge the limitations of existing waste classification models driven by DL and underscore the imperative for heightened accuracy and runtime improvements for precise outcomes, nevertheless. They argue that precise waste categorization holds paramount importance for numerous reasons, including facilitating recycling, safeguarding both the environment and human health, and minimizing costs associated with waste management. The existence of varied waste types, ranging from non-biodegradable and hazardous to industrial, municipal solid, agricultural waste, and necessitates a comprehensive approach.

In addition, Wilts et al. (2021) focus on the automation of municipal waste-sorting plants using AI and robotics. Their research explores the practical application of AI in waste management, highlighting its potential to improve recycling rates and working conditions for employees. AI has become integral to enhancing waste management, capable of handling vast amounts of data, automating processes, and yielding reliable results. Various AI techniques, such as Linear Regression (LR), Support Vector Machines (SVM), Decision Trees (DT), Genetic Algorithms (GA), and Artificial Neural Networks (ANN), have been integrated across different facets of waste management (Bibri et al. 2023b; Fang et al. 2023; Olawade et al. 2024). These methods are applied in activities ranging from waste collection and sorting to treatment and planning. LR is useful for predicting waste quantities and planning, while SVM excels in large dataset analysis for bin-level detection and sorting. DT is suitable for optimizing waste predictions and planning. GA applies principles of natural selection to optimize routes and management plans, proving valuable in refining efficiency and operational strategies in waste management systems. These AI techniques contribute significantly to the advancement of waste management, promoting more sustainable and efficient practices within the sector. Lastly, ANN are effective in identifying complex patterns and enhancing sorting processes. For example, these techniques aid in tracking the end-of-life status of materials and support predictions related to the purchase of new products (Schöggl et al. 2023).

Moreover, AI-driven circular economy initiatives play a key role in safeguarding our planet's ecosystem. They actively combat resource exploitation, championing the conservation of natural resources by closing the loop to curtail excessive resource consumption. Among the notable systems contributing to this cause is Artificial Intelligence for Ecological Services, a rule-based system renowned for its efficient modeling of ecosystem services. Beyond these applications, AI models can be effectively harnessed for predictive maintenance in waste management facilities. By scrutinizing patterns in datasets, AI models can proactively identify potential equipment malfunctions, enabling timely interventions and minimizing operational disruptions. Moreover, AI-powered decision support systems can seamlessly integrate diverse data streams, such as population density and waste composition. This holistic approach empowers waste management stakeholders with actionable insights and recommendations, facilitating the formulation and execution of effective strategies.

On the whole, the studies discussed highlight the transformative impact of AI and AIoT on waste management, aligning closely with the principles of circular economy. They emphasize how these technologies optimize waste collection, categorization, and resource recovery, resulting in enhanced efficiency and sustainability. Predictive maintenance capabilities ensure operational continuity, while decision support systems enable comprehensive waste management strategies in circular systems. AI-driven solutions contribute to reduced transportation distances, accurate carbon emissions estimation, and efficient energy conversion, all of which are central to circular economy goals. This underscores their crucial role in reshaping waste management practices to support circular economy principles, benefit the environment, and reduce operational costs.

8.4.5 Artificial Intelligence and Artificial Intelligence of Things Solutions for Circular Economy in Smarter Eco-Cities

The circular economy model has opened up new avenues for innovation, particularly in the realm of "eco-innovation." Additionally, eco-innovation is identified as crucial for transitioning from a linear to a circular production and consumption model. De Jesus (2018) suggests that transitioning to circular economy through eco-innovation requires tailored solutions, termed "clean congruence," and systemic eco-innovation that integrates technological advancements with innovative service and organizational models.

AI and AIoT play a central role in optimizing natural resource utilization and expanding the potential for pollution prevention and environmental regeneration in emerging smarter eco-cities (Bibri et al. 2023b). These technologies also contribute to the shift from a linear to a circular model of the economy by facilitating the design, development, and maintenance of circular products and materials; the creation of resilient and sustainable products; the advancement of sustainable business models; and the establishment and optimization of the comprehensive infrastructure necessary to amplify circular practices (Lekan and Oloruntoba 2020; Roberts et al. 2022). AI, IoT, and Big Data play an important role in closing material loops by providing precise information regarding product availability and location, thereby significantly enhancing resource efficiency in circular business models, as evidenced by multiple case studies (Ranta et al. 2021).

Consequently, researchers are now exploring the concept of building circular smart sustainable cities and hence circular smarter eco-cities, capitalizing on the applications of AI and AIoT. Lampropoulos et al. (2024) highlight the critical role of AI and IoT in achieving the UN' SDGs and advancing the circular economy. They introduce the concept of AIoT, which combines AI with IoT to enhance sustainable practices. Ali et al. (2023) explore the integration of AI into circular economy practices and assess the current waste situation in the context of circular economy and AI-driven circular economy. The authors highlight the potential of AI in enhancing resource management and its importance in achieving sustainability goals. They identify the need for critical measures to promote these practices on broader scales, such as cities and regions. The value lies in expanding the scope of AI-driven circular economy research and underscores the significance of AI in facilitating the transition to a circular economy. For example, ANN techniques enhance circular economy by emphasizing material circularity through prediction and execution, employing ML to forecast the quantities of reusable, residual, and recyclable materials (Ali et al. 2023).

Furthermore, Akinode and Oloruntoba (2020) investigate the potential of AI in the transition to circular economy. The authors emphasize the role of AI in optimizing circular processes by reducing traffic congestion, optimizing energy usage, and enhancing collaboration in different areas. Their study highlights the initial steps in understanding AI's contributions to a circular economy. D'Amore et al. (2022) analyze the role of AI in the Water-Energy-Food nexus and its connection to the SDGs. The authors emphasize the importance of AI as a technology for promoting the SDGs and suggest the need for a multi-stakeholder approach to address these nexus challenges. Their findings highlight AI's relevance in enhancing resource management for urban sustainability. Jose et al. (2020) advocate the use of AI and related technologies to enhance energy efficiency, facilitate carbon trading, and achieve the vision of circular economy, thereby mitigating climate change risks. From a different perspective, Ghoreishi and Happonen (2020) explore the key enablers for incorporating AI techniques into circular product design. They emphasize the optimization and real-time data analysis benefits of AI in circular economy solutions. This study offers insights into how AI can increase productivity and sustainability in manufacturing industries. From a general perspective, Agrawal et al. (2021) explore the applications of AI techniques in enhancing the adoption and implementation of circular economy practices across various domains. Salvador et al. /2021) apply fuzzy logic approach to analyze the data on circular economy strategies. The authors identify focus areas where industries need to focus more for successful implementation of circular economy practices.

Overall, these studies collectively demonstrate the diverse innovative applications of AI in advancing circular economy practices, optimizing resource management, and supporting sustainability goals. While some studies focus on AI's role in circular economy implementation, others explore its connections with SDGs, circular product design, and waste management. These contributions highlight AI's potential to enhance decision-making, improve efficiencies, and contribute to the circular economy in smarter eco-cities across various domains.

Furthermore, it is important to acknowledge that much of the concept of circular smart cities and circular smart sustainable cities remains idealistic or theoretical in nature. This is similar to the concept of eco-cities as ideal urban models (Roseland 1997), which, as Register (1987) noted since the inception of the notion of sustainable development in 1987, do not exist. Nevertheless, as Rapoport (2014) suggests, while the ability of eco-cities to fully realize their utopian aspirations may be constrained by the realities of operating within profit-driven planning environments, they still serve a valuable role as experimental grounds and aspirational ideals. Similarly, circular smart sustainable cities and hence circular smarter eco-cities emerge as sites for experimentation and innovation, where new ideas can be tested and lofty visions can be pursued. With that in mind, smarter eco-cities, with the aid of AIoT technologies, aim to harness the benefits of the synergistic potential of eco-cities, smart cities, and circular cities by optimizing resource usage, reducing waste, and promoting circularity while enhancing the overall quality of life for urban residents. Drawing on the strategies of the circular economy, coupled with the previously reviewed and discussed studies, AIoT can significantly bolster each of these strategies as part of building smarter eco-cities of the future (Table 8.4).

In a nutshell, the role of AIoT in smarter eco-cities is integral to advancing the principles of the circular economy. AIoT is key to advancing the development of smarter eco-cities through real-time monitoring, predictive analytics, timely data-driven insights, and intelligent decision-making and automation capabilities that enhance resource efficiency, minimize waste, and promote sustainable practices. This aligns with the principles of circular economy. Accordingly, AIoT serves as a linchpin technology

TABLE 8.4 The Role of Artificial Intelligence of Things in Enhancing Circular Economy Strategies in Smarter Eco-Cities

Resource management optimization	AIoT offers real-time resource tracking, providing valuable data on resource availability and consumption. This data-driven approach transforms procurement processes, preventing overstocking and reducing waste generation. AI algorithms use this information to predict demand accurately, empowering smarter eco-cities to optimize raw material use and minimize waste generation through design improvements and efficient resource allocation. By leveraging AIoT's capabilities, these cities can move closer to circularity, where resources are efficiently managed, reused, and repurposed, fostering sustainable urban environments while minimizing the ecological footprint.
Product lifespan extension	AIoT contributes significantly to extending the lifespan of products and materials. It leverages real-time data analytics, predictive maintenance, and condition monitoring to facilitate recovery processes. The synergy between IoT and AI promotes sustainable product lifecycles, reducing waste and resource consumption. Specifically, AIoT enables the tracking of products and materials throughout their lifecycle. By embedding smart sensors, it becomes possible to monitor usage, condition, and location. This information is critical for recovering products at the end of their life. Also, AIoT can predict when a product or component is likely to fail, allowing for proactive maintenance or repair. AI-driven maintenance schedules ensure interventions occur at optimal times, preventing premature disposal and extending product lifespans. The use of AIoT informs and guides design decisions toward products that prioritize ease of disassembly, repairability, and recyclability.
Resource efficiency	AIoT, through real-time monitoring and adaptive control, brings unparalleled efficiency to resource-intensive processes. By identifying bottlenecks and performance gaps, AIoT optimizes operations, reducing energy consumption and minimizing resource waste. These systems ensure that resources are used judiciously, helping organizations achieve sustainability goals while enhancing productivity and cost-effectiveness. In industries, AIoT's ability to fine-tune processes fosters resource conservation and promotes eco-friendly practices, thereby aligning with the principles of sustainability and responsible resource management.

(*Continued*)

TABLE 8.4 (*Continued*)   The Role of Artificial Intelligence of Things in Enhancing Circular Economy Strategies in Smarter Eco-Cities

Recycling	AI-powered sorting systems in recycling facilities can identify and separate recyclable materials more efficiently. AIoT-enabled devices can sort and categorize waste more efficiently, increasing the quality and quantity of materials that can be recycled. This can lead to better separation of materials, reducing contamination, and enhancing the circularity of materials. Moreover, AI algorithms can help optimize recycling collection routes, reducing fuel consumption and hence emissions.
Reuse	IoT sensors can track the usage and condition of reusable items by providing real-time data, making it easier to manage and allocate them efficiently. They can be integrated into various products and assets, ranging from electronic devices to consumer goods. By continuously monitoring the status of these items, IoT sensors enable businesses and organizations to gain valuable insights into their lifecycle, usage patterns, and potential maintenance needs. AI can analyze the usage patterns of products and identify those that are still in excellent condition and suitable for resale or rental. This ensures that products reach their full lifecycle potential, reducing waste and the need for new manufacturing. AI can also suggest ways to reuse items effectively.
Remanufacturing	AIoT advances revolutionize remanufacturing by continuously monitoring product condition and pinpointing components suitable for refurbishment or replacement. This real-time tracking ensures that products are remanufactured with precision and resource efficiency. AIoT's analytical capabilities optimize the remanufacturing process, from disassembly to quality control, reducing waste and conserving resources. AIoT contributes significantly to sustainable practices in the circular economy by streamlining remanufacturing operations and extending the lifecycle of products.
Reduction, disposal, and treatment	AIoT can monitor waste generation in real-time and provide insights into waste patterns. AIoT devices can monitor waste management processes, such as landfill operations and waste-to-energy conversion. By analyzing data on waste composition and decomposition rates, AIoT systems can optimize disposal methods and minimize environmental impacts. Smarter eco-cities can use various forms of data to implement targeted waste reduction, treatment, and disposal strategies. For instance, changes in industrial processes can be suggested to minimize material scrap or propose more efficient packaging designs.

(*Continued*)

TABLE 8.4 (*Continued*) The Role of Artificial Intelligence of Things in Enhancing Circular Economy Strategies in Smarter Eco-Cities

Closed-loop systems	AI can help design products with circularity in mind. AI plays a crucial role in the design phase by offering insights into material selection and product design. It evaluates the recycleability and reusability of materials, facilitating the creation of closed-loop systems in which products are designed to minimize waste and maximize resource efficiency. By simulating different design scenarios and assessing their environmental impact, AI empowers manufacturers to make informed decisions that align with circular economy principles, ultimately fostering sustainable and eco-friendly product development.
Consumer awareness	AI can personalize sustainability recommendations for residents by leveraging data on individual preferences and behaviors. It can suggest eco-friendly product choices, tailor recycling guidance to specific needs, and propose transportation alternatives aligned with personal preferences and environmental goals. This level of customization enhances sustainability adoption by making eco-conscious decisions more accessible and relevant to individuals, thus contributing to more sustainable and environmentally responsible cities.
Decision and policy support	AIoT systems can provide real-time data-driven insights that guide decision-making, supporting policymakers and industries. For instance, businesses can use AIoT analytics to determine optimal routes for collection and transportation of recyclable materials, minimizing energy consumption and reducing GHG emissions. Industries can use AI to optimize supply chains by identifying opportunities to reuse packaging materials, shipping containers, or other logistical assets. This minimizes the environmental impact of transportation and packaging waste. Moreover, AIoT can monitor compliance with sustainability regulations and track the environmental impact of various initiatives.

that underpins the transition to circular economies in smarter eco-cities. By integrating this advanced technology into the urban fabric, existing smart eco-cities can achieve higher levels of environmental sustainability and circularity while enhancing the quality of life for their residents.

8.4.6 The 4R Framework and Artificial Intelligence: Their Contribution to Circular Economy and Tripartite Sustainability in Smarter Eco-Cities

In the context of smarter eco-cities, the concept of circular economy significantly contributes to addressing and integrating the three dimensions of sustainability. Circular economy strategies, which primarily align with environmental sustainability goals, contribute to social sustainability by enhancing the quality of life for urban residents. Economically, the circular economy model promotes resource efficiency, cost savings, and the creation of new business opportunities, strengthening economic sustainability in the process. Therefore, the integration of circular economy principles in smarter eco-cities is a holistic approach that fosters a symbiotic relationship between sustainability dimensions, promoting a sustainable urban future. For example, the 4R can be closely related to the three dimensions of sustainability in the context of smarter eco-cities (Table 8.5).

In the context of smarter eco-cities, integrating the 4R framework with AIoT can further enhance sustainability efforts. AIoT systems can provide real-time data on resource usage and waste generation, allowing these cities to make informed decisions to balance economic, social, and environmental dimensions of sustainability more effectively. Lampropoulos et al. (2024) underscore the essential roles of AIoT in meeting the UN's SDGs and promoting circular economy. The authors propose extending the traditional 3R principles to include 'Rethink,' forming a 4R sustainability framework. They evaluate this framework using a STEEPLED analysis—considering Sociocultural, Technical, Economic, Environmental, Political, Legal, Ethical, and Demographic factors. The study highlights the importance of education and mindset shifts toward sustainability and identifies AI as a powerful technology in optimizing resource efficiency and fulfilling SDGs for a sustainable future.

Indeed, circular economy practices align with numerous SDGs, including clean water and sanitation, affordable and clean energy, reduced inequalities, quality education, climate action, life on land and below water, sustainable consumption and production, sustainable cities and communities. Agrawal et al. (2021) propose a research framework for the future that explores the relationship between sustainable development, AI techniques, and the TBL—encompassing environmental, economic, and social benefits. Engaging in the circular economy is crucial for promoting sustainability, which requires a comprehensive systems-level approach that considers these three aspects and assesses their complex interactions.

TABLE 8.5 4R Framework for Circular Smarter Eco-Cities

4R	Economic Dimension	Environmental Dimension	Social Dimension
II. Reduce	Reducing resource consumption and waste generation can lead to cost savings for businesses and individuals. It promotes economic sustainability by optimizing resource use and minimizing production costs.	Reducing resource extraction and consumption directly contributes to environmental sustainability by lowering the ecological footprint of cities. It minimizes habitat destruction and resource depletion.	Resource reduction can lead to improved public health and quality of life in cities. For example, reduced air pollution from fewer manufacturing processes can lead to better respiratory health for citizens.
Reuse	Reusing products or materials can stimulate economic growth through the development of reuse markets, repair services, and refurbishment industries. It can also create jobs and boost local economies.	Reusing products reduces the need for new manufacturing and reduces waste generation, which is environmentally beneficial. It conserves resources and reduces energy and water consumption.	Promoting reuse can provide affordable goods and services to underserved communities, contributing to social equity. It can also foster a sense of community through sharing and collaborative consumption.
Recycle	Recycling can stimulate economic growth by creating markets for recyclable materials and reducing the costs of raw material acquisition.	Recycling conserves natural resources, reduces energy consumption, and mitigates pollution. It plays a crucial role in environmental sustainability by diverting waste from landfills and reducing the ecological impact of resource extraction.	Recycling programs create jobs in waste collection, processing, and recycling industries, contributing to social sustainability by reducing unemployment rates.
Recover	Energy and value recovery from waste, such as waste-to-energy processes, can generate revenue and reduce waste disposal costs for cities.	Waste recovery reduces the environmental impact of waste disposal methods like landfills and incineration. It contributes to cleaner air and water and overall environmental sustainability.	Proper waste recovery practices can minimize health risks associated with landfills and incineration, benefiting public health.

In pursuit of the circular economy vision, numerous industries are actively investing in and embracing the potential of AI technologies. This transition represents a fundamental shift toward a more sustainable future, one that demands a collaborative ecosystem comprising the collective efforts of all sectors. This holistic approach benefits individual sectors and has broader implications for an entire nation's economic and environmental landscape.

Focusing on the 4R Sustainability framework in terms of the enabling role of emerging technologies in transitioning to a sustainable and circular economy, Siakas et al. (2023) focus on the transition from a linear to circular economy model with an emphasis on resource regeneration and ecosystem restoration. The authors highlight the role of AI, IoT, and Big Data technologies in facilitating this transition. They argue that these technologies play a vital role in achieving a sustainable and closed-loop circular economy. The study contributes to the understanding that emerging technologies are tools, not magic solutions, and that a fundamental social and cultural shift is essential for a genuine transition toward a cleaner environment in line with the 4R framework.

Economic benefits are significantly enhanced through the combination of circular economy strategies and AI technologies. This synergy not only reduces the need for resource extraction and consumption but also improves the efficiency of resource use and boosts production capabilities, all of which have a positive impact on the economy. In particular, unsustainable consumption, which leads to greater extraction of raw materials and boosts production, is accelerating environmental degradation and climate change. The incorporation of AI into circular economy practices supports more sustainable economic activities and promotes environmental stewardship. Furthermore, this integration plays a crucial role in improving various aspects of sustainable smart cities, enhancing living conditions, mobility, environmental sustainability, and overall economic performance, thus creating a more holistic urban development approach (Liu et al. 2021).

The use of AI within the framework of circular economy also drives economic growth by enhancing production processes and facilitating smart urban development. It contributes to the creation of new job opportunities and fosters an environment ripe for innovation. This dynamic leads to a proliferation of economic activities and a boost in job creation, positioning AI-powered circular economy models as pivotal in the transformation toward more adaptive and innovative urban ecosystems (Rathore and

Malawalia 2021). This progression revitalizes local economies and sets a foundation for long-term economic resilience and sustainability, reflecting a shift toward more forward-thinking and eco-conscious urban planning and development.

AI-driven the circular economy approaches effectively utilize rural resources and agricultural waste, converting them into valuable energy outputs like heat and electricity, which can help bridge the energy gap in rural areas (Ngan et al. 2019). This transformation capitalizes on underutilized resources and contributes to the diversification of energy sources, which is vital for sustainable development. Additionally, these AI-enhanced circular economy strategies provide numerous social benefits. They lead to heightened customer satisfaction by offering products that are both environmentally friendly and innovative, aligning with the growing consumer preference for sustainable goods. This shift helps reduce the overall cost of products due to more efficient use of materials and energy, which, in turn, enhances the profitability of businesses engaged in these practices. Moreover, the implementation of AI in circular economic models has a direct positive impact on public health. By reducing waste and optimizing resource use, these models decrease environmental pollutants, thus contributing to cleaner air and water. This reduction in pollution leads to better health outcomes for communities, especially in areas affected by industrial waste. The synergy between AI and the circular economy fosters a healthier environment and supports broader societal well-being by promoting sustainable living and consumption patterns.

AI plays a role in the creation of circular components, materials, and products that support the principles of reuse, repair, and refurbishment, alongside critical attributes such as disassembly, recycled content, and upgradability. AI amplifies the R-principles for establishing circular economy (Agrawal et al. 2021; Akinode and Oloruntoba 2020; Lampropoulos et al. 2024; Wilts et al. 2021), including Reuse, Reduce, Recycle, Regenerate, Restore, and Renew. With respect to the latter, for example, AI possesses predictive capabilities essential for providing solutions related to renewal. It plays an instrumental role in transitioning away from waste and fossil fuels by facilitating the generation of renewable energy. Furthermore, AI actively collects real-time data on weather conditions and energy demand patterns to make short-term predictions related to renewable energy generation, energy consumption, and weather patterns.

The core tenet of circular economy underscores the importance of ensuring that materials and products are not needlessly discarded but are instead employed in continuous industrial processes (Jude et al. 2021). Consequently, the establishment of a robust infrastructure for extracting valuable nutrients from biological waste becomes imperative. AI leverages advanced visual recognition techniques and methodologies to fine-tune the components within mixed streams, thereby enabling the optimal utilization of materials and products.

The symbiotic relationship between circular economy and AI underscores the interconnectedness of modern industries, emphasizing that the pursuit of sustainability is not the sole responsibility of any single sector. Instead, it necessitates a concerted, cross-sectoral effort to create a harmonious ecosystem where resources are utilized efficiently, waste is minimized, and economic growth is aligned with environmental preservation. This collaborative endeavor fosters innovation and enhances the resilience of industries in the face of the evolving global challenges.

Furthermore, the advantages of this collective approach extend beyond individual sectors. A sustainable circular economy model not only reduces the environmental footprint but also promotes economic stability and resilience by decreasing resource dependency and minimizing waste-related costs. As industries transition toward circular economy principles supported by AI, they unlock new opportunities for growth, foster innovation, and contribute to the overall prosperity of the nation.

In conclusion, the convergence of circular economy principles and AI technologies heralds a transformative era where industries recognize their interconnectedness and their shared responsibility toward sustainability. By working together within a collaborative ecosystem, industries can collectively strive toward a common sustainability goal that benefits their individual operations but also has far-reaching positive impacts on the economic, environmental, and societal well-being of the entire nation.

8.4.7 Exploring the Interrelationships among Circular Economy, SDGs, and Smart Cities

By extending the lifespan of materials, products, and services, the circular economy model serves as a fundamental principle in accomplishing the United Nations' SDGs in smart sustainable cities. In recent years, the global discourse on urban sustainability has been marked by a growing emphasis on circular economy—a transformative concept that transcends the linear model of production and consumption within cities. In this evolving landscape, scholars, practitioners, and policymakers alike have sought to unravel the intricate relationship between the circular economy model and the broader canvas of sustainability, as encapsulated by the United Nations' SDGs. A constellation of studies, each offering unique insights and perspectives, has endeavored to illuminate this nexus, shedding light on how circular economy practices align with and contribute to these goals. These studies form a tapestry of research that collectively enhances our understanding of the interplay between circular economy strategies, environmental factors, and urban sustainability in the context of SDGs and smart cities. This synthesis embarks on a comparative journey to discern their commonalities and differences, and thus weaving together a more comprehensive understanding of the intricate relationship between the circular economy model and the tripartite sustainability framework, particularly in this context[[Tab]].

Concerning the link between circular economy and SDGS, Trica et al. (2019) focus on the environmental factors and sustainability of the circular economy model in the European Union. Their study aims to present "the economic factors of the sustainable development of a circular economy," emphasizing the importance of environmental sustainability and recycling. Their findings highlight the significance of investing in recycling infrastructure and innovative resources for sustainable economic growth and environmental protection. Schroeder et al. (2019) explore the relevance of circular economy practices to SDGs. Their literature review and matching exercise reveal that circular economy practices can directly contribute to achieving several SDG targets, such as clean water and sanitation (SDG6), affordable and clear energy (SDG7), decent work and economic growth (SDG8), and responsible consumption and production (SDG 12). They emphasize the potential for creating synergies among these goals through circular economy practices, as well as the need for empirical research to determine the implementation means required for adopting circular economy practices in the context of SDGs. Rodríguez-Anton et al. (2019) investigate the relationship between circular economy initiatives in the European Union and compliance with the SDGs. Their exploratory factor analysis and correlation analysis aim to determine if there is a statistically significant relationship between circular economy initiatives and SDG compliance. They also explore homogeneous groups of European countries in terms of SDG compliance.

Regarding the relationship between circular economy and smart cities, Formisano et al. (2022) examine the interconnection between the concepts of smart sustainable city and circular economy. The authors reveal a significant positive relationship between smart cities and circular cities, emphasizing the importance of waste management in promoting sustainable urban smartness. Waste management emerges as the most influential element of the circular economy for sustainable smart cities. It is worth noting that operationalizations rely on the relative mentions of specific EU cities within scholarly documents in this regard. The study is one of the early endeavors to quantify the relationship between smart sustainable cities and circular economies, shedding light on their systemic nature. It adds to the interplay of these two domains, serving as a foundation for future research in this area.

Dincă et al. (2022) investigate the factors supporting the development of a smart environment in smart cities, with a focus on reducing air pollution. They assess the impact of circular economy, fiscal, and environmental factors on air pollution reduction using regression models. Their findings emphasize the effectiveness of renewable energy, environmental education, circular economy implementation, and suitable policies in reducing air pollution. Additionally, a significant benefit lies in supplying data regarding the evaluation of local air quality and advocating for relevant policies to accomplish two primary goals: enhancing well-being and fostering sustainable urban areas.

The study by Rosa et al. (2023) dovetails with the study by Dong et al. (2021), which addresses the relationship between circular economy practices and their alignment with SDGs in urban contexts. Specifically, Dong et al. (2021) match circular economy with SDGs and explore how circular economy can contribute to realizing 17 SDGs in cities, considering both positive and negative impacts. Their study highlights the role of the circular economy in mitigating urban sustainability issues and provides insights into potential trade-offs. The study conducted by Rosa et al. (2023) provides an overview of the main research trends at the intersection of circular economy and the SDGs by analyzing the scientific literature. The findings offer valuable insights for advancing sustainability research aligned with 17 SDGs, emphasizing the role of the circular economy as a burgeoning area for sustainable solutions. The study's contribution lies in its novel approach, offering a comprehensive analysis that aids both the academic community and practitioners in navigating this complex and crucial intersection of themes. This contribution should be included in the synthesis of studies exploring the relationship between circular economy, sustainability goals, and smart cities, enriching the understanding of research trends and directions in this field. While Rosa et al. (2023) provide a comprehensive overview of research trends and interdisciplinary aspects, Dong et al.

(2021) delve into the specific impacts of circular economy strategies on the achievement of individual SDGs in urban settings. Together, these two studies offer a holistic perspective on how circular economy practices can contribute to sustainable urban development and the realization of SDGs.

Overall, these studies collectively highlight the importance of circular economy practices in advancing sustainability, especially within the framework of the SDGs and smart cities. They offer valuable theoretical and empirical insights into how circular economy initiatives can contribute to achieving sustainable development objectives while acknowledging potential challenges and trade-offs. They also offer practical guidance for policymakers and city planners in devising effective strategies to enhance the circular economy and tripartite sustainability practices.

8.4.8 Smarter Eco-Cities as an Emerging Approach to Data-Driven Smart Sustainable Cities

In the quest for a more sustainable future, the concept of smarter eco-cities has emerged as a beacon of holistic urban development, transcending the confines of traditional eco-cities and building on smart cities. While environmental sustainability remains its core focus, this emerging paradigm of urbanism extends its ambition to encompass a broader spectrum of sustainability facets beyond the responsible and sustainable management of the natural environment, ecosystems, and biodiversity. Accordingly, it advocates for the harmonious coexistence of economic vitality, social equity, and ecological stewardship within urban landscapes. This analysis focuses on the multifaceted nature of emerging smarter eco-cities, exploring their ability to address not only environmental sustainability but also the intertwined dimensions of economic resilience and social inclusivity. This comprehensive approach marks a pivotal shift in urban planning and policy, striving to create cities that thrive in three dimensions of sustainability in line with smart sustainable cities (Bibri et al. 2023b, 2024a) as an umbrella term for smarter eco-cities and with support of AI and AIoT technologies.

8.4.8.1 Evolution of Environmental Urban Sustainability Paradigms: Interlinked Development of Eco-Cities and Smart Cities

Since the early 2000s, eco-cities have been at the forefront of sustainable urbanism. However, it was not until the mid-2010s that smart cities began gaining momentum in pursuit of environmental sustainability (Bibri et al. 2023a). Interestingly, both eco-cities and smart cities have influenced each other (Bibri and Krogstie 2020a; Noori et al. 2020), with eco-cities incorporating advanced technologies from smart city initiatives to enhance sustainability, while smart cities increasingly adopt eco-city principles to promote environmental resilience and sustainable urban development. This interplay has driven innovation in both areas, fostering cities that are both technologically advanced and environmentally conscious. Eco-cities began benefiting from the innovative technologies of smart cities around the mid-2010s, leveraging advancements in data analytics, IoT, and smart infrastructure to enhance sustainability efforts. Conversely, smart cities started incorporating environmental technologies and practices from eco-cities in the late 2010s, integrating green solutions such as renewable energy, sustainable urban design, and circular economy principles to address environmental challenges and promote long-term resilience. The shift toward environmental sustainability in smart cities is discussed in a recent review conducted by Stübinger and Schneider (2020), where the focus on minimizing energy consumption, water usage, waste generation, and pollution has become increasingly significant.

Moreover, the United Nations' SDGs report in 2015 played a pivotal role in the transformation of eco-cities into smart sustainable cities, as well as smart cities into sustainable smart cities, giving rise to smart eco-cities. Indeed, SDG 11—Sustainable Cities and Communities—underscores the importance of advanced ICT in enhancing resource efficiency, reducing pollution, advancing urban infrastructure, addressing climate change, fostering socioeconomic development, and promoting knowledge dissemination (United Nations 2022). The contribution of advanced ICT is associated with the synergy and integration of smart city technologies and solutions and SDGs (Grossi and Trunova 2021; Mishra et al. 2022; Sharifi et al. 2024). This growing interest in innovative technologies, coupled with their capacity to address complex sustainability challenges, has driven research and performance improvements in smart sustainable cities, particularly in the fields of environmental sustainability (Bibri et al. 2023a).

The focus on environmental sustainability has shifted from primarily being associated with eco-cities to becoming a significant driving force for the development of smart cities. This evolution has led to various environmentally driven approaches to sustainable smart cities, encompassing various aspects of urban life and technology integration (e.g., Al-Dabbagh, 2022; Kim et al., 2021; Mishra et al. 2022; Sharifi et al. 2024; Shruti et al. 2019; Saravanan and Sakthinathan 2021; Tripathi et al. 2022). These approaches underscore a move toward holistic urban development that integrates environmental concerns with technological advancements while still aiming further for sustainable outcomes across economic and social dimensions. To better understand the intricate and nuanced evolution of these urban approaches and on the rapidly evolving areas in the field, Bibri et al. (2023a) provide comprehensive insights into the key underlying models—smart cities and sustainable cities. The authors explore how these models have co-evolved and transitioned, driven by the convergence of emerging technologies and their applied solutions, including Big Data, IoT, and AI. Their research emphasizes the transformative potential of these technologies in enhancing urban resilience, efficiency, and sustainability, thereby paving the way for more adaptive and responsive urban environments.

In tracing the evolution of urban paradigms from eco-cities to smart cities, it becomes evident that their mutual influence has significantly shaped their trajectories. Both have undergone a shift toward a shared focus on environmental sustainability, which has been amplified by advanced ICT. This transition has particularly given rise to environmentally focused approaches to the development of smarter eco-cities, fueled by AI and IoT and their impactful convergence—AIoT (Bibri et al. 2023b, 2024a). However, it is crucial to recognize that while environmental sustainability takes precedence, these cities continue to strive for a harmonious balance among the three dimensions of sustainability: environmental, social, and economic, especially smart eco-cities. To explore this intricate evolution and gain comprehensive insights into these evolving urban models, the subsequent discussion addresses the interconnected themes, trends, and key models, propelled by AI and AIoT technologies and solutions in light of the three dimensions of sustainability.

8.4.8.2 Evolving Emphasis on Environmental Sustainability in Current Smart Eco-Cities through Policy Instruments

While eco-cities promote the concept of balancing the three dimensions of sustainability, they have historically placed a focus on environmental sustainability. This focus often overlooks the equally important social and economic dimensions, which are essential for fostering holistic and resilient urban environments (Holmstedt et al. 2017; Kenworthy 2006; 2019; Rapoport and Vernay 2011). This narrow emphasis extends to smart eco-cities, which incorporate IoT and Big Data technologies (Bibri 2021a, b; Caprotti 2020; Caprotti et al. 2017; Späth 2017) and further to emerging smarter eco-cities that are now integrating AI and AIoT technologies (Bibri et al. 2023b, 2024a).

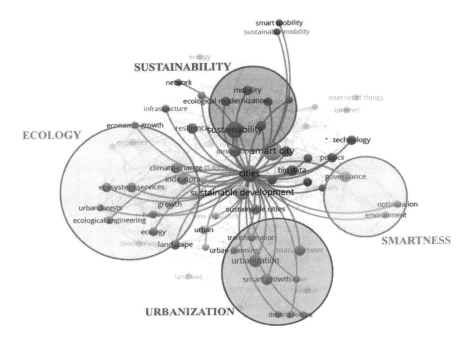

FIGURE 8.3 A bibliometric analysis for building a conceptual framework for smart eco-cities.

The dominance of environmental sustainability is evidenced in a recent bibliometric study conducted by Tarek (2023) on smart eco-cities. This research addresses the need for guidelines that connect urban design principles with smart-ecological indicators for sustainable city planning. It formulates a conceptual framework that merges smart and ecological dimensions, ultimately contributing to the achievement of SDGs across diverse contexts. As depicted in Figure 8.3, the conceptual framework is constructed around four core clusters: sustainability, ecology, smartness, and urbanization. These clusters encompass various interrelated themes, including ecological modernization, climate change, ecosystem services, ecological engineering, smart and sustainable mobility, environmental governance, resource optimization, urban planning, urban management, urban transformation, IoT, and Big Data. In smart eco-cities, the interconnectedness of sustainability, ecology, smartness, and urbanization is readily apparent. Sustainability serves as the overarching objective, guiding the integration of ecological principles, advanced technologies, and urbanization strategies. Ecology highlights the importance of harmonious coexistence between humans and the natural environment, steering eco-friendly practices, and resource preservation. Smartness involves the integration of advanced technologies and data-driven solutions to optimize urban systems and enhance the quality of life. Meanwhile, urbanization encompasses the development and transformation of cities, where smart and ecological approaches converge to create sustainable urban environments. Collectively, these components synergize to promote environmentally sustainable practices and enhance the well-being of both citizens and the surrounding ecosystem in smart eco-cities.

Smart eco-cities still strive and remain dedicated to adapting to and fostering environmental, social, and economic changes. They can achieve this by efficiently managing available resources to enhance economic and societal outcomes securely and sustainably. Bibri (2021d) develops an integrated model for strategic sustainable urban development—smart eco-cities—based on case studies. This model merges the leading global paradigms of urbanism—eco-cities, data-driven smart cities, and environmentally data-driven smart sustainable cities—by integrating their strategies and solutions. The study offers an innovative approach to constructing future models for sustainable urban development and provides practical insights into developing strategic planning processes for transformative change toward sustainability based on integrated approaches. The proposed model aims to assist cities that are branding or regenerating themselves as eco-cities, or planning to evolve into smart eco-cities in the era of big data in balancing the long-term goals of sustainability.

Bibri (2021c) investigates the emerging concept of "data-driven smart sustainable cities," which integrates sustainable urbanism and eco-urbanism, harnessing data-driven technologies to improve and balance the three dimensions of urban sustainability. Based on case study research, the author identifies and enumerates the advantages of smart eco-cities across these dimensions and assesses the institutional transformations necessary to support this integration. The findings underscore the synergistic potential of combining smart urbanism and eco-urbanism for advancing the environmental, social, and economic goals of sustainability. The study contributes to the development of a novel model for data-driven smart sustainable cities and highlights the transformative impact of data-driven technologies on urban institutions and competences.

In their study, Bibri and Krgostie (2020b) explore the development and integration of a novel model for data-driven smart sustainable cities. This model combines several existing and emerging urban paradigms, including eco-cities, data-driven smart cities, and environmentally focused smart sustainable cities. The integration of these models for sustainable urban development is based on findings from four case studies of six leading European cities known for their ecological and technological advancements. The authors argue that this model has significant potential to enhance the environmental, social, and economic goals sustainability in future cities by leveraging the synergistic benefits of data-driven technologies and solutions.

Important to note is that smart eco-cities are fundamentally shaped by the confluence of technological progress, environmental imperatives, and strategic policy instruments. Technological advancements equip these sites of urban experimentation with the means to optimize diverse systems in pursuit of sustainability. Simultaneously, environmental concerns propel them to embrace smart, eco-conscious solutions. However, it is through the prism of policy instruments that smart eco-cities find their direction and evolve. Policy frameworks, regulations, and incentives establish an enabling context for sustainable practices and innovative technologies (Joss and Cowley 2017). They facilitate the seamless integration of smart solutions into urban landscapes, ensuring their efficacy. Targeted policies are imperative for propelling progress across different facets of environmental sustainability and for effective implementation of data-driven innovations. Overall, policy instruments serve as the backbone of smart eco-city realization, providing a structured framework for planning, execution, and regulation. Among the array of policy instruments facilitating the realization of smart eco-cities are government regulations, mandating sustainability standards; financial incentives, including subsidies; public-private partnerships, which leverage expertise and resources; open data initiatives, promoting information sharing; and standards and certification programs, ensuring interoperability and reliability (Bibri and Krogstie 2020a; Caprotti et al. 2017; Cugurello 2018; Kramers et al. 2016; Noori et al. 2020; Pasichnyi 2019; Spath et al. 2017).

8.4.8.3 Smart Sustainable Cities and Their Evolving Priorities: Leveraging AI and AIoT for Advancing Tripartite Sustainability in Smarter Eco-Cities

While various approaches to smart sustainable cities aim to incorporate advanced technologies and data-driven solutions of smart cities to support or realign with the goals of environmental, social, and economic sustainability (Alagirisamy and Ramesh 2022; Bibri 2021b, c; Corsi et al. 2022; Rietbergen 2021), it is notable that the environmental and economic considerations often take precedence over the social aspects. Furthermore, the dominance of certain approaches can vary based on their discursive influence, universal acceptance, and practical feasibility. These approaches encompass smart and sustainable cities (Martin et al. 2018; Yigitcanlar and Cugurullo 2020), smart sustainable cities (Corsi et al. 2022; Alagirisamy and Ramesh 2022), data-driven smart sustainable cities (Bibli 2021b, c), smart eco-cities (Caprotti et al. 2017; Späth 2017), data-driven smart eco-cities (Bibri 2021a), and smart cities for integrated SDGs (Mishra et al. 2022; Sharifi et al. 2024; Visvizi and del Hoyo 2021). Regardless, the foundational principles of smart sustainable cities revolve around the pursuit of environmental, economic, and social sustainability.

Furthermore, there is a growing trend of integrating AI and AIoT systems to enhance efficiency, sustainability, equity, resilience, and the overall quality of life as part of the effort to rebalance the three dimensions of sustainability. This perspective aligns with one of the thematic groups in the global UN initiative "United for Smart Sustainable Cities" known as "AI in Cities," and "City Platforms" which emphasizes the implementation of efficient and sustainable AI solutions in smart sustainable cities to achieve the environmental, social, and economic targets of SDGs. In essence, novel strategies need to be devised to amplify the sway of social and economic objectives over the blueprint and growth of smart cities, an effort that resonates in the aspiration of fostering smarter and more sustainable eco-cities. This trajectory aligns with Corsi et al.'s (2022) findings that this paradigm primarily benefits the social axis, followed by the environmental axis, with comparatively less impact on the economic axis. Complementing this trajectory, Yigitcanlar and Cugurullo (2020) delve into AI's sustainability within the context of smart sustainable cities, unraveling potential synergies between this paradigm and large-scale AI.

Recent research underscores that leveraging advanced technologies and data-driven strategies in smart cities can play a pivotal role in driving the transition toward achieving SDGs. Thamik et al. (2024) explore the applications of AIoT in achieving SDGs, highlighting its potential to advance specific goals related to environmental protection, energy efficiency, renewable energy, water management, agriculture, and smart sustainable urban development, among others. Mishra and Singh (2023) investigate the role of AI technologies in boosting the economy, improving citizens' lives, and assisting government functions, contributing to safer and more livable cities. Their examination focuses on sustainable development in smart cities through AI applications in areas such as smart transport, healthcare, homes, industries, energy, waste, agriculture, governance, and education.

Tripathi and Saxena (2024) assess how AI impacts the UN' 17 SDGs, which encompass economic, social, and environmental objectives. The authors highlight that AI has the potential to enhance global productivity, equality, and environmental outcomes. However, it also poses risks that could lead to economic instability, environmental degradation, and social unrest. The study stresses the importance of evaluating both the positive and negative impacts of AI on sustainable development, identifying which SDGs are likely to benefit from AI advancements and which may face challenges. The authors argue that a thorough understanding of these dynamics is essential for leveraging AI to foster a sustainable future while mitigating its adverse effects. In their analysis of AI's impact on achieving SDGs, Vinuesa et al. (2020) reveal that AI can facilitate 134 targets across various goals spanning the three dimensions of sustainability, while potentially impeding 59 others. Joshi et al. (2024) explore AI's potential to affect individuals and the environment positively and negatively. The authors introduce two theoretical models to assess how AI aligns with the SDGs, emphasizing the risks of focusing too narrowly on AI applications without considering broader international research and development goals. They provide current information on AI's role in various sectors, such as food and energy, and suggest that while AI can significantly contribute to achieving the SDGs, more precise and coordinated efforts are necessary. They demonstrate how different AI applications can play crucial roles in advancing SDGs across various contexts, such as cities, regions, and nations, underscoring the need for tailored AI solutions to address diverse sustainability challenges.

AIoT has shown great potential to advance the environmental, social, and economic goals of sustainability by integrating AI with IoT to optimize resource management, enhance efficiency, and improve the quality of life. In this context, by harnessing the cognitive capabilities of AI and the interconnectedness of IoT devices and sensors, these technologies can transform how smart sustainable cities can be environmentally, economically, and socially managed, planned, and governed. They facilitate the creation of eco-conscious, efficient, resilient, and inclusive urban environments while addressing pressing challenges. Environmentally, AIoT demonstrates applications across numerous areas of environmental sustainability and climate change (Bibri et al. 2023b, 2024a, b; El Himer et al. 2022; Dheeraj et al. 2020; Popescu et al. 2024; Puri et al. 2020; Rane et al. 2024; Sleem and Elhenawy 2023; Thamik et al. 2024; Tomazzoli et al. 2020; Samadi 2022). These areas include resource management, energy efficiency, renewable energy deployment, waste handling, water management, transportation management, environmental monitoring, resilience against environmental threats, and climate mitigation and adaptation.

Socially, it enhances public services, healthcare, and education through real-time data and intelligent systems, promoting equity and well-being (Koffka 2023; Ishengoma et al. 2022; Mishra 2023): Economically, AIoT drives innovation and efficiency across industries, reducing costs and fostering sustainable growth (Koffka 2023; Parihar et al. 2023; Seng et al. 2022). Through these multifaceted applications, AIoT supports the holistic achievement of sustainability goals. Al-Turjman et al. (2021) present a comprehensive exploration of AI-IoT, providing a holistic understanding by integrating both theoretical and practical perspectives of the AI paradigm for IoT. The authors address a range of AI-enabled IoT applications across various domains, including the social and economic implications of AI-IoT. They cover applications in different aspects of sustainability, security, privacy, and ethics, highlighting the broad and impactful scope of AI-IoT technologies.

Overall, the role of AI and AIoT in advancing the three dimensions of sustainability is significant, offering a range of opportunities and benefits. These technologies have shown great potential to address various social challenges, promote inclusivity, improve access to essential services, and enhance the overall well-being of individuals and communities. They also drive efficiency, innovation, and growth across various economic domains. Additionally, they facilitate the creation of intelligent, interconnected systems that can address a wide range of environmental challenges and improve urban life.

The dynamic interplay between smarter eco-cities and smart sustainable cities with respect to the three dimensions of sustainability involves integrated approaches that lead to synergistic effects that promote sustainable development. To further advance the concept of smarter eco-cities, it is crucial to shift the focus from environmental sustainability to embrace the economic and social dimensions of sustainability through the utilization of AI and AIoT technologies. While environmental sustainability is a key driver in the evolution of smarter eco-cities, integrating social and economic aspects can create more comprehensive and balanced approaches to sustainable urban development. Utilizing AI and AIoT can optimize resource management, enhance economic efficiency, and improve public services, fostering social equity and well-being.

Embracing a multidimensional approach allows smarter eco-cities to achieve SDGs in a more balanced and comprehensive manner. By leveraging AI and AIoT technologies, these cities can enhance their capacity to address complex sustainability challenges, fostering innovation and collaboration across various domains. This integrated approach not only promotes environmental stewardship but also drives economic prosperity and social inclusivity, ensuring that the benefits of sustainable development are shared broadly and equitably. In other words, this holistic approach ensures that the benefits of sustainable development are widely distributed, promoting inclusive growth and resilient urban environments.

8.5 A NOVEL FRAMEWORK FOR INTEGRATING CIRCULAR ECONOMY, METABOLIC CIRCULARITY, AND TRIPARTITE SUSTAINABILITY THROUGH ARTIFICIAL INTELLIGENEC OF THINGS IN SMARTER ECO-CITIES

The integration of circular economy, metabolic circularity, and tripartite sustainability in the dynamic context of smarter eco-cities can be facilitated by recent advancements in AIoT. AIoT significantly enhances the TBL benefits associated with circular and metabolic processes while advancing sustainability goals, thereby generating value across the environmental, economic, and social spheres of smarter eco-cities. This framework outlines the key components and relationships involved in leveraging AIoT to foster sustainable urban development. It is derived from the insights gleaned from the reviewed studies on the relationship between AIoT and each of the three frameworks in question. These insights highlight how AIoT can be strategically applied to optimize resource use, enhance resilience, and achieve comprehensive sustainability. The framework illustrates how AIoT can align these frameworks to achieve balanced sustainability in smarter eco-cities. It operates through the following key components and subcomponents:

Data integration and analysis: The AIoT infrastructure integrates data from a multitude of sources within the smarter eco-city environment. These sources include energy consumption data from smart meters, material usage statistics from industrial IoT devices, waste generation records from smart waste management systems, transportation operations data from connected vehicle networks, mobility patterns from GPS and mobile sensors, and pollution levels from environmental monitoring stations. Advanced AI models and ML algorithms process and analyze this vast array of data in real-time. Through techniques such as predictive analytics and pattern recognition, AI can derive actionable insights that inform decision-making processes. These insights help optimize resource allocation, improve operational efficiency, and enhance urban sustainability by enabling proactive measures to address emerging issues.

Circular economy principles: The circular economy model forms the backbone of this framework, emphasizing principles such as resource regeneration, resource extraction reduction, waste minimization, product life extension, and closed-loop systems. These principles guide decision-making throughout smarter eco-cities. In the context of AIoT-enhanced smarter eco-cities, all these principles are crucial. For example, in resource optimization, AI algorithms can predict material usage patterns and optimize resource allocation, reducing waste; in closed-loop systems, IoT sensors track material flows, ensuring that products are reused or recycled efficiently; and in product lifecycle management, AI enables predictive maintenance and end-of-life management, extending product lifecycles and ensuring proper disposal or recycling.

Metabolic circularity: Metabolic circularity focuses on optimizing the flow and reuse of materials and energy within smarter eco-cities to create efficient, closed-loop systems that mimic natural metabolic processes. It ensures that resources are continuously cycled through the urban system, reducing waste and resource consumption. It focuses on creating synergies between different urban domains to enhance sustainability. AI analyzes data from IoT sensors to understand and optimize the flow of resources such as water, energy, and waste within the city. AIoT enables the identification and creation of synergies between different domains (e.g., waste-to-energy initiatives) to maximize resource efficiency. Real-time data from IoT devices provides continuous feedback, allowing urban systems to adapt and optimize their metabolic processes dynamically. Prior to this, SUM analyzes and manages urban resource flows using AIoT to monitor and optimize the city's energy, water, and material usage. It aims to create a balanced and sustainable urban ecosystem by ensuring that resource inputs and outputs are efficiently managed.

Tripartite sustainability: Tripartite sustainability encompasses the environmental, economic, and social dimensions of urban development. It is used to comprehensively assess the sustainability performance of smarter eco-cities. AIoT tools provide real-time data and analytics to measure and report on sustainability indicators, ensuring continuous improvement and adaptation to evolving sustainability targets. Integrating AIoT with this framework enhances the ability to achieve balanced sustainability.

AI-powered environmental monitoring systems track air and water quality, energy consumption, and waste generation, providing actionable insights to reduce ecological footprints. AIoT-driven optimization of resource usage and operational efficiency leads to cost savings and economic resilience. AI enhances urban planning by incorporating social data to improve public services, transportation systems, and community engagement, ensuring equitable access to resources and opportunities.

AIoT-enabled optimization, intelligence, and collaboration: AIoT continuously monitors and optimizes resource usage, waste management, energy consumption, and other relevant factors. It dynamically adjusts city operations to reduce resource waste, enhance energy efficiency, and promote sustainable practices. IoT sensors are deployed throughout the urban environment to collect real-time data on resource usage, environmental conditions, and human activities. AI algorithms are utilized to analyze these data, identify patterns, and generate predictive models for resource optimization, waste management, and energy efficiency. Moreover, AIoT-enabled smart grids are implemented to optimize energy distribution, integrate renewable energy sources, and reduce energy waste. AIoT systems are used to monitor waste generation, optimize collection routes, and improve recycling processes. They are also used to monitor and manage water resources, ensuring efficient usage and minimizing wastage.

Furthermore, integrated urban data platforms are created to aggregate data from various AIoT systems, providing a holistic view of the city's metabolism and resource flows. Various stakeholders, including government agencies, businesses, and communities, are engaged through collaborative platforms that facilitate data sharing, decision-making, and co-creation of sustainable solutions. AIoT facilitates real-time communication, allowing stakeholders to participate in sustainability initiatives and provide valuable feedback.

Policy and regulations: The framework advocates for the development and implementation of policies and regulations that support circular economy, metabolic circularity, and tripartite sustainability practices in smarter eco-cities. These policies also support the integration of AIoT in urban planning and management, ensuring ethical use, data privacy, and security. Furthermore, policies should provide incentives and investments for the development and adoption of AIoT technologies to promote circular economy, metabolic circularity, and tripartite sustainability. Initiatives for educating and raising awareness among citizens and businesses about the benefits of circular economy, metabolic circularity, and tripartite sustainability are integrated into the framework. Continuous research and innovation efforts are encouraged by these policies to identify emerging AIoT technologies and best practices for further enhancing the benefits of the circular economy, metabolic circularity, and sustainability outcomes.

As illustrated in Figure 8.4, the framework provides a clear and comprehensive visual representation of the relationships among its essential components. AIoT, an integration of AI with IoT, serves as the central enabler, connecting to circular economy, metabolic circularity, and SUM. Circular economy is linked to, metabolic circularity, showing the flow from resource regeneration to material and energy reuse, leading to waste minimization and environmental footprint reduction. Metabolic circularity is connected to SUM, emphasizing the optimization of urban resource flows and the reuse of materials and energy. SUM, which analyzes and manages urban resource flows for sustainability and resilience, feeds back into metabolic circularity, which in turn connects to the circular economy, creating a closed-loop system. Tripartite sustainability encompasses all components, highlighting the integration and balance of economic, social, and environmental dimensions of sustainability. SDGs provide the overarching goals that guide all efforts within the framework. Smarter eco-cities are the central focus connecting all the components of this framework, demonstrating their collective contribution to achieving SDGs.

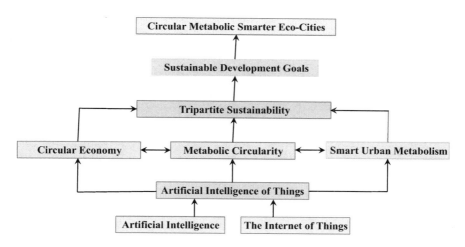

FIGURE 8.4 A novel framework for integrating circular economy, metabolic circularity, and tripartite sustainability through Artificial Intelligence of Things in smarter eco-cities.

Circular metabolic smarter eco-cities utilize the integration of AIoT to optimize resource regeneration and minimize waste, while continuously managing and analyzing urban resource flows. This holistic approach ensures a balance of economic, social, and environmental sustainability, contributing effectively to achieving SDGs. By creating a closed-loop system, these cities enhance efficiency and sustainability, paving the way for resilient and intelligent urban environments.

The framework highlights the transformative potential of AIoT in fostering sustainable urban development by integrating circular economy, metabolic circularity, and tripartite sustainability. By addressing the key components, it provides a comprehensive approach to creating smarter, more sustainable eco-cities. This integrated approach not only advances sustainability goals but also enhances the quality of life. It addresses environmental and economic challenges and ensures social equity and resilience, paving the way for sustainable urban ecosystems. By harmonizing the three frameworks through AIoT, it creates a dynamic, data-driven, responsive, and inclusive approach to advancing the sustainability of smarter eco-cities, ultimately contributing to the balanced progress of its dimensions in the pursuit of an efficient, sustainable, resilient, and prosperous future.

8.6 DISCUSSION

The adoption and integration of circular economy, metabolic circularity, and tripartite sustainability practices in smarter eco-cities have been progressing at a gradual pace. This slow advancement is primarily due to the hurdles and bottlenecks arising from the complex dynamics of sustainability transitions and the innovations required to achieve a unified and holistic perspective. In the pursuit of advancing these practices, the role of AI and AIoT has emerged as a transformative force. While the potential for positive transformation is immense, the journey is not without its challenges and barriers. This discussion delves into the complex landscape where AI and AIoT converge with the ideals of circular economy, metabolic circularity, tripartite sustainability practices, and critically examines the multifaceted aspects of incorporating these technologies into smarter eco-cities and their implications for these practices. Additionally, it highlights gaps in current knowledge and outlines future research directions. This aims to pave the way for more resilient, efficient, intelligent, and sustainable urban ecosystems.

8.6.1 Key Findings and Their Interpretation

The findings of this study illuminate the transformative potential of AI and AIoT in advancing the development of smarter eco-cities, offering a promising pathway toward achieving urban sustainability goals. The integration of AI and IoT into AIoT has demonstrated substantial potential in optimizing urban resource management and promoting sustainable development. By facilitating real-time data collection and analysis, AIoT enables more informed decision-making processes, which are crucial for the effective implementation of circular economy, metabolic circularly, and tripartite sustainability. This capability not only enhances resource regeneration and waste minimization but also supports the creation of closed-loop systems that emulate natural metabolic processes. Additionally, AIoT promotes tripartite sustainability by balancing environmental, economic, and social dimensions, ensuring a holistic approach to urban development. The practical applications of AIoT in urban settings are manifold and contribute to the broader sustainability goals by promoting resource efficiency and reducing environmental impacts.

8.6.2 Circular Economy and Artificial Intelligence of Things in Smarter Eco-Cities

8.6.2.1 Challenges of Implementing Circular Economy in Smarter Eco-Cities

The synergy between circular economy and AIoT in the realm of smarter eco-cities represents an exciting frontier, but it comes with its fair share of complexities. Therefore, it is important to unravel the multifaceted challenges that lie ahead in the quest to create more circular and smarter eco-cities. Some of these challenges are distilled and elaborated on based on the recent research addressing circular economy and its relation to urban sustainability and SDGs (e.g., Busu and Trica 2019; Dincă et al. 2022; Dong et al. 2021; Kirchherr et al. 2017; Merli et al. 2018; Rodríguez-Anton 2019; Rosa et al. 2023; Schroeder et al. 2019; Smol et al 2021; Trica et al. 2019; Velte et al. 2018).

Behavioral change: One of the fundamental challenges facing the transition to a circular economy in smarter eco-cities is the need to induce behavioral change among residents and businesses. Individuals and organizations are often accustomed to linear consumption patterns, making it difficult to shift from such entrenched habits toward recycling, reusing, and reducing waste. Overcoming this challenge necessitates comprehensive public awareness campaigns, education programs, and incentives to motivate the adoption of circular practices. Citizens and businesses must recognize the long-term benefits and positive environmental impact of circularity.

Infrastructure and technology: Establishing the requisite infrastructure for effective recycling, waste separation, and material reuse presents a substantial challenge. The development and maintenance of advanced recycling facilities, waste-sorting centers, and collection networks demand significant financial investments. Smarter eco-cities need to allocate resources to build and upgrade these facilities while integrating cutting-edge technologies that enhance efficiency in material recovery and processing.

Regulatory frameworks: Developing and enforcing regulatory frameworks that promote circularity is a complex endeavor. It involves the formulation of stringent standards for waste reduction, sustainable product design, and responsible recycling practices. Achieving consensus among stakeholders, including government bodies, industries, and environmental organizations, can be challenging. Striking the right balance between incentivizing circular practices and imposing necessary regulations is crucial to success.

Economic viability: Circular economy initiatives often require substantial initial investments, which can deter both businesses and municipalities. Convincing stakeholders that embracing circular practices can yield long-term cost savings and economic benefits is a formidable task. Demonstrating the financial viability of circularity through case studies and tangible examples is essential to garnering buy-in from decision-makers and investors.

Consumer awareness: Educating citizens about the advantages of circular economy practices and their role in contributing to sustainability is of paramount importance. Without widespread awareness and understanding, the adoption of circular practices remains limited. Smarter eco-cities must embark on comprehensive awareness campaigns, engaging the public through schools, community programs, and digital media to foster a sense of responsibility and commitment to circularity.

8.6.2.2 Challenges of Integrating Circular Economy with AI and AIoT in Smarter Eco-Cities

The integration of AI and AIoT with the circular economy presents complex challenges, as highlighted by numerous recent studies (e.g., Agrawal et al. 2021; Akinode and Oloruntoba 2020; Ali et al. 2023; Alonso et al.

2021; Cîmpeanu et al. 2022; Lampropoulos et al. 2024; Roberts et al. 2022; Ramadoss et al. 2018; Rathore and Malawalia 2021; Sertyesilisik 2023; Siakas et al. 2023).

Data management: The proliferation of AI and AIoT technologies in smarter eco-cities generates an immense volume of data. Effectively managing and harnessing these data for optimizing circular practices, such as waste collection and resource allocation, is a multifaceted challenge. These cities must develop advanced data management systems capable of handling and analyzing these vast datasets efficiently. This includes implementing data storage solutions, data integration platforms, and real-time analytics tools to derive actionable insights for informed decision-making.

Privacy and security: Robust data security and privacy measures are imperative in AI and AIoT systems related to circular economy practices. Safeguarding these data against potential breaches and misuse is paramount. Smarter eco-cities must implement robust cybersecurity measures, data encryption protocols, and stringent access controls to protect against data breaches and ensure residents' trust in the use of AI and AIoT in circular economy practices. Blockchain can also be employed to enhance data security and traceability in supply chains.

Complexity: The integration of circular economy principles into AI and AIoT systems adds layers of complexity to smarter eco-cities. Designing models and algorithms that factor in metabolic circularity alongside considerations such as efficiency and cost-effectiveness presents a formidable challenge. Striking the right balance between these competing priorities and ensuring that metabolic circularity remains a central focus in the decision-making process requires innovative solutions and multidisciplinary collaboration.

Interoperability: Seamless interoperability between different AI and AIoT devices and systems remains a technical hurdle yet pivotal for the success of both metabolic and circular practices in smarter eco-cities. Incompatibility and fragmentation issues can hinder the efficiency and effectiveness of these initiatives, as well as the optimization of urban functions, necessitating standardization protocols and communication interfaces. Establishing these interfaces is essential to enable the integration of diverse technologies and ensure that they work together harmoniously for achieving better outcomes.

Scalability: Adapting circular practices across an entire city while integrating AI and AIoT necessitates meticulous planning and scalability considerations. Solutions that prove effective in small-scale pilot projects at the district level may not be readily scalable to the broader urban environment. Identifying scalable models and strategies that can be applied city-wide is crucial for achieving meaningful metabolic circularity at scale. Furthermore, scalability in AIoT involves designing systems that can handle a growing number of connected devices and data points. Implementing micro services-based architectures and utilizing cloud-based IoT platforms can help scale AIoT solutions efficiently. Moreover, adherence to standardized communication protocols and device management frameworks enhances scalability.

Human-centric design: Circular economy initiatives should align with the needs and behaviors of residents. AI and AIoT systems must be designed to interact with people in ways that not only encourage but also facilitate circular practices. User-friendly interfaces, clear communication, and incentives for citizen engagement are essential components of human-centric design to ensure that circularity becomes an integral part of daily life.

Regulatory frameworks: Creating regulatory frameworks for AI and AIoT in circularity and sustainability requires collaboration between technical experts and policymakers. Regulations may encompass data governance, environmental certifications, and ethical guidelines for AI models and algorithms. Compliance can be ensured through auditing mechanisms and certification processes.

Cross-disciplinary collaboration: Coordinating circular economy efforts among diverse stakeholders requires the development of data-sharing standards and interoperability protocols. Technical experts from various industries should collaborate to establish these standards. For example, the development of open data formats and Application Programming Interfaces for AIoT devices can facilitate data exchange and interoperability between different systems.

Investment: Integrating AI and AIoT technologies to advance circular economy goals demands substantial upfront investment. Convincing stakeholders, including government bodies and investors, of the long-term benefits and return on investment can be challenging. This return is often calculated based on cost savings, resource optimization, and revenue generation. As these are typically quantified, the benefits of circular economy might be less immediate and tangible, requiring multiple tracking metrics that can be complex to quantify. Smarter eco-cities must provide compelling evidence through comprehensive cost-benefit analyses and case studies to secure the necessary funding for these transformative initiatives.

Demonstrating the economic advantages of circular practices and their positive impact on resource efficiency is essential for garnering financial support and investment.

Addressing these challenges will be critical in realizing the potential of circular economy principles integrated with AI and AIoT technologies in smarter eco-cities. Collaboration among governments, businesses, technology providers, communities, and citizens is essential to overcome these obstacles and advance sustainability goals. Moreover, addressing the technical challenges and open issues associated with AIoT in metabolic circularity and tripartite sustainability requires an multidisciplinary approach. It involves technological innovation, policy development, and industry collaboration.

8.6.2.3 Future Research Directions

The challenges associated with implementing circular economy principles and integrating them with AI and AIoT technologies in the context of smarter eco-cities open the door to a host of intriguing research avenues. In navigating the dynamic landscape of urban sustainability and technological innovation, addressing these challenges presents an opportunity for scholars and practitioners to explore novel solutions and contribute to the advancement of smarter and more sustainable cities. Next is an outline of the key promising avenues for future research in response to the existing knowledge gaps, shedding light on key directions that can shape the discourse surrounding circularity, AI, and AIoT in future urban environments.

- Develop effective strategies for promoting behavioral change among residents and businesses in smarter eco-cities. This includes understanding what motivates individuals and induces businesses to embrace circular practices and designing targeted measures and campaigns accordingly.
- Investigate and develop innovative circular economy business models suitable for smarter eco-cities, considering economic viability and incentives for businesses to participate.
- Analyze the effectiveness of regulatory frameworks in promoting circularity. Comparative studies of different cities' regulatory approaches can provide insights into best practices.

- There is a lack of comprehensive research on the cost-effectiveness and scalability of circular infrastructure in smarter eco-cities. Studies should focus on evaluating the long-term benefits of circular systems and technologies.
- Conduct comprehensive impact assessments of circular economy initiatives in smarter eco-cities to evaluate their environmental, economic, and social outcomes.
- Develop standardized metrics and assessment tools to measure the circularity level of smarter eco-cities. This would allow for benchmarking and tracking progress over time.
- Explore AI-driven tools and algorithms for circular product and infrastructure design, considering factors such as material selection, recyclability, and reusability.
- Improve interoperability among various AI and AIoT systems to enable seamless integration of circular practices with other urban operations and functions.
- Investigate robust privacy and security measures for AIoT devices and AI systems used in circular economy applications to address potential data breaches and privacy concerns.
- Explore effective models of multi-stakeholder collaboration involving governments, businesses, technology providers, and citizens to drive circular economy initiatives in smarter eco-cities.
- Develop frameworks and policies for responsible data governance in circular economy applications, ensuring transparency, accountability, and data ethics.

Exploring these research avenues can contribute to the successful implementation of circular economy principles in smarter eco-cities while leveraging the potential of AI and AIoT technologies.

8.6.3 Artificial Intelligence of Things in Smarter Eco-Cities

8.6.3.1 Challenges of Incorporating and Balancing the Three Dimensions of Sustainability

In the context of emerging smarter eco-cities, there are several key challenges and barriers associated with incorporating and balancing the three dimensions of sustainability and utilizing AI and AIoT technologies to enable this incorporation and balance. The following challenges are distilled from recent research addressing the relationship between tripartite sustainability and smart cities and smarter eco-cities (e.g., Bibri et al. 2023b; Bibri et al. 2024b; Iris-Panagiota and Egleton 2023; Mishra and Singh 2023; Sharifi et al. 2024; Thamik et al. 2024; Vinuesa et al. 2020; Yigitcanlar and Cugurullo 2020; Yigitcanlar et al. 2020; Zaidi et al. 2023):

Competing priorities: Striking a balance between environmental, social, and economic sustainability is a persistent challenge in smarter eco-cities. Often, these dimensions present conflicting priorities. For instance, economic growth, driven by industrial expansion, may conflict with environmental conservation efforts aimed at reducing pollution and preserving natural habitats. Navigating these tensions and finding ways for these dimensions to complement rather than hinder each other is a complex task for policymakers and city planners.

Resource scarcity: Many existing smart eco-cities grapple with resource scarcity, adding complexity to sustainability efforts. The need to ensure economic prosperity and social well-being while managing limited resources like energy and raw materials can be a daunting challenge. Balancing these demands, especially in rapidly growing urban environments, requires innovative strategies and technologies that can maximize resource efficiency and minimize waste.

Behavioral change: Encouraging behavioral change among citizens and businesses to adopt sustainable practices across all three dimensions is a formidable task. People often resist changes that impact their daily lives or economic interests. It can be an uphill battle to engage and persuade individuals to embrace practices, such as energy conservation and sustainable transportation. This necessitates effective communication, education, and incentives. Fostering a culture of sustainability within smarter eco-cities requires comprehensive and targeted strategies.

Policy integration: Local governments often maintain separate policies and regulations for each dimension of sustainability. Aligning these policies and integrating them effectively to create a cohesive sustainability framework poses a significant barrier. Policymakers must work to harmonize regulations, streamline administrative processes, and promote cross-sector collaboration to achieve synergistic outcomes across the environmental, social, and economic realms.

Data integration: The complexity of collecting, managing, and integrating data related to environmental, social, and economic factors can impede holistic decision-making. Smarter eco-cities rely on data-driven insights to inform their sustainability strategies, but different data sources, formats, and quality standards can hinder the seamless integration of these data. Developing robust data infrastructure and analytics capabilities to overcome these challenges is vital for effective urban planning and decision support.

Trade-offs: Trade-offs between sustainability dimensions, in some instances, may be necessary, introducing another layer of complexity. For example, prioritizing economic sustainability through technological development may temporarily impact environmental sustainability by increasing emissions or resource consumption. Balancing these trade-offs requires careful consideration of short-term versus long-term benefits and the development of strategies to mitigate negative impacts.

Measuring impact: Establishing standardized metrics and assessment methods to measure the impact of sustainability initiatives across dimensions is a daunting task. Quantifying social and environmental outcomes can be particularly challenging due to the interconnected nature of these dimensions. Smarter eco-cities need reliable tools and indicators to evaluate the effectiveness of their policies and initiatives comprehensively. This necessitates interdisciplinary research and the development of robust evaluation frameworks tailored to the unique characteristics of smarter eco-city contexts.

8.6.3.2 Challenges of Utilizing AI and AIoT for Tripartite Sustainability in Smarter Eco-Cities

Implementing AI and AIoT in smarter eco-cities involves navigating complex challenges that can hinder their effectiveness and wide adoption. Additionally, integrating these advanced technologies into existing urban frameworks requires careful planning and adaptation to ensure they contribute positively to sustainability goals. A set of key challenges is outlined below, drawing on insights from many studies (e.g., Ahmed et al. 2020; Bibri et al. 2023b, 2024b; Brevini 2020; Dauvergne et al. 2021; Lannelongue et al. 2021; Lutz 2019; Paracha et al. 2024; Raman et al. 2024; Vinuesa et al. 2020; Wazir et al. 2022; Yigitcanlar et al. 2020).

Data privacy and security: As AI and AIoT systems amass vast quantities of data, concerns about data privacy and security intensify. Striking a balance between extracting meaningful insights from these data and safeguarding sensitive information is a critical challenge. Additionally, robust identity and access management systems can help control data access and prevent unauthorized use of AIoT systems. Technical solutions to address privacy in AI algorithms involve privacy-preserving ML techniques, such as federated learning and homomorphic encryption.

Digital divide: Bridging the digital divide to ensure equitable access to AI and AIoT technologies is a multifaceted challenge. Achieving social sustainability in this regard hinges on providing all citizens, regardless of socioeconomic status, with equal opportunities to benefit from these innovations. It also involves providing affordable, Internet-connected devices to different user groups and communities. Technological solutions such as low-cost IoT sensors and community-based Wi-Fi networks can play a key role in bridging this gap. Smarter eco-cities must develop inclusive strategies to address disparities in digital access and literacy, ensuring that citizens and communities do not fall further behind in the digital age.

Algorithm bias: Addressing bias in AI and AIoT systems remains a persistent challenge, as biased algorithms can exacerbate social inequality. To achieve social sustainability, smarter eco-cities must continuously work to identify and rectify bias in ML/DL algorithms used for decision-making, ensuring fairness, inclusivity, and transparency in various domains.

Regulatory frameworks: The legal and regulatory landscape for AI and AIoT is still evolving, contributing to uncertainty and challenges related to data ownership, liability, and ethical usage of AI. Smarter eco-cities must actively engage in shaping the regulatory frameworks that govern these technologies to ensure that they align with their sustainability goals.

Environmental impact: AI and AIoT's impacts on environmental sustainability can be significant. They can be energy-intensive in terms of computational resources associated with sensing, processing, analysis, and communication, potentially conflicting with environmental sustainability objectives. Finding eco-friendly solutions, applying eco-design principles, and promoting energy-efficient AI and AIoT technologies are essential to minimize these impacts. The latter can include using low-power hardware components, efficient data transmission protocols, and advanced power management techniques. AI algorithms can also be designed to operate more efficiently, reducing computational demands and energy consumption. E-waste management is another significant concern. AIoT devices may have a short lifespan. Ensuring responsible disposal and recycling of these devices requires incorporating design principles for easy disassembly and recycling. It also involves the development of AIoT devices with longer lifespans and upgradability options, reducing the need for frequent replacements.

Community engagement: Engaging citizens in the utilization of AI and AIoT technologies for sustainability goals can be met with skepticism. Overcoming these challenges requires proactive communication, public education campaigns, and mechanisms for soliciting and incorporating community input. Ethical Concerns: Addressing the ethical implications of AI and IoT technologies, such as surveillance, data privacy, and algorithmic decision-making, is essential for maintaining social trust and equity. Smarter eco-cities must establish robust ethical guidelines and mechanisms for transparent and accountable technology usage.

Job displacement: AI and AIoT pose concerns due to automation. To mitigate this, reskilling and workforce development programs can be implemented. These programs should focus on teaching digital literacy, programming, and data analysis skills. Technical training institutions and online platforms can provide accessible resources for up-skilling the workforce.

Ultimately, incorporating and balancing the three dimensions of sustainability in emerging smarter eco-cities while utilizing AI and AIoT technologies is a complex endeavor. Overcoming these challenges will be essential for creating and achieving smarter eco-cities of the future.

8.6.3.3 Existing Gaps and Avenues for Future Exploration

While the integration and balance of the three dimensions of sustainability are fundamental to the concept of smarter eco-cities, knowledge gaps continue to persist in our understanding and practical application. Moreover, AI and AIoT further complicate this intricate puzzle. To navigate the complex landscape of emerging smarter eco-cities successfully, it is essential to identify and address these gaps while charting robust avenues for future research. Next is a set of relevant suggestions of promising directions for research, shedding light on how AI and AIoT can catalyze the pursuit of tripartite sustainability in urban environments.

- Research should explore the development of integrated policy frameworks that prioritize sustainability in its tripartite composition. Strategies might include policy coherence and the creation of multidisciplinary teams to oversee sustainability initiatives.
- Investigate community-centric approaches that empower citizens to actively participate in sustainability efforts. Strategies could include the co-creation of solutions and incentives for sustainable behavior.

- Develop cross-disciplinary metrics and assessment methods that can comprehensively measure the impact of sustainability initiatives across dimensions. This can involve collaboration between environmental scientists, economists, and social scientists.

- Investigate the holistic impact of the interplay between smarter eco-cities and smart sustainable cities in terms of the underlying multifaceted interactions, synergies, and trade-offs. A potential research avenue is to conduct empirical research that examines how this interplay influences the environmental, social, and economic aspects of sustainability. This avenue could involve conducting case studies, quantitative analysis, or qualitative research to explore the interconnectedness, feedback loops, and potential conflicts between these two related paradigms of urbanism.

- Develop strategies to ensure that sensitive data is protected while still enabling data-driven decision-making processes related to tripartite sustainability.

- Study strategies for ensuring fairness and social equality in AI and AIoT systems. This can include bias mitigation techniques, algorithm auditing, and transparent decision-making processes.

- Develop regulatory frameworks for AI and AIoT and advocate for clear guidelines on data ownership, liability, and ethical AI usage at local and regional levels.

- Explore interoperability standards for AIoT devices and AI systems to ensure they can seamlessly work together based on collaboration with industry and standards organizations.

- Investigate strategies for sustainable or green AI and AIoT, including energy-efficient hardware and algorithms. Promote responsible AI development practices to minimize environmental impact.

- Create digital platforms that facilitate community engagement in AI and AIoT initiatives. These platforms can encourage participation, feedback, and co-design of solutions.

- Promote ethical considerations as part of the design process for AI and AIoT systems. Encourage research on ethical frameworks and guidelines for technological development.

These gaps and research avenues provide a roadmap for addressing the challenges and barriers while harnessing the potential of AI and AIoT in achieving tripartite sustainability in emerging smarter eco-cities. Collaboration between academia, industry, government, and communities will be crucial in advancing research and implementing effective strategies to create urban environments that are environmentally sound, economically viable, and socially beneficial in line with the fundamentals of smart sustainable cities.

8.7 CONCLUSION

This study has delved into the transformative role of AI and AIoT in advancing the development of smarter eco-cities. By leveraging the synergistic potential of circular economy, metabolic circularity, and tripartite sustainability frameworks, AIoT can significantly enhance the sustainability and resilience of urban environments.

The aim of this study was to explore the transformative role of AI and AIoT technologies in harnessing the synergies of circular economy, metabolic circularity, and tripartite sustainability to advance the development of smarter eco-cities toward achieving SDGs. The findings of this study indicate that AI and AIoT can profoundly impact data-driven decision-making processes, leading to smarter eco-cities that are more aligned with SDGs. AI and AIoT technologies facilitate real-time data collection, analysis, and interpretation, supporting the circular economy's principles of resource regeneration and waste minimization. Additionally, they bolster urban metabolic circularity by optimizing the flow, reuse, and closed-loop systems of materials and energy within smarter eco-cities. Moreover, AI and AIoT's ability to support urban metabolic circularity is particularly noteworthy. These technologies enable cities to emulate natural metabolic processes, creating more resilient and adaptive urban systems. This approach minimizes environmental impacts and enhances the overall quality of life for urban residents. Furthermore, AI and AIoT technologies promote tripartite sustainability by balancing environmental, economic, and social dimensions, ensuring a holistic approach to urban development.

The contributions of this study are multifaceted. Firstly, it provides a comprehensive examination of AI and AIoT's capabilities in fostering sustainable urban development. By highlighting the practical applications and benefits of these technologies, the study offers valuable insights for researchers, practitioners, and policymakers. These insights are crucial for advancing the knowledge base in the field of sustainable urban development and for guiding future research endeavors toward achieving urban sustainability goals. Secondly, the study presents actionable guidance for urban planners and policymakers. By showcasing the potential of AI and AIoT to address urban challenges, it underscores the need for strategic deployment of these technologies. Policymakers are encouraged to support the integration of AI and AIoT in urban planning and governance initiatives to create smarter, more sustainable cities.

The implications of this research are profound. The strategic deployment of AI and AIoT technologies can lead to more sustainable, resilient, and efficient urban environments. This can help cities to meet and exceed their sustainability targets. Furthermore, the insights gained from this study can inform policy and regulatory frameworks, ensuring that the adoption of AIoT is aligned with broader sustainability objectives. By emphasizing the importance of circular economy and metabolic circularity, this research also highlights the need for a systemic and holistic approach to urban development. This approach should integrate environmental, economic, and social dimensions, thereby achieving the tripartite sustainability that is essential for the long-term viability of urban areas.

While this study provides a foundational understanding of the role of AI and AIoT in advancing smarter eco-cities, it also identifies several avenues for future research. There is a need for more empirical studies that examine the real-world applications and impacts of AI and AIoT in diverse urban contexts. Additionally, research should explore the potential challenges and barriers to the implementation of these technologies, including issues related to data privacy, cybersecurity, digital divide, fairness, interoperability, and scalability.

Furthermore, future studies should investigate the long-term effects of AI and AIoT on urban sustainability, considering both positive outcomes and potential unintended consequences. By addressing these research gaps, scholars can contribute to a more comprehensive understanding of how AIoT can be harnessed to create truly sustainable and resilient cities.

In summary, this study underscores the transformative potential of AI and AIoT in advancing the development of smarter eco-cities. These technologies can significantly advance urban sustainability by leveraging the synergistic potential of the circular economy, metabolic circularity, and tripartite sustainability frameworks. The findings and insights provided herein offer valuable guidance for researchers, practitioners, and policymakers, contributing to the advancement of knowledge in sustainable urban development. As cities continue to grow and evolve, the strategic deployment of AI and AIoT technologies will contribute to the achievement of global sustainability goals.

REFERENCES

Abou Baker, N., Szabo-Müller, P., & Handmann, U. (2021). A feature-fusion transfer learning method as a basis to support automated smartphone recycling in a circular smart city. In: S. Paiva, S. I. Lopes, R. Zitouni, N. Gupta, S. F. Lopes, & T. Yonezawa (Eds.), *Science and Technologies for Smart Cities. SmartCity360° 2020.* Lecture Notes of the Institute for Computer Sciences, Social Informatics and Telecommunications Engineering (vol. 372). Springer https://doi.org/10.1007/978-3-030-76063-2_29

Agrawal, R., Wankhede, V. A., Kumar, A., Luthra, S., Majumdar, A., & Kazancoglu, Y. (2021). An exploratory state-of-the-art review of artificial intelligence applications in circular economy using structural topic modeling. *Operations Management Research, 15*(4), 609–626. https://doi.org/10.1007/s12063-021-00212-0.

Ahmad, M. A., Teredesai, A., & Eckert, C. (2020). Fairness, accountability, transparency in AI at scale: Lessons from national programs. In: Proceedings of the 2020 Conference on Fairness, Accountability, and Transparency, Barcelona, Spain, 27–30 January 2020 (pp. 690–699).

Ahvenniemi, H., Huovila, A., Pinto-Seppä, I., Airaksinen, M., & Valvontakonsultit Oy, R. (2017). What are the differences between sustainable and smart cities? *Cities, 60*, 234–245. https://doi.org/10.1016/j.cities.2016.09.009

Akinode, J. L., & Oloruntoba, S. A. (2020). Artificial intelligence in the transition to circular economy. *American Journal of Engineering Research (AJER), 9*, 185–190.

Alagirisamy, B., & Ramesh, P. (2022). Smart sustainable cities: Principles and future trends. In I. Pal & S. Kolathayar (Eds.), *Sustainable Cities and Resilience* (Vol. 183, Lecture Notes in Civil Engineering). Springer. https://doi.org/10.1007/978-981-16-5543-2_25.

Alcayaga, A., Wiener, M., & Hansen, E. G. (2019). Towards a framework of smart-circular systems: An integrative literature review. *Journal of Cleaner Production, 221*, 622–634. https://doi.org/10.1016/j.jclepro.2019.02.085

Al-Dabbagh, R. (2022). Dubai, the sustainable, smart city. *Renewable Energy and Environmental Sustainability, 7*, 3.12. https://doi.org/10.1051/rees/2021049

Ali, Z. A., Zain, M., Pathan, M. S., & Mooney, P. (2023). Contributions of artificial intelligence for circular economy transition leading toward sustainability: An explorative study in agriculture and food industries of Pakistan. *Environmental Development and Sustainability, 26*, 19131–19175. https://doi.org/10.1007/s10668-023-03458-9

Allam, Z., Ayyoob, S, Bibri, S., & Jones, D. (2022). The metaverse as a virtual form of smart cities: Opportunities and challenges for environmental, economic, and social sustainability in urban futures. *Smart Cities, 5*, 771–801. https://doi.org/10.3390/smartcities5030040

Arbolino, R., De Simone, L., Carlucci, F., Yigitcanlar, T., & Ioppolo, G. (2018). Towards a sustainable industrial ecology: Implementation of a novel approach in the performance evaluation of Italian regions. *Journal of Cleaner Production, 178*, 220–236. https://doi.org/10.1016/j.jclepro.2017.12.183

Backmann, A, Ericsson, H, Eriksson, T, Jarnhammar, M., Klasson, K., Larsson, S., & Ranhagen, U. (2014). SymbioCity PROCESS GUIDE In search of synergies for sustainable cities. In: *Swedish Association of Local Authorities and Regions (SALAR), SKL International and Sida*. Swedish Development Cooperation Agency.

Bag, S., Pretorius, J. H. C., Gupta, S., & Dwivedi, Y. K. (2021). Role of institutional pressures and resources in the adoption of big data analytics powered artificial intelligence, sustainable manufacturing practices and circular economy capabilities. *Technological Forecasting and Social Change, 163*, 120420.

Bai, X. (2007). Industrial ecology and the global impacts of cities. *Journal of Industrial Ecology, 11*(2):1–6.

Bibri, S. E. (2019). Smart sustainable urbanism: Paradigmatic, scientific, scholarly, epistemic, and discursive shifts in light of big data science and analytics. In: Big Data Science and Analytics for Smart Sustainable Urbanism. Advances in Science, Technology & Innovation. Springer. https://doi.org/10.1007/978-3-030-17312-8_6

Bibri, S. E. (2020). Data-driven environmental solutions for smart sustainable cities: Strategies and pathways for energy efficiency and pollution reduction. *Euro-Mediterranean Journal for Environmental Integration, 5*(66). https://doi.org/10.1007/s41207-020-00211-w

Bibri, S. E. (2021a). Data-driven smart eco-cities and sustainable integrated districts: A best-evidence synthesis approach to an extensive literature review. *European Journal of Futures Research, 9*(1), 16. https://doi.org/10.1186/s40309-021-00181-4

Bibri, S. E. (2021b). Data-driven smart sustainable cities of the future: An evidence synthesis approach to a comprehensive state-of-the-art literature review. *Sustainable Futures, 3*, 100047. https://doi.org/10.1016/j.sftr.2021.100047

Bibri, S. E. (2021c). A novel model for data-driven smart sustainable cities of the future: The institutional transformations required for balancing and advancing the three goals of sustainability. *Energy Informatics, 4*(4). https://doi.org/10.1186/s42162-021-00138-8

Bibri, S. E. (2021d). Data-driven smart eco-cities of the future: An empirically informed integrated model for strategic sustainable urban development. *World Futures, 79*(7–8), 703–746. https://doi.org/10.1080/02604027.2021.1969877

Bibri, S. E., & Huang, J. (2025). Artificial intelligence of sustainable smart city brain and digital twin: A pioneering framework for advancing environmental sustainability. *Environmental Technology and Innovation* (in press).

Bibri, S. E., & Krogstie, J. (2020a). Environmentally data-driven smart sustainable cities: Applied innovative solutions for energy efficiency, pollution reduction, and urban metabolism. *Energy Informatics, 3*(1), 29. https://doi.org/10.1186/s42162-020-00130-8

Bibri, S. E., & Krogstie, J. (2020b). Data-driven smart sustainable cities of the future: A novel model of urbanism and its core dimensions, strategies, and solutions. *Journal of Futures Studies, 25*(2), 77–94. https://doi.org/10.6531/JFS.202012_25(2).0009

Bibri, S. E., Alahi, A., Sharifi, A., & Krogstie, J. (2023a), Environmentally sustainable smart cities and their converging AI, IoT, and big data technologies and solutions: An integrated approach to an extensive literature review. *Energy Information, 6*, 9. https://doi.org/10.1186/s42162-023-00259-2

Bibri, S. E., Huang, J., Jagatheesaperumal, S. K., & Krogstie, J. (2024b). The synergistic interplay of artificial intelligence and digital twin in environmentally planning sustainable smart cities: A comprehensive systematic review. *Environmental Science and Ecotechnology, 20*, 100433. https://doi.org/10.1016/j.ese.2024.100433

Bibri, S. E., Huang, J., & Krogstie, J. (2024a). Artificial intelligence of things for synergizing smarter eco-city brain, metabolism, and platform: Pioneering data-driven environmental governance. *SustainableCities and Society*, 105516. https://doi.org/10.1016/j.scs.2024.105516

Bibri, S., Krogstie, J., Kaboli, A., & Alahi, A. (2023b). Smarter eco-cities and their leading-edge artificial intelligence of things solutions for environmental sustainability: A comprehensive systemic review. *Environmental Science and Ecotechnlogy, 19*, https://doi.org/10.1016/j.ese.2023.100330

Brevini, B. (2020). Black boxes, not green: Mythologizing artificial intelligence and omitting the environment. *Big Data & Society, 7*(2), 2053951720935141. https://doi.org/10.1177/2053951720935141

Busu, M., & Trica, C. L. (2019). Sustainability of circular economy indicators and their impact on economic growth of the European Union. *Sustainability, 11*(5481). https://doi.org/10.3390/su11195481

Campbell-Johnston, K., Cate, J., Elfering-Petrovic, M., & Gupta, J. (2019). City level circular transitions: Barriers and limits in Amsterdam, Utrecht, and The Hague. *Journal of Cleaner Production, 235*, 235–1239. https://doi.org/10.1016/j.jclepro.2019.06.317

Caprotti, F. (2020). Smart to green: Smart eco-cities in the green economy. In: K. S. Willis & A. Aurigi (Eds.), *The Routledge Companion to Smart Cities* (pp. 200–209). Routledge.

Caprotti, F., Cowley, R., Bailey, I., Joss, S., Sengers, F., Raven, R., Spaeth, P., Jolivet, E., Tan-Mullins, M., & Cheshmehzangi, A. (2017). *Smart Eco-City Development in Europe and China: Policy Directions*. University of Exter (SMART-ECO Project, Exter.

Cîmpeanu, I.-A., Dragomir, D.-A., & Zota, R. D. (2022). Using artificial intelligence for the benefit of the circular economy. *Proceedings of the International Conference on Business Excellence, 16*(1), 294–303. https://doi.org/10.2478/picbe-2022-0029.

Cooper, H. M. (2017). *Research Synthesis and Meta-Analysis: A Step-by-Step Approach*. SAGE.

Corsi, A., Pagani, R., e Cruz, T. B. R., de Souza, F. F., & Kovaleski, J. L. (2022a). Smart sustainable cities: Characterization and impacts for sustainable development goals. *Revista de Gestão Ambiental e Sustentabilidade, 11*(1), 20750.

Cugurullo, F. (2018). The origin of the Smart City imaginary: From the dawn of modernity to the eclipse of reason. In: C. Lindner & M. Meissner (Eds.), *The Routledge Companion to Urban Imaginaries*. Routledge. https://www.researchgate.net/publication/325474312_The_origin_of_the_Smart_City_imaginary_from_the_dawn_of_modernity_to_the_eclipse_of_reason

Cui, X. (2022). A circular urban metabolism (CUM) framework to explore resource use patterns and circularity potential in an urban system. *Journal of Cleaner Production, 359*, 132067. https://doi.org/10.1016/j.jclepro.2022.132067

D'Amico, G., Arbolino, R., Shi, L., Yigitcanlar, T., & Ioppolo, G. (2021). Digital technologies for urban metabolism efficiency: Lessons from urban agenda partnership on circular economy. *Sustainability, 13*(11), 6043.

D'Amico, G., Arbolino, R., Shi, L., Yigitcanlar, T., & Ioppolo, G. (2022). Digitalisation driven urban metabolism circularity: A review and analysis of circular city initiatives. *Land Use Policy, 112*, 105819. https://doi.org/10.1016/j.landusepol.2021.105819

D'Amico, G., Taddeo, R., Shi, L., Yigitcanlar, T., & Ioppolo, G. (2020). Ecological indicators of smart urban metabolism: A review of the literature on international standards. *Ecological Indicators, 118*, 106808.

D'Amore, G., Di Vaio, A., Balsalobre-Lorente, D., & Boccia, F. (2022). Artificial intelligence in the water–energy–food model: A holistic approach towards sustainable development goals. *Sustainability, 14*(2), 867.

da Rosa, L. A. B., Cohen, M., Campos, W. Y. Y. Z., Ávila, L. V., & Rodrigues, M. C. M. (2023). Circular economy and sustainable development goals: Main research trends. *Revista De Administração Da UFSM, 16*(1), e9. https://doi.org/10.5902/1983465971448

Dauvergne, P. (2021). The globalization of artificial intelligence: Consequences for the politics of environmentalism. *Globalizations, 18*(2), 285–299. https://doi.org/10.1080/14747731.2020.1785670

De Jesus, A., Antunes, P., Santos, R., & Mendonça, S. (2018). Eco-innovation in the transition to a circular economy: An analytical literature review. *Journal of Cleaner Production, 172*, 2999–3018. https://doi.org/10.1016/j.jclepro.2017.11.111

De Jong, M., Joss, S., Schraven, D., Zhan, C., & Weijnen, M. (2015). Sustainable-smart-resilient low carbon-eco-knowledge cities; making sense of a multitude of concepts promoting sustainable urbanization. *Journal of Cleaner Production, 109*, 25–38.

Dheeraj, A., Nigam, S., Begam, S., Naha, S., Jayachitra Devi, S., Chaurasia, H. S., Kumar, D., Ritika, Soam, S. K., Srinivasa Rao, N., Alka, A., Sreekanth Kumar, V. V., & Mukhopadhyay, S. C. (2020). Role of artificial intelligence (AI) and internet of things (IoT) in mitigating climate change. In: Ch. Srinivasa, R. T. Srinivas, R. V. S. Rao, N. Srinivasa Rao, S. Senthil Vinayagam, & P. Krishnan

(Eds.), *Climate Change and Indian Agriculture: Challenges and Adaptation Strategies* (pp. 325–358). ICAR-National Academy of Agricultural Research Management.

Dincă, G., Milan, A.-A., Andronic, M. L., Pasztori, A.-M., & Dincă, D. (2022). Does circular economy contribute to smart cities' sustainable development? *International Journal of Environmental Research and Public Health*, *19*(13), 7627. https://doi.org/10.3390/ijerph19137627

Dixon-Woods, M., Agarwal, S., Jones, D., Young, B., & Sutton, A. (2005). Synthesizing qualitative and quantitative evidence: A review of possible methods. *Journal of Health Services Research & Policy*, *10*(1), 45–53.

Dong, L., Liu, Z., & Bian, Y. (2021). Match circular economy and urban sustainability: Re-investigating circular economy under sustainable development goals (SDGs). *Circular Economy and Sustainability*, *1*(4), 243–256. https://doi.org/10.1007/s43615-021-00032-1

El Himer, S., Ouaissa, M., Ouaissa, M., & Boulouard, Z. (2022). Artificial intelligence of things (AIoT) for renewable energies systems. In: S. El Himer, M. Ouaissa, A. A. A. Emhemed, M. Ouaissa, & Z. Boulouard (Eds.), *Artificial Intelligence of Things for Smart Green Energy Management. Studies in Systems, Decision and Control* (vol. 446). Springer.

Esposito, M., Tse, T., & Soufani, K. (2017). Is the circular economy a new fast-expanding market? *Thunderbird International Business Review*, *59*, 9–14.

Evans, J., Karvonen, A., Luque-Ayala, A., Martin, C., McCormick, K., Raven, R., & Palgan, Y. V. (2019). Smart and sustainable cities? Pipedreams, practicalities and possibilities. *Local Environment*, *24*(7), 557–564. https://doi.org/10.1080/13549839.2019.1624701

Fang, B., Yu, J., Chen, Z., Osman, A. I., Farghali, M., Ihara, I., Hamza, E. H., Rooney, D. W. & Yap, P.-S. (2023). Artificial intelligence for waste management in smart cities: A review. *Environmental Chemistry Letters*. https://doi.org/10.1007/s10311-023-01604-3

Floridi, L., Cowls, J., King, T. C., & Taddeo, M. (2020). How to design AI for social good: Seven Essential factors. *Science and Engineering Ethics*, *26*, 1771–1796.

Formisano, V., Iannucci, E., Fedele, M., & Bonab, A. B. (2022). City in the loop: Assessing the relationship between circular economy and smart sustainable cities. *Sinergie Italian Journal of Management*, *40*(2), 147–168.

Ghoreishi, M., & Happonen, A. (2020). Key enablers for deploying artificial intelligence for circular economy embracing sustainable product design: Three case studies. *AIP Conference Proceedings*, *2233*(1), 050008.

Ghosh, R., & Sengupta, D. (2023). Smart urban metabolism: A big-data and machine learning perspective. In: R. Bhadouria, S. Tripathi, P. K. Singh, P. K. Joshi, & R. Singh (Eds.), *Urban Metabolism and Climate Change*. Springer. https://doi.org/10.1007/978-3-031-29422-8_16

Gough, D, Oliver, S., & Fdixon, J. (2017). *An Introduction to Systematic Reviews.* SAGE.

Grossi, G., & Olga, T. (2021). Are UN SDGs useful for capturing multiple values of smart city? *Cities*, *114*, 103193. https://doi.org/10.1016/j.cities.2021.103193

Hannan, M. A., Begum, R. A., Al-Shetwi, A. Q., Ker, P. J., Al Mamun, M. A., Hussain, A., Basri, H., & Mahlia, T. M. I. (2020). Waste collection route optimisation model for linking cost saving and emission reduction to achieve sustainable development goals. *Sustainable Cities and Society, 62*, 102393. https://doi.org/10.1016/j.scs.2020.102393

Holmstedt, L., Brandt, N., & Robèrt, K.-H. (2017). Can Stockholm Royal Seaport be part of the puzzle toward global sustainability? From local to global sustainability using the same set of criteria. *Journal of Cleaner Production, 140*, 72–80. https://doi.org/10.1016/j.jclepro.2016.07.019

Husain, Z., Maqbool, A., Haleem, A., Pathak, R. D., & Samson, D. (2021). Analyzing the business models for circular economy implementation: A fuzzy TOPSIS approach. *Operations Management Research, 14*(3–4), 256–271. https://doi.org/10.1007/s12063-021-00197-w.

Ingrao, C., Messineo, A., Beltramo, R., Yigitcanlar, T., & Ioppolo, G. (2018). How can life cycle thinking support sustainability of buildings? Investigating life cycle assessment applications for energy efficiency and environmental performance. *Journal of Cleaner Production, 201*, 556–569. https://doi.org/10.1016/j.jclepro.2018.08.080

International Telecommunication Union (ITU) (2022). The United for Smart Sustainable Cities (U4SSC). The International Telecommunication Union. https://u4ssc.itu.int (accessed 10 December 2022).

Iris-Panagiota, E., & Egleton, T. E. (2023). Artificial intelligence for sustainable smart cities. In: B. K. Mishra (Eds.), *Handbook of Research on Applications of AI, Digital Twin, and Internet of Things for Sustainable Development* (pp. 1–11). IGI Global. https://doi.org/10.4018/978-1-6684-6821-0.ch001

Ishengoma, F. R., Shao, D., Alexopoulos, C., Saxena, S., & Nikiforova, A. (2022). Integration of artificial intelligence of things (AIoT) in the public sector: Drivers, barriers and future research agenda. *Digital Policy, Regulation and Governance, 24*(5), 449–462.

Jose, R., Panigrahi, S. K., Patil, R. A., Fernando, Y., & Ramakrishna, S. (2020). Artificial intelligence-driven circular economy as a key enabler for sustainable energy management. *Materials Circular Economy, 2*, 1–7.

Joshi, R., Pandey, K., & Kumari, S. (2024). Artificial intelligence for advanced sustainable development goals: A 360-degree approach. In: P. K. Prabhakar & W. Leal Filho (Eds.), *Preserving Health, Preserving Earth*. World Sustainability Series. Springer. https://doi.org/10.1007/978-3-031-60545-1_16

Jude, A. B., Singh, D., Islam, S., Jameel, M., Srivastava, S., Prabha, B., & Kshirsagar, P. R. (2021). An artificial intelligence based predictive approach for smart waste management. *Wireless Personal Communications, 127*, 15–16.

Kennedy, C., Cuddihy, J., & Engel-Yan, J. (2007). The changing metabolism of cities. *Journal of Industrial Ecology, 11*(2), 43–59. https://doi.org/10.1162/jie.2007.1107

Kennedy, C., Pincetl, S., & Bunje, P. (2011). The study of urban metabolism and its applications to urban planning and design. *Environmental Pollution, 159*(8–9), 1965–1973.

Kenworthy, J. R. (2006). The eco-city: Ten key transport and planning dimensions for sustainable city development. *Environment and Urbanization, 18*(1):67–85 https://doi.org/10.1177/0956247806063947

Khan, I. S., Ahmad, M. O., & Majava, J. (2021). Industry 4.0 and sustainable development: A systematic mapping of triple bottom line, circular economy and sustainable business models perspectives. *Journal of Cleaner Production, 297*, 126655.

Kierans, G., Jüngling, S. & Schütz, D. (2023). Society 5.0 2023. *EPiC Series in Computing, 93*, 82–96.

Kim, H., Choi, H., Kang, H., An, J., Yeom, S., & Hong, T. (2021). A systematic review of the smart energy conservation system: From smart homes to sustainable smart cities. *Renewable and Sustainable Energy Reviews, 140*, 110755. https://doi.org/10.1016/j.rser.2021.110755

Kirchherr, J., Reike, D., & Hekkert, M. (2017). Conceptualizing the circular economy: An analysis of 114 definitions. *Resources, Conservation and Recycling, 127*, 221–232. https://doi.org/10.1016/j.resconrec.2017.09.005

Koffka, K. (2023). Intelligent Things: Exploring AIoT Technologies and Applications. Kinder edition.

Kramers, A., Höjer, M., Lövehagen, N., & Wangel, J. (2014). Smart sustainable cities – Exploring ICT solutions for reduced energy use in cities. *Environmental Modelling & Software, 56*, 52–62. https://doi.org/10.1016/j.envsoft.2013.12.019

Lampropoulos, G., Rahanu, H., Georgiadou, E., Siakas, D., & Siakas, K. (2024). Reconsidering a sustainable future through artificial intelligence of things (AIoT) in the context of circular economy. In: S. Misra, K. Siakas, & G. Lampropoulos (Eds.), *Artificial Intelligence of Things for Achieving Sustainable Development Goals* (vol. 192). Springer. https://doi.org/10.1007/978-3-031-53433-1_1

Lannelongue, L., Grealey, J., & Inouye, M. (2021). Green algorithms: Quantifying the carbon footprint of computation. *Advanced Science, 8*, 2100707. https://doi.org/10.1002/advs.202100707

Lantto, R., Järnefelt, V., Tähtinen, M., Jääskeläinen, A.-S., Laine-Ylijoki, J., Oasmaa, A., Sundqvist-Andberg, H., & Sözer, N. (2019). Going beyond a circular economy: A vision of a sustainable economy in which material, value and information are integrated and circulate together. *Industrial Biotechnology, 15*(1), 12–19.

Lanzalonga, F., Marseglia, R., Irace, A., & Biancone, P. P. (2024). The application of artificial intelligence in waste management: Understanding the potential of data-driven approaches for the circular economy paradigm. *Management Decision.* https://doi.org/10.1108/MD-10-2023-1733

Lekan, A. J., & Oloruntoba, A. S. (2020). Artificial intelligence in the transition to circular economy. *Am. J. Eng. Res. 9*(6), 185–190.

Levoso, S. A., Gasol, C. M., Martínez-Blanco, J., Gabarell Durany, X., Lehmann, M., & Farreny Gaya, R. (2020). Methodological framework for the implementation of circular economy in urban systems. *Journal of Cleaner Production, 248*, 119227. https://doi.org/10.1016/j.jclepro.2019.119227

Li, J., Sun, W., Song, H., Li, R., & Hao, J. (2021)Toward the construction of a circular economy eco-city: An emergy-based sustainability evaluation of Rizhao city in China, *Sustainable Cities and Society*, 71, 102956, https://doi.org/10.1016/j.scs.2021.102956.

Lieder, M., Asif, F. M., & Rashid, A. (2020). A choice behavior experiment with circular business models using machine learning and simulation modeling. *Journal of Cleaner Production*, 258, 120894.

Lim, Y., Edelenbos, J., & Gianoli, A. (2019). Identifying the results of smart city development: Findings from systematic literature review. *Cities*, 95, 102397, ISSN 0264-2751.https://doi.org/10.1016/j.cities.2019.102397

Liu, Y., Wood, L. C., Venkatesh, V. G., Zhang, A., & Farooque, M. (2021). Barriers to sustainable food consumption and production in China: A fuzzy DEMATEL analysis from a circular economy perspective. *Sustainable Production and Consumption*, 28, 1114–1129.

Lucertini, G., & Musco, F. (2020). Circular urban metabolism framework. *One Earth*, 2(2), 138–142. https://doi.org/10.1016/j.oneear.2020.02.004

Lutz, C. (2019). Digital inequalities in the age of artificial intelligence and big data. *Human Behavior and Emerging Technologies*, 1(2), 141–148. https://doi.org/10.1002/hbe2.140

Ma, S., Zhang, Y., Yang, H., Lv, J., Ren, S., & Liu, Y. (2020). Data-driven sustainable intelligent manufacturing based on demand response for energy-intensive industries. *Journal of Cleaner Production*, 247, 123155.

Mahdavinejad, M. S., Rezvan, M., Barekatain, M., Adibi, P., Barnaghi, P., & Sheth, A. P. (2018). Machine learning for internet of things data analysis: A survey. *Digital Communications and Networks*, 4(3), 161–175. https://doi.org/10.1016/j.dcan.2017.10.002

Maranghi, S., Parisi, M. L., Facchini, A., Rubino, A., Kordas, O., & Basosi, R. (2020). Integrating urban metabolism and life cycle assessment to analyse urban sustainability. *Ecological Indicators*, 112, 106074. https://doi.org/10.1016/j.ecolind.2020.106074

Marin, J., & De Meulder, B. (2018). Interpreting circularity. Circular city representations concealing transition drivers. *Sustainability*, 10(5), 1310. https://doi.org/10.3390/su10051310

Martin, C. J., Evans, J.,& Karvonen, A. (2018). Smart and sustainable? Five tensions in the visions and practices of the smart-sustainable city in Europe and North America. *Technological Forecasting and Social Change*, 133, 269–278. https://doi.org/10.1016/j.techfore.2018.01.005

Mastorakis, G., Mavromoustakis, C. X., Batalla, J. M., & Pallis, E. (Eds.). (2020). *Convergence of Artificial Intelligence and the Internet of Things*. Springer.

Merli, R., Preziosi, M., & Acampora, A. (2018). How do scholars approach the circular economy? A systematic literature review. *Journal of Cleaner Production*, 178, 703–722. https://doi.org/10.1016/j.jclepro.2017.12.112

Mishra, P., & Singh, G. (2023). Artificial intelligence for sustainable smart cities. In: *Sustainable Smart Cities*. Springer. https://doi.org/10.1007/978-3-031-33354-5_6

Mishra, R. K., Kumari, C. L., Chachra, S., Krishna, P. S. J., Dubey, A., & Singh, R. B. (Eds.). (2022). Smart cities for sustainable development. *Advances in Geographical and Environmental Sciences*. Springer. https://doi.org/10.1007/978-981-16-7410-5_1

Mostafavi, M., & Doherty, G. (2010). *Eco-urbanism*. Lars Muller.

Nañez Alonso, S. L., Forradellas, R. F. R., Morell, O. P., & Jorge-Vazquez, J. (2021). Digitalization, circular economy and environmental sustainability: The application of artificial intelligence in the efficient self-management of waste. *Sustainability, 13*(4), 2092.

Nasir, I., & Aziz Al-Talib, G. A. (2023). Waste classification using artificial intelligence techniques: Literature Review. *Technium: Romanian Journal of Applied Sciences and Technology, 5*, 49–59. https://doi.org/10.47577/technium.v5i.8345

Ngan, S. L., How, B. S., Teng, S. Y., Promentilla, M. A. B., Yatim, P., Er, A. C., & Lam, H. L. (2019). Prioritization of sustainability indicators for promoting the circular economy: The case of developing countries. *Renewable and Sustainable Energy Reviews, 111*, 314–331.

Nishant, R., Kennedy, M., & Corbett, J. (2020). Artificial intelligence for sustainability: Challenges, opportunities, and a research agenda. *International Journal of Information Management, 53*, 102104.

Niza, S., Rosado, L., & Ferrão, P. (2009). Urban metabolism: Methodological advances in urban material flow accounting based on the Lisbon case study. *Journal of Industrial Ecology, 13*(3), 384–405.

Núñez-Cacho, P., Maqueira-Marín, J. M., Rata, M., & Molina-Moreno, V. (2022). Building a model for the predictive improvement of air quality in Circular Smart cities. In: 2022 IEEE International Smart Cities Conference (ISC2) (pp. 1–5). https://doi.org/10.1109/ISC255366.2022.9922373

Olawade, D. B., Fapohunda, O., Wada, O. Z., Usman, S. O., Ige, A. O., Ajisafe, O., & Oladapo, B. I. (2024). Smart waste management: A paradigm shift enabled by artificial intelligence. *Waste Management Bulletin, 2*(2), 244–263.

Oluleye, B. I., Chan, D. W. M., & Antwi-Afari, P. (2022). Adopting Artificial Intelligence for enhancing the implementation of systemic circularity in the construction industry: A critical review. *Sustainable Production and Consumption, 35*, 509–524. https://doi.org/10.1016/j.spc.2022.12.002.

Paracha, A., Arshad, J., Ben Farah, M., & Ismail, K. (2024). Machine learning security and privacy: A review of threats and countermeasures. *EURASIP Journal on Information Security, 10*. https://doi.org/10.1186/s13635-024-00158-3

Parihar, V., Malik, A., Bhawna, Bhushan, B., & Chaganti, R. (2023). From smart devices to smarter systems: The evolution of artificial intelligence of things (AIoT) with characteristics, architecture, use cases, and challenges. In: *AI models for Blockchain-Based Intelligent Networks in IoT Systems: Concepts, Methodologies, Tools, and Applications* (pp. 1–28). Springer.

Pasichnyi, O., Wallin, J., Levihn, F., Shahrokni, H., & Kordas, O. (2019). Energy performance certificates—new opportunities for data-enabled urban energy policy instruments? *Energy Policy 127*, 486–499.

Peponi, A., Morgado, P., & Kumble, P. (2022). Life cycle thinking and machine learning for urban metabolism assessment and prediction. *Sustainable Cities and Society, 80*, 103754. https://doi.org/10.1016/j.scs.2022.103754

Popescu, S. M., Mansoor, S., Wani, O. A., Kumar, S. S., Sharma, V., Sharma, A., Arya, V. M., Kirkham, M. B., Hou, D., Bolan, N., & Chung, Y. S. (2024). Artificial intelligence and IoT driven technologies for environmental pollution monitoring and management. *Frontiers in Environmental Science, 12*, 1336088. https://doi.org/10.3389/fenvs.2024.1336088

Potting, J., Hekkert, M., Worrell, E., & Hanemaaijer, A. (2017). Circular Economy: Measuring Innovation in the Product Chain. Planbureau voor de Leefomgeving, issue 2544 (Report).

Puri, V., Jha, S., Kumar, R., Priyadarshini, I., Son, L. H., Abdel-Basset, M., Elhoseny, M., & Long, H. V. (2019). A hybrid artificial intelligence and Internet of Things model for generation of renewable resource of energy. *IEEE Access, 7*, 111181–111191.

Ramadoss, T. S., Alam, H., & Seeram, R. (2018). Artificial intelligence and Internet of Things enabled circular economy. *The International Journal of Engineering and Science, 7*(9), 55–63.

Raman, R., Pattnaik, D., Lathabai, H. H., Kumar, C., Govindan, K., & Nedungadi, K. C.. (2024). Green and sustainable AI research: An integrated thematic and topic modeling analysis. *Journal of Big Data 11, 55* . https://doi.org/10.1186/s40537-024-00920-x

Rane, N., Choudhary, S., & Rane, J. (2024). Artificial Intelligence and machine learning in renewable and sustainable energy strategies: A critical review and future perspectives. *SSRN*. https://ssrn.com/abstract=4838761, https://doi.org/10.2139/ssrn.4838761

Ranta, V., Aarikka-Stenroos, L., & Väisänen, J.-M. (2021). Digital technologies catalyzing business model innovation for circular economy: Multiple case study. *Resources, Conservation and Recycling, 164*, 105155.

Rapoport, E. (2014). Utopian visions and real estate dreams: The eco–city past, present and future. *Geogr Compass, 8*, 137–149

Rapoport, E., & Vernay, A.-L. (2011). Defining the eco-city: A discursive approach. In: Paper Presented at *Management And Innovation For A Sustainable Built Environment Conference, International Eco-Cities Initiative*, Amsterdam, The Netherlands.

Rathore, A. S., & Malawalia, P. (2021). Deploying artificial intelligence for circular economy and its link with sustainable development goals. *International Journal, 1*(4).

Register, R. (1987). *Ecocity Berkeley: Building Cities for a Healthy Future*. North Atlantic Books.

Register, R. (2006). *Ecocities: Rebuilding Cities in Balance with Nature*. New Society Publishers.

Riesener, M., Dölle, C., Mattern, C., & Kreß, J. (2019, October). Circular economy: Challenges and potentials for the manufacturing industry by digital transformation. In: 2019 IEEE International Symposium on Innovation and Entrepreneurship (TEMS-ISIE) (pp. 1–7). IEEE.

Rietbergen, M. G., Velzing, E.-J., & van Stigteds, R. (2021). *Smart Sustainable Cities: A Handbook for Applied Research*. HU University of Applied Science.

Roberts, H., Zhang, J., Bariach, B., Cowls, J., Gilburt, B., Juneja, P., Tsamados, A., Ziosi, M., Taddeo, M., & Floridi, L. (2022). Artificial intelligence in support of the circular economy: Ethical considerations and a path forward. *AI & Society*. https://doi.org/10.1007/s00146-022-01596-8

Rodríguez-Anton, J. M., Rubio-Andrada, L., Calemín-Pedroche, M. S., & Alonso-Almeida, M. D. M. (2019). Analysis of the relations between circular economy and sustainable development goals. *International Journal of Sustainable Development & World Ecology*, 26(8), 708–720. https://doi.org/10.1080/13504509.2019.1666754

Roseland, M. (1997). Dimensions of the eco-city. *Cities*, 14 (4): 197–202. https://doi.org/10.1016/s0264-2751(97)00003-6

Rudskaya, E. N., Eremenko, I. A., & Yuryeva, V. V. (2019). Eco-cities in the paradigm of a circular economy and a comprehensive internet. *IOP Conference Series: Materials Science and Engineering 698*, 077023 https://doi.org/10.1088/1757-899X/698/7/077023

Rusch, M., Schöggl, J.-P., & Baumgartner, R. J. (2023). Application of digital technologies for sustainable product management in a circular economy: A review. *Business Strategy and the Environment*, 32(3), 1159–1174.

Salvador, R., Barros, M. V., Freire, F., Halog, A., Piekarski, C. M., & Antonio, C. (2021). Circular economy strategies on business modeling: Identifying the greatest influences. *Journal of Cleaner Production*, 299, 126918.

Samadi, S. (2022). The convergence of AI, IoT, and big data for advancing flood analytics research. *Frontiers in Water*, 4, 786040. https://doi.org/10.3389/frwa.2022.786040

Sanchez, T. W. (2023). Planning on the verge of AI, or AI on the verge of planning. *Urban Science*, 7, 70. https://doi.org/10.3390/urbansci7030070

Saravanan, K., & Sakthinathan, G. (2021). *Handbook of Green Engineering Technologies for Sustainable Smart Cities*. CRC Press.

Sastry, M. K. S. (2020). Circular economy in energizing smart cities. In: N. Baporikar (Ed.), *Handbook of Research on Entrepreneurship Development and Opportunities in Circular Economy* (pp. 251–269). IGI Global. https://doi.org/10.4018/978-1-7998-5116-5.ch013

Schmidt, S., Laner, D., Van Eygen, E., & Stanisavljevic, N. (2020). Material efficiency to measure the environmental performance of waste management systems: A case study on PET bottle recycling in Austria, Germany, and Serbia. *Waste Management*, 110, 74–8.

Schöggl, J.-P., Rusch, M., Stumpf, L., & Baumgartner, R. J. (2023). Implementation of digital technologies for a circular economy and sustainability management in the manufacturing sector. *Sustainable Production and Consumption*, 35, 401–420. https://doi.org/10.1016/j.spc.2022.11.012

Schroeder, P., Anggraeni, K., & Weber, U. (2019). The relevance of circular economy practices to the sustainable development goals. *Journal of Industrial Ecology*, 23, 77–95.

Seng, K. P., Ang, L. M., & Ngharamike, E. (2022). Artificial intelligence Internet of Things: A new paradigm of distributed sensor networks. *International Journal of Distributed Sensor Networks*, *18*(3). https://doi.org/10.1177/15501477211062835

Sertyesilisik, B. (2023). Circular, smart, and connected cities: A key for enhancing sustainability and resilience of the cities. In: *Research Anthology on BIM and Digital Twins in Smart Cities*, edited by Information Resources Management Association (pp. 230–243). IGI Global. https://doi.org/10.4018/978-1-6684-7548-5.ch012

Sertyesilisik, B. (2023). Circular, smart, and connected cities: A key for enhancing sustainability and resilience of the cities. In: *Research Anthology on BIM and Digital Twins in Smart Cities*, edited by Information Resources Management Association (pp. 230–243). IGI Global. https://doi.org/10.4018/978-1-6684-7548-5.ch01

Shaamala, A., Yigitcanlar, T., Nili, A., & Nyandega, D. (2024). Algorithmic green infrastructure optimization: Review of artificial intelligence driven approaches for tackling climate change. *Sustainable Cities and Society*, *101*, 105182. https://doi.org/10.1016/j.scs.2024.105182

Shahrokni, H., Årman, L., Lazarevic, D., Nilsson, A., & Brandt, N. (2015). Implementing smart urban metabolism in the Stockholm Royal Seaport: Smart city SRS. *Journal of Industrial Ecology*, *19*(5), 917–929.

Shahrokni, H., Lazarevic, D., & Brandt, N. (2015). Smart urban metabolism: Towards a real–time understanding of the energy and material flows of a city and its citizens. *Journal of Urban Technology*, *22*(1), 65–86.

Sharifi, A., Allam, Z., Bibri, S. E., & Khavarian-Garmsir, A. R. (2024). Smart cities and sustainable development goals (SDGs): A systematic literature review of co-benefits and trade-offs. *Cities*, *146*, 104659. https://doi.org/10.1016/j.cities.2023.104659

Shruti, S., Singh, P. K., & Ohri, A. (2019). Toward developing sustainable smart cities in India. *International Journal of Engineering and Advanced Technology (IJEAT)*, *9*, 2.

Shruti, S., Singh, P. K., Ohri, A., & Singh, R. S. (2022). Development of environmental decision support system for cities in India. *41*(5), e13817. https://doi.org/10.1002/ep.13817

Siakas, D., Lampropoulos, G., Rahanu, H., Georgiadou, E., & Siakas, K. (2023). Emerging technologies enabling the transition toward a sustainable and circular economy: The 4R sustainability framework. In: M. Yilmaz, P. Clarke, A. Riel, & R. Messnarz (Eds.), *Systems, Software and Services Process Improvement. EuroSPI 2023. Communications in Computer and Information Science* (vol. 1891). Springer. https://doi.org/10.1007/978-3-031-42310-9_12

Sihvonen, S., & Ritola, T. (2015). Conceptualizing ReX for aggregating end-of-life strategies in product development. *Procedia CIRP*, *29*, 639–644.

Singh, S., Singh, P. K., Ohri, A., & Singh, R. (2022). Development of environmental decision support system for sustainable smart cities in India. *Environmental Progress & Sustainable Energy*, *41*(5), e13817. https://doi.org/10.1002/ep.13817

Sleem, A., & Elhenawy, I. (2023). Survey of Artificial Intelligence of Things for Smart Buildings: A closer outlook. *Journal of Intelligent Systems and Internet of Things*, 8(2), 63–71. https://doi.org/10.54216/JISIoT.080206

Smol, M., Marcinek, P., & Koda, E. (2021). Drivers and barriers for a circular economy (CE) implementation in Poland—A case study of raw materials recovery sector. *Energies, 14, 2219*, https://doi.org/10.3390/en14082219

Son, T. H., Weedon, Z., Yigitcanlar, T., Sanchez, T., Corchado, J. M., & Mehmood, R. (2023). Algorithmic urban planning for smart and sustainable development: Systematic review of the literature. *Sustainable Cities and Society, 94*, 104562

Späth, P., Hawxwell, T., John, R., Li, S., Löffler, E., Riener, V., & Utkarsh, S. (2017). *Smart-Eco Cities in Germany: Trends and City Profiles*. University of Exeter Press.

Stapper, E. W., & Duyvendak, J. W. (2020). Good residents, bad residents: How participatory processes in urban redevelopment privilege entrepreneurial citizens. *Cities, 107*, 102898. https://doi.org/10.1016/j.cities.2020.102898

Stübinger, J., & Schneider, L. (2020). Understanding smart city: A data-driven literature review. *Sustainability, 12*(20), 8460. https://doi.org/10.3390/su12208460

Stuiver, M., & O'Hara, S. (2022). Envisioning the symbiotic city in 2050: Two visions of Washington DC and the Netherlands. In: *The Symbiotic City: Voices of Nature in Urban Transformations* (pp. 63–81). Wageningen Academic Publishers. https://doi.org/10.3920/978-90-8686-935-0_3

Tarek, S. (2023). Smart eco-cities conceptual framework to achieve UN-SDGs: A case study application in Egypt. *Civil Engineering and Architecture, 11*(3), 1383–1406. https://doi.org/10.13189/cea.2023.110322

Thamik, H., Cabrera, J. D. F., & Wu, J. (2024). The digital paradigm: Unraveling the impact of artificial intelligence and internet of things on achieving sustainable development goals. In: S. Misra, K. Siakas, & G. Lampropoulos (Eds.), *Artificial Intelligence of Things for Achieving Sustainable Development Goals*. Lecture Notes on Data Engineering and Communications Technologies (vol. 192). Springer. https://doi.org/10.1007/978-3-031-53433-1_2

Tomazzoli, C., Scannapieco, S., & Cristani, M. (2020). Internet of Things and artificial intelligence enable energy efficiency. *Journal of Ambient Intelligence and Humanized Computing, 14*, 4933–4954. https://doi.org/10.1007/s12652-020-02151-3

Trencher, G. (2019). Towards the smart city 2.0: Empirical evidence of using smartness as a tool for tackling social challenges. *Technological Forecasting and Social Change, 142*, 117–128.

Trica, C. L., Banacu, C. S., & Busu, M. (2019). Environmental factors and sustainability of the circular economy model at the European Union Level. *Sustainability, 11*, 1114.

Tripathi, P., & Saxena, P. (2024). An assessment of the role of artificial intelligence on sustainable development goals. In: L. V. Sannikova (Ed.), *Digital Technologies and Distributed Registries for Sustainable Development*. Law, Governance and Technology Series (vol. 64). Springer. https://doi.org/10.1007/978-3-031-51067-0_1

Tripathi, S. L., Ganguli, S., Magradze, T., & Kumar, A. (2022). Intelligent Green Technologies for Sustainable Smart Cities. Scrivener Publishing, Beverly, MA. https://doi.org/10.1002/9781119816096

United Nations (2022). The Sustainable Development Goals Report 2022 (SDG Report 2022 Issue). https://unstats.un.org/sdgs/report/2022/

Velte, C. J., Scheller, K., & Steinhilper, R. (2018). Circular economy through objectives–Development of a proceeding to understand and shape a circular economy using value-focused thinking. In: Proceedings of the Procedia CIRP 69–25th CIRP Life Cycle Engineering (LCE) Conference, Copenhagen, Denmark, 30 April–2 May 2018 (pp. 775–780).

Venkata Mohan, S., Amulya, K., & Modestra, J. A. (2020). Urban biocycles – Closing metabolic loops for resilient and regenerative ecosystem: A perspective. *Bioresource Technology*, 306, 123098. https://doi.org/10.1016/j.biortech.2020.123098

Verrest, H., & Pfeffer, K. (2019). Elaborating the urbanism in smart urbanism: Distilling relevant dimensions for a comprehensive analysis of Smart City approaches. *Information, Communication & Society* 22:1328–1342. https://doi.org/10.1080/1369118X.2018.1424921

Vinuesa, R., Azizpour, H., Leite, I., Balaam, M., Dignum, V., Domisch, S., Felländer, A., Langhans, S. D., Tegmark, M., & Fuso Nerini, F. (2020). The role of artificial intelligence in achieving the sustainable development goals. *Nature Communications*, 11(1), 233. https://doi.org/10.1038/s41467-019-14108-y

Vishkaei, B. M. (2022). Metaverse: A new platform for circular smart cities. In: P. De Giovanni (Eds.), *Cases on Circular Economy in Practice* (pp. 51–69). IGI Global. https://doi.org/10.4018/978-1-6684-5001-7.ch003

Visvizi, A., & del Hoyo Raquel, P. (2021). *Smart Cities and the UN SDGs*. Elsevier. https://doi.org/10.1016/C2020-0-01556-2

Wazid, M., Das, A. K., Chamola, V., & Park, Y. (2022). Uniting cyber security and machine learning: Advantages, challenges and future research. *ICT Express*, 8(3), 313–321. https://doi.org/10.1016/j.icte.2022.04.007

Weisz, H., & Steinberger, J. K. (2010). Reducing energy and material flows in cities. *Current Opinion in Environmental Sustainability*, 2(3), 185–192.

Williams, K. (2009). Sustainable cities: Research and practice challenges. *International Journal of Urban Sustainable Development*, 1(1), 128–132.

Wilts, H., Garcia, B. R., Garlito, R. G., Gómez, L. S., & Prieto, E. G. (2021). Artificial intelligence in the sorting of municipal waste as an enabler of the circular economy. *Resources*, 10(4), 28.

Yao, T., Huang, Z., & Zhao, W. (2020). Are smart cities more ecologically efficient? Evidence from China. *Sustainable Cities and Society*, 60, 102008. https://doi.org/10.1016/j.scs.2019.102008

Yigitcanlar, T., & Cugurullo, F. (2020). The sustainability of artificial intelligence: An urbanistic viewpoint from the lens of smart and sustainable cities. *Sustainability*, 12(20), 8548.

Yigitcanlar, T., Desouza, K. C., Butler, L., & Roozkhosh, F. (2020). Contributions and risks of artificial intelligence (AI) in building smarter cities: Insights from a systematic review of the literature. *Energies*, 13(6), 1473.

Yigitcanlar, T., David, A., Li, W., Fookes, C., Bibri, S. E., & Ye, X. (2024a). Unlocking artificial intelligence adoption in local governments: Best practice lessons from real-world implementations. *Smart Cities*, 7(4), 1576–1625. https://doi.org/10.3390/smartcities7040064

Yigitcanlar, T., Senadheera, S., Marasinghe, R., Bibri, S. E., Sanchez, T., Cugurullo, F., & Sieber, R. (2024b). Artificial intelligence and the local government: A five-decade scientometric analysis on the evolution, state-of-the-art, and emerging trends. *Cities*, 152, 105151. https://doi.org/10.1016/j.cities.2024.105151

Yin, P. Y., Chen, H. M., Cheng, Y. L., Wei, Y. C., Huang, Y. L., & Day, R. F. (2021). Minimizing the makespan in flowshop scheduling for sustainable rubber circular manufacturing. *Sustainability*, 13(5), 2576.

Zaidi, A., Ajibade, S.-S. M., Musa, M., & Bekun, F. V. (2023). New insights into the research landscape on the application of artificial intelligence in sustainable smart cities: A bibliometric mapping and network analysis approach. *International Journal of Energy Economics and Policy*, 13(4), 287–299. https://doi.org/10.32479/ijeep.14683

Zhang, J., & Tao, D. (2021). Empowering things with intelligence: A survey of the progress, challenges, and opportunities in artificial intelligence of things. *IEEE Internet of Things Journal*, 8(10), 7789–7817. https://doi.org/10.1109/JIOT.2020.3039359

Index

Note: **Bold** page numbers refer to tables and *italic* page numbers refer to figures.

AC-GAN *see* Auxiliary Classifier GAN (AC-GAN)
ACO *see* ant colony optimization (ACO)
actuation system 210, 281
adaptation policy 144
adaptive urban planning and management 311
advanced AIoT solutions, for waste management 537–540
aggregative synthesis 424, 520
AIoE *see* Artificial Intelligence of Everything (AIoE)
AIoSSCT *see* Artificial Intelligence of Smart Sustainable City Things (AIoSSCT)
AIoT *see* artificial intelligence of things (AIoT)
Alesund City, Norway 225, 297
algorithm bias 575
Amsterdam Smart City (ASC) 237, 238, 388
analytics module, AI and 298
ANN *see* Artificial Neural Networks (ANN)
ant colony optimization (ACO) 60
APIs *see* Application Programming Interfaces (APIs)
Application Programming Interfaces (APIs) 214, 301, 306, 315
artificial intelligence (AI) 58–62
    and AIoT and smart city relationship between 136–137, **136**
    systems and domains 154–159

    and AIoT harnessing synergies among 240–243
    algorithms 7–10, 62, 63, 79, 114, 156, 278–279, 309, 511
    applications 437–463
        computer vision 441–443
        deep learning 439–440
        generative artificial intelligence 453–456
        generative models for urban planning and design 444–447
        machine learning 438–439
        natural language processing 443–444
        urban digital twin 449–453
    and climate change, empirical research on relationship between 133, **133–135**, *135–136*, 142–147
    environmental challenges of 166, **167**, 168
    and environmental sustainability
        empirical research on relationship between 130–132, **131**, 137, **138–141**, *139*, *141*
        theoretical and literature research on relationship between **133**, 132–133
    ethical, societal, and regulatory challenges of 168–170, **168–169**
    evolution of 21, 159, **167**, 170, 172, 401
    governance **169**
    groundbreaking convergence of 38–39
    integration of 115, 144, 155, 236

595

artificial intelligence (AI) (*cont.*)
  IoT devices, integration 154
  layer 289
  models 59, 62, 278, 539
    in UDT systems 419
  powered models 122
  powered UDT 296
  powered user interfaces 8
  tools 146
  utilization and integration of 130
artificial intelligence-driven circular economy 539
artificial intelligence-driven middleware 8
artificial intelligence-driven UDT 229–231, 296
artificial intelligence-enabled Mobile Marine Protected Areas (MMPAs) 141
artificial intelligence-enabled sensors 8
Artificial Intelligence of Everything (AIoE) 307–310
  emergence of 308
  primary objective of 307
"Artificial Intelligence of Smart City Things for Transportation Safety" 215
Artificial Intelligence of Smarter Eco-City 14
Artificial Intelligence of Smart Sustainable City Things (AIoSSCT) 210–218, *215*, 240, 246, 273, 274, 286, **287**, 308, 312, *313*, 320
  architecture 315–316
  benefits and opportunities 322–323
  challenges, barriers, and limitations 325–327
  comparative analysis 321–322
  findings and interpretation of results 320
  framework 284, *284*, 322, 324, 325, 327
  implications for research, practice, and policymaking 324–325
  suggestions for future research directions 327–328
artificial intelligence of things (AIoT) 3, 63, 66–68, 277–280, 359
  in advancing metabolic circularity in smarter eco-cities 534–537

  applications and development of smarter eco-cities **174**
  architectural layers of 304–307, 362
  City Brain as large-scale system of 364–365
  cyber-physical systems and 7, 280–282, **283**
  data-driven approaches 4
  in digital platforms 236
  dynamic interplay of 307–312
  environmental challenges of 166, *167*, 168
  environmental governance 355–356, *356*
  in environmental sustainability 157
  ethical, societal, and regulatory challenges of 168–170, **168–169**
  evolution 170, 172, 214, 286, 329, 400
    and pervasive integration 159
  evolving nature of 398
  foundational elements of synergistic interplay of 310–312
  functionalities in UB 270
  functions, platform urbanism and 378–380, **379**
  fundamental layers of 307
  governance systems 345, 355, 397
  groundbreaking convergence of 38–39
  grounded in data science cycle 304–307
  importance of 156
  integration 9, 54, 115, 211, 401
    AIoT in digital platforms 380
    AIoT with CPSoS 15
    in environmental governance 400
  interplay of 12
  large-scale system 80
  as natural evolution of the internet of things 361–362
  optimization, intelligence, and collaboration 564
  pillars 118–120, *119*
    decision-making 120
    learning 119–120
    perceiving 119
    sensing 119
    visualizing 120
  potential of 117
  powered solutions 5

Index ■ 597

powered systems 205
role 7, 8, 12, 319
    in operation and functioning of UB 309
    as pivotal enabler 395
serves as driving catalyst 5
smart city for circular economy framework 535
in smarter eco-cities 572
in SUM 79
for Sustainable Development Goals 113
synergistic potential of 500
system 118
    in smarter eco-city 120
tools 389
transformative potential of 279–280, 350
urban system 290
Artificial Neural Networks (ANN) 73, 74, 155, 157, 440, 539, 541
ASC *see* Amsterdam Smart City (ASC)
atmosphere, greenhouse gases (GHG) into 121
autoencoders **76**
autonomy 9, 67, 118, 277, 500
Auxiliary Classifier GAN (AC-GAN) 457

backpropagation algorithm 62
Backtracking Search Algorithm (BSA) 157
behavioral change 567–568, 573
bias and fairness **168**
big data 426
    for City Brain 368
    layer 288
    utilization and integration of 130
BigGAN 433
biodiversity conservation 124
Blue City Project 225–226, 296, 463–466
BSA *see* Backtracking Search Algorithm (BSA)

capsule networks (CapsNets) 75, **76**
carbon emissions **167**
    forecasting 143
CCPA *see* Climate-Calibrated Precipitation Analysis (CCPA)
cGAN-Flood 459
ChatGPT 430, 460–463

circular cities
    integration of 534–537
    urban metabolism of 529
circular economy
    and AI 550
    with AI and AIoT in smarter eco-cities, challenges of integrating 568–571
    core tenet of 549–550
    exploring the interrelationships among 550–553
    framework, AIoT-based smart city 535
    fusion of 534
    integration 220
    novel framework for integrating 562–566, *565*
    principles 529, 536
    and SDGs 551
    in smarter eco-cities 540–542, **543–545**, 546, **547**, 548–550
        challenges of implementing 567–568
    and urban circularity, commonalities and differences 504–507
circular metabolic smarter eco-cities 566
"Circular Resource Efficiency Management Framework" 509
circular smarter eco-cities, 4R framework for **547**
city agents 46, 212
city-as-a-platform, evolution of 355
City Brain 5, 6, 10, 78, 79, 89
    big data for 368
    case study analysis 218–219
    challenges and complexities surrounding the integration of 250, **251–252**, 252
    cloud computing for 368
    collaborative contributions of 393, **394**
    collaborative integration of 243
    complementary capabilities of 40
    construction 367
        and implementation 79
    emerging data-driven governance systems 383–386
    in environmental governance 384–385
    functioning of 9
    harnessing synergies among 240–243

City Brain (cont.)
   integration of 242
   internet neural network for 368–369
   interplay of 12
   IoT for 368
   as large-scale system of artificial intelligence of things 364
   operational urban management 219–221
   pillars based on data science cycle 365–367
   role in environmental governance 243–244
   synergistic integration of 396
   technical components of 367–369
   transformative capabilities of 7
   urban planning and governance 221–223
City Information Models (CIMs), to urban digital twin 227
city operation 52, 78, 164, 230–231, 271, 272, 296, 307–308, 355, 367, 425, 564
city platforms 14, 16, 559
City Protocol 238
civic engagement and collaboration 238
clean congruence 540
climate action 122, 130, 142, 394
Climate-Calibrated Precipitation Analysis (CCPA) 459
climate change 56, 57, 113, 120–122
   AIoT applications for 129–130, 146
   artificial intelligence for 123–124, 129–130, 146
   convergence of AI and IoT in 150–151, **152–153**, 153, 154
   empirical research on relationship between AI and 133, **134–135**, 135–136, 142–147
   environmental sustainability and 55–58
   generative artificial intelligence for 431–460
   mitigation and adaptation 151
   role of AI and IoT technologies in mitigating 153
climate disruption 56, 126

climate mitigation, biodiversity conservation for 124
CLIP *see* Contrastive Language-Image Pre-training (CLIP)
closed-loop systems 499, 506, 510
cloud computing 216, 363
   for City Brain 368
   layer 288–289
   models **64**
cloud infrastructure 314
CNNs *see* convolutional neural networks (CNNs)
collaboration and enhancement, synergizing framework components through 315–320, *316*
co-locating industries 532
communication
   infrastructure 215
   network 280
   protocols 298
community engagement 576
   in environmental stewardship 518
competing priorities 573
computer vision (CV) 60, 430
configurative synthesis 287, 359, 424, 521
consumer awareness 506, 568
Contrastive Language-Image Pre-training (CLIP) 431
convolutional neural networks (CNNs) 59, 74, 75, 429, 431, 538
CPSoS *see* cyber-physical systems of systems (CPSoS)
CPSs *see* cyber-physical systems (CPSs)
"cradle-to-cradle" model 500
cross-disciplinary collaboration 570
cutting-edge technologies 43
cyber-physical systems (CPSs) 4, 6, 210, 218, 271–272, 276
   and artificial intelligence of things 280–282, **283**
   efficacy of 210
   interplay of 12
   leverages IoT 9
      for deployment 309
   real-time data analyze 7
   and SUM, integration 12

Index ■ 599

cyber-physical systems of systems (CPSoS)
    8, 9, 12, 13, 210, 272, 273,
    282, 454
  context of 10
  for data-driven environmental
    management and planning
    310–312
  dynamic interplay of 307–312
  fosters 311
  framework 317, 319
  integration of 311–312
    AIoT with 15
cybersecurity measures 281

data acquisition 298, 301
data analytics 363
  models 8
  and predicative modeling 216
  and predictive modeling 314
data-driven decision-making 216–217, 310, 314–315
data-driven environmental
    governance 356
  in smarter eco-cities, pioneering
    framework 390–395, *391*, **394**
data-driven environmental management
    and planning 270
  cyber-physical systems of systems
    310–312
data-driven management and planning
    240–243
data-driven policy decisions 387
"data-driven smart sustainable
    cities" 557
data-driven technology 43, 237–238,
    275, 532
  rapid advancement of 113, 130
  of smart cities 164–166
data-driven urban governance 39
data-driven urban management 39
Data Hub Manager 237–238
data integration 573–574
  and analysis 562
  and interoperability 216, 314
data management 569
data science cycle 306–307
  artificial intelligence of things
    grounded in 304–307

City Brain pillars based on
    365–367, *366*
data utilization 67
data visualization 305
decision-making 85, 364
  AIoT pillars 120
  data-driven 216–217, 310, 314–315
  hub, real-time 317
decision trees (DTs) 10
DECODE project 238
deep learning (DL) 118, 278, 429, 432
  framework for load forecasting 156
  realm of 433
  as subfields of artificial
    intelligence 73–75, 77–78
  techniques, tasks, and applications **76**
Deep Neural Networks (DNN) 278
deep reinforcement learning (DRL) 75, **76**
deforestation 57, 143
deployment, CPS leverages IoT for 309
differential evolution (DE) 60
digital divide 575
digital mediation 236
"digital mediation of cities" 379
digital models 83
digital platforms
  AIoT in 236
  platform urbanism leverages 15
digital twin, urban *see* urban digital twin
    (UDT)
Disaster City DT concept 450
disaster management, virtual earthquake
    scenarios for 457
disaster resilience and management 142
disaster response and management 143
DL *see* deep learning (DL)
DRL *see* deep reinforcement learning
    (DRL)
drought tolerance determination 143
DT-based deep reinforcement learning
    approach 230
dynamic 3D city models 297
dynamic simulation and modeling
    310–311

eco-cities
  integration of 534–537
  interlinked development of 553–555

eco-cities (cont.)
    metabolic and circular futures in smarter 522, **523–526**, 527–530
    to smarter eco-cities 517–519
eco-innovation 540
ecological degradation 56, 113, 345, 349
ecological footprint 3, 242, 352, 425, 508, 532
economic sustainability **514**
economic viability 568, 571
ecosystem changes monitoring 124
edge computing **64**, 216, 313–314, 363, 368
Edge Computing Layer 289
emerging data-driven governance systems
    City Brain 383–386
    in enhancing environmental governance processes 380–390
    platform urbanism 388–390
    smart urban metabolism 386–388
emerging technological integration 24
emissions
    carbon **167**
    carbon emissions forecasting 143
    greenhouse gases (GHGs) 57, 121, 124, 142, 499, 501
energy consumption **167**
energy demand prediction 143
energy management, DL-based framework for load forecasting in 151
enviro-economic sustainability 516
environmental and technological trajectory perspective
    from eco-cities to smarter eco-cities 517–519
environmental challenges 24
    of AI and AIoT Technologies 166, **167**, 168
environmental degradation 88
environmental footprint 113
environmental governance 213, 345, 381
    AIoT integration in 400
    challenges 393
    City Brain in 384–385
    institutions 382
    mechanisms in 47–48
    obstacles in 247–248
    role in 48
    in smarter eco-cities 381–383
    strategies 48, 392
    structures of 48
environmentally smart sustainable urbanism
    smart eco-cities and symbiotic cities as prominent approaches to 530–534, **533**
environmental monitoring 385
environmental stewardship, community engagement in 518
environmental sustainability 2, 24, 44, 56–57, 120–122, 311, **514**
    AIoT technologies in 129–130, 156
    artificial intelligence for 123–124, 129–130
    challenges 56
    mitigation 245
    and climate change 55–58
    convergence of AI and IoT in 150–151, **152–153**, 153, 154
    in current smart eco-cities 555–558, 556
    dominance of 556
    empirical research on relationship between AI and 130–132, **131**, 137, **138–141**, 139, 141
    generative artificial intelligence for 456–460
    overarching goal of 57–58
    strategies 56
    theoretical and literature research on relationship between AI and 132–133, **133**
environmental urban governance 381
environmental urban management, landscape of 54
environmental urban planning 51
    and governance, novel conceptual framework 210–216
environmental urban sustainability paradigms, evolution of 553–555
European Union's scientific DECODE project 238
evolutionary computing (EC) 60
evolution of smart eco-cities 114
E-waste generation **167**

Index ■ 601

explainable AI (XAI), challenges, issues, and considerations of 170–172, **171**
Extract, Transform, Load (ETL) processes 451

federated learning **72**
feedback mechanism 298
  robust 245
fenerative models 460
few-shot learning **71**
flood prediction and protection 143
flow-based models 433
fog computing 216, 363
  models **64**
footprint, ecological 3
forest degradation 143
fossil fuels, burning of 57
foundation models, pre-trained 431–436
foundations and frameworks, smarter eco-cities 42–45, *43*
4R framework and artificial intelligence 546
4R framework for circular smarter eco-cities **547**
4R sustainability framework 546
fuzzy logic (FL) 60

GANs *see* generative adversarial networks (GANs)
GenAI *see* generative artificial intelligence (GenAI)
GenDT *see* Generative Digital Twin (GenDT)
generative adversarial networks (GANs) 75, **76**, 429, 431, 457–458
generative artificial intelligence (GenAI) 418, 420, 431–436, 453–456
  challenges of 474–477
  for environmental sustainability and climate change 456–460
  models 418, 419, 458
    advent of 460–461
    high energy consumption of 476
  potential of 459–460
  strengths of 458
  sustainable smart city, planning and design 466–469, *467*
  for urban planning and design 460–463
  utilization of 472
Generative Digital Twin (GenDT) 457
generative models, for urban planning and design 444–447
Generative Spatial AI (GSAI)
  model 464–466
  into policy frameworks 473
  for Lausanne city digital twin 463–466
  sustainable smart city 421
Generative Urban Design (GUD) 422–423, 461
genetic algorithms (GAs) 60, 157
geo-foundation model 434–435
geographically diverse models 435
geographic information systems (GIS) 434
GHGs *see* greenhouse gases (GHGs)
GIS *see* geographic information systems (GIS)
global warming 121
GNNs *see* graph neural networks (GNNs)
governance
  and eco-cities 46
  platforms 244–245
  platform urbanism and 378
  systems, integration with 387
GPT 433
gradient descent 62
graph neural networks (GNNs) **140**, 461
greenhouse gases (GHGs)
  accumulation of 57
  into atmosphere 121
  emissions 57, 121, 124, 142, 499, 501
groundbreaking approach 204
GSAI *see* Generative Spatial AI (GSAI)
GUD *see* Generative Urban Design (GUD)

Hangzhou City Brain 291, 384
Helsinki Digital Twin project 226, *227*
Helsinki Smart City 238
  development 389
  DT project 296
High-Performance Computing platform 221
holistic approach 4, 16
  of sustainable smart cities 5
  to urban management and planning 12
human-centric design 570

ICT *see* Information and Communication Technologies (ICT)
IML *see* interpretable ML (IML)
inductive approach 359
industrial symbiosis fosters 531
Information and Communication Technologies (ICT) 275, 554, 555
infrastructure and technology 568
infrastructure urban planning 51
innovative data-driven management, and planning framework 312–315
integrated framework, essential components of proposed 350–353, *351*
integration layer, data acquisition and 298
integration of software 281
integration with governance systems 387
intelligent transportation systems (ITS) 5, 66, 132, 144, 165, 221
interconnected systems 310
interdisciplinary approach 24–25
international intergovernmental organizations 81
International Telecommunication Union (ITU) 81
Internet Neural Network Layer 289
Internet of Everything (IoE) 14
Internet of Things (IoT) 3, 62–63
  devices 118
  enabled devices, remote monitoring capabilities 151
  groundbreaking convergence of 38–39
  layer 288
  potential of 115
  sensors 301
  and devices 363
  for smart cities 124–127, **125–126**
  synergy with AI 151
  technical and practical capabilities of IoT 151
  technologies, integration 154
  utilization and integration of 130
interoperability 569
interpretable ML (IML), challenges, issues, and considerations of 170–172, **171**

IoE *see* Internet of Everything (IoE)
IoT *see* Internet of Things (IoT)
ITS *see* intelligent transportation systems (ITS)
ITU *see* International Telecommunication Union (ITU)

job displacement 576
joint foundation models 435

Key Performance Indicators (KPIs) 220, 235, 301, 303
k-means clustering 62
knowledge-sharing mechanisms 302

lack of system integration 24
Lancaster City Information Model (LCIM) prototype 227
land-use urban planning 51
large language models (LLMs) 431–436, 457, 460, 461, 463
Latent Diffusion Model (LDM) 458
Lausanne city digital twin, generative spatial artificial intelligence for 463–466
LCIM prototype *see* Lancaster City Information Model (LCIM) prototype
LCZs *see* local climate zones (LCZs)
LDM *see* Latent Diffusion Model (LDM)
learning
  AIoT pillars 119
  capability 67
leveraging data-driven insights 534
leveraging digital technologies 515
leveraging DL techniques 439
Life Cycle Thinking 158, 234, 304, 528
linchpin technology 270, 350, 542
load forecasting, DL-based framework for 156
local and regional urban planning 51
local climate zones (LCZs) 457
long short-term memory networks (LSTMs) 75, **76**, 430–431

machine learning (ML) 68–69, 428–429
  algorithms 118
  models 8, 59, 158, 164, 278

as subfields of artificial
   intelligence 73–78
techniques 61
and common tasks **69–72**
MADDPG algorithm *see* Multi-Agent Deep Deterministic Policy Gradient (MADDPG) algorithm
Marine Protected Areas (MPAs), identification and management of 141
Mask-R-CNN *see* Mask Region-Based Convolutional Neural Networks (Mask-R-CNN)
Mask Region-Based Convolutional Neural Networks (Mask-R-CNN) 459
metabolic circularity 510
   novel framework for integrating 562–566, 565
   in smarter eco-cities 522–530, **523–526**
   urban 502
Metabolic Performance Metrics (MPMs) 233, 303
"Metabolism of Cities Data Hub" 233–234
meta-governance 46
methodological stages of study 207–210, 208
mitigating climate change, role of AI and IoT technologies in 153
Mobile Marine Protected Areas (MMPAs), AI-enabled 141
model-based reinforcement learning 61
modern urban environments, intricacy of 420
monitoring environmental impact 387
MPMs *see* Metabolic Performance Metrics (MPMs)
MSA model *see* Multi-sectoral Systems Analysis (MSA) model
Multi-Agent Deep Deterministic Policy Gradient (MADDPG) algorithm 295
multilayer perceptron (MLP) network 158, 234, 304
Multi-sectoral Systems Analysis (MSA) model 233, 303
multi-stakeholder platforms 389
multitasking learning **72**

municipal waste-sorting plants, automation of 539

natural computing (NC) 60
natural language processing (NLP) 59, 60, 430–431, 443–444
natural resources, depletion of 269
NC *see* natural computing (NC)
Near Zero Energy Buildings (NZEBs) 450–451
network governance 46
Neural Network for City Brain 368–369
neural network (NN) models 61
NLP *see* natural language processing (NLP)
non-governmental organizations (NGOs) 48
novel conceptual framework, for data-driven environmental management, planning, and governance 88, 88–90
NZEBs *see* Near Zero Energy Buildings (NZEBs)

online learning **71**
operational city management, domain of 219
operational urban management, City Brain 219–221
optimization, waste management 385

particle swarm optimization (PSO) 60, 157
PEDs *see* Positive Energy Districts (PEDs)
perceiving, AIoT pillars 119
pinpointing critical areas 387
planning framework, innovative data-driven management and 312–315
planning systems, data-driven environmental management and 270
platformization 84, 345
platform urbanism 5, 12, 84–86, 89, 345
   and AIoT functions 378–380, **379**
   case study analysis 236–240
   challenges and complexities surrounding the integration of 250, **251–252**, 252

platform urbanism (cont.)
    collaborative contributions of 393, **394**
    emerging data-driven governance systems 388–390
    and governance 378
    interconnectedness of 393
    leverages digital platforms 15
    progression of 375
    and smart urbanism, comparative dimensions of governance 374–378, **375–377**
    from smart urbanism to 373–374
    synergistic integration of 396
    technical logic behind the integration of 245–248, *247*
    in urban governance 389
policy and industry support 506
policy and regulations 564–566
policy decisions, data-driven 387
"policy-driven mode of governance" 53
policy instruments 558
policy integration 573
policymakers 25, 93, 399, 472
pollution control 385
Positive Energy Districts (PEDs) 449
precision agriculture harnesses cutting-edge technologies 146
predictive analytics 363, 385
Preferred Reporting Items for Systematic Reviews and Meta-analyses (PRISMA) approach 127, *128*, 284, *284*, 285, 357, *357*
ProcessGPT modeling 456
product lifespan, extension 506
PSO *see* particle swarm optimization (PSO)
public engagement 386

rapid urbanization 37, 88, 113, 345, 349
raw material use 505
real-time decision-making hub 317
real-time resource tracking 220
recurrent neural networks (RNNs) 59, 75, **76**, 431
recycling 506
regulatory frameworks **169**, 568, 570, 575
reinforcement learning 61, **70**
    algorithms 438
    model-based 61

Reinforcement Learning Empowered Digital Twins (RLEDs) 230, 295, 296
remanufacturing 506
resource depletion **167**
resource efficiency 435, 506
resource optimization 385
resource scarcity 573
reuse of products and components 506
RLEDs *see* Reinforcement Learning Empowered Digital Twins (RLEDs)
RNNs *see* recurrent neural networks (RNNs)
robust data security and privacy 569
robust feedback mechanism 245
R-principles 507

SAR *see* Synthetic Aperture Radar (SAR)
scalability 569–570
scenario planning 292, 293, 302, 321, 387
ScienceDirect 208, 285, 357, 424
Scopus 127, 208, 284, 357, 423
security infrastructure 298
security issues **168**
self-supervised learning **70**
semantic knowledge representation 298
semi-supervised learning **70**
sensing, AIoT pillars 119
sensors
    AI-driven 154
    AI-enabled 8
    IoT 301
    networks and data collection 215
simulation engine, modeling and 298
smart cities
    artificial intelligence and the internet of things for 124–127, **125–126**
    data-driven technologies of 164–166
    exploring the interrelationships among 550–553
    integration of 534–537
    interlinked development of 553–555
"Smart City Brain for Circular Economy" 220, 290–291
smarter eco-cities
    advanced AIoT solutions for waste management in 537–540

Index ■ 605

artificial intelligence of things in 534–537, 572
circular economy in 540–542, **543–545**
artificial intelligence of things and 567–568
challenges of implementing 567–568
tripartite sustainability 546–550, **547**
development, novel conceptual framework 161–163, *162*
distinctive features of 45
dynamics of 87, 165, 166, 170
to eco-cities 517–519
as emerging approach to data-driven smart sustainable cities 553–562
environmental governance in 380–383
evolving landscape of 173
foundations and frameworks 42–45, *43*
4R framework for circular **547**
interconnected concepts shaping the development of 85–87
landscape and dynamics 164–166
leveraging AI and AIoT for advancing tripartite sustainability in 558–562
metabolic and circular futures in 522–530, **523–526**
models and principles **533**
multifaceted examination of 90
pioneering framework for data-driven environmental governance in 390–395, *391*, **394**
realm of 346, 500
and symbiotic cities as prominent approaches 530–534, **533**
tripartite sustainability in 574–576
smart grids 512
smart metabolism circularity, analytical framework of 232
smart sustainable cities 203
dimensions of sustainability within 517
and evolving priorities 558–562
and sustainable cities (eco-cities) 513
tripartite sustainability perspective 512–517, **514**, *516*

smart sustainable indicators 81
smart urbanism and platform urbanism, comparative dimensions of governance 374–378, **375–377**
smart urban metabolic circularity 510, 511
and tripartite sustainability 510–512
smart urban metabolism (SUM) 5, 6, 78–81, 86, 158, 299–304, 318, 369–373
AIoT functionalities in 270
architecture 300, 302
case study analysis 231–236, **232**
challenges and complexities surrounding the integration of 250, **251–252**, 252
circularity, circular economy, and 507–510
collaborative contributions of 393, **394**
collaborative integration of 243
common steps of 371–372
complementary capabilities of 40
comprehensive data sources and management for developing **523–526**
and CPS, integration 12
dynamic visualization platform 242
emerging data-driven governance systems 386–388
essential characteristics of 370
functioning of 9
general and contextual approaches to **371**
harnessing synergies among 240–243
integrates computational modeling 11
integration of AI and AIoT in 79
interplay of 12
key features of 371
role 13–14
and advantages of artificial intelligence of things powered 372–373
in environmental governance 243–244
in Stockholm Royal Seaport project 302–303
sustainable urban development through 302
synergistic integration of 396

606 ■ Index

smart urban metabolism (SUM) (*cont.*)
    technical logic behind the integration of 245–248, *247*
    transformative capabilities of 7
smart water infrastructure, urban digital twin for 231
social inclusion 237
social sustainability 277, **514**
socioeconomic sustainability 515
socio-environmental sustainability 517
spatial urban planning 51
stakeholder collaboration 220
stakeholder engagement 388
STEEPLED analysis 546
Stockholm Royal Seaport (SRS) project 233, 371
    SUM in 302–303
strategic urban planning 51
sub-grid processes modeling 144
SUM *see* smart urban metabolism (SUM)
supervised learning **69**
support vector machines (SVMs) 59, 539
sustainability
    challenges of incorporating and balancing three dimensions of 572–574
    dimensions of *516*
    environmental *see* environmental sustainability
    metrics 387
    smart urban metabolic circularity and tripartite 510–512
    social realm of 279
sustainable circular economy model 550
sustainable cities (eco-cities), and smart sustainable cities 513
Sustainable Development Goals (SDGs) 81, 324, 348, 425, 463, 500, 515, 541, 542
    AIoT for 113
    circular economy and 551
    exploring the interrelationships among 550–553
    potential of AIoT in advancing 3
    United Nations 57–58
sustainable smart cities 4, 5, 44, 275–277, 418
    emergence of 276
    evolving landscape of 428
    holistic approach of 5
    and planning and design processes 425–427
    rapid evolution of 271
    and smarter eco-cities 5, 9
    to urban management and planning 12
Sustainable Smart City CPSoS 14
sustainable transportation, domain of 132
sustainable urban development 2, 503–504
    through SUM 302
sustainable urban planning 51, 385
Suzhou City, urban socioeconomic metabolism 233, 303
SVMs *see* support vector machines (SVMs)
swarm learning **72**
SymbioCity model 530
symbiotic cities 532
    and smart eco-cities as prominent approaches 530–534, **533**
synergistic integration of AI 241
Synthetic Aperture Radar (SAR) 459

TBL *see* Triple Bottom Line (TBL)
tendency of smart city initiatives 43–44
thematic synthesis 287, 359
3-D vision transformer (ViT) 458–459
traditional urban metabolism, evolution from 299
transduction learning **71**
transfer learning 61, **71**, 435
    with pre-trained models 75, **76**
transformative trajectory, in urban development 203
transportation urban planning 51
tripartite sustainability
    challenges of utilizing AI and AIoT for 574–576
    novel framework for integrating 562–566, *565*
    in smarter eco-cities 546–550, **547**, 558–562
    smart sustainable cities 512–517, *516*
    smart urban metabolic circularity and 510–512
Triple Bottom Line (TBL) 501

UAPCE *see* Urban Agenda Partnership on Circular Economy (UAPCE)
UB *see* urban brain (UB)
UDT *see* urban digital twin (UDT)
United for Smart Sustainable Cities (U4SSC) 515
United Nations Economic Commission for Europe–ITU (UNECE–ITU) 80
Unreal Engine 456
unsupervised learning **69**
urban
  context 510
  data overload 24
  design 427
  ecology 518
Urban Agenda Partnership on Circular Economy (UAPCE) 232, 303
urban brain (UB) 270, 271, 282, 288–291
  AIoT functionalities in 270
  AIoT role in operation and functioning of 309
  predictive analytics 321
urban circularity and circular economy, commonalities and differences 504–507
urban computing and intelligence 290
  data-driven approach in 535
urban development 3
  evolving landscape of 43, 512
  landscape of 166
  realm of 517
  sustainable 2, 503–504
  trajectory of 518
  transformation in 38
  transformative trajectory in 203
urban digital twin (UDT) 5, 9, 81–84, 87, 89, 318
  AI-driven 229–231
  AI models in 419
  AIoT functionalities in 270
  applications
    data augmentation 454
    data synthesis and scenario generation 455
    generative urban design and optimization 455
    3D building generation 455

    3D city modeling 455
    3D street environment construction 455
  architecture of 297
  case study analysis 223–229
  challenges and complexities surrounding the integration of 250, **251–252**, 252
  City Information Models (CIMs) to 227
  collaborative integration of 243
  complementary capabilities of 40
  development and implementation of 419
  frameworks 419
  functioning of 9
  harnessing synergies among 240–243
  interplay of 12
  role in environmental governance 243–244
  in smart cities 11
  for smart water infrastructure 231
  synergistic integration of AI with 437
  and synergistic integration with artificial intelligence 436–437
  technical logic behind the integration of 245–248, *247*
  transformative capabilities of 7
  in urban planning and design 437–463
urban environments 270, 518
  intricacy of modern 420
urban governance 46–47, 53–55, 212, 353
  data-driven 39
  environmental 380
  platform urbanism in 389
urbanization 2, 3
  pace of 417
  rapid 37, 269
urban management 52–55
  data-driven 39
  and planning 11, 12, 279
urban managers 52
urban metabolic circularity 502
urban metabolism (UM) 78, 80, 369
  of circular cities 529
  and circularity 508, 509
  complex nature of 299
  evolution from traditional 299
  smart 80–81

"urban nervous system" 373
urban operational management 219
urban planners 25
urban planning 49–55, 212, 427
    and governance 291
        City Brain 221–223
        integration of 53
        landscape of 204
        sustainable 385
        transformative potential of AI in 422
urban planning and design 422
    generative artificial intelligence for 460–463
    generative models for 444–447
    urban digital twin in 437–463
urban platforms 13
    in Finland's largest cities 239
urban policy 46
urban socioeconomic metabolism, Suzhou City 233, 303
urban sustenance lab 238
urban systems, impact of AI on 115
urban transportation, domain of 165
user interface and visualization 217, 290, 298, 301, 315

VAEs *see* variational autoencoders (VAEs)
variational autoencoders (VAEs) 75, 432
versatility 436

virtual earthquake scenarios, for disaster management 457
virtual reality platform 297
virtual simulation algorithms 10
vision transformer (ViT), 3-D 295
visualization
    AIoT pillars 119
    and user interface 290, 298, 301
    user interfaces and 217, 315
volunteered geographic information (VGI) 225

waste management
    advanced artificial intelligence of things solutions for 537–540
    optimization 385
waste reduction 506
water resource conservation, exploration of 132
water utilization management 143
Web Geographic Information System 221
Web of Science (WoS) 127, 208, 284–285, 357, 423–424
wildfire-related applications 143
wind generation prediction 143
Wireless Sensor Networks (WSN) 276

XAI *see* explainable AI (XAI)

Printed in the United States
by Baker & Taylor Publisher Services